Plant Breeding and Cultivar Development

Plant Breeding and Cultivar Development

Dhan Pal Singh
G.B. Pant University of Agriculture and Technology, Uttarakhand, India

Asheesh K. Singh
Department of Agronomy, Iowa State University, Iowa, United States

Arti Singh
Department of Agronomy, Iowa State University, Iowa, United States

ELSEVIER

ACADEMIC PRESS
An imprint of Elsevier

Academic Press is an imprint of Elsevier
125 London Wall, London EC2Y 5AS, United Kingdom
525 B Street, Suite 1650, San Diego, CA 92101, United States
50 Hampshire Street, 5th Floor, Cambridge, MA 02139, United States
The Boulevard, Langford Lane, Kidlington, Oxford OX5 1GB, United Kingdom

British Library Cataloguing-in-Publication Data
A catalogue record for this book is available from the British Library

Library of Congress Cataloging-in-Publication Data
A catalog record for this book is available from the Library of Congress

ISBN: 978-0-12-817563-7

For Information on all Academic Press publications
visit our website at https://www.elsevier.com/books-and-journals

Publisher: Charlotte Cockle
Acquisitions Editor: Nancy Maragioglio
Editorial Project Manager: Billie Jean Fernandez
Production Project Manager: Joy Christel Neumarin Honest Thangiah
Cover Designer: Miles Hitchen

Typeset by MPS Limited, Chennai, India

Contents

Preface

The motivation of this book is to provide a comprehensive coverage on numerous facets important in plant breeding, particularly cultivar development. This book provides a one-stop learning resource, featuring an optimal balance between classical and modern plant breeding tools and techniques. The authors have aimed to create this work to cater the needs of undergraduate and graduate students, practicing plant breeders interested in germplasm and variety development, and to a broader audience interested in plant breeding. The book encompasses important insights on the translational shift from traditional to technologically enhanced—both genomics and phenomics—plant breeding, addressing a wide range of topics.

This book systematically builds through its 28 chapters to educate readers on well-established principles and theories to modern tools and technology; and presents this complex subject with its numerous intricacies in a straightforward and practical way. The contents are organized methodically to lay the foundations of plant biology, genetics (Mendelian, population, quantitative), plant genetic resources, wide hybridization, and genetic variability. The book describes the methods and approaches in the breeding of self-pollinating, cross-pollinating, and clonal crops; along with a primer on experimental designs and statistical analysis applicable to cultivar development. Additional topics include intellectual property rights and protections, participatory breeding, and ideotype-based breeding. Fundamental of genomics, phenomics, high-throughput phenotyping, and advanced data analytics including machine learning and deep learning are covered, along with examples of their potential utilization in breeding schemes.

The authors have attempted clarity and readability so that plant breeders in public and private sector working on a range of crops find it informative and applicable to their programs. Authors have tried to capture information from multiple crops, bringing in their previous experiences as they have worked on several crops in their careers; including barley, *Brassica*, chickpea, durum wheat, field pea, lentil, maize, mung bean, oat, pigeon pea, rice, soybean, urd bean, and common wheat. Authors have developed commercially grown cultivars in chickpea, durum wheat, field pea, lentil, mung bean, pigeon pea, soybean, urd bean, and common wheat.

The body of work in plant breeding is very extensive; therefore, authors have tried to include papers that laid the foundation for more extensive work in plant breeding. The inclusion of work from these pioneers is also a tribute to them.

Acknowledgments

D.P. Singh sincerely acknowledges the support and encouragement of his wife Asha Rani and support of Shailendra, Shubra, Amit, Kanchan, Ipshita, Harshita, Yashovardhan, and Kaushiki.

A.K. Singh acknowledges his mentors, colleagues, and team members who have enabled his career as a plant breeder and expresses sincere gratitude to his teachers and mentors at G.B. Pant University of Agriculture, University of Saskatchewan, and University of Guelph. He sincerely appreciates collaborations with scientists and staff while working at the Swift Current Research and Development Center, Agriculture and Agri-Food Canada, and at Iowa State University, United States. He is very thankful to his breeding and research group at Iowa State University, for their support. The constant love, support, encouragement, and inspiration of authors' parents, Dr. Harpal Singh and Ms. Indra Singh, were vital for the completion of this book. He sincerely appreciates the support from his sister Dr. Ashima Chandra, her husband Mr. Sahil Chandra, and their children Aryaman Chandra and Satyaarth Chandra. He is thankful to his aunt Mrs. Vimla Sisodia and uncle Dr. C.S. Sisodia for their encouragement and motivation.

A. Singh thanks members of her research team at Iowa State University and research mentors in Canada (Dr. R. Knox and Dr. R. DePauw) and the United States (Dr. K. Lamkey, Dr. B. Ganapathysubramanian, and Dr. G Tylka) for their support. She is very thankful to her parents and siblings for always being so encouraging and supportive.

Authors truly appreciate the efforts of Dr. R. Knox for reviewing the section on molecular markers, Dr. J. Zhang for reviewing the GWAS and GS sections, Dr. S. Whitham for reviewing the plant transformation and gene editing sections, Dr. Z. Jubery and Mr. K. Nagasubramanian for reviewing the phenomics and ML chapter, Dr. S. Sarkar and Dr. B. Ganapathysubramanian for reviewing the smart breeding section, Mr. M. Carroll for assisting with ground sampling distance calculations, and Dr. K. Parmley for assisting with the development of Fig. 28.6. Authors are very appreciative of the assistance provided by Ms. K. Vasquez in preparing figures.

A. Singh and A.K. Singh feel privileged by the love and support of their children, Ms. Anneka G. Singh and Mr. Arron J. Singh. They were instrumental in the completion of this book with all the big and small things they did, enabling us to work long hours to write this book. Their maturity and patience defies their young age, and their support and encouragement motivated us to complete this book.

CHAPTER

Plant breeding: past, present, and future perspectives

1

Abstract

This chapter focuses on the past, present, and future of plant breeding particularly geared toward cultivar development. Plant breeders have made tremendous progress in the genetic enhancement of yield, adaptation traits, disease–pest resistance, abiotic stress tolerance, and nutritional and/or end-use quality, increasing crop production and the productivity (yield per unit area) of major crops. With the increasing needs of human population and an ever-increasing climatic variability, plant breeding becomes more important than ever to respond to production challenges to ensure a stable food, feed, and fiber supply. The recent research advances in the areas of biotechnology, genomics, and phenomics provide sophisticated tools to plant breeders to build improved cultivars. Major topics covered in this chapter include crop domestication, pre-Mendelian and Mendelian plant breeding, and plant breeding in the 20th and 21st centuries, which include details on green revolution, genetically modified crops, molecular marker tools, image-based phenotyping, machine learning methods, relationship between private and public sector plant breeding organization, and future plant breeding efforts.

Plant breeding began with the domestication of wild plants and the cultivation of plants, which were useful to man. The process of selecting plant types that provided a dependable source of food, feed, fiber, oil, and other useful products was a part of domestication; therefore, selection became the first method of plant breeding. Selection depends on practitioner's knowledge of plant itself and its response to its environment. The experience plays an important role that is why plant breeding was considered as an art rather than science; and selection continued for thousands of years until human could learn about the structure and function of male and female flowers and their role in the formation of seed. Hybridization was used in certain species to develop superior strains without the knowledge of inheritance of traits (i.e., genetics). The discoveries of cell and their organelles, Mendel's laws and their rediscovery, their application in plant breeding, and infusion of learnings from various scientific disciplines started to make plant breeding more as a science and less of an art discipline.

Plant Breeding and Cultivar Development. DOI: https://doi.org/10.1016/B978-0-12-817563-7.00004-0

Plant breeding deals with the inheritance of qualitative and quantitative traits and includes the genetic improvement in existing genotypes for specific trait(s), and the creation of altogether new genotype with new gene combination called recombinant. Present day plant breeding also involves integration of alien gene(s) from plant kingdom or beyond. Through plant breeding, plants have been improved for their productivity, quality traits (physical and chemical), and also for various other desirable traits, such as resistance to environmental stresses and harmful parasite and for their suitability to mechanical harvesting. Plant breeding is also referred to as plant/crop improvement. In view of these considerations, plant breeders need to be trained not only in genetics (population, quantitative, and Mendelian) but also in related disciplines, including agronomy, biotechnology, entomology, food science, molecular genetics, pathology, physiology, and soil science and many more. A plant breeder is required to have general understanding of the crop(s) including cultivars being grown by the farmers and their problems in different cropping systems and niches, needs of consumers, and industries. Therefore, it is imperative that a plant breeder has a general understanding and knowledge of related disciplines such as crop physiology, crop protection, crop husbandry, biochemistry, and statistics including field plot techniques and analysis. With the advent on modern genomic technologies, phenomics, and mechanization/computerization, they also need to be aware or moderately experienced in computer programing.

In this chapter, historical development of plant breeding, role of plant breeding in the welfare of mankind, its present scenario, and future perspectives are described in brief.

Primitive agriculture and crop domestication

Archeological evidence suggests that angiosperms (higher plants, also called flowering plants) appeared at the end of Mesolithic age and human beings invented agriculture during Neolithic age. They used stone implements for cultivation that was the beginning of primitive agriculture, which was a switch off from the lifestyle of hunting and gathering. Until the Neolithic period, humans devoted a great deal of time and energy in search of food that was largely through hunting and gathering of animals and plants. This lifestyle changed in due course when they started domesticating plant species some 10,000 years ago. It was a gradual and slow process of change where plant species generated enormous variability through mutations, cross-pollination, and recombination (through natural hybridization). Of these variations, variations only favorable (i.e., advantageous) for adaptation to the environmental conditions were primarily selected by nature and others not fit for survival were lost. It is believed that the beginning of agriculture and plant domestication of our present-day plant species occurred thousands of years ago at different times and places. These places were geographically isolated and inhabited by different societies around the world, and were called as centers of origin (and later named centers of diversity). The currently cultivated crop species were domesticated from wild types at these centers, and from these centers domesticated plants first spread. Probably the first development in plant breeding occurred with the domestication of wild plants such as wheat, barley, and lentils around 13,000–10,000 years ago. It is suggested that perhaps the first successful selections were made in the Fertile Crescent in Eurasia between 9000 and 7000 BCE along the Tigris and Euphrates rivers in the present day Iraq extending into Iran and Southeast Turkey on the north, and south into Lebanon, Israel, and Southern highlands. This area shows a rich diversity of cereals and pulses.

The wild species progenitors have traits for adaptations, for example, the wild species of cereals were tall, branched, or bushy, and in case of wild species of legumes the plants were trailing type. In general, the wild species produced small seeds with undesirable appearance and hard seed coat. The spikes of cereals and pods of legumes were prone to shattering. In nature, these traits were helpful for seed dispersal and perpetuation from year to year. Other traits that were altered during domestication included size and shape of foliage, tubers, berries, fruits, and grains that increased yield, give better nutritional ability, and other desirable traits. Molecular analysis has revealed that key factors that distinguished a domesticated plant from its wild type are often in the genes that encode transcription factors, proteins that regulate the expression of many other genes (Doebley et al., 2006). Mutation in the genes coding for protein of certain biosynthetic pathways give rise to difference in nutrient composition among varieties of the same crop. These include mutations in genes that created sweet corn varieties, lower lignin for improved digestibility for livestock, and improved fatty acid profile in oilseed crops.

Of the thousands of plant species, only few plant species were domesticated for food, fiber, oil, and other human use. Of about 250,000–300,000 plant species, 10,000–12,000 (4% of total plant species) are edible but less than 200 are used. Furthermore, rice, wheat, and maize contribute ~60% of the calories and protein consumed by humans from plant sources (FAO 1999, 2004), and about 30 crop species provide calories and nutrients that humans need every day.

Pre-Mendelian plant breeding

The Assyrians and Babylonians in ancient times recognized that date palms were of two types, male and female. As early as 700 BCE reportedly people in these civilizations hand pollinated date palms, which was one of the oldest domesticated fruit trees. Selection of superior seed was mentioned during Roman times. First real description of sex and plant hybridization was made by Rudolph Camerer (also known as Camerarius) (1665–1721) at the University of Tübingen in Germany. He conducted experiments on mulberry, castor bean, spinach, and maize and showed that pollen produced on male flowers was indispensable to fertilization and seed development on female plants. He, however, did not work with plants of both sex organs in a single flower. Botanist Richard Bradley (1688–1732) used controlled experiments to observe that pollen had a role in fertilization. He made this conclusion after observing that if he removed pollen producing stamen of the flower, in isolation no viable seed was produced. He also reported controlled (and not random) cross-fertilization as a path of fruit improvement. It has been reported that in the 17th century in China, the Emperor Khang Hi (1662–1723) selected an early maturing rice plant from his garden, which was multiplied and cultivated widely because of desirable quality and was named Imperial Rice.

Thomas Fairchild an Englishman in 1717 produced an interspecific cross between Sweet William (*Dianthus barbatus* L.) as female and Carnation (*Dianthus caryophyllus*) as pollen parent and the product was known as Fairchild's Sweet William (also called Fairchild's Mule), which was sterile. In 1716, an American, Cotton Mather observed that ears from yellow corn grown next to blue or red corn had blue or red kernels in them, suggesting the occurrence of natural cross-pollination. Jean-Baptiste van Mons (1765–1842) observed that seed of the oldest trees (with good fruits) generally produce inferior progeny, and seed of the young trees (even with bad fruit) produce superior progeny (with good fruits). Van Mons developed a lot of varieties in fruit trees (particularly pears) by sowing seed from younger trees (with good fruits), selecting the best progenies (seedlings) to obtain new seed (fruits), and repeating the process to develop superior varieties (and after four to five generations, nearly all progenies were considered superior). However, the decline in older varieties is probably attributed to diseases and other factors, and not all hybridizations were planned. Thomas Knight (1759–1838) is mostly known for his work on phototropism and geotropism, and on fruit trees, strawberry, and pea breeding. He made controlled crosses (in late 1700s) and kept record of progeny performance to develop varieties. However, these efforts were led by individuals. Philippe-Victoire Levêque de Vilmorin and his wife revived the plant and seed company of her parents Claude Geoffroy and Pierre Andrieux, which later became the Vilmorin-Andrieux seed company in the mid-18th century. The Vilmorin breeding efforts in France included work on individual plant selection based on progeny test (was used for improving sugar content in sugar beet), which was called Vilmorin Isolation Principle (however, this method was not effective in sugar beet, a cross-pollinated crop).

The Swedish Botanist Carlous Linnaeus (1753) published Species Plantarum and established the binomial nomenclature, which consisted of generic name and a species name, for example, *Zea mays* for maize. This work helped in the understanding of the plant species as the system used number and arrangement of reproductive organs (stamens and pistils) as the critical characteristics for nomenclature. The first systematic investigations into hybridization were conducted in many plant species (54 species belonging to 13 genera) by Joseph Koelreuter (between 1760 and 1766) in Germany, who promoted theory of fertilization and artificial hybridization. Most of the crosses resulted in sterile progenies, which suggested that hybridization in general is possible between closely related plants. He produced fertile hybrid in *Nicotiana* (e.g., *Nicotiana paniculata* and *N. rustica*) and demonstrated that hybrid progeny received traits from both parents and were intermediate in most traits. He also reported varying degrees of sterility from his experiment involving interspecific observed luxuriance, and postulated commercial potential of hybrid vigor. Christin Konrad Sprengel (1750–1816) identified self- and cross-fertilized plants. This provided better understanding about floral morphology and mode of pollination, and that the function of flowers was to attract insects for pollination.

Thomas Andrew Knight (1759–1838) in late 18th and early 19th century effectively selected superior cultivars of fruits and vegetables through crossing to improve useful food plants for better quality traits. Superior progenies of a cross were identified and were asexually propagated, thereby overcoming subsequent segregation and reselection. Patrick Shirreff, a Scotsman, began his work on wheat and oat crops in 1819. He practiced selection in these crops and developed superior pure line cultivars. For the first time he evaluated his selections along with other in plots, which were sampled and data were recorded on height, maturity, and other agronomic traits. Henry Zimmerman in 1837 in Maryland selected pure line cultivar in wheat after noticing three ears of singular appearance near the edge of his wheat fields. These were grown for seed increase and after 6 years, seed was sold to the public as "Zimmerman" wheat (white kernel wheat). John Le Coutier (1836) published his results on selection in wheat and reported that progenies from single grain or ear were more uniform and pure. David Fife, a farmer from Ontario, Canada, made a selection named Red Fife in 1842, which came from a seed lot by Fife from Poland (via Scotland) in 1841. From the winter wheat lot, only one plant produced spikes as the rest did not produce spikes because they were spring seeded and therefore, did not have a winter cold treatment for vernalization. Fife harvested the seed from this single spring wheat plant and increased them to develop the cultivar Red Fife, which possessed excellent bread making qualities and it credited to have ushered in an exponential wheat growth in Canada. It was the first truly hard red spring wheat in North America that made substantial contribution to agriculture as it became the predominant parent in the development of improved cultivars including Marquis spring wheat, which compared with Red Fife was short statured, less prone to shattering, and higher yielding with similar bread making qualities.

In middle of the 19th century, Hallett in England carried out his work in what is known as "head-row method" and he believed that selection within a pure line can affect improvement if practiced over multiple generations. He practiced single plant selections in wheat, oat, and barley and developed several commercial varieties. William James Farrer (1845–1906), a famous Australian wheat breeder, is best recognized for his early maturing wheat variety Federation for its general resistance to rusts, grain quality, and adaptation traits. He is credited for visualizing ideotype of wheat including shorter stature, stiff straw, and narrow leaves.

It should be mentioned that much of his work was conducted prior to the universal understanding of principles of genetics and helped establish a flourishing wheat breeding industry in Australia. Luther Burbank (1849–1926) made outstanding contributions to crop breeding, and probably he can be considered the most successful plant breeder. He was responsible for the development of 800 strains of asexually propagated plant species including Burbank and Russet Burbank potatoes, "spineless" cactus, walnuts, berries, ornamentals, flowers, vegetables, stoneless plum and semi- and free-stone plums, plumcot (cross between plum and apricot), and many more crop plants. Russet Burbank potato cultivar became the world's predominant potato in the food processing industry.

It may be emphasized that prior to 1900 much of the plant breeding was practiced not by the trained geneticists and plant breeders but by farmers, amateurs, and non-scientist who selected seed or clone from preferred plant types of land races or populations for subsequent sowing. This period was also devoted to the understanding of the floral structure and function of both self- and cross-fertilized species. Some of the workers also made successful crosses among closely related species and less closely related species. The crosses were fertile between closely related species, whereas less related species produced sterile F_1 hybrids. However, the mechanism of segregation and genetic principles was not universally understood (or mainstream).

Mendelian plant breeding

During the 18th century, hybridizations were made by several researchers and interesting results were obtained. New varieties were produced from such crosses; however, mechanism of fertilization, that is, the development of pollen tube and its approach to the ovule was discovered around 1833. Robert Brown an English botanist did considerable work on this problem. The cell theory first put forward convincingly in 1839 by German microscopists Schleiden and Schwann proposed that all plants and animals are constructed from small fundamental units called cells. The cell contains nucleus and cell organelles, which are surrounded by the cell membrane. The discovery of strand-like structure, that is, chromosomes by C.W. Nageli in 1842, was followed by the discoveries of the cell divisions (mitosis and meiosis) by work of several researchers (including Nageli and Robert Remak). This helped in the understanding that the cells arise from other cells by the process of cell divisions and the traits are transmitted from generation after generation.

During this period, attempts were being made to use hybridization to develop superior cultivars both in cross- and self-fertilized crop species. Gartner in Germany made extensive crosses involving some 700 species and obtained 250 hybrids, and this work was published in 1849. In 1847–48, American farmer Robert Reid in Illinois crossed a semigourd seed dent with little yellow to create Reid Yellow Dent. Robert Reid and his son James Reid improved the variety for ear selection, higher kernel rows, brighter yellow color, etc. Reid yellow dent is reported to be in the background of about half of modern maize hybrids in the United States. However, after native Americans, the first variety development records suggest that Noyes Darling (1844) developed a white seeded sweet maize variety called Indian Corn using systematic breeding approaches (Singleton WR, 1944). For hybridization of maize, detasseling of female plants was done to allow male plants (yellow corn) to pollinate female plants (sweet corn). The earliest recorded attempt to develop barley cultivars using hybridization was made by F.H. Horsford in 1879 in the United States.

The Scientific foundation for evolution was laid during the second half of 19th century through the writings of an English scientist Charles Darwin. He described the theory of evolution in his book "Origin of Species by Natural Selection." His theory was based on few important features: (1) overproduction of offspring and consequent struggle for existence, and that there are some forms that will be more successful at surviving and reproducing than other forms in a given environment, (2) variations and their inheritance, such that with any population there will be variation in morphology, physiology, etc., among individuals, (3) elimination of unfavorable variations (i.e., survival of the fittest), and (4) offspring resemble their parents more than they resemble unrelated individuals. He proposed that dissimilarities seen in cultivated plants from their wild relatives were due to selection pressures imposed by early man. Darwin's theory formed consciously or subconsciously approaches for crop improvement in the late 1800s along with Mendel's law of inheritance.

Gregor Johann Mendel (1822–84), an Austrian monk, formulated the fundamental laws of inheritance (i.e., the law of segregation and the law of independent assortment). The Mendelian laws of genetics form the basis of classical genetics. Mendel performed his scientific investigations for 8 years (1857–1965) in the monastery garden at Brunn (Austria) on hybridization of plants, particularly garden peas, a self-fertilized plant. He presented results of his experiments before the Natural History Society of Brunn in 1865. While describing the results Mendel used the term "factors" for the unit of inheritance, which were later on called "genes" by Johannsen. The proceedings of the society were published in the *Transactions of the Brunn Natural History Society* but had limited circulation. Therefore, the work of Mendel remained unnoticed for 35 years until three distinguished botanists DeVries in Holland, Tschermak in Austria, and Correns in Germany working independently rediscovered the Mendel's laws in 1900, which provoked a tremendous scientific interest in genetics and had great influence on the development of plant breeding.

Plant breeding in the 20th and 21st centuries

The rediscovery of Mendel's laws opened up a new era in plant breeding. Before these principles were understood, plant breeding was more an art. However,

Mendelian principles accelerated the exploitation of naturally occurring genetic variability and in the generation and manipulation of combination of new variability, and as a result, plant breeding became more of science and less an art.

Johannsen (1903), the Danish botanist, developed pure line theory of selection in Princess bean (*Phaseolus vulgaris* L.) (pure line is a progeny of a single self-fertilizing individual). Nilsson-Ehle (1909) proposed the existence of polymeric (multiple) factors for inheritance of seed coat color in wheat. He gave scientific background to transgressive breeding, fully explaining the advantage of recombination breeding. Wilhelm Rimpau (1842–1903) was one of the earliest to develop a partially fertile hybrid from wheat/rye cross and this work led to the later development of triticale crop species. East (1908) working at the Connecticut Experiment Station, in the United States, published his work on inbreeding in maize. Shull (1908, 1909) at Cold Spring Harbor conducted extensive research to develop homozygous lines and the best two lines were used to develop hybrids in maize. The discoveries of East and Shull in maize were instrumental for the wide application of hybrid vigor in other crop plants including field crops, vegetables, and horticultural plants. Jones (1918) suggested the production of double cross hybrids in maize using four inbreds, and the commercial seed was produced on high-yielding single cross parents, as at that time single cross commercial hybrid proposed by Shull was considered infeasible for seed quantity production.

Nikolai I. Vavilov (1887–1943) is recognized as the foremost plant geographer, botanist, and geneticist of contemporary times. He was ahead of his time; as he was a pioneer in genetic resource management. After several years of investigations, he proposed now the familiar eight centers of origin. He anticipated genetic erosion and therefore, he organized and took part in more than 100 germplasm collecting missions in the major agricultural areas of the world and concluded that the greatest species variation occurred in certain restricted areas (centers of diversity), which he believed identified the centers of origin of those species. He insisted on a systematic classification, evaluation, and utilization of vast assembled collections. He is credited for creating one of the world's largest seed repository with over 250,000 accessions. Vavilov also proposed law of homologous series in hereditary variation.

Muller (1927) for the first time demonstrated the ability of X-rays to induce genetic variation in fruit fly (*Drosophila melanogaster*). This work was used in crop plants (barley) by Stadler (1928). The discovery of X-rays as mutagen and its use to induce variation had initiated a new field, that is, induced mutagenesis, which was used as an important tool in basic studies (i.e., locating genes on chromosomes, studying gene structure, expression, and regulation, and for exploring genomes). Plant Breeders across the globe started to investigate the use of radiation-induced mutations. More than 2252 mutant varieties have been released (Maluszynski et al., 2000). Of these 1585 were crop varieties released as direct mutant, and 667 were derived through crosses with induced mutants; however, more varieties have been developed since the publication. Mutation breeding approach has been used in sexually propagated crops with small flowers which are difficult to cross and in asexually vegetatively propagated crops. In general, mutation breeding approach has been to improve the well-adapted plant varieties by altering one or few major traits.

During the 20th century, plant breeders and geneticists were trained in the colleges and universities and several experiment stations were established across the world. Most of plant breeding was directed toward collection of the indigenous variability and introductions of exotic materials. The improved cultivars of self-pollinated crops were developed using pure line selection. The plant breeders developed specialized techniques of hybridization and used diverse lines to cross and generate segregating generations to select superior segregants. The mass selection and its modifications were used to improve the cultivars of cross-pollinated crops which were largely populations, but hybrids were rapidly gaining popularity in some regions (e.g., the United States). The inbred lines were produced through artificial selfing and superior inbred lines were used to develop hybrids. Clonal selection was used to develop superior clones in vegetatively propagated crops. The field plot techniques were developed for testing the improved genotypes. The use of field designs and statistical analysis of agronomic and other traits data added precision in varietal evaluation. To develop varieties with stable performance for adoption in wider areas, multilocation testing was used. The data were recorded not only for agronomic traits but also for resistance to biotic and abiotic stresses. Genotypes with (significantly) higher yield when compared with previously cultivated check cultivar were promoted for further testing and commercialization. The line(s) which possessed resistance to one or two major stresses of the area/region were released for commercial cultivation. This period also witnessed the development of network for seed multiplications of cultivars/hybrids, and as a result several private companies came into existence.

Green revolution

Since the time immemorial, the scarcity of food has always been a problem for the human civilization. Food shortages have caused several famines; some of these are well documented (e.g., the Irish famine of 1840s, and Bengal famine of 1943 in India). Moreover, the human population in certain regions of the world grew at a much faster rate and food production increased at a slower rate, leading to insufficient food supply necessitating steps to be taken at the National and International levels. In view of the food shortages, serious efforts were made in the 20th century to achieve yield increases in major food crops including cereals, such as maize, wheat, rice, sorghum, and pearl millet. The remarkable events were the shift toward single-cross maize hybrids, use of cytoplasmic genetic male sterility in the production of hybrid varieties in maize, sorghum, and pearl millet, and semidwarf varieties of rice and wheat. The genesis of semidwarf wheat and rice and their cultivation were responsible in quantum jump in yield gains of these crops in several countries and was termed "Green Revolution."

In the middle of 1920s H.A. Wallace founded pioneer Hi-Bred Seed Company and produced hybrid corn; however, due to the disastrous economic depression of 1930s, the use of hybrids did not really take off until the early 1940s. By the mid-1950s, double cross hybrid dominated US maize production and open-pollinated varieties had virtually disappeared. Later on in the United States, the double cross hybrids were replaced by single cross hybrids further increasing yields in 1960s. H.A. Wallace, who himself was a breeder, became Secretary of Agriculture and the US Vice-President under Roosevelt. He interacted with J.A. Ferrell from the Rockefeller foundation to promote agriculture initiative in Mexico and launched a program of crop breeding in Mexico. The Mexican Government in collaboration with the Rockefeller foundation launched a project in 1943, to set up the Mexican Agriculture Program, which is credited with ushering in the green revolution. N.E. Borlaug, a plant pathologist, joined this project in 1944 where he along with his team of dedicated scientists developed higher yielding stem rust-resistant varieties of wheat and appropriate practices including fertilizer, irrigation, tillage, disease, and pest management for optimizing the yield potential of these varieties. The concept of shuttle breeding was also initiated at this time (1945) to reduce the time to develop varieties; and shorter statured varieties responsive to irrigation and fertilizers began to be developed. In view of encouraging results from this program, the Rockefeller Foundation decided to establish an independent institute, which was named, International Maize and Wheat Improvement Centre (CIMMYT) headquarter in Mexico. CIMMYT was created (1966) by merging several regional and country research programs initiated by the Rockefeller Foundation with a focus on international agricultural research and programs tackling global food security. Rockefeller Foundation and Ford Foundation joined hands with the Philippines government to establish the International Rice Research Institute (IRRI) near Manila (Los Banos) in 1962. IRRI focused on research and breeding of rice, which was staple food across the world, to mirror the wheat crops' transformation and adoption.

During the 1960s, efforts were being made to introduce green revolution programs in developing countries including, India, Pakistan, Philippines, and Indonesia, through partnerships between federal governments and the international programs of IRRI and CIMMYT. The semidwarf varieties of wheat developed at CIMMYT by Borlaug and the wheat breeders, and rice varieties developed by H.M. Beachell and the rice breeders at IRRI spread in developing nations and became very popular in 1960–70s in the wheat and rice growing reasons in several countries including India and Pakistan. The new varieties, along with improved management practices revolutionized the wheat and rice production in these countries and saved millions of people from starvation. Borlaug was awarded the Nobel Peace Prize in 1970 for this work.

The traditional cultivars of wheat and rice grown prior to Green Revolution were tall, leafy with weak stems that were prone to lodging; and non-responsive to high yield conditions and environments. Such varieties had a harvest index of 0.3. Contrary to this, the semidwarf varieties of wheat and rice had a harvest index of 0.5 that gave at least two to three times higher yield than the local varieties. *Norin 10* variety from Japan was the source of semidwarf wheat that channeled the wheat green revolution. Two US varieties, Fultz and Turkey red, are in the lineage of Norin 10. Glassy Fultz was developed using pedigree method and was crossed as a male with Daruma as a female (donor for semidwarf trait, contained recessive dwarfing genes *rht1* and *rht2*) to create Fultz-Daruma. In 1924, Fultz-Daruma was crossed with Turkey Red (as female), and F_3 selection named Tohoku No. 34 was named Norin 10. Norin 10 was released in 1935. Norin 10 and other similar Japanese lines showed short stiff straw with higher tillering capacity. S.C. Salmon brought Norin 10 and 15 other varieties from Japan to the United States and was used in breeding of the semidwarf winter wheat variety Gaines by Orville Vogel of USDA and released in 1961. Gaines was developed from a multiple parent cross. The original cross made in 1949 was Norin 10 × Brevor, and selection from this cross was crossed with a selection from a cross of Orfred and sister selection of Brevor. The selection from this cross was crossed with Burt in 1952.

The semidwarf winter wheat cultivar Gaines was used by Borlaug in hybridization with Mexican strains to develop semidwarf spring wheat cultivars. He used shuttle breeding, where in different generations were grown at two diverse environments, that is, sites which were located at different latitudes with changing day length and altitudes by 2600 m and were different in terms of soil temperature, rainfall, and photoperiod. In summer, wheat crop was grown at high altitude near Mexico City, while in winter season they used the Yaqui Valley of northern Mexico (under irrigation plots, Sonoran Desert). Only the varieties that withstood the rigors of both environments were advanced in the breeding program. At CIMMYT, cultivars Hope (a stem rust-resistant wheat from the United States) and Frontana (a leaf-resistant Brazilian cultivar) were used to transfer their durable resistance (*Sr2* gene complex from Hope, and *Lr34* gene complex from Frontana). The improved varieties developed using the above methodology along with appropriate agronomic packages resulted in fourfold increased productivity in Mexico. Some of these varieties were introduced into other parts of the world, including India, Pakistan, and Turkey in 1966 with spectacular results.

Similar approach was used in rice at IRRI, and semidwarf varieties were released in mid-1960s to farmers in the Philippines. Of these, IR8 was most successful; it was semidwarf, early maturing, non-photosensitive, and nitrogen responsive with high yield potential. Due to its earlier maturity, it was suitable for double cropping in certain rice growing regions. IR8 was developed from a cross of Indonesian variety Peta (from a cross between Indian variety Latisail and the Chinese variety Cina) as female and a semidwarf Taiwan cultivar Dee-Geo-Woo-Gen (DGWG) was used as male. DGWG possessed a single recessive semidwarf gene *sd1*. IR8 had stiff straw, was resistant to lodging and photoperiod insensitive making it widely adapted. IR8 and several other varieties with IR series were introduced in other rice growing countries of Southeast Asia where these varieties gave higher and stable yields, generating a name of miracle rice. The higher yields were obtained due to their responsiveness to higher doses of fertilizers and other inputs in contrast to the local varieties that were prone to lodging.

The green revolution was successful in increasing the world production of cereals and thereby provided availability of food. However, the achievements of green revolution have been criticized by some that it has benefited the rich and large farmers in areas where irrigation facilities were available, and the technology was not environment friendly as excessive use of agrochemicals is responsible for soil degradation.

The next revolution that has ushered in increased crop productivity relates to biotechnology innovations. Biotechnology and molecular tools have been used to boost agriculture production, and these include genetically modified (GM) crops and marker-enabled breeding to develop higher yielding varieties with desirable traits.

Genetically modified crops

The GM or transgenic plant is one that contains a gene or genes which have been introduced artificially into the plant's genetic make-up using a set of biotechnology techniques collectively known as recombinant DNA (rDNA) technology. Transgene(s) are transferred using either particle bombardment or plasmid-mediated transformation. Once stably transformed, the transgene(s) is inherited along with the rest of the plant genes. The offspring (plants) carrying transgene(s) are also transgenic plants. Major examples of crops grown with Genetically Modified Organism (GMO) technology and grown on farmer fields, include insect-resistant (IR) "*Bt*" cotton, maize, and canola; herbicide-resistant soybean, cotton, and maize; viral disease-resistant papaya, etc.

The GM revolution started with discoveries in the 1980s by M. Montagu, J. Schell, and M. Chilton who discovered a natural gene transfer mechanism between *Agrobacterium tumefaciens* and plants. The tumor causing genes of *Agrobacterium* could be replaced by other genes not present endogenously in the crop species to create new crop trait package. The GM cultivars were developed in 1990s and were first commercialized in the United States in 1996. The two major types of GM crops in the US row crops are herbicide tolerant (HT) and IR.

The HT traits include tolerance against herbicides, such as glyphosate, glufosinate, and dicamba. These expand farmer's ability to control weed in their field more effectively. As per the ERS-USDA (2019), the HT soybean acres in the United States have expanded from 17% in 1997 to 94% in 2014, and in 2019 it has grown to 98%. Similarly GM cotton with HT trait has also grown from 10% in 1997 to 98% in 2019. In 2019 maize HT trait crops were grown in 92% of the total acres. The IR traits include primarily genes from the soil bacterium *Bacillus thuringiensis* "*Bt*." The bacteria produces crystalline protein called Cry protein that has a high target specificity to control insect–pest (Cry toxins only impact insects of specific taxonomic order and are shown to be safe to mammals). Some of the destructive insects controlled by the *Bt* toxin are European corn borer, rootworm, corn earworm, and bollworm. In the United States, IR *Bt* corn acres have grown from 8% in 1997 to 83% in 2019. The IR *Bt* crops have grown from 15% in 1997 to 92% in 2019 (ERS-USDA, 2019). The increase in crop acre percentage is likely due to the newer IR stacked maize hybrids that have IR traits that provide protection against both corn borer and corn rootworm (additionally corn earworm and fall armyworm); while the earlier hybrids only provided protection against the European corn borer). The Bt technology in cotton varieties has upgraded from a single-gene trait to multigene trait packages. The first-generation *Bt* cotton had a single *Bt* gene, the second generation had two *Bt* genes (producing two toxin proteins), and more recently to a third generation has a three *Bt* gene package combining crystal (Cry) and vegetative insecticidal proteins (Vip). *Bt* cotton can control bollworm, tobacco budworm, pink bollworm, and (Beet and Fall) armyworm infestations.

The global crop hectarage of GM crops continued to grow from 1.7 m ha in 1996 to 191.7 m ha in 2018. These crops were grown by 17 million farmers in 26 countries (ISAAA, 2018). The adoption rate of GM crops speaks of sustainability and benefits it delivers to the growers and consumers. In 2017, the United States had maximum area under GM crops (75 m ha), followed by Brazil (51.3 m ha), Argentina (23.9 m ha), Canada (12.7 m ha), and India (11.6 m ha). The area in India of GM crop is only from cotton. Soybean are the most widely grown GM crop worldwide, with 50% (95.9 m ha) of the global biotech crops and 78% of the total global soybean production (ISAAA, 2018). In maize, 31% of the global maize production is with GM crops.

The major GM crops are soybean, maize, cotton, and canola, while the other crops with GM grown varieties include alfalfa, papaya, eggplant, potato, safflower, apple, and sugarcane (ISAAA, 2018). Not all GM varieties are developed by transgene DNA sequences. For example, Arctic apple was created by introducing antipolyphenol oxidase (anti-PPO) RNA [using RNA interference (RNAi) technology]. The RNAi is used to lower PPO levels in apple by introducing RNA sequences that cause the degradation of PPO RNA, therefore, no PPO enzymes are produced leading to non-browning apples when cut.

As per ISAAA (2016), since 1996, in the United States 195 single trait events in 20 crop species have been approved: maize (43), potato (43), cotton (28), soybean (24), canola (20), tomato (8), alfalfa (3 events), apple (3), chicory (3), papaya (3), rice (3), sugar beet (3), rose (2), squash (2), creeping bentgrass (1), flax (1), melon (2), plum (1), tobacco (1), and wheat (1). The examples of GM crops with their traits are presented in Table 1.1. The traits such as insect resistance with *Bt* genes in maize, cotton, rice, tomato, and potato and herbicide tolerance (glyphosate tolerance) with *A. tumefaciens* strain CP4 [gene product CP4 EPSPS giving herbicide tolerant form of 5-enolpyruvulshikimate-3-phosphate synthase (EPSPS) enzyme] in soybean and alfalfa have resulted in substantial yield gains and benefits to farmers.

Table 1.1 Examples of GM traits and crops (either grown commercially or not grown commercially).

GM trait	Crop(s)
Insect resistance	Cotton, eggplant, maize, potato, poplar, sugarcane, soybean, tomato, and rice
Herbicide tolerance	Alfalfa, canola, chicory, cotton, creeping bentgrass, maize, rice, soybean, sugarbeet, and wheat
Virus resistance	Papaya and potato
Disease resistance	Potato
Altered oil composition (i.e., improved oil quality)	Canola, safflower, and soybean
Male sterility and restorer genes	Canola, chicory, and maize
Drought tolerance	Maize and rice
Nutritional or quality trait	Rice, tomato, and potato
Altered lignin	Alfalfa

This list is not exhaustive.
GM, Genetically modified.

The GM technology has been developed for value addition and nutritional quality of crop plants. Initial GM varieties had a single trait focus, which evolved to multiple genes (e.g., insect tolerance through multiple Cry proteins). In high value crops, insect and herbicide tolerance was developed. The current trend of biotech crops and traits is better stacked traits (e.g., multiple modes of herbicide resistance for a more comprehensive weed control, multiple insect species protection). This trend will likely continue and deploy multiple modes of simultaneous resistance to pests and diseases and perhaps include additional traits such as quality and climate resiliency. The current stacked trait example includes glufosinate–dicamba HT, glyphosate–dicamba HT, glyphosate–2,4-D choline tolerant, and Bt–HT. Triple herbicide tolerance is available in soybean varieties (Xtendflex) against dicamba, glyphosate, and glufosinate herbicides, Enlist E3 against glyphosate, glufosinate, and 2,4-D herbicides, and LibertyLink GT27 against Liberty, glyphosate, and 4-hydroxyphenylpyruvate dioxygenase (HPPD) inhibitors based herbicide. In maize, an eight-stack hybrid variety (SmartStax) is available with protection against corn earworm, European corn borer, southwestern corn borer, fall armyworm, corn earworm, black cutworm, along with herbicide tolerance against glyphosate, and glufosinate. However, the cost of seed generally includes the cost of technology.

The land races, wild, and weedy relatives need to be collected, conserved, and preserved; because GM technology does not replace the essence of plant breeding: useful genetic variation. The GM technology is also competing with newly emerged gene editing technique (see Chapter 27: Molecular Tools in Crop

Improvement and Cultivar Development), which provide more precision in genetic engineering without being classified as GM. The most well-known current gene editing technology is labeled clustered regularly interspaced short palindromic repeats/CRISPR-associated protein 9 (CRISPR/Cas9) system due to the protein complex used for editing DNA sequences. This technology has generated excitement due to the prospects in application and rapid acceptance as the most effective technology to change the DNA of an organism, including addition, deletion, or altering nucleotides at specific locations in the genome. The CRISPR/Cas9 system for gene editing is lower cost, faster, and higher efficiency than other existing gene editing methods, therefore, is expected to spur innovations in plant breeding.

Molecular markers in plant breeding

In the past 15–20 years molecular markers usage by plant breeders has become mainstream to introgress genes and or quantitative trait loci (QTL) in germplasm and variety development. This technique is called marker-assisted breeding (MAB) [also called marker-assisted selection (MAS)] for application in forward breeding or marker-assisted backcross breeding (MABB) [also called marker-assisted backcrossing (MABC)] to pyramid few genes/QTLs for different traits without changing the original genetic constitution of the parental cultivar. MAB is now routinely used in private and public breeding programs, for example, in most wheat breeding programs in Canada (DePauw et al., 2011). MAB has been used for traits controlled by major genes such as disease and insect resistance, for quality traits such as low cadmium content in durum wheat and high grain protein content (Table 1.2). Significant efforts have also been made in area of rice molecular breeding for drought, submergence, and salinity, the three most devastating abiotic stresses. QTLs and genes controlling these abiotic stresses have been identified and introgressed into high-yielding varieties through MAS and MABB (see Chapter 21: Breeding for Resistance to Abiotic Stresses, and Chapter 27: Molecular Tools in Crop Improvement and Cultivar Development).

For this technology to be successful, the development of markers that are in the gene (i.e., perfect markers) or very tightly linked to the gene (i.e., linked markers) and marker validation is imperative. Meta-analysis (comparing and validating QTL across genetic backgrounds and populations) to identify QTL that are relevant in breeding programs, stable in different background, and devoid of linkage drag is essential prior to their usage in plant breeding programs. This method has resulted in significant economic incremental benefit in major food crops, such as, wheat, maize, and rice. With the ongoing mapping of agronomically important genes and the development of high-throughput marker systems, the use of molecular markers in cultivar development would continue to be important in future plant breeding efforts.

Table 1.2 Examples of Western Canadian wheat cultivars registered using marker-assisted breeding (DePauw et al., 2011).

Market class	Cultivar	Registered in	Breeding institution	DNA marker or gene and trait
CWRS	Lillian	2003	AAFC, Swift current and Winnipeg	*Gpc-B1/Yr36*, grain protein
CWRS	Somerset	2004	AAFC, Winnipeg	*Gpc-B1*, grain protein
CWES	Burnside	2004	AAFC, Winnipeg	*Gpc-B1*, grain protein
CWRS	Goodeve	2007	AAFC, Swift Current	*Sm1*, midge resistance
CWES	Glencross	2008	AAFC, Winnipeg	*Sm1*, midge resistance
CWAD	Brigade	2008	AAFC, Swift Current	*Cdu1*, grain cadmium uptake
CWAD	Verona	2008	CDC, University of Saskachewan	*ScOpc20*, *Usw15*, grain cadmium uptake

Since the publication DePauw et al. (2011), numerous other cultivars have been developed using marker-assisted selection in Canada.
AAFC, Agriculture and Agri-Food Canada; *CDC*, Crop Development Centre; *CWRS*, Canada Western Red Spring; *CWES*, Canada Western Extra Strong; *CWAD*, Canada Western Amber Durum.

The current technology of next-generation sequencing (NGS) has made DNA sequencing high-throughput and cheaper. NGS provides extensive and detailed genome coverage and easier to explore and subsequently exploit the genetic and phenotypic relationships and diversity. In addition to the identification of polymorphic SNPs for MAS or MAB approaches, haplotype blocks are generated that are correlated with trait variation (quantitative). Marker saturation also allows for more powerful genome-wide association studies to identify novel genes (and markers for MAS and MAB) from diverse association panels (cultivated and landraces for identification of rare alleles) through marker trait associations. Sufficient marker numbers also facilitate genomic prediction and selection, which is a new and powerful tool in plant breeders toolkit to make selection purely on genotypic information (without phenotypic assessment) allowing selection for complex and quantitative traits (e.g., yield) in earlier generations (single plants and single short rows) minimizing time and gaining selection and breeding efficiencies. In genomic prediction, contrary to MAS or MAB approaches, all molecular markers are used to determine the individual's breeding value. The advent of NGS is also beneficial for targeted genome editing with CRISPR/Cas9 type of systems, which require knowledge of the gene sequence to be edited. This approach of genome editing is currently not considered GMO in some countries, therefore pointing to their wide application for targeted gene-trait improvement (to remove deficiency or improve trait, as the need may be). Some of the newer upcoming molecular techniques attempt to increase genetic recombination to create more genetic diversity, as well as gamete selection to make selection of gametes (based on genomic prediction) and only advancing those which meet the breeder-imposed selection criteria. These are ongoing or upcoming technologies but point to the innovative approaches in plant breeding.

Advances in image based and high-throughput phenotyping

Phenotyping remains integral to plant breeding efforts, despite the development of genetics and genomics tools. Therefore, plant breeding requires advances in phenotyping especially for traits that are difficult, time consuming, or costly to phenotype. Phenotyping involves a broad range of traits including yield, days to maturity, plant height, stem lodging, disease and insect resistance, responses to abiotic factors such as water stress (drought, flooding), heat stress, cold stress, salt tolerance, quality, and many more. The list of traits is long and can be unique to each crop species and breeding program. However, phenotyping for these traits requires experienced researchers who spend countless hours sometimes on routine repetitive tasks. Circumventing the challenges of phenotyping will allow plant breeders to work on more traits, collect timely ratings, increase population sizes, increase trait measurement accuracy, and reduce the cost of the breeding program.

With these motivations, there have been a rapid development and deployment of sensors and carrying platforms (aerial and ground) for remote and proximal sensing. These sensors include cameras that can capture more accurate and precise information at a larger scale, in a time-series, and on numerous genotypes with the advanced data analytics to tease out important insights and information (this constitutes the field of phenomics). The higher throughput phenotyping (HTP) involves the usage of sensors to rapidly collect data in an automate manner. Image-based phenotyping is one of the major methods to collect HTP data, but many other types of sensors are also available that capture information in wavelengths outside of human eye capability, generating very useful data. Sensors are deployable on both unmanned ground vehicle and unmanned aerial systems.

The cost of sensors and unmanned system is continually decreasing; therefore, these are now accessible to a larger plant breeding community. However, the adoption will depend on cost, simplicity, data accuracy, and statistical analysis capabilities (especially due to the data size and complexities). The most exciting development in data analysis domain related to plant breeding and phenotyping is machine learning (ML) (Singh et al., 2016, 2018). The ML methods are able to handle data deluge coming from digitalization of plant breeding data, as it can extract features using automated approaches. These allow generation of information at a level and scale that was previously not possible. A subset of ML tool is deep learning, which have transformed numerous things around us, including in the areas of consumer analytics, diagnostics, autonomous vehicles, and financial management. Phenomics and ML methods are revolutionizing the plant breeding pipeline, and details are provided in Chapter 28, Phenomics and Machine Learning in Crop Improvement.

Contributions of plant breeding to the world agriculture

Plant breeders have made enormous contributions to increased food production in the 20th and 21st centuries throughout the world. Along with genetic improvement of yielding ability, efforts were made to reduce maturity duration of the cultivars to fit them in multiple cropping; modify plant architecture for pure and mixed culture; insulate the cultivars against harmful parasite and extreme weather conditions (e.g., drought), soil-related stresses (e.g., salinity); and genetic enhancement of nutritional quality. For this success, both cultivated and diverse accessions (related species) have been used. This has provided stability to the yielding ability with a suite of other traits. After the World War II, plant breeders quickly responded to the human requirements and developed higher yielding cultivars and hybrids, which were cultivated using appropriate crop husbandry including the mechanization of agriculture (from seeding to harvesting) and the use of agrochemicals has accelerated the production of major field crops in the world.

From 1964 to 2014, the total area production worldwide has seen 182%, 622%, 68%, and 76% increase for cereals, oilseeds, pulses, roots, and tubers crops, respectively (source: FAOSTAT). While the yield (kg/ha) has increased 161%, 192%, 48%, and 35% for cereals, oilseeds, pulses, roots, and tubers crops, respectively, from 1964 to 2014 (Fig. 1.1, data source: FAOSTAT). The world productivity (kg/ha) of cereals, oilseeds, pulses, and roots and tubers is presented in Fig. 1.1.

In case of cereals, the increase in area during the past 50 years was only marginal (8%) but the increment in productivity was spectacular (161%) which was partly due to the large-scale cultivation of semidwarf varieties of rice and wheat, and single cross hybrids and GM cultivars in the case of maize. The oilseed crops have a noticeable rise in area seeded (147.4%) and productivity (192%), attributed to dramatic increase in soybean and other oilseed crops for their oil and meal (protein). This includes a large increase of GM soybean in the United States, Brazil, and of canola in Canada. The HT GM cultivars and non-GM cultivars may not differ significantly for their yield potential; however, GM cultivars are preferred by large acreage farmers due to their advantage in weed management practices and improving farm production systems. Although pulse crops area only increased 13.5% in 50 years (1964–2014), the yield per area increased 48% in the same period. This is due to the fact that pulse crops across the world have gained lesser emphasis of breeders and geneticists and of the funding agencies. Moreover, the pulse crops are grown largely under rainfed cropping systems, and limited private sector investments are ongoing for pulse crops. The crop species

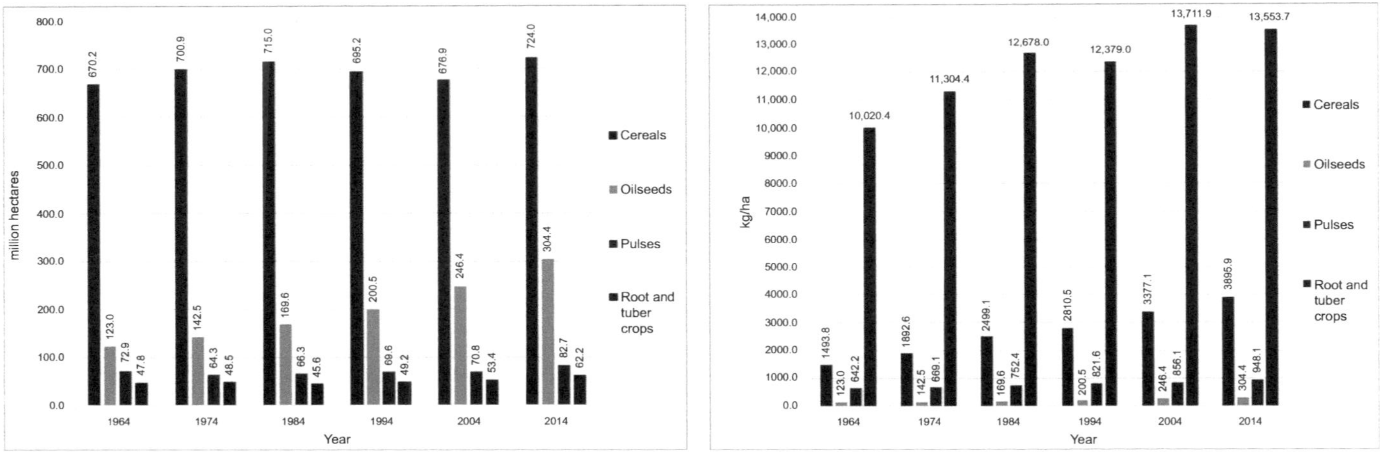

FIGURE 1.1

World production (Left graph) and productivity (Right graph) of cereals, oilseed, pulses, and root–tuber crops.

FAOSTAT, 2020. <http://www.fao.org/faostat/en/#data/QC> (accessed 05.06.20).

in category of roots and tubers group have shown an increase in area (30.3%) and productivity (35%). It is estimated that on an average a 70% yield improvement in cereals is due to genetic gains from breeding and remainder is attributed to improved farming practices, although some researchers attribute more equitable contributions to breeding and non-breeding factors. Irrespective, the role of breeding is fundamental to meet the needs of the human population in near and distant future.

Complementary role of public and private sector plant breeding

The public and private sector plant breeding have coexisted in industrialized countries over the past 100 years and both have contributed in increasing the crop yield. Public sector institutions, in general, have made contributions in the areas of basic and applied plant breeding for the benefit of society and without economic considerations. These efforts were not limited to the genetic improvement in food crops but all important crop and tree species useful to the welfare of mankind, including cereals, grain legumes, oilseeds, fiber crops, medicinal and aromatic plants, horticultural plants (vegetables, flowering plants, and fruit trees), timber trees, etc. Germplasm collection, evaluation, maintenance, preservation, and their utilization were given priority to achieve the gains. The germplasm with useful traits was made available on request to other breeders. The plant breeders and geneticists used knowledge of both the Mendelian and molecular genetics for the genetic enhancement of cultivars. Basic seeds of crop cultivars were provided to seed producing agencies for multiplication and distribution. Plant breeders in collaboration with the scientists of other disciplines have developed cultivation packages for each crop for achieving higher yields. In addition, public sector has devoted considerable resources for educating and training of plant breeders.

Private sector have worked selectively on some of the crops based on their importance in certain areas/regions and countries and profitability, for example, maize, soybean, canola, and cotton breeding is important for private sector in the industrialized world particularly the United States. Wheat is important for R&D of private companies in Western Europe, United States, Canada, and Australia, whereas canola is a priority in Canada, and cotton in some cotton growing countries. Rice is a priority in some countries of South-East Asia. Although not universally true, private sector has preferred to focus their largest R&D investments where they can protect their variety (generally GM) through a patent to ensure a more steady and protected revenue stream. These were possible through patents granted on GM traits and for crops that are not human mainstay diets.

Depending on the country and region, the relative roles of public and private plant breeding have varied. Intellectual property protection, globalization, and constraints in the funding for innovations of public sector plant breeding during the past 40–50 years have witnessed acceleration of the privatization of plant breeding, particularly in the developed countries. The discoveries in biotechnology and molecular biology disciplines, for example, increased emphasis on GM technology and molecular breeding, have reduced the activities of conventional public sector plant breeding which is somewhat concerning as they too play an important role in upstream work and product development for public good. Some of the multinational companies have made heavy investments to strengthen their capabilities in these areas and have an edge in the development of biotech cultivars of important crop species.

Future plant breeding

Population is expected to add of another 2–4 billion people to the planet within next 3–4 decades. This dictates the need to increase food production. According to Food and Agriculture Organization (FAO) estimate mostly in developing countries, food production must increase by 60% between now and 2050 to meet the demands. However, the amount of arable land is not likely to increase much in future because of increasing urbanization, salinization, and desertification. Average temperatures and dry land areas are predicted to expand in arid and semiarid tropics, which occupies about 40% of land area. In future, freshwater availability for irrigation would also be scarce. Global climatic changes can affect the crop productivity. An optimum temperature range is required for an optimum growth and development of the crop as well as for sink to accumulate starches, proteins, and fats. Schlenker and Roberts (2009) reported that yields increase with temperature up to 29°C for corn, 30°C for soybeans, and 32°C for cotton in the United States, but they would decline precipitously at higher temperatures. They predicted that yields of these crops in the current growing areas will decline by 30–46% by the end of the 21st century under most moderate climate change and by 63–82% under most warming scenario. It is estimated that even an increase of 1–2°C temperature would have dramatic deterioration in yields of the food crops and most dramatic impacts are expected in the arid zones. Flood prone regions of South-East Asia would also be worstly affected due to climatic changes. In view of above reasons, plant breeding priorities need to be redefined and strategized to be future looking. An important strategy would be to increase yielding abilities of crops vertically and to continue doing so, not only of major food crops but also of so-called minor crops especially the ones that were ignored during the era of green revolution. The ways and means of increasing the food production and the productivity of the food crops through plant breeding research are briefly described herewith.

One of the negative impacts of green revolution was widespread cultivation of only few crop species and their fewer cultivars which resulted in the genetic homogenization. Millets, small millets, and pulse crops are grown traditionally by small and marginal farmers under rainfed conditions which are stressful environments where major cereals (rice, wheat, and maize) fail. These crops were less attended by plant breeders and were somewhat ignored by policy makers and funding agencies during and after the green revolution era. Therefore, these crops need to be revisited for their genetic improvement. Major millets, which include sorghum (*Sorghum bicolor*) and pearl millet (*Pennisetum glaucum*), and small millets, which include finger millet (*Eleusine coracana*), foxtail millet (*Setaria italica*), kodo millet (*Paspalum scrobiculatum*), proso millet (*Panicum miliaceum*), barnyard millet (*Echinochloa esculenta* and *Echinochloa frumentacea*), and little millet (*Panicum sumatrense*), are valued for their nutritional quality and health benefits. In general, small millets provide reasonable and assured harvest under low inputs and most adverse climatic conditions. Millets are C_4 crops and hence are climate change compliant (NAAS, 2013). In Table 1.3 important traits of adaptation to adverse climatic and soil conditions of small millets are presented. Small millets, in general, are suitable for rainfed areas and dry land agriculture. Some of the cultivars of these millets are early maturing, suitable as catch crop.

Grain legumes (pulse crops; family—Leguminosae) occupy a unique position in the world agriculture by virtue of their 2-3x higher seed protein content than cereals. These crops have wider adaptation under stress prone marginal conditions, due to their capacity to fix atmospheric nitrogen (30–150 kg/ha) symbiotically in association with *Rhizobium* bacteria (Table 1.4). For many of the developing countries in the world, pulses constitute a cheap and readily available source of dietary protein, and is produced in an environmentally sustainable manner. Therefore, the only practical avenue to solve protein malnutrition problem in these countries is to significantly increase the productivity and total production of these crops. In the developed world, soybean [*Glycine max* L. (Merr.)] is an indirect source of protein as they are used in animal feed. Although soybean and peanuts belong to the family Leguminosae, as per the FAO, they are classified as oilseed crops but both are desirable for their oil (cooking and industrial applications) and protein (feed applications and food source) content. Pulse crops do not include legume crops that are

Table 1.3 Small millets and their adaptation ability traits to extreme climatic and adverse soil conditions.

Crop	Adaptation ability to abiotic stresses
Barnyard millet	It is able to grow on poor soil and is tolerant to cold.
Finger millet	It is tolerant to drought and salt stresses.
Foxtail millet	It is tolerant to drought and salt stresses.
Kodo millet	It is tolerant to drought and can be grown on poor soils from gravelly to clay.
Little millet	It is well adapted to regions with less rainfall and is tolerant to drought, pests, and salts.
Proso millet	It is well adapted to clay sandy soils.

Table 1.4 Adaptation ability of important pulse crops and soybean to climatic and edaphic conditions.

Crop	Adaptation ability to abiotic stresses
Chick pea	Tolerant to drought and low management, large seed variation.
Cow pea	It is grown on wide range of soil types (sandy to heavy clay).
Lentil	It can be grown on poorer soils and even on soils of moderate alkalinity. It can tolerate drought conditions.
Mung bean, Urd bean	These are grown widely in various agroclimatic zones and cropping systems.
Peas	It thrives well in places with cool climate and is grown on wide variety of soils types.
Pigeon pea	Better adapted to marginal climatic conditions. Its roots open up the soil and improve the soil structure.
Soybean	Very wide adaptation from tropics to temperate regions, and large variation in days to maturity.

harvested green (such as green peas, green beans, and edamame as these are classified as vegetable crops), and crops that are planted exclusively for sowing (cover crops) purposes (such as clover and alfalfa).

Pulse crops, soybean, and peanuts (*Arachis hypogaea*) can meet the needs of a growing population. Important pulse crops that need immediate attention of plant breeders are chickpea (*Cicer arietinum*), pigeon pea (*Cajanus cajan*), urd bean (*Vigna mungo*), mung bean (*Vigna radiata*), cow pea (*Vigna unguiculata*), lentil (*Lens culinaris*), peas (*Pisum sativum*), common bean (*P. vulgaris*), and faba bean (*Vicia faba*). Similarly, breeding efforts and resources are needed in soybean and peanut. These crops have a rich genetic diversity. The genetic enhancement in the yielding ability and production of these crops that would facilitate their easy availability to poor people which in turn would help in decreasing the number of malnourished people of rapidly growing populations in underdeveloped and developing countries. With a rapidly emerging interest in plant protein diets, and moving away from more energy intensive animal protein, pulse crops are expected to play a huge role in human diet in coming years.

The genetic improvement of millets and pulses would pave the way for ushering the second green revolution by developing higher yielding and climate resilient varieties of these crops. Pulse, oilseed, and millet crops need R&D investment by federal, state, private, and foundations, with special emphasis on crop breeding and related research.

Plant breeders have made tremendous progress during the past 75 years in the genetic enhancement of yield traits, nutritional quality, disease, and insect–pest resistance, which increased crop production and productivity of major food crops. However, rarely cultivars are insulated against the adverse environmental stresses that are now increasingly becoming more important due to climate change. The climatic changes would not only affect the yield *per se* but also the end-use quality. It will affect the prevalence and severity of harmful parasites and in turn the host–parasite interactions. Therefore, both biotic and abiotic stresses are to be identified together and given due attention. There is an urgent need to search genetic variation in crop species for agroclimatic factors (e.g., drought, heat, cold, and flood) and soil-related stresses (e.g., mineral deficiencies and toxicities, saline, and acidic soils). The recent research advances in the areas of biotechnology, genomics, and phenomics will not replace conventional plant breeding in public and/or private sector. Instead, these are tools that will aid in the development of superior cultivars of all food, feed, fiber, and fuel crops with multiple modes of resistance to pests/diseases and to abiotic stressors, improved nutritional quality (for food crops), and high yield potential with plasticity (stable and responsive against myriad conditions).

CHAPTER

Mode of reproduction in crop plants

2

Abstract

This chapter broadly covers the modes of reproduction in crop plants. The knowledge of the reproductive biology, life cycle, and type of cultivars are important for plant breeders to develop appropriate breeding strategies. Information on sexual and asexual modes of reproduction and propagation in crop plants are covered. The chapter contains concepts on mechanisms that promote self-fertilization and cross-fertilization, as well as genetic consequences of self- and cross-fertilization. Details on male sterility (cytoplasmic, genetic, and cytoplasmic–genetic male sterility) and maintenance of these genetic materials are included. Concepts and manipulation of self-incompatibility (heteromorphic and homomorphic—gametophytic and sporophytic) are described and their applications in plant breeding are provided. This chapter establishes the foundation of competency in familiarity of crop plants.

The knowledge about the duration of growth cycle, types of flowers, structure of flowers, mode of reproduction, and genetic consequences of reproductive systems are important to plant breeders for the understanding of mechanisms of heredity and in the development of superior cultivars. The duration of the plant growth cycle in flowering crop species (*angiosperms*) is classified into three categories (i.e., *annuals*, *biennials*, and *perennials*). Plants that complete their life cycle in one growing season are called annuals, for example, wheat, rice, grain legumes, etc. Plant that complete their life cycle in two growing seasons (complete reproductive growth) are called biennials, for example, carrot, onion, sugar beet, which grow vegetatively in the first season, and during winter it is vernalized to start reproductive phase in the spring season. Plants that persist for several years are called perennials—they may be herbaceous, which continue to grow with underground vegetative parts (for eg., rhizomes) or above ground parts (for eg., stolons). However, certain plants which are natural biennials or perennials are cultivated by producers as annuals for their roots, or for breeding purposes can bolt to produce flowers for crossing and to produce hybrid seed. Based on the frequency of flowering, plant species can be classified as monocarpic or polycarpic. Monocarpic plants produce flowers and seed only once in the lifecycle, while polycarpic plants produce flowers

Plant Breeding and Cultivar Development. DOI: https://doi.org/10.1016/B978-0-12-817563-7.00002-7

and fruits numerous times (for example, yearly). Bamboo is an example of a monocarpic plant, while most of the perennials are polycarpic including perennial crops and trees species.

The reproductive system of the cultivated species is of two types, that is, *sexual* (are largely propagated through seeds) and *asexual* reproduction (are largely propagated through vegetative parts or through apomixis). The sexually propagated plants are divided into two groups: largely self-pollinated (*autogamous*) and largely cross-pollinated (*allogamous*). The genetic constitution of the species depends on the mode of reproduction. The differences between the genetic constitution of self- and cross-pollinated groups are related to the influence of inbreeding and outbreeding. In cross-pollinated crops, enforced selfing is required to maintain purity, which leads to a general deterioration of vigor and other adverse effects. On the other hand, selfing is not required to maintain purity in self-pollinated crops, as it happens naturally. The asexually propagated species are highly heterozygous and segregate widely upon sexual reproduction, for example, potato and sugarcane.

The population of self-pollinated crops (e.g., wheat and soybean) is homozygous and homogeneous, whereas the population of cross-pollinated crops is heterozygous and heterogeneous, for example, maize. Clonal, synthetic, and hybrid cultivars are heterozygous but homogenous. Pure line cultivars are homozygous. In autogamous species, self-pollination (generation advancement = seed to seed in successive generations) is used to achieve homozygosity. In allogamous species, inbred lines (which are used as parents to develop a hybrid) are developed through forced self-pollination. In allogamous species, hybrid production is achieved by crossing different genotypes (i.e., inbred lines) that creates a heterozygous cultivar (in this case, a hybrid). For a gene A, alleles A1 and A2 (assuming a diploid plant species):

- Homozygous = A1A1 or A2A2; alleles at a locus are the same
- Heterozygous = A1A2; alleles at a locus are different
- Homogeneous = all genotypes are A1A1 or all genotypes are A2A2; all plants are genetically identical
- Heterogeneous = genotypes are A1A1 and A2A2 in a population; plants are genetically different

Briefly some examples of populations are (1) a commercial maize hybrid cultivar (Allogamous), which is homogeneous (for a single cross hybrid) and heterozygous, (2) a commercial soybean pure line cultivar (Autogamous), which is homogeneous and homozygous, (3) a commercial maize synthetic cultivar, which is heterogeneous and heterozygous, and (4) a commercial potato cultivar (Clonal), which is homogeneous and heterozygous.

Depending on the mode of reproduction, appropriate breeding methods are used to develop superior varieties. Therefore, an understanding of the details of pollination and fertilization of a crop is a prerequisite to develop an effective and efficient breeding procedure.

Sexual reproduction

In sexual reproduction specialized reproductive cells, the male and female gametes are formed by the process called *gametogenesis*. The fusion of male and female gametes leads to the formation of zygote from which embryo develops that ultimately gives rise to seed. The formation of gametes onto seed takes place in the flowers that are of two types: *complete* flowers (e.g., peas, cotton, tomato, etc.)—contain all the four floral parts (sepals or calyx, petals or corolla, stamens or androecium, and pistil or gynoecium) and *incomplete* flowers (e.g., rice, maize, etc.)—lack one or more of these floral parts.

Of the four floral parts, the stamens and pistil are essential to produce gametes. The flowers that contain both organs are called *perfect* flowers, which are also *bisexual* or *hermaphrodite* (e.g., wheat, rice, grain legumes, forage grasses, etc.). The flowers that lack one of these essential organs are called *imperfect* flowers, which are *unisexual*. Such flowers could be either *staminate* (have no pistils) or *pistillate* (have no stamens). When the *staminate* and *pistillate* flowers are born on the same plant but on different places the condition is called monoecy and plants are *monoecious* (e.g., maize) and if they are present on different plants, the condition is called dioecy and plants are *dioecious* (e.g., papaya).

The alternation of generation

The sexual reproduction involves two phases or generations in the life of a plant. The *sporophytic* generation is a non-sexual generation (i.e., the formation of a diploid *sporophyte*) which starts from fertilized egg to the formation of seed, seedling and fully developed sporophyte or plant. The *gametophytic* generation is a haploid generation, which is contained wholly within the reproductive structures of the sporophyte (Fig. 2.1).

Flowers produce two types of spores through meiosis. The formation of *microspores* and *megaspores* from their mother cell, that is, *microspore mother cell* and *megaspore mother cell* by meiosis is called *microsporogenesis* and *megasporogenesis*, respectively. The development of one cell structure (pollen grain) to the two nuclei cell and the development of *embryo sac* from megaspore by *mitosis* is called *microgametogenesis* and *megagametogenesis*, respectively. The *microgametophyte* has two sperm nuclei and one tube nucleus, whereas the female *gametophyte* is eight nucleate. The pollen grain germinates on stigma and produces pollen tube that penetrates down through the style to reach *micropylar* end of the embryo sac. One sperm nucleus combines with an egg cell nucleus to produce the diploid embryo and another sperm nucleus fertilizing an embryo sac nucleus (formed by the fusion of polar nuclei) to produce a triploid, that is, *primary endosperm*. This is called *double fertilization*.

The fertilized egg (*zygote*) develops into an *embryo* (a miniature plant), which on germination of the seed develops into a plant. The primary endosperm nucleus after many divisions forms numerous nuclei. These are enclosed by cell walls to

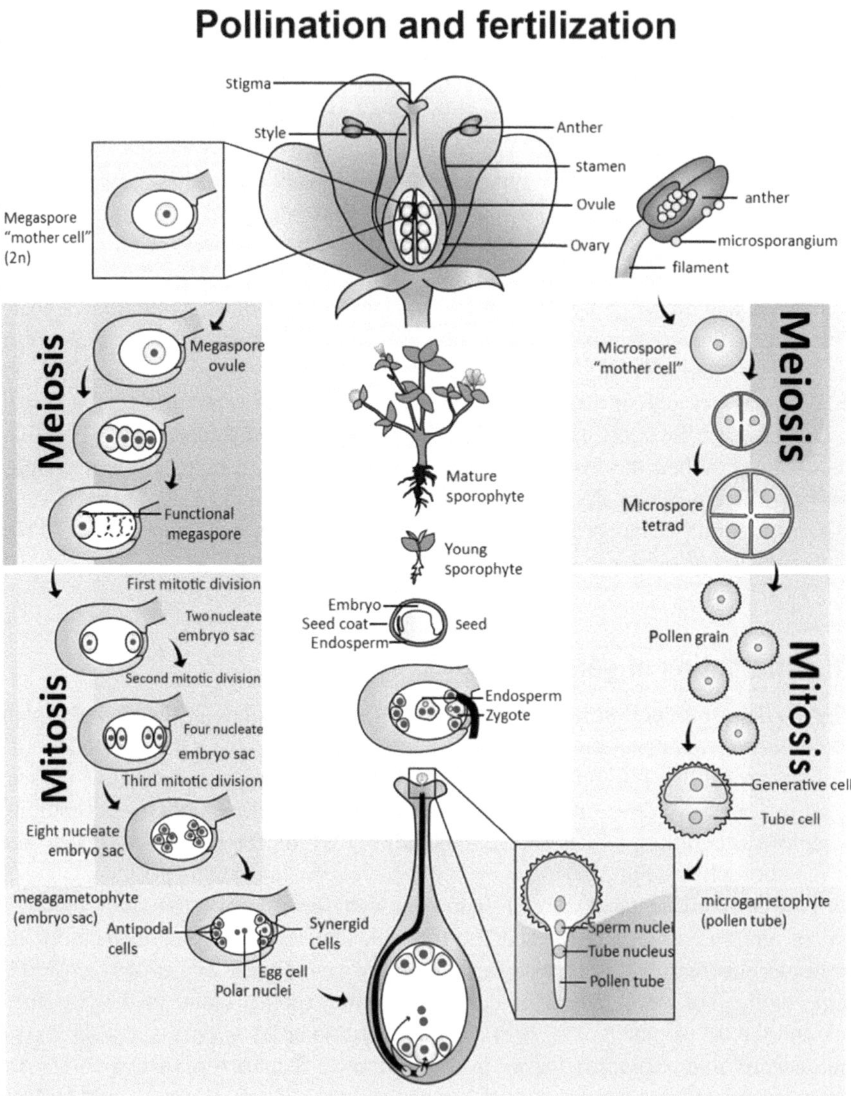

FIGURE 2.1

Generalized diagram for the formation of spores, gametes, embryo, and endosperm.

form endosperm tissue—storage tissue of starch, oil, and protein. The stored food is used by the germinating embryo and seedlings in early stages. In cereals, the larger parts of the seed are *endosperm* (i.e., endospermous or albuminous seeds), while in grain legumes the endosperm is absorbed by the developing embryo (i.e., non-endospermous or exalbuminous seeds) and the food materials for energy source to young seedlings are stored in cotyledons. The seed coat is formed from integuments surrounding the ovule.

Self- and cross-fertilization: anther contains pollen, and movement of pollen to the female part (stigma) is called *pollination*. The fusion or union of male and female gametes is *fertilization*. The process of pollination and fertilization may occur within the same flower which is called self-fertilization (i.e., *autogamy*) or between flowers of different plants that is called cross-fertilization (i.e., *allogamy*). The pollen grain may fall itself on the stigma as in self-fertilized cereals and grain legumes or may be transferred to the stigma as in cross-fertilized crops by wind (i.e., *anemophilous*), for example, maize, or by insects (i.e., *entomophilous*), for example, cotton.

Based on pollination mechanisms, the plant species have been grouped into two categories: predominantly self-pollinated and predominantly cross-pollinated species. Genetic information (i.e., DNA) from both the male and female parents is present in a seed produced by fertilization. It is this union of sperm and egg cell that results in the creation of genetic variation (if the male and female gametes possess different genetic information). Offspring that result from the union of gametes from male and female plants with dissimilar genotypes are known as hybrids. Most cultivated varieties (i.e., cultivars) are created by a process that involves at least one generation of cross-pollination by plant breeders in both self- and cross-pollinated species. Plant breeders select genotypes (i.e., male and female parents) that complement each other to combine the positive (or desirable) traits from each parent, remove the undesirable traits, and develop a progeny (cultivar = cultivated variety) that will be commercially grown.

Self-pollinated species

In self-pollinated species, self-pollination is predominant, however, cross-pollination may occur up to 4%–5% (Table 2.1). It should be noted that there may be large differences in the amount of out-crossing among different species of this group. In barley and soybean, the cross-pollination is generally about 0.5%, whereas in wheat, rice, and tobacco it may be up to 2%–3%. The stigma in the flower of cotton is exposed, due to which insect may cause 5%–25% cross-pollination. In sorghum, 5% or higher cross-pollination may result due to early exposure of stigma before the anthers dehisce. Insects, mainly bees, are attracted to pigeon pea flowers, due to which 5%–15% cross-pollination may occur.

Mechanisms that promote self-fertilization

The anthers and stigmas of a bisexual flower mature at the same time (i.e., *homogamy*) thus effecting self-pollination. In certain species, flowers fail to open, and pollen shed occurs before the flower opens (i.e., *cleistogamy*). The flowers of some species open only after the pollination has taken place (i.e., *chasmogamy*). In tomato, stamens form a cone enclosing the stigma in such a way that self-pollination is

Table 2.1 Few examples of predominantly self-pollinated crop species.

Common name	Botanical name
Barley	*Hordeum vulgare*
Black gram	*Vigna mungo*
Chickpea	*Cicer arietinum*
Common bean	*Phaseolus vulgaris*
Cotton[a]	*Gossypium* spp.
Cowpea	*Vigna unguiculata*
Eggplant (Brinjal)	*Solanum melongena*
Flax	*Linum usitatissimum*
Jute	*Corchorus capsularis*
Lentil	*Lens culinaris*
Lettuce	*Lupinus* sp.
Mung bean	*Vigna radiata*
Oat	*Avena sativa*
Pea nut (Ground nut)	*Arachis hypogaea*
Peach	*Pinus persica*
Peas	*Pisum sativum*
Pigeon pea[a]	*Cajanus cajan*
Rice	*Oryza sativa*
Sorghum[a]	*Sorghum bicolor*
Soybean	*Glycine max*
Tomato	*Lycopersicon esculentum*
Wheat	*Triticum aestivum*

[a]*Frequently > 10% outcrossed.*

ensured. The presence of a sticky exudate on the stigmatic surface due to which pollen readily adheres to this exudate, promoting autogamy.

Genetic consequences of self-fertilization

The self-pollination leads to the highest degree of inbreeding. It promotes homozygosity at all loci and traits of the plant. The important genetic consequences are as follows:

1. Genotypes of a sporophyte are homozygous, and the genotypes of gametes of a plant are all the same.
2. Population of self-pollinated species comprises of a mixture of homozygous lines, which do not show inbreeding depression and loss of vigor on selfing.
3. As a result of selfing in such species, in nature, the recessive deleterious alleles are exposed and are removed in due course.
4. New genes may arise through spontaneous mutations, which are usually recessive and are exposed through homozygosity and could be selected if useful.

5. Even very low level of out-crossing permits appearance of new gene combinations, which leads to the origin of a mixture of different homozygous types.
6. The breeding methods frequently used for self-pollinated crops include pure line selection, pedigree, bulk, and backcross breeding.

Cross-pollinated species

In such crop species, cross-pollination is a rule; however, self-pollination may be 5% or higher (Table 2.2). Cross-pollination may be brought about by wind, water, or insects.

Mechanisms that promote cross-fertilization

Imperfect flowers: Imperfect flowers are missing either stamens or pistils (i.e., *unisex* flowers). Unisex flowers either occur on the same plant (i.e., monoecious) or different

Table 2.2 Few examples of predominantly cross-pollinated crop species.

Common name	Botanical name
Alfalfa	*Medicago sativa*
Annual ryegrass	*Lolium multiflorum*
Asparagus	*Asparagus officinalis*
Banana	*Musa spp.*
Cabbage	*Brassica oleracea*
Carrot	*Daucus carota*
Cassava	*Manihot esculentum*
Castor	*Ricinus communis*
Coconut	*Cocos nucifera*
Cucumber	*Cucumis sativus*
Fescue	*Festuca spp.*
Kentucky bluegrass	*Poa pratensis*
Maize	*Zea mays*
Musk melon	*Cucumis melo*
Onion	*Allium spp.*
Pearl millet	*Pennisetum typhoides*
Pepper	*Capsicum spp.*
Pineapple	*Ananas bracteatus*
Potato	*Solanum tuberosum*
Radish	*Raphanus sativus*
Rye	*Secale cereale*
Sugar beet	*Beta vulgaris*
Sunflower	*Helianthus annuus*
Sweet potato	*Ipomea batatas*
Watermelon	*Citrullus lanatas*

plants (i.e., dioecious). In some species, a plant has flowers of only one sex, either male or female and in such crops, cross-pollination is obligatory. The condition is *dioecy* and such plants are called *dioecious*, for example, papaya, hemp, hop, asparagus, pistachio, etc. Some cross-pollinated species bear the male and female sexual organs at separate locations of the same plants that favor cross-pollination. This condition is called *monoecy* and such plants are referred as *monoecious*, for example, maize, wild rice, castor, and walnut. In maize, the tassel bears staminate flowers and pistillate flowers are born on the shoot. Although cross-pollination is a rule in maize, but some self-pollination may also occur as pollen is wind borne. In castor, pollen transfer is both by wind and insects. In wild rice (*Zizania palustris*), male and female flowers are born in the same panicle but on different branches. The cross-fertilization is promoted usually from adjacent plant before the staminate flowers on the same tillers are exposed.

Dichogamy: In some species, although the flowers are *bisexual* and perfect, cross-fertilization occurs. This could be when stamens mature before pistil matures (i.e., *protandry*) or when pistil matures before the stamens mature (i.e., *protogyny*). Pearl millet has both perfect and staminate flowers. The stigmas emerge several days before the anthers and normally pollinated before the anthers dehisce. Cassava, avocados, and walnuts are examples of *protogynous* species (i.e., stigma receptive first). Sunflower, coconut, carrots, and raspberries are examples of *protandrous* species (i.e., pollen shed first). In case of alfalfa and narrow leaf trefoil (*Lotus tenuis*) a mechanical obstruction to self-pollination in the form of a protective film over stigmatic surface inhibits pollen germination on stigma and helps to ensure cross-fertilization. The film is ruptured through activity by an insect vector, such as, bees, and cross-pollination takes place.

Significant differences in the length of stamens and pistils (i.e., *heterostyly*) also promote cross-fertilization, for example, buckwheat. In some crop species, the flowers are *hermaphrodite* but due to genetic mechanisms self-fertilization is not possible, for example, *self-incompatibility* and *male sterility* that promotes cross-fertilization. Self-incompatibility refers to the inability of viable pollen to fertilize flowers of the same or similar genotype. Self-pollen is rejected on the surface of the stigma or in the style while foreign pollen is unaffected and can germinate, grow, and fertilize the egg cell. Self-incompatibility prevents self-pollination and enforces cross-pollination. These genetic mechanisms are of great practical significance to the plant breeder and growers, hence, will be given a detailed consideration separately in this chapter itself. Male sterility is caused by the formation of non-functional pollen grains, which prevents self-pollination and promotes cross-pollination. Two main types of male sterility are: *cytoplasmic male sterility* (CMS), which is caused by mitochondrial genes interacting with nuclear genes, and *genic male sterility* (GMS), which is caused by nuclear genes alone. These phenomena can be exploited for hybrid seed production.

Genetic consequences of cross-fertilization

The cross-fertilization leads to heterozygous plants and populations of such crops are heterozygous and heterogeneous. The genetic consequences of cross-fertilization are as follows:

1. Genotype of the sporophytic generation is heterozygous and the genotypes of gametes of a plant are all different.
2. Cross-pollination preserves and promotes heterozygosity in a population and heterozygosity is the characteristic feature of cross-pollinated species.
3. Self-incompatibility often occurs in such crops.
4. Selfing leads to mild to severe inbreeding depression and expression of deleterious alleles. However, there are exceptions to this feature. Cucurbits are monoecious but do not show inbreeding depression. This may happen if only small populations of such plants have existed in nature and cultivation.
5. Crossing of genetically distinct inbred lines shows considerable heterosis in the F_1 generation. Heterosis has been exploited in cross-pollinated crops to develop cultivars.
6. Breeding methods frequently used by the plant breeders are mass-selection, recurrent selection, synthetics/composites, and hybrid breeding. However, in maize, breeding methods applicable to self-pollinated species are used to develop inbred lines that are subsequently crossed to develop hybrid cultivar.

Asexual reproduction

Asexual reproduction generates individuals (also called as clones) that are genetically identical to the mother plant. In asexual reproduction, fusion of male and female gametes does not happen; however, new plants are produced from (1) the vegetative parts of the plant (i.e., *vegetative propagation*) and/or (2) from embryos that develop without fertilization (i.e., *apomixis*). Asexual reproduction also includes *clonal propagation*. Vegetative asexual reproduction (or more appropriately, propagation) occurs via tubers, stem cuttings, runners, bulbs, and so on. Examples include sugarcane (stem cuttings), potato (tubers), fruit and nut trees (stem grafts), alfalfa (stem cuttings), and bermudagrass (stolons). Apomictic reproduction is the asexual production of seeds formed by one of the several means that either bypasses meiosis or has a meiotic failure. Examples of apomictic crops include citrus. Examples of major clonal crops (asexual mating system) in the world include potatoes, sugarcane, banana, cassava, and sweet potatoes.

Vegetative propagation

A group of plants that are propagated vegetatively from a single plant are *clones*. Vegetative propagation includes all the methods of multiplication through vegetative parts. For example, potato, cassava, sugarcane, roses, grapes, and some perennial grasses are frequently propagated by stem cuttings; Johnson grass, brome grass are propagated by rhizomes or underground stems; onions or tulip are propagated by bulbs; and gladioli, potato, strawberry, and mints are propagated by corms, tubers, runners or stolon, and bulbils, respectively. The list of plant species propagated using vegetative parts (cuttings) includes most of the fruit plants, ornamentals, some of the important food crops, and perennial grasses. Vegetative propagation can also be practiced in some species that are even normally seed producing crops because many new plants can be obtained in a short period. Rice may also be propagated vegetatively on small scale (non-commercial) by separation and rooting of individual tillers. Grafting and budding are used for propagating tree crops. Large number of plantlets may be produced from small pieces of vegetative materials like a segment of leaf through *micropropagation*. Tissue culture has been used to rapidly multiply planting materials under aseptic conditions.

Apomixis

It is a form of *asexual reproduction* where seed is produced without fertilization. The process through which embryo develops without fusion of male and female gametes is called *agamospermy*. Apomixis is of two types: *facultative apomixis* (i.e., when both sexual and apomictic seeds are formed), for example, bluegrass (*Poa pratensis*) and *obligate apomixis* (i.e., when sexual reproduction is absent or nearly so), for example, bahia-grass (*Paspalum notatum*). There are different ways in which apomixis happens. A cell of integuments or nucellus grows into an embryo (i.e., *adventitious embryony*), for example, citrus and mango. It may originate from haploid embryo sac (i.e., *apospory*), for example, *Hieraceum*, *Malus*, etc., or from diploid megaspore mother cell (i.e., *diplospory*). The development of an egg cell into an embryo is called *parthenogenesis*, which could be *haploid parthenogenesis* (e.g., *Solanum* and *Nicotiana*) or *diploid parthenogenesis* (e.g., grasses). When an embryo develops from synergids or antipodal cells it is called *apogamy* which could be either haploid or diploid type depending on the state of embryo sac. The haploid seed embryo may develop from sperm nucleus itself (i.e., *androgenesis*) or both the sperm and egg nucleus develop independently into seed embryo without uniting (i.e., *semigamy*). The use of apomixis in breeding and its benefits to the plant breeders are further described in Chapter 20, Breeding Methods Used in Asexual Crops.

Genetic consequences of asexual reproduction

The varieties of asexually propagated species are highly heterozygous and vigorous. They segregate widely on sexual reproduction; however, through asexual reproduction, the same genotype of an individual can be perpetuated with great precision. A *clone* selected from a heterozygous population or even a hybrid or a desirable recombinant obtained following sexual hybridization and the desirable mutants (sports) selected in natural or induced after mutagen treatment could be multiplied and their 'true to type' are maintained indefinitely through vegetative propagation.

Determination of mode of reproduction

The examination of the flower and its morphology is an initial step to understand the mode of reproduction. A bisexual flower alongwith any of the mechanisms that hinder cross-fertilization (e.g., cleistogamy or chasmogamy) suggests that the plant is self-pollinated; whereas if the plant has any of the mechanisms that hinder self-fertilization (e.g., imperfect flower, self-incompatibility, sterility, etc.) suggests that the plant has propensity to be cross-pollinated. Failure of seed set under isolation in space or using selfing bags indicate the plant (species) is allogamous (cross-pollinating). The seed set in space isolation (required isolation distance to prevent cross-pollination) suggest that the plant (species) is autogamous (self-pollinating). Selfing does not have any adverse effect in autogamous species, whereas cross-pollinators show adverse effects of inbreeding including loss of vigor and reduced productivity.

The predominantly self-pollinated species have negligible cross-pollination. The extent of cross-pollination is studied by interplanting the strain with homozygous recessive marker gene as female and the homozygous dominant allele as male, preferably for seedling or seed characteristics. It is desirable that such genotypes flower at the same time. The seeds are collected from the recessive type (female parent) and the progenies are grown next season. At a particular location, the female and male types are planted at several spots. The amount of natural crossing is recorded by counting the proportion of dominant types of progenies. If such studies are conducted over years (seasons) and location, then inferences about the effect of temperature, insect vector, and direction of wind as well as their interaction on natural crossing can be made. Different types of designs such as rectangular (Fig. 2.2A) and circular (Fig. 2.2B) have been used. In these design, female and male lines with recessive and dominant gene markers are interplanted. Additionally, the plants with dominant markers are planted all around the experiment. However, circular designs which provide plants with dominant marker in the outer circle and the recessive marker in the inner circle may be preferred.

Selfing and crossing

The selfing and crossing are essential processes in plant breeding and depends on a species, that is, structure of flowers and normal manner of pollination. These aspects are understood in almost all major crop species of economic importance; however, if unknown, breeder should learn these before starting crop breeding. Selfing is practiced to prevent natural cross-pollination and to maintain the purity of a variety. It is

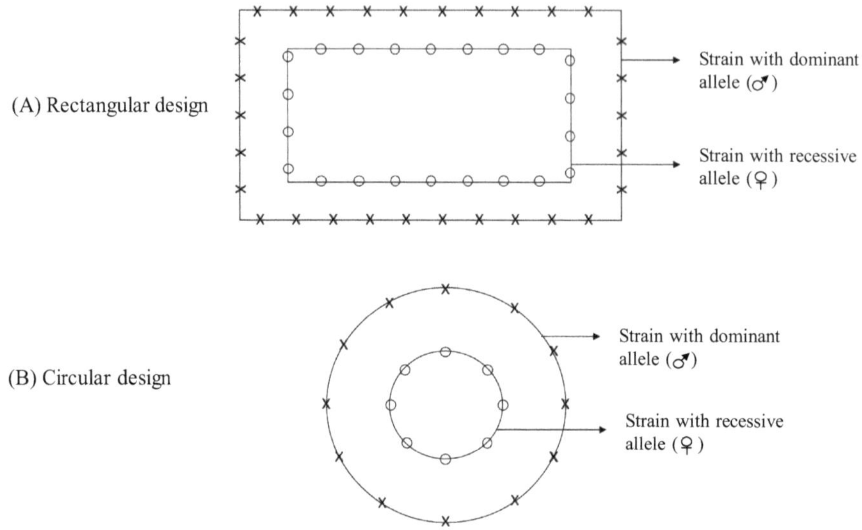

FIGURE 2.2

(A) The rectangular and (B) circular design for studying the extent of natural crossing. The distance between female and male strains could be varied. The seed is harvested from the female strain.

also used to develop true breeding lines and to increase the homozygosity. In self-pollinated crops, selfing has no adverse effects on vigor or fertility as it is natural mode of pollination in such crops and no separate (human interference) operation is needed (since selfing is automatically performed). However, to initiate new breeding population or family, manual crossing is made by emasculation or other methods to combine the male and female gametes of two or more parents. All plant breeders working in self-pollinating crops create artificial pollination with two or more parents to create new genetic combinations (F_1) followed by natural selfing to advance generations to homozygosity and select superior pure lines.

To perform self-pollination in cross-pollinated species, breeders use paper or cloth bags to enclose single flower or whole inflorescence (e.g., pearl millet, sorghum, maize, wheat, castor, etc.). Cages are useful for covering whole plants in crops such as pigeon pea, tomato, onion, and cumin. Space isolation is very effective to maintain genetic purity during commercial seed production in self- and cross-pollinated crops. The isolation distance of 3–5 m is recommended in predominantly self-pollinated crops such as wheat, peanut; while in often-cross-pollinated crops such as cotton, 30–50 m of isolation distance is required. In cross-pollinated crops requiring selfing, space isolation would depend on the mode of cross-pollination. In sorghum and pearl millet, an isolation distance of 150–200 m is required. The pollen movement in these crops is through wind, however, when cross-pollination occurs through insect vector an isolation distance of 400 m is required (e.g., lucerne or alfalfa). For making artificial crosses, generally bagging is done a day before opening of flowers and care may be taken to remove all unwanted flowers (i.e., flowers which have already opened). Bagging is also required after emasculation and pollination to prevent pollen from reaching the stigma of the emasculated flowers. On the other hand, in cross-pollinated crops with bisexual flowers or with male or female flowers in a single inflorescence, bagging of entire inflorescence or sometimes entire plant is practiced.

In cross-pollinated crops, bagging is required to enforce selfing or inbreeding to produce inbred lines. Inbred lines are produced by continuous selfing in a single plant or by continuous sib-mating between two closely related plants. The inbred lines are used to produce hybrid seed (single crosses, double crosses, etc.). Selfing reduces vigor or productivity, which is restored in the F_1 generation. The crossing of two genetically dissimilar plants or lines is called hybridization. In this process, pollen grains of desirable male parent are transferred to the stigma of a selected female parent. It is practiced to create genetic variability and to combine the desirable traits into a single variety. It is also used to study the inheritance of a character, to study combining ability, to exploit heterosis, and to produce hybrid varieties.

The crosses involve plants or parents of two varieties of the same species (i.e., intervarietal or intraspecific) or between different species (i.e., interspecific, e.g., crosses in cotton—*Gossypium hirsutum* and *Gossypium barbadense*, soybean—*Glycine max* and *Glycine sojae*, chick peas—*Cicer arietinum* and *Cicer reticulatum*), or between different genera (i.e., intergeneric, e.g., wheat × rye to produce triticale).

Methods of emasculation

Generally crossing is accomplished by removing the stamens of the female parent or seed parent, this process is called *emasculation*. The stigma of the female parent is pollinated with the pollen from the male parent or pollen parent, the process is called *pollination*. The emasculation is required in the bisexual plants and is not required in monoecious or dioecious plants. However, it is necessary to protect the pistillate flowers before it becomes receptive and is pollinated with pollen collected from the desired parent. In bisexual flowers, the emasculation is completed before anthers ripen and self-pollen reaches the stigma.

The exact procedure for emasculation must be learned for the species with which the breeder is working. The time of emasculation and pollination would vary from crop to crop. In some species morning emasculation and evening pollination or evening emasculation and morning pollination is followed. In wheat, barley, and rice, bracts may be cut back without adverse effect, however, same treatment in oats would upset the seed setting. In grain legumes, the undehisced stamens of the selected buds are removed with forceps by gently slit opening the keel tube. The pollen grains are dusted on the stigma of the emasculated flower(s). The flowers are bagged after pollination and record details are kept with the flower (parent ID, cross combination, date, etc.). Emasculation could also be circumvented by some procedure that would ensure an acceptable degree of cross-pollination. Some of these are described as follows:

1. *Removal of the anthers:* the anthers could be removed with the help of forceps or suction before pollen is shed. Hand emasculation is most commonly used method in crops, for example, wheat, rice, barley, oats, grasses, soybean, chickpea, and many other grain legumes, linseed, sugar beet, and tobacco. Small forceps and scissors are required. Suction has been used successfully to emasculate small flowered legumes. The emasculation is generally done in the late evening or in early morning (1 day before the anthers are expected to dehisce and stigma is fully receptive) but may be different based on crop, geography, etc. In few crops, earlier stigma receptivity can be used effectively (e.g., soybean).
2. *Use of hot or cold water treatment:* the hot water (45°C–48°C) is taken in a thermoflask to greenhouse/field and flowers are immersed (for periods ranging from 1 to 10 minutes) depending upon species. This practice has been reported to kill the pollen in sorghum, rice and grasses. Similar technique with cold water (around freezing temperature) is reported to kill the pollen (e.g., wheat, rice). After hot/cold water treatment anthers are removed with forceps. The flowers may be pollinated within 30 minutes.
3. *Use of gametocide:* ethyl alcohol (57%) has been found effective to kill pollen in alfalfa. The flowers are immersed in ethyl alcohol for a period of 10 minutes. The chemicals have also been used at large scale to produce commercial hybrid seed. The spray of chemicals at the time of flowering induces male sterility.
4. *Use of male sterility and self-incompatibility:* these systems have been described in more details subsequently.

Precautions during emasculation

- Normal time and pattern of blooming of the flowers must be understood.
- It may be determined which flower(s) produce better seed.
- Buds of proper size may be selected for emasculation.
- Selected flower buds are retained and rest (including opened flowers, damaged flowers, and fruits) may be removed.
- Anthers may be removed by holding filaments with forceps. Emasculation may be done at the time when stigma is not receptive. Care must be taken to not damage inflorescence parts during emasculation.
- If required, magnifying lens should be used to confirm that any piece of anther is not left in the emasculated flower bud.
- Do not remove surrounding flower parts (petals or glumes) and flag leaf unless required. The emasculated flower/inflorescence may be bagged to prevent foreign pollen.

Pollination

Pollens are collected in petri dish (e.g., in maize and pearl millet) from the bagged flower of male parent and dusted with brush on to the female part. The pollen may be dusted directly from the male flower on the stigma (e.g., wheat, peas, soybean, cotton, brinjal, okra, etc.). After pollination, the flowers are to be bagged and labeled. Care must be taken in hot dry environments to not leave out anthers for too long. Cryopreservation techniques have been created, and this enables hybridizations between parents with completely dissimilar flowering dates. It is more common in horticulture crops, but can be used in different crops.

Precautions during pollination

- Pollination is done usually in morning or during the time of day when the stigma is receptive. The buds of the pollen parent should be enclosed. This may vary for crop, environment, geography, and needs to be understood by the breeder for their location.
- The tags/labels should be light weight so that flower or inflorescence does not break with a jerk of wind.
- Care must be taken to not damage inflorescence parts during pollination.

Male sterility and self-incompatibility

Among all the mechanisms responsible for cross-fertilization, male sterility and self-incompatibility are the most important as these have contributed toward development of hybrid cultivars in different species.

Male sterility

Sterility is characterized by non-functional gametes. Sterility can be male or female sterility, and is caused by chromosomal irregularities during meiosis, presence of Mendelian genes for male sterility, and non-nuclear factors that lead to abortion or modification of entire flowers or their male/female organs that hinder self-fertilization. In case of male sterility, the pollen grains (male gametes) are non-functional while the female gametes are functional. The male sterile plants may show the following modifications.

- stamens may be malformed or may abort before the pollen formation.
- anthers do not dehisce due to the presence of hard outer layer as a result pollen grains are not released.
- pollen grains are not viable. Such pollen grains do not show stain or stain poorly.

The male sterility can be classified into four categories: GMS, CMS, cytoplasmic–genetic male sterility (CGMS), and environmental-induced male sterility [thermosensitive (TGMS) and photosensitive (PGMS)].

Genetic male sterility: it is governed by a single recessive gene (*ms ms* in diploid plant organisms). GMS is widespread in several crop plants (e.g., barley, soybean, pigeon pea, tomato, etc.). The male sterility alleles arise due to spontaneous mutations, which are deleterious from the evolution point of view; however, are useful to the plant breeders as a substitute to emasculation. The GMS could also be induced artificially using mutagens. In some crops, GMS is conditioned by a dominant gene (e.g., in safflower), which cannot be used as the F_1 hybrid as it would be male sterile. In alfalfa, male sterility is reported due to two genes that are independently inherited. The recessive genetic male sterility has been frequently and successfully used in self-pollinated crops hence is described.

For the maintenance of male sterile line, A (*ms ms*) is crossed with a heterozygous male fertile line B (*Ms ms*) (i.e., sib-mating). Such a mating produces 50% male sterile (*ms ms*) plant and 50% male fertile (*Ms ms*) plants (Fig. 2.3). Only seed from the male sterile plants is harvested. During the hybrid seed production, heterozygous male plants must be identified and rouged out before flowering. This is a tedious job. However, if a marker gene is linked with GMS (preferably for seedling character), it would facilitate the rouging of male fertile plants before anthesis. For the commercial hybrid seed production, the male sterile (*ms ms*) line is crossed with another homozygous fertile line (C). The GMS system has been used effectively in hybrid seed production of castor and pigeon pea.

Cytoplasmic male sterility: it is controlled by factors present in the cytoplasm (mitochondrial gene). The plants have male sterile cytoplasm and will produce seed if pollinators (i.e., plants with fertile factor in cytoplasm) are present. The F_1

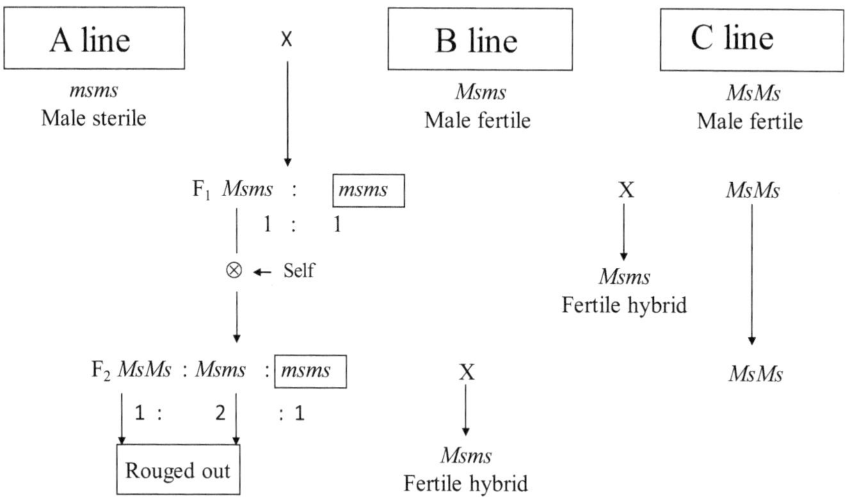

FIGURE 2.3

Inheritance of genetic male sterility, its maintenance, and use in hybrid seed production.

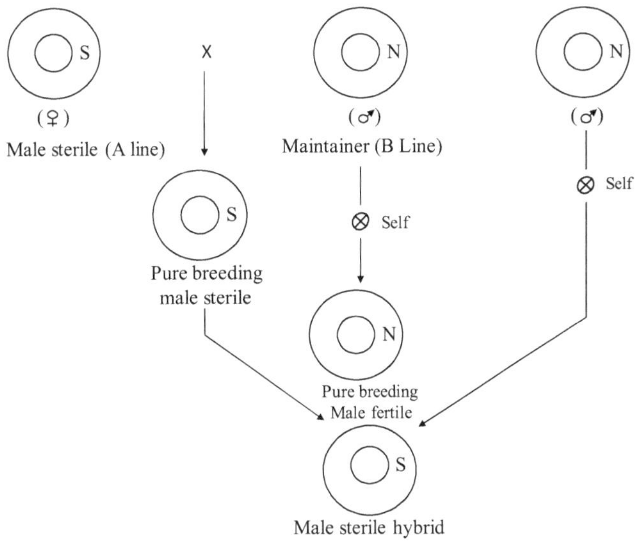

FIGURE 2.4

Inheritance of cytoplasmic male sterility, its maintenance, and use in hybrid seed production. The S and N are sterile and fertile cytoplasm, respectively.

produced by crossing between male sterile and male fertile plants would always produce male sterile plants because their cytoplasm is derived entirely from the female gamete. The male sterile line is known as A-line which is maintained by crossing it with its male fertile counterpart, known as B line or maintainer (as it is used to maintain A-line). Both lines are isogenic lines (Fig. 2.4), which need to be maintained.

The CMS character is transferred to a given strain by using that strain as pollinator (recurrent parent) in successive generations (6–7) of backcross program would recover almost all the nuclear genes of new strain with the cytoplasm of male sterile line as a result new male sterile would be similar to that of the pollinator strain (Fig. 2.5). The CMS system is highly stable and is not influenced by environmental conditions such as temperature and day length. As F_1 is male sterile, therefore, such male sterility is not useful for seed crop (except if hybrid seed is commercially sold), but is useful in ornamental species and in crop species of which vegetative parts are used as fodder or for any other use because plants remains fresh and in bloom for a longer time. The CMS system is reported in onion, forage crops, etc.

Cytoplasmic–genetic male sterility: this system is conditioned by an interaction between nuclear gene(s) (mostly recessive, e.g., in onion or dominant, e.g., in most grasses) and cytoplasmic or non-nuclear factor. Jones and Clarke (1943) reported inheritance of CGMS in onions. They crossed single plants of the Italian

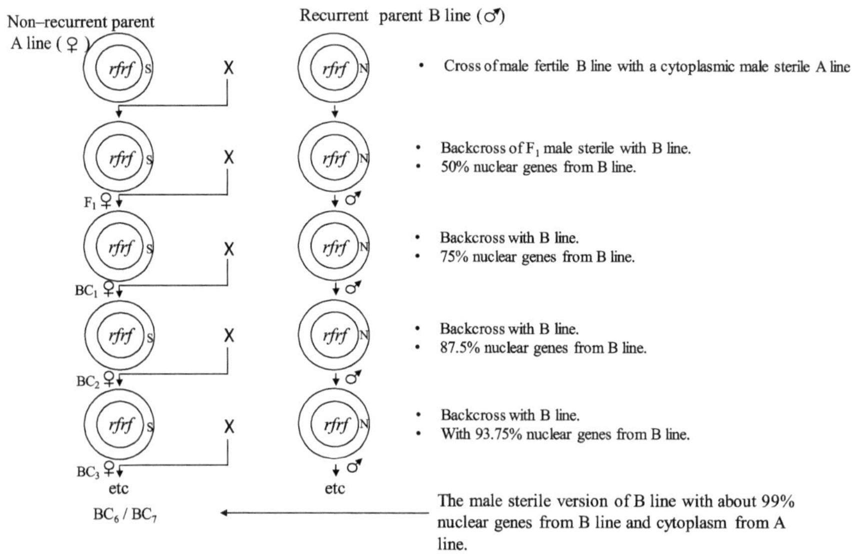

FIGURE 2.5

Development of new cytoplasmic male sterile line. A-line is used as non-recurrent parent and B line is a recurrent parent in the backcross program. B line serves as maintainer for new male sterile line. N is normal cytoplasm and S is sterile cytoplasm.

Red 13–53 onion variety with various male fertile plants and studied the genetic behavior of crosses and backcrosses based on the following assumptions:

1. male sterile condition is due to an interaction between a nuclear gene and a non-nuclear or cytoplasmic factor.
2. two types of cytoplasm are: normal type of cytoplasm (N) and sterile type of cytoplasm (S). Such factor is inherited only through egg and not through the male parent.
3. gene for male sterility (*rf*) also influenced pollen development with S cytoplasm but has no effect when carried by plants with N cytoplasm, this showed that there is no natural selection for or against the recessive gene *rf*, when it is present in N cytoplasm.

Therefore, the male sterile plants belonged to the genotype S-*rf*-*rf*, and fertile plants belonged to N-*Rf*-*Rf* or N-*Rf*-*rf* or N-*rf*-*rf* genotypes. S-*Rf*-*Rf* or S-*Rf*-*rf* would also be fertile although they carry S cytoplasm because they also carried dominant gene *Rf*.

The cross, S-*rf*-*rf* × N-*Rf*-*Rf* gave all F_1 plants with S-*Rf*-*rf* which were fertile due to *Rf* gene, despite the S cytoplasm (Fig. 2.6A). The cross, S-*rf*-*rf* × N-*rf*-*rf* produced all male sterile plants in F_1 with S-*rf*-*rf* genotype (Fig. 2.6B), and the cross S-*rf*-*rf* × N-*Rf*-*rf* segregated into male fertile (S-*Rf*-*rf*) and

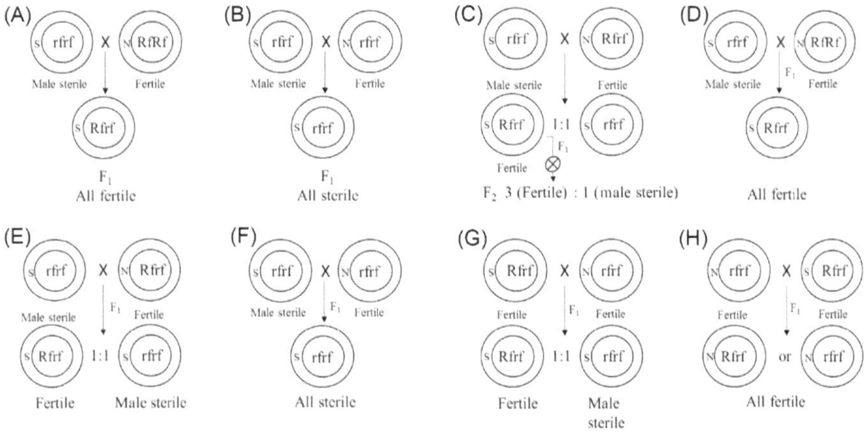

FIGURE 2.6

Inheritance of cytoplasmic–genetic male sterility. See text for details on (A)–(H).

Modified after Jones, H.A., Clarke, A.E., 1943. Inheritance of male sterility in the onion and the production of hybrid seed. Proc. Amer. Soc. Hort. Sci. 43, 189–194.

sterile (S-*rf*-*rf*) in 1:1 ratio. The F_2 generation from a self-fertile plant (S-*Rf*-*rf*) segregated in 3 (fertile or normal): 1 (male sterile) (Fig. 2.6C). All the male sterile F_1 plants were S-*rf*-*rf* and when this F_1 was backcrossed with three types of male fertile plants, the outcomes were: for N-*Rf*-*Rf*, all fertile (Fig. 2.6D); for N-*Rf*-*rf*, 1 (fertile): 1 (sterile) (Fig. 2.6E), and for N-*rf*-*rf*, all sterile progenies (Fig. 2.6F). When fertile F_1 with S-*Rf*-*rf* genotype was used as female and backcrossed to N-*rf*-*rf*, a 1 (fertile): 1 (sterile) segregation was observed (Fig. 2.6G); however, in the reciprocal backcross (N-*rf*-*rf* backcrossed to S-*Rf*-*rf*) all fertile progenies were noted since all carried the N cytoplasm (Fig. 2.6H). The production of 100% male sterile progenies in a backcross involving male sterile F_1 (S-*rf*-*rf*) and fertile/normal line (N-*rf*-*rf*) suggested that through repeated backcrossing (6–7) with N-*rf*-*rf*, male sterile lines could be produced (Fig. 2.7).

Maintenance of cytoplasmic–genetic male sterility: the male sterile line is known as A-line, which is cytoplasmic male sterile. The A-line (S-*rf*-*rf*) is maintained by crossing with its counterpart, male fertile maintainer line or B line (N-*rf*-*rf*). The product of this cross would also be male sterile (S-*rf*-*rf*). The cross of male sterile line (female parent, A) of S-*rf*-*rf* genotype with male fertile line (male parent B) N-*rf*-*rf* would provide male sterile F_1 hybrid which is useful in crops like onion (Fig. 2.7).

The cross of male sterile line of S-*rf*-*rf* genotype with fertile line N-*Rf*-*rf* would segregate in 1 (fertile): 1 (sterile), hence is not very useful for hybrid seed production. The third type of cross which would be suitable for seed crops, is between a male sterile A line (S-*rf*-*rf*) with a fertile line of N-*Rf*-*Rf* genotype (called restorer or R-line), which would produce all fertile F_1 plants.

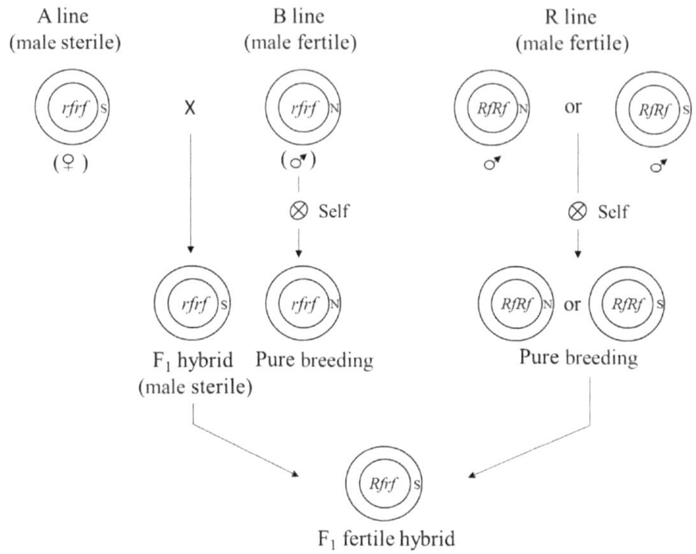

FIGURE 2.7

Maintenance and production of fertile F_1 hybrid using cytoplasmic–genetic male sterility system. The cross between A- and B-lines is useful for the production of F_1 hybrid where commercial product is a vegetative part of the plant.

The A-lines stocks may contain different sources of male sterility. Breeders focus on developing improved (for traits of interest, e.g., yield, other agronomic, and quality traits) B-lines in their programs and R-line germ pool. Once an elite B-line is developed, A-line (from A-line stock) can be used to convert the B-line into A-line through backcrossing. The three-line system (i.e., A, B, and R) could be used for commercial seed production as is being followed in self-pollinated crop such as rice and cross-pollinated crops such as maize, sorghum, pearl millet, sunflower, etc. In this system, A-line is cytoplasmic male sterile (with S cytoplasm) and is crossed with its counterpart B line (N cytoplasm for normal fertility in the male maintainer, and does not restore fertility in the progeny when crossed to the female A-line). The resultant F_1 hybrid is crossed with R-line that restores the fertility in the F_1 hybrid (Fig. 2.7). The restorer line carries a dominant gene which overcomes the effect of male sterile cytoplasm on pollen sterility, hence restores the fertility of F_1 hybrid. See Chapter 19, Hybrid Varieties, for more details on hybrid cultivars.

Environmental-induced genetic male sterility: The environmental-induced GMS system is reported to be successfully used in rice. Both thermosensitive GMS and photosensitive GMS lines are identified. Such male sterility system is controlled by one or two nuclear genes. These systems are called two-line male sterility systems because male sterile line becomes fertile under specific environments; hence there is no need of a maintainer line. Only two lines: male sterile and restorer lines are used. The utility of male sterility to develop hybrid cultivars in self-pollinated crops is described in Chapter 8, Hybridization and selection in self-pollinated crops, and Chapter 19, Hybrid Varieties.

Self-incompatibility

The self-incompatibility refers to the inability of a plant with functional pollen grains to produce seeds when self-pollinated. It is also referred as the hindrance to self-fertilization. However, the pollen grains are fully functional on another genotype. The self-incompatibility systems may be classified into heteromorphic and homomorphic.

1. *Heteromorphic incompatibility:* it is due to differences in the floral morphology (i.e., the length of stamens and style) and is called heterostyly, for example, *Primula*, the pin and thrum types. In pin type of flowers, the styles are long, and anthers are short. In thrum type of flowers, the styles are short, and anthers are long. This type of difference is known as *distyly*. The pin and thrum are conditioned by genotypes ss and Ss, respectively, where Ss produces thrum type and ss produces pin type of flowers. Allele S is dominant to s. The pin flower produces gametes of s type as well as phenotype of s type, and thrum type of flower produces two type of gametes; S and s type but all of them are phenotypically S type. This shows that incompatibility reaction of pollen is determined by the genotype of plant producing them. The cross between pin × pin and thrum × thrum are incompatible whereas cross between pin × thrum and thrum × pin are compatible (Table 2.3). The tristyly is known in some plants like *Lythrum* in which the style of the flower may be short, long, or medium length.

   Mather (1950) reported inheritance of this heterostyly (in *Primula sinensis*) to be a complex character involving several variations, including long, medium, and short styles along with long, medium, and short stamens. Buckwheat (*Fagopyrum* spp.) displays heteromorphic incompatibility and is normally cross-fertilized species. However, self-fertile flowers with styles and stamens of the same length have been reported.

Table 2.3 The four types of crosses in *Primula* showing heteromorphic incompatibility system.

Phenotype of crosses	Genotypes of crosses	Progenies
Pin × Pin	ss × ss	Incompatible (no progeny)
Pin × Thrum	ss × Ss	1 Thrum: 1 Pin
Thrum × Pin	Ss × ss	1 Thrum: 1 Pin
Thrum × Thrum	Ss × Ss	Incompatible (no progeny)

2. *Homomorphic incompatibility:* the system is called homomorphic as it is not associated with the differences in the morphology of the flowers. It is of two kinds: gametophytic and sporophytic.
   a. *Gametophytic incompatibility:* in this system, the ability of the pollen to function is determined by its own genotype and not by the plant (sporophyte) on which it is produced. The incompatibility reaction is controlled by a single gene S having multiple alleles (S_1, S_2, ..., S_n). It was originally called as oppositional factor system by East and Mangelsdorf (1925). The alleles of incompatibility gene(s) act individually in the style and they exhibit no dominance. The incompatible pollen is inhibited in the style, which is diploid and contains two incompatibility alleles in heterozygous condition at S locus (e.g., S_1S_2 and S_3S_4). The incompatible reactions would occur if both pollen and style have identical alleles. Through this system of incompatibility, only heterozygous for S alleles are produced. In a single gene system this type of incompatibility reaction gives three types of mating (Table 2.4). It is reported to occur in red clover, white clover, alsike clover, and yellow sweet clover.

      This form of incompatibility is more common than sporophytic self-incompatibility system. With gametophytic incompatibility, the genotype of the haploid pollen grain determines its incompatibility reaction and occurs in pollen grains having one generative nucleus at the time of release from the anthers. In most plant families with gametophytic incompatibility, the generative cell does not divide until after germination of the pollen tube.
   b. *Sporophytic incompatibility:* in this system, the ability of the pollen to function is determined by the plant (sporophyte) that produces it and not by the genotype of the pollen, while that of style is governed by its own genotype (Table 2.5). It may have individual action in both pollen and style; however, dominance is determined by the pollen parent ($S_1 > S_2 > S_3 > S_4$... etc.). Incompatible pollen may be inhibited on the stigmatic surface for instance in a plant with S_1S_2 genotype (S_1 is dominant to S_2). The pollen will function like S_1. The S_1 pollen would be rejected by S_1 style. Therefore, homozygote for S alleles is possible.

Incompatibility reaction is associated with pollen having two generative nuclei when the pollen is shed from the anther. This is reported to occur in Composite, Cruciferae, and Rubiaceae families and is important in the breeding of cabbages.

Table 2.4 Incompatibility reaction in three types of mating system in gametophytic self-incompatibility.

Types of mating	Progeny
$S_1S_2 \times S_1S_2$	Incompatible (no progeny)
$S_1S_2 \times S_1S_3$	Partial compatible (S_1S_3 and S_2S_3)
$S_1S_2 \times S_3S_4$	Fully compatible (S_1S_3, S_1S_4, S_2S_3, and S_2S_4)

Table 2.5 Incompatibility reaction in three types of mating system in sporophytic self-incompatibility.

Types of mating	Incompatibility reaction
$S_1S_2 \times S_1S_2$	Complete incompatibility
$S_1S_2 \times S_1S_3$	Complete incompatibility
$S_1S_2 \times S_3S_4$	Complete compatibility (S_1S_3, S_1S_4, S_2S_3, and S_2S_4)

Table 2.6 The differences between the gametophytic and sporophytic self-incompatibility.

Gametophytic	Sporophytic
• Pollen's incompatibility reaction is determined by its genotype. The system has independent action in both pollen and style.	• Pollen's incompatibility reaction is determined by the genotype of the plant (sporophyte) where pollen is produced and not by its own genotype while that of style is governed by its own genotype.
• The genetic control is by a single locus (S) with multiple alleles without dominance in the pollen and by its own genotype in the style generally by two alleles which do not show dominance but show codominance.	• The genetic control is by a single locus (S) with multiple alleles showing dominance ($S_1 > S_2 > S_3 > S_4$...Sn), codominance (individual action) or competition in the pollen. No dominance in style and independent action. There are frequent reciprocal differences.
• The incompatibility reaction takes place in the style tissues.	• The site of incompatibility reaction is the stigmatic surface.
• The stigma surface is plumose having elongated receptive cells, known as wet stigma. The pollen grain germinates once it reaches stigma. The incompatibility reaction takes place at later stage.	• The stigma is papillate and dry, and is covered with pellicle (hydrated layer of protein), which is involved in the incompatibility reaction.
• Pollen tube in binucleate.	• Pollen tube is trinucleate.
• Homozygosity through normal pollination is not possible.	• Homozygosity is a normal part of the system.

It was reported by Hughes and Babcock in 1950 in *Crepis foetida*. Sometimes incompatible crosses can result in fertilization due to incomplete retardation of the pollen tube growth. For better understanding of the gametophytic and sporophytic incompatibility their contrasting features are presented in Table 2.6 and Fig. 2.8.

There is a continual research initiative to identify and clone male sterility genes. For example, *MS1* in wheat and *OsPKS2* in rice. *Pms1* gene has been studied for its role in two-line hybrid rice that is built on photoperiod-sensitive male sterility—male sterile under long-day conditions, male fertile under short-day conditions, along with *ZmMs33* in maize, and *MSH1* in *Brassica juncea*.

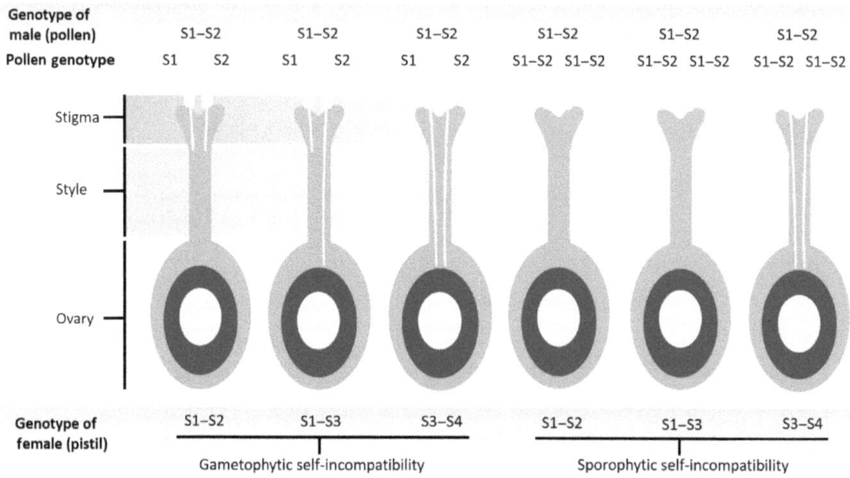

FIGURE 2.8

Examples of gametophytic and sporophytic self-incompatibility. In gametophytic self-incompatibility, $S_1S_2 \times S_1S_3$ is partial compatible, and $S_1S_2 \times S_3S_4$ are fully compatible. In sporophytic self-incompatibility, only $S_1S_2 \times S_3S_4$ are fully compatible. See Tables 2.4 and 2.5 for further details.

Modified from Claessen, H., Keulemans, W., Van de Poel, B., De Storme, N. (2019). Finding a Compatible Partner: Self-Incompatibility in European Pear (Pyrus communis); Molecular Control, Genetic Determination, and Impact on Fertilization and Fruit Set. Front. Plant Sci., https://doi.org/10.3389/fpls.2019.00407.

Role of self-incompatibility in plant breeding

In self-incompatibility the male and female gametes are normal but non-functional due to some physiological processes. Therefore, it can serve as a strong mechanism of cross-fertilization, hence useful in hybrid breeding. However, production of inbred lines by overcoming the incompatibility barriers is a prerequisite in the exploitation of heterosis and for hybrid seed production. Different methods have been used to overcome self-incompatibility barriers are described below:

1. *Bud pollination*: it consists of pollinating immature buds with mature pollens (i.e., 2–4 days before anthesis) for the production of self-seed. It is useful both in gametophytic and sporophytic self-incompatibility systems. In young buds, the stigmas lack adhesivity which could be achieved if the stigma is smeared with the stigmatic fluids (exudates) from an open flower (i.e., mature flower). Such an approach is used for the development of inbred lines for seed production is *Brassica* species.
2. *Delayed pollination:* in some species pollination at delayed stage (i.e., aged pistils) with normal incompatible pollens resulted in some degree of self-fertilization, for example, in *Brassica* and *Linum*.

3. *Removal of stigmatic surface:* the stigmatic surface is site of the pollen inhibition in self-incompatibility. The removal of stigmatic surface followed by application of pollen, results in certain amount of seed production. It is effective in *B. oleracea* but not in *B. campestris*. In this species, mutilation of stigma with steel brush during pollination is useful.
4. *High temperature:* treatment of style at high temperature than normal range (up to 60°C) that inactivates enzyme or substances, which results in the breakdown of gametophytic single locus self-incompatibility in certain plant species (i.e., pseudofertility) such as rye.
5. *In vitro fertilization:* in this method, the pollen grains are placed in direct contact of ovule resulting in production of seed, for example, in *Petunia auxiliaries*. This method excludes the effect of interference in self-compatibility by the stigma, style, or ovary.
6. *Irradiation:* the irradiation of style with X-rays (in *Petunia*) or with gamma-rays (in *Lycopersicon peruvianum* and *Nicotiana alata*) resulted in breakdown of self-incompatibility and production of self-seed.

Utilization of self-incompatibility in plant breeding

Self-incompatibility is a hindrance in self-fertilization and promotes cross-fertilization. It is useful in the production of hybrids and in combining desirable genes from different sources.

1. *Production of hybrids:* self-incompatibility used in the production of hybrids in radish, cabbage, cauliflower, broccoli, Brussels sprouts, and kale. Inbred lines which are self-incompatible are produced by five to six generations of selfing through bud pollination. During the process of selfing, only the plants with desirable traits and strong level of self-incompatibility are selected. The self-incompatible but cross-compatible inbred lines such as S_1S_1 and S_2S_2 are planted in alternate rows in recommended isolation distance to produce single cross. Seed obtained from both the inbred lines would be hybrid seed. Alternatively, a self-incompatible line and self-compatible lines are interplanted in isolation. Self-incompatibility is used to produce hybrid seed at commercial level in cruciferous crops in Japan. The seed from self-incompatible line would be hybrid.

   There are limitations of using self-incompatibility in hybrid seed production.

   a. Production and maintenance of inbred lines using hand pollination is a tedious and time consuming as well as costly.
   b. Continuous selfing may result in reduced level of self-incompatibility. Such an effect is more pronounced in gametophytic system.

c. Environmental factors including high temperature and high humidity may also produce self-seed. The insect pollinators may increase the proportion of selfed seed.

2. *Combining desirable genes:* using self-incompatibility system, the desirable genes from two or more sources can be combined in a single genotype through natural cross-pollination. The knowledge of self-incompatibility in fruit crop would help to increase the yields by providing suitable pollinators.

CHAPTER

Genetics in relation to plant breeding

3

Abstract

This chapter focuses on genetic principles as they relate to plant breeding. Chapter contents include the basic concepts of cell structure, cell division (mitosis and meiosis), chromosome and overall genome, Mendelian inheritance principles including the law of segregation and the law of independent assortment, epistasis and pleiotropy, genetic linkage, and cytoplasmic inheritance. Principles of quantitative inheritance and its importance to plant breeding are described. These contents are intended to establish the foundation of these basic genetic principles for plant breeding activities.

Plant breeding applies the principles of genetics for crop improvement. The knowledge of principles of genetics provide the information about expression of characters, that is, genes, their location, number, structure and their transmission from generation after generation. Therefore, it is essential for a plant breeder to understand the expression and transmission of genes. In this chapter, the basics of genetics, genes, their function and role in the hereditary of the plants are briefly described.

Cell structure (typical plant cell structure)

The continuity of life depends on the transmission of genetic information from the parents to offspring that are organized in all forms above the parasitic viruses into cells. The organisms could be divided into two categories; one celled (i.e., unicellular) or multi-celled (i.e., multicellular). Simple unicellular organism, such as bacterium or blue green algae, do not have membrane bound nucleus (no nuclear membrane, no membrane bound organelles) and are called *prokaryotes* (*karyote* = *nucleus*). Organism with cells that have membrane bound nucleus and several other membranes enclosed organelles are called *eukaryotes*.

The cell consists of denser body, the *nucleus*, which contains *deoxyribose nucleic acid (DNA)* and proteins. Chromatin is a complex of DNA and proteins that forms condensed strands called *chromosomes*. In addition to chromatin, one or more dark bodies per nucleus are present which are called *nucleolus or nucleoli* (plural). They are rich in histones, a low molecular weight protein in *ribose nucleic acid (RNA)*.

Plant Breeding and Cultivar Development. DOI: https://doi.org/10.1016/B978-0-12-817563-7.00001-5

The remainder of the nucleus is filled with a fluid called nuclear sap. The nucleus is enclosed in its own membrane called nuclear membrane.

The extra-nuclear region is called *cytoplasm*, which is enclosed by a semipermeable membrane called the *plasma membrane* through which metabolic exchanges occur. In plant cells, there is additionally a cell wall of cellulose or related substances. The cytoplasm contains the mitochondria, golgi apparatus, chloroplasts, and vacuoles. *Mitochondria* are minute structures lying free (randomly distributed) in the cytoplasm of nondividing cells and taking the form of rods or spherical granules. These are the sites of enzyme activity in oxidative metabolism and contains DNA. The *ribosomes* that contain RNA are sites of protein synthesis. *Chloroplasts* are found within the cell of plants and contains DNA and chlorophyll and are site of photosynthesis. *Plastids*, other than chloroplasts also exist. The leucoplasts are found in tissues that are not exposed to light, for example, in potato tuber. The *vacuoles* are conspicuous in the cytoplasm of plant cells, less so in animal cells which contain a fluid called cell sap. Vacuoles help to regulate water pressure in the cell and maintain rigidity. *Golgi apparatus* is chemically lipoprotein in structure and has a role in cell wall formation (synthesis of complex polysaccharides of the plant cell wall). The Golgi apparatus functions as a factory in which proteins (received from the endoplasmic reticulum) are further processed and sorted for transport to lysosomes, plasma membrane, or secretion.

Cell division

The process of formation of new cells from the existing cell is referred to as cell division (i.e., the cell arises from other cell by process of cell division). There are two types of cell division; mitosis and meiosis. During the cell division, *cytokinesis* (i.e., division of cell cytoplasm) follows *karyokinesis* (i.e., division of nucleus). Typically, the somatic cells of a sexually reproducing species contain diploid chromosomes, that is, chromosome occur in pairs or two copies, called homologous chromosomes. The gametic cells (the male and female gametes) contain half as many chromosomes as the somatic cells, that is, haploid number (n) of chromosomes. One member of each pair enters each gamete in the process of chromosome reduction or *meiosis*. The union of sperm and egg during fertilization results in a fertilized egg *(zygote)* containing one homologous chromosome from male parent and another from female parent. Thus the normal diploid chromosome number ($2n$) constitution is maintained. The somatic cells are formed and maintained by the process of equatorial division at *mitosis*. The chromosomes present in somatic cells are called somatic chromosomes.

Mitosis

The term mitosis was coined by Walther Fleming (German biologist, considered founder of cytogenetics) in 1882 and provided the detail of chromosomal movements in the process of mitosis. *Mitosis* is a type of nuclear division that produces two identical daughter cells from a mother cell. The daughter nuclei normally receive duplicates of each chromosome originally present in the nucleus of the parent cell. Mitosis occuring in somatic cells and is responsible for an increase of cell number and size. Cell division is regulated by cells communicating with each other using chemical signals (using proteins called cyclins), and these signals act as switches to start or stop cell division. Since the daughter cells have exact copies of their parent cell's DNA, no genetic diversity is created through mitosis and mutations happening in single cells are confined to single cell or a localized region. *Mitosis is good for basic growth, repair, and maintenance*. For example, growth happening in fast growing tissues called meristematic tissues—located mainly at stem apex and root apex. Mitosis is responsible for the replacement of old tissues and organs as well as for the production of new tissues and organs. It is the only form of cell division associated with asexual reproduction.

Meiosis

The term meiosis was coined by British biologists/botanists Farmer and Moore (1905) (Farmer and Moore had chosen the word "maiosis" that was soon changed to "meiosis" through discussion among researchers across the world). The meiosis involves two nuclear and two cell divisions. The first division is known as *reductional division*. It involves separation of chromosomes. During the reductional division, the chromosome number is halved but the centromeres of homologous chromosomes do not separate. At the end of the first meiotic division each daughter cell has n chromosomes, consisting of one bipartite chromosome of each pair. The second division is called as *equational division*. It involves separation of chromatids. At the end of second meiotic division, each of the four daughter cells has only single stranded chromosomes with the haploid number of chromosome number. The process of meiosis occurs in reproductive organs of plants/animals. If an organism has somatic chromosome number $2n = 20$, then at the end of meiosis the resultant four daughter cells (haploid) would have $n = 10$. *It facilitates crossing over, segregation and independent assortment of chromosomes and genes and is the basis of genetic variation, therefore, making meiosis fundamental for crop improvement programs (as it creates genetic variation, in progenies compared to parents, through recombination)*. It helps in maintaining the chromosome number in a species from one generation to the next generation produced by sexual reproduction. Meiotic irregularities include failure of pairing or partial pairing, lagging and unequal distribution of chromosomes, etc.

The differences between mitosis and meiosis are presented in Table 3.1.

Table 3.1 Comparison of mitosis and meiosis.

Mitosis	Meiosis
It occurs in somatic cells.	It occurs in reproductive organs.
It has one division per cycle, that is, one cytoplasmic division per equational division (sister chromatids separates).	It has two divisions per cycle. First division is reductional division (homologous chromosomes separates) and second division is equational division (sister chromatids separates). However, meiosis II, an equational division, is physically the same as mitosis. Although the genetic composition in the nuclei are different due to crossing over and recombination.
Since the chromosomes fails to synapse, there is no exchange of genetic material between homologous chromosomes.	As the chromosome synapse, the genetic exchange occurs between the chromatids of homologous chromosomes.
Two daughter cells are produced per cycle.	Four daughter cells are produced per cycle.
The chromosome numbers of daughter cells and genetic content of mitotic products are identical.	The chromosome number of meiotic products is half to that of mother cell.

Chromosome

Karl Wilhelm von Nägeli in 1842 discovered strands, which were later named chromosomes. Plant characters are controlled by genes (DNA sequences), located in a linear nucleoprotein body of the cell nucleus (i.e., chromosomes). Genes are transmitted according to *Mendelian inheritance* through the process of nuclear division. A species is characterized by a set of chromosomes per cell. The shape, size, and number of chromosomes vary from species to species. The characteristic features of the species are represented by a diagram known as *ideogram*. The normal structure of a chromosome may change, which could change the gene number in the chromosome (i.e., *deletion* and *duplication*) or the sequence of genes in the chromosome (i.e., *translocation* and *inversion*). Deletion is a loss of a portion or segment either from terminal or from intercalary portion of chromosome and leads to loss of the gene or genes in that chromosomal segment. Duplication is the presence of a segment twice in the same or sometimes different chromosomes (or genome), and lead to addition of some genes in population and can result in an additive effect. Translocation is the exchange of segment between non-homologous chromosomes and is useful in locating the position of genes, centromere, and genetic markers on the chromosomes and for the transfer of desirable characters from one species to another. In case of inversion, the chromosome segment is oriented in a reverse order and plays an important role in the evolution of new species by changing the karyotype of an individual. The number of chromosomes are changed due to *polyploidy*, namely, *autopolyploidy* (is the multiplication of the chromosomes of a single species) and *allopolyploidy* (originates from complete chromosome set from two species).

Novel genetic variation and genes are produced by several methods, commonly through the chromosomal aberrations described above and also through mutation. Lethal mutations do not carry their germline forward; however, non-lethal mutations over generations accumulate within the gene pool leading to an increase in genetic variation. Natural selection can reduce the abundance of some genetic changes within the gene pool, while adaptive changes can happen due to accumulation of favorable mutations. A germline mutation gives rise to a constitutional mutation (i.e., a mutation that is present in every cell) in the offspring and is passed on to descendants through reproductive cells. Somatic mutations on the other hand, involve non-reproductive system cells and are therefore, not usually transmitted to descendants.

Gene

A certain section on the DNA that provides the cell with the information to produce a protein or a functional RNA. The DNA is composed of nitrogenous bases: the purines (adenine = A, and guanine = G) and pyrimidines (cytosine = C, and thymine = T), deoxyribose sugar and phosphate group. When a nitrogen base in linked to a sugar the product is called *nucleoside,* and when a nucleoside is linked with a phosphate it is called *nucleotide*. Nucleotides are bound together to form a *nucleic acid* or *polynucleotide chain* (Fig. 3.1A) in which the

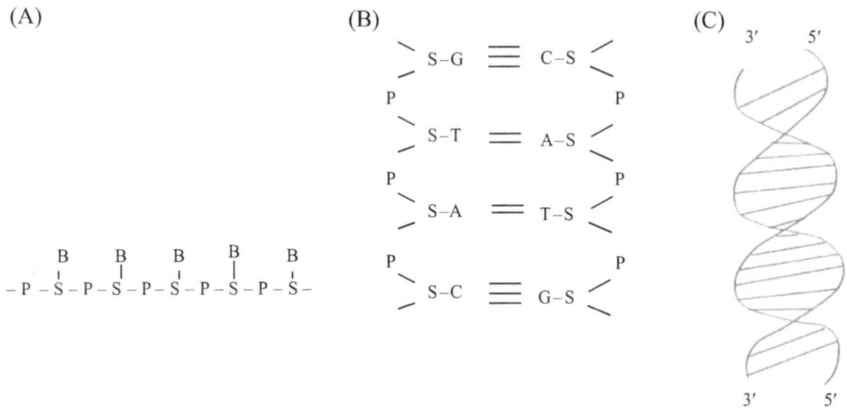

FIGURE 3.1

(A) Formation of nucleic acid or polynucleotide. (B) Double helical structure of DNA comprising a sugar-phosphate backbone, horizontal rungs of nitrogenous base connected with hydrogen bonds. (C) The polynucleotide chains are antiparallel.

sugar of one nucleotide is linked to the phosphate group of another and so on (Fig. 3.1B). The DNA is made up of a double helical structure of antiparallel polynucleotides with the base pairs connected by hydrogen bonds (Fig. 3.1C). The pairing of nitrogen bases is always specific. The adenine with thymine and guanine with cytosine are connected by hydrogen bonds. The change in any of the base pairs (from $A = T$ to $G \equiv C$) will result to a change in the DNA or gene. The gene controls the expression of a character through its mechanism of controlling protein synthesis. As per the central dogma – DNA encodes RNA, and RNA encodes protein, consequently building instructions for protein production are stored on DNA.

RNA is the hereditary material in viruses. It is very similar to DNA, however the difference is that it contains: (1) uracil instead of thymine; and (2) ribose sugar instead of deoxyribose sugar. RNA is predominantly single stranded (except in some viruses); and mRNA (messenger RNA), tRNA (transfer RNA), and rRNA (ribosomal RNA) fulfills various functions in the production of proteins.

Genome

A genome is defined as the set of chromosomes within a gamete of a species. The DNA (acronym for deoxyribonucleic acid) is the hereditary material, and as previously described, most of it is present in nucleus of the plants in the form of chromatin strands called chromosomes. DNA is also found in certain cytoplasmic organelles such as plastids (chloroplasts in green cell) and mitochondria. The nuclear genes show *Mendelian inheritance* whereas cytoplasmic organelles show *cytoplasmic inheritance*, which is also called *extra-chromosomal* or *extra-nuclear* inheritance. Plant cells contain large amounts of DNA, the amount being very variable between species from the smallest (0.5 picogram = pg) in *Arabidopsis thaliana* to 107 pg per haploid genome in *Viscum album*. Both chromosomal or extra-chromosomal genes have been found useful in plant breeding.

Mendelian inheritance

Gregor Johann Mendel (1822–84), called as the father of genetics, was an Austrian monk. He studied crosses in garden pea (*Pisum sativum* L.) between 1856 and 1865. He presented the result of his experiments in 1865 in the Natural History Society at Brunn, Austria, and the results were published in 1866. However, the research findings of Mendel were not appreciated until 1900, when the principles of genetics were rediscovered independently by three scientists: Hugo de Vries (in Holland), Carl Correns (in Germany), and Erich von Tschermak-Seysenegg (in Austria).

Mendel confined his attention to a single character at a time, for example, flower color. When the behavior of single traits was established, he then studied two traits together such as flower color and vine height. He counted the number of each type of progeny that resulted from the cross. Mendel's procedure was to cross two plants differing in a pair of contrasting characters and observe first hybrid (F_1) generation. He then crossed the hybrid plants together (or allowed them to self-fertilization) and raised the second generation (F_2). The F_2 offsprings were counted and number of plants in each contrasting category in which parental generation differed were classified. The simple data he used to formulate hypotheses or theoretical explanation (i.e., the pattern of inheritance of traits he studied), which were confirmed by Mendel himself and by others in plants and animals, were established as Mendel's laws of inheritance or Mendelian inheritance or Mendelism. Mendel's pairs of factors were later called as *genes* (by Johannsen) and each factor pair (AA or aa) is referred as an *allele* (i.e., A or a, the alternative form of a gene). A gene has a specific location on the chromosome, which is called *locus*. William Bateson started the practice of choosing letters mnemonically as gene symbols. He symbolized the dominant allele by capital letter. He also coined the words genetics, alleleomorph (which was shortened to allele), homozygote, and heterozygote, and he introduced the practice of denoting the generations in a breeding scheme as parental, first filial, and second filial generation as P_1, F_1, F_2, respectively, and so forth.

The two major principles: (1) law of segregation, and (2) independent assortment of genes were established. These laws permitted the prediction of traits among offspring according to the laws of chance. The genetic analysis of a cross can be explained with the help of a genetic grid called a Punnett square or checkerboard. Alternatively, the genetic analysis of a cross is also explained by the branch diagram or forked line method.

Law of segregation

When two plants with a contrasting trait (i.e., differing in one character) and each trait is governed by a dominant and recessive factors are crossed, in the F_1 hybrid one of the traits is expressed, which is called *dominant* trait and the other trait that is suppressed is called *recessive*. When the two F_1 plants were crossed (or selfed) and resultant seeds were planted, Mendel discovered that both traits appeared in the second generation (F_2) in a 3 (dominant): 1 (recessive) ratio, suggesting that the recessive factor governing it was not lost or modified. Based on these results, Mendel concluded that each parent contributed one factor for a character to the F_1 hybrid, which has two factors, one for each character. The two factors separate from each other, then the gametes of F_1 hybrid are formed.

By assuming that the tall peas (with dominant gene, AA) and dwarf peas (with recessive gene, aa) of the parental generation (P generation) are crossed. The F_1 generation were all tall (Aa, *heterozygous*) because A was dominant to a. The hybrid seed of F_1 plants (selfed seed) would separate to haploid gametes (A and a), which according to the laws of probability and random union would give a genotypic ratio of 1 AA: 2 Aa: 1 aa in the F_2 generation. Since the heterozygous Aa would show

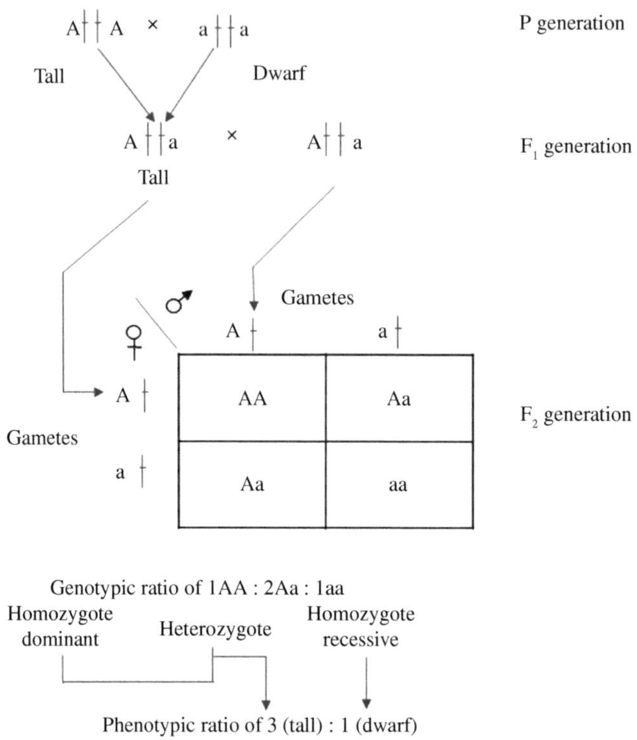

FIGURE 3.2

Law of segregation explained in a mono-hybrid cross that explains random union between two types of gametes formed in F_1 heterozygote and produced genotypes which showed a phenotypic ratio of 3 (dominant):1 (recessive) in F_2 of the Punnett square.

same phenotypic expression as the homozygous AA, the phenotypic ratio of 3 (tall):1 (dwarf) is observed (Fig. 3.2). Mendel studied following seven characters, one of each pair of contrasting traits appearing to be dominant and the other as recessive.

1. round vs wrinkled seed,
2. yellow vs green cotyledon color,
3. inflated vs constricted form of pods,
4. green vs yellow color of unripe pods,
5. axillary vs terminal position of the flowers,
6. red vs white flower color, and
7. tall vs dwarf habit of vine.

Such crosses where parents differ in one pair of alternative characters are known as *monohybrids*. To test the validity of F_2 genetic ratio, a test cross was developed by Mendel. He crossed the heterozygous F_1 plants to the homozygous recessive type and equal numbers of two kinds of offspring in 1:1 ratio were obtained. To distinguish between heterozygous and homozygous individuals in a segregating population, sometimes it is necessary to know the exact genotype of a plant. It can be known by progeny test of the F_2 plants showing dominant trait. For this, each F_2 plant is bagged separately then harvested seed is subsequently planted as plant to progeny rows. In the F_3 generation, a homozygous dominant plant would produce progenies that are uniform for the trait and the heterozygous plant will produce segregating progenies.

Law of independent assortment

When the genetic behavior of each single character was established, Mendel considered two characters simultaneously for which he crossed two plants differing in two characters. He observed that the factors (genes) for different characters (for each pair of traits) assort independently of each other during gamete formation. Such crosses, where parents differ in two pairs of alternative characters are known as dihybrid crosses and the resulting hybrids are known as *dihybrids*. Backcrosses (test crosses) of the heterozygous F_1 plants to the homozygous double recessive verified the correctness of Mendel's assumption.

One of his crosses was between a pea plant with yellow color and round shape and one with green color and wrinkled shape (Fig. 3.3). The F_1 hybrid had all yellow round seeds, and this was due to the dominance of yellow color over green, and round shape over wrinkled. When F_1 hybrid plants were crossed to each other or allowed to self-fertilize, the F_2 generation was obtained consisting of 556 seeds, which were classified for each character, seed color and seed shape separately as well as for both the characters simultaneously to prove his hypotheses. Among 556 seeds, the expected segregation in the ratio of 3 (dominant: 1 (recessive) was observed for both seed shape (423 round seeds: 133 wrinkled seed) and seed color (416 yellow seeds: 140 green seeds), which showed a good fit in monohybrid ratio. He studied the simultaneous segregation of 556 seeds for both characters and obtained 315 yellow round seeds, 108 green round seeds, 101 yellow wrinkled seeds, and 32 wrinkled green seeds. This fitted well in a dihybrid (9:3:3:1) ratio. Mendel found that the segregation of seed color is independent of the seed shape and that both the parental and new combinations of the characters appear in the F_2 offspring. The dihybrid segregation (9:3:3:1) explained with two genes can also be obtained by applying the law of probability (i.e., product rule), which states that two or more events are occurring independently. The chances of events occurring together is the product of probabilities of individual events.

1. *Segregation for first pair of characters (gene pair) in the* F_2
   - The probability of getting plants with yellow seed color = ¾.
   - The probability of getting plants with green seed color = ¼.
2. *Segregation for second pair of characters (gene pair) in the* F_2
   - The probability of getting plants with round seed shape = ¾.
   - The probability of getting plants with wrinkled seed shape = ¼.

Based on the probability applicable for two characters the joint probabilities of occurrence of different possible events would be as follows.

- Yellow round = ¾ × ¾ = 9/16.
- Yellow wrinkled = ¾ × ¼ = 3/16.
- Green round = ¼ × ¾ = 3/16.
- Green wrinkled = ¼ × ¼ = 1/16.

The F_2 phenotypic and genotypic segregation are presented through Punnett square (Fig. 3.3). It can also be presented using branch diagram or forked line method (Fig. 3.4).

The test cross is useful to test the validity of assumption that different genes in a dihybrid cross show independent assortment can be proved by crossing F_1 heterozygote (YyWw) with yellow round phenotype backcrossed (test crossed) to the double

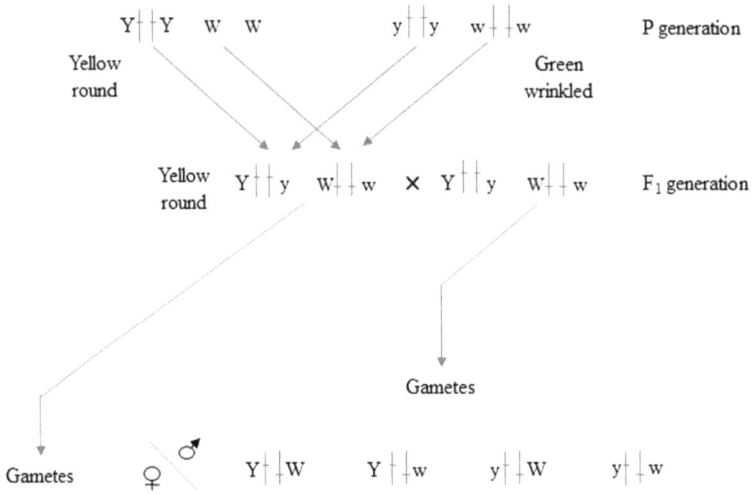

♀ \ ♂	Y W	Y w	y W	y w
Y W	YYWW	YYWw	YyWW	YyWw
Y w	YYWw	YYww	YyWw	Yyww
y W	YyWW	YyWw	yyWW	yyWw
y w	YyWw	Yyww	yyWw	yyww

F_2 generation

FIGURE 3.3

Law of independent assortment explained (in a dihybrid cross of pea) for two pairs of seed traits (yellow round and green wrinkled) which shows complete dominance. The results of random union between the four types of gametes formed in the F_1 heterozygote and produced genotypes that showed 9:3:3:1 phenotypic ratio in the F_2 Punnett square.

3/4 Y_ Yellow	3/4 W_ Round	=	9/16 Y_W_ Yellow round
	1/4 ww Wrinkled	=	3/16 Y_ww Yellow wrinkled
1/4 yy Green	3/4 W_ Round	=	3/16 yyW_ Green round
	1/4 ww Wrinkled	=	1/16 yyww Green wrinkled

FIGURE 3.4

Forked diagram used to predict the phenotypic ratios in the F_2 population with two dominant genes at both loci.

recessive (yyww) parent with green wrinkled seeds. The F_1 heterozygote would produce four kinds of gametes with genes, YW, Yw, yW, yw and the double recessive parent would produce only one kind of gamete, yw. The expected four kinds of genotypes: YyWw, Yyww, yyWw, and yyww should be obtained in equal proportion that is in a ratio of 1:1:1:1. Such results were actually obtained in experiment in pea, which confirmed Mendel's theory. When lines differing in three or more independently inherited characters are crossed, the principle of independent assortment still holds true.

Gene interactions

In earlier examples interaction between alleles of the same gene has been described. The relationship between the number of genes (with complete dominance and independent assortment) in a cross and their resultant gametes in the F_1 heterozygote, the different kinds of genotypes and phenotypes and all possible combinations in the F_2 generation are presented in Table 3.2.

In a typical digenic F_2 segregation, 16 genotypes with four kinds of phenotypic classes are observed (Table 3.3). The non-allelic gene interactions do occur, and this phenomenon is called *epistasis*. The term epistasis was used for the first time by William Bateson in 1909 who described interaction between different genes, that is, when allele at one locus masks or inhibits or masks the expression of an allele at

Table 3.2 Number of genes with complete dominance and independent assortment involved in the heredity of a character and number of possible F_2 combinations.

Number of genes involved	Kinds of gametes produced by F_1 heterozygote	Kinds of F_2 genotypes	Kinds of F_2 phenotypes	Number of all possible combinations in the F_2
1	2	3	2	4
2	4	9	4	16
3	8	27	8	64
4	16	81	16	256
5	32	243	32	1024
10	1024	59,049	1024	1,048,576
n	2^n	3^n	2^n	4^n

Table 3.3 The F_2 digenic segregation in a typical hypothetical example with complete dominance.

Phenotypic classes	Number in each class	Genotypic classes	Number in each class
A_B_ (AB)	9	AABB	1
		AABb	2
		AaBB	2
		AaBb	4
A_bb (Ab)	3	AAbb	1
		Aabb	2
aaB_(aB)	3	aaBB	1
		aaBb	2
aabb (ab)	1	aabb	1

another locus (i.e., genic interaction may occur between genes located in the same or different chromosomes). The gene that masks the expression of another gene (i.e., non-allelic gene) is called *epistatic gene*. The gene which is masked is called *hypostatic gene*. In case of non-allelic dihybrid interaction, the F_1 resembles neither parent and in the F_2 generation apparently new characters appear because some genes work together while other gene prevent the expression of other genes. The typical digenic F_2 phenotypic interaction (9:3:3:1) would be modified with different kinds of non-allelic interaction (Table 3.4); however, the phenotypic classes in F_2 of epistatic interaction would be less than four kinds. The genotypic ratios are not affected. The epistatic digenic interactions with the help of two non-allelic genes are described as follows.

Dominant epistasis (12:3:1): in this case dominant gene A is epistatic to B and b, and does not allow B and b to express. In the absence of A (i.e, aaB_),

Table 3.4 Modifications of the phenotypic segregation of independent assortment in digenic epistatic gene interactions.

Gene interaction	**Phenotypic classes**			
	A_B_	**A_bb**	**aaB_**	**aabb**
Independent assortment	9	3	3	1
Dominant epistasis or masking interaction	12		3	1
Recessive epistasis or supplementary gene interaction	9	3	4	
Duplicate genes with additive effect	9	6		1
Duplicate dominant epistasis or duplicate gene interaction	15			1
Duplicate recessive epistasis or complementary genes	9	7		
Dominant and recessive epistasis or inhibitory gene interaction	13		3	

B express and produces different phenotype. Similarly, in aabb, A is absent; therefore, bb expresses and produces different phenotype.

Recessive epistasis (9:3:4): in this case, aa is epistatic to B and b resulting in 3 (aaB_) and 1 aabb as one phenotypic class of four individuals and two dominant A_B_ and one dominant gene A_bb produce two different phenotypic classes as a result 9:3:4 ratio is obtained.

Duplicate genes with additive effects (9:6:1): the genotypes with one dominant gene produce one unit of the particular substance and two dominant genes produce cumulative effect and double recessive aabb produce no substance.

Duplicate dominant epistasis (15:1): the dominant alleles of both loci produce the same phenotype whether alone or in combination. Only aabb will produce another phenotypic class.

Duplicate recessive epistasis (9:7): two dominant genes at different loci of the same chromosome or different chromosomes, interact with each other to produce a character. However, in the absence of the other, neither of them produces that character. The action of these genes is complementary.

Dominant and recessive epistasis (13:3): one dominant gene (A) is epistatic to other dominant gene (B) and recessive allele b and bb is epistatic to the other gene A in dominant condition. Thus 9 A_B_, 3 A_bb, and 1 aabb produce one phenotype and aaB_ produce another phenotype in a 13:3 ratio.

Chi-square goodness of fit test

Chi-square is sum of square of differences (deviation) of observed frequency and expected frequency divided by corresponding expected frequency of all the classes. It is useful to test the validity of the deviation between the hypothesis (i.e., expected one) and the actual field data in the genetic experiments (such as inheritance studies). Thus the chi-square test is a statistical tool to determine whether the deviation is within the limits of experimental error or large enough to reject the hypothesis (null hypothesis). The null hypothesis states that observed data are in agreement with the expected ratio. In other words, deviations if any between the observed data and expected ratio is not real but due to chance only.

$$\text{Chi-square } (\chi^2) = \sum [(O-E)^2/E]$$

where O is the observed frequency, E is the expected frequency, and Σ is the summation over all classes.

The value is calculated for $n-1$ degree of freedom. The n is the number of phenotypic classes in the data. Thus in tests of 1:1 or 3:1 ratios, there is one degree of freedom. A test of 1:1:1:1 or 9:3:3:1 would have three degrees of freedom. Yates correction should be applied if sample size in any class is less than five or whenever only one degree of freedom exists.

$$\text{Chi-square } (\chi^2) \text{ corrected} = \sum [((O-E)-0.5)^2/E]$$

The calculated value of χ^2 from the observed data is compared with table value at desired degree of freedom with 5% and 1% level of significance. If calculated χ^2 value is less than the table χ^2 value of 5% or 1% chosen level of significance, the results are non-significant. This suggests that null hypothesis is accepted, and it is concluded that the observed data are in agreement with the expected data, that is, fit is good. On the other hand, if calculated χ^2 value is more than table χ^2 value at 5% level of significance, the results are significant (assuming a type I error rate of 5% is used). This suggests that null hypothesis is rejected, that is, the observed data are not in agreement with expected data (there is a significant difference between observed and expected frequency), that is, fit is not good. The chi-square test is used only when the number of observations are sufficiently large (at least 50) and is applicable only to the numerical data themselves and not to percentages or ratios derived from frequencies.

Linkage and crossing over

The number of genes of an organism exceeds the number of its chromosomes, hence a single chromosome carries a large number of genes. The tendency of two or more genes remain together in the same chromosome during inheritance is called *linkage*. It denotes the tendency of genes to be inherited in groups. The genes in a single chromosome constitute a *linkage group*, although large chromosome can be broken into more than one linkage group. The number of genetic linkage groups in a species corresponds to the haploid number of chromosomes, for example, in peas (*Pisum sativum* L.) the haploid (n) number of chromosomes are seven, hence, peas has seven linkage groups.

It is quite often said that Mendel was very fortunate not to run into complication of linkage because he used genes each of which were located on seven different chromosomes of peas. This is, however, an over-simplification. The actual situation was analyzed by Blixt (1975) (Table 3.5). Mendel worked with three genes on chromosome IV, two genes on chromosome I, and one in each of chromosomes V and VII. Genes "*A*" and "*I*" on chromosome I are so distantly located that no linkage is normally detected. The same is true for genes "*V*" and "*Le*" on one arm, and "*Fa*" on the other arm of chromosome IV. This leaves "*V*" and "*Le*," which ought to have shown linkage; however, Mendel did not publish on this particular combination.

The genes on different chromosomes as well as on same chromosome can assort independently. An exception to law of independent assortment was discovered in 1906 by Bateson and Punnett. They found two pair of alleles in sweet pea (*Lathyrus odoratus*), which did not assort independently. They coined terms *coupling* and *repulsion*. The coupling phase is due to linkage of two or more dominant or recessive genes (AB/ab) and the repulsion phase is due to linkage to a dominant gene with recessive gene (Ab/aB).

If there is no recombination between A and B on a chromosome, they are transmitted intact from one generation to the next. Instead of nine genotypes in the F_2 only three types are produced in a 1 AAbb: 2 AaBb: 1 aaBB ratio, that is a ratio expected from a single gene segregation. Of these, two genotypes are parental types and other genotype is product of non-crossover gametes, it is referred as *complete linkage*. In *incomplete linkage*, genes separate due to crossing over. If linkage is incomplete between A and B, new genotypes are produced as a result of crossing over.

Crossing over is the exchange of corresponding segments between non-sister chromatids of homologous chromosomes during meiotic prophase. The crossing over leads to recombination between linked genes. The parental combinations represent non-crossover types and the new recombination are cross-over types. The frequency of crossing over depends on the distance [recombination frequency (RF)] between two gene loci. The chief effect of linkage is to reduce the frequency of recombination between linked genes. Genes located at opposite end of chromosomes are likely to

Table 3.5 Chromosome locations of traits and genes studied by Mendel.

Character pair used by Mendel	Alleles	Chromosome
Seed color: yellow–green	I–i	I
Seed coat and flowers: colored–white	A–a	I
Mature pods: smooth expanded–wrinkled indented	V–v	IV
Inflorescence: from leaf axils–umbellate in top of plant	Fa–fa	IV
Plant height: >1 m to around 0.5 m	Le–le	IV
Unripe pods: green–yellow	Gp–gp	V
Mature seeds: smooth–wrinkled	R–r	VII

From Blixt, S., 1975. Why didn't Gregor Mendel find linkage? Nature 256, 206.

Table 3.6 Effect of linkage on the recovery of AB/AB genotypes in F_2 from double heterozygote.

Recombination value (*p*)	Percent of AB/AB individuals in F_2, in double heterozygote F_1	
	AB/ab	**Ab/aB**
0.50 (independent assortment)	6.25	6.25
0.25	14.06	1.56
0.10	20.25	0.25
0.01	24.50	0.0025
Formula	$100 \times [0.25 \times (1-p)^2]$	$100 \times (0.25\,p^2)$

separate by crossing over. Crossing over is less frequent near the centromere, and more common in the gene rich regions and telomere. Usually crossing over may vary from <1% to almost 50% of the total. The crossing over is also useful for locating genes in the chromosome and in the preparation of *linkage map.*

The linkage between two or more genes controlling desirable characteristics is an aid to the plant breeder because it tends to hold the favorable combination of characters. However, linkage between desirable and undesirable genes is a hindrance to wide segregation (Table 3.6). However, segregants with desirable alleles could be obtained if large segregating populations are grown and/or through the irradiation of male gametes (example of mutation breeding; covered in Chapter 14: Mutation Breeding). The linked characters show high value of genetic correlation and coheritability, that is, two different desirable characters would be improved simultaneously.

Detection of linkage

When two genes are located on different chromosomes, they assort independently and a typical F_2 ratio of 9:3:3:1 and a typical back cross (test cross) ratio of 1:1:1:1 is observed. The failure to the expression of these ratios is an indication of abnormal segregation at either loci or involvement of linkage. The linkage can be detected either from F_2 or test cross data.

1. *Detection of linkage using F_2 data*: when the F_2 data fails to fit in a 9:3:3:1 ratio using chi-square test, the chi-square is partitioned into various components. This is to ascertain, if linkage is involved. It is explained as follows.

Phenotypes =	A_B_	A_bb	aaB_	aabb
Expected ratio =	9	3	3	1
Observed values =	a_1	a_2	a_3	a_4

a. Chi-square due to Segregation of A locus $(\chi^2_A) = (a_1 + a_2 - 3a_3 - 3a_4)^2/3n = 1DF$ (i.e., one degree of freedom).
b. Chi-square due to Segregation of B locus $(\chi^2_B) = (a_1 - 3a_2 + a_3 - 3a_4)^2/3n = 1DF$ (i.e., one degree of freedom).
c. Chi-square due to joint segregation or linkage (χ^2_L) is obtained by multiplication of coefficients in (a) and (b) $= (\chi^2_L) = (a_1 - 3a_2 - 3a_3 + 9a_4)^2/9n = 1DF$ (i.e., one degree of freedom).

A significant chi-square shows linkage.

If two genes are studied, the outcomes in F_2 generation at different recombination frequencies are described below in Table 3.7.

2. *Detection using test cross data*: when the test cross data do not show good fit in a 1:1:1:1 ratio using chi-square test then the chi-square is partitioned as follows.

Phenotypes =	A_B_	A_bb	aaB_	aabb
Expected ratio =	1	1	1	1
Observed values =	a_1	a_2	a_3	a_4

a. Chi-square due to Segregation at A $(\chi^2_A) = ((a_1 + a_2 - a_3 - a_4)^2/n) = 1DF$ (i.e., one degree of freedom).
b. Chi-square due to Segregation at B locus $(\chi^2_B) = ((a_1 - a_2 + a_3 - a_4)^2/n) = 1DF$ (i.e., one degree of freedom).
c. Chi-square due to joint segregation or linkage (χ^2_L) is obtained by multiplication of coefficients in (a) and (b) $= ((a_1 - a_2 - a_3 + a_4)^2/n) = 1DF$ (i.e., one degree of freedom).

Table 3.7 Explanation of computation of genotypic ratio with complete linkage and independent assortment with two genes segregating.

Class	Formula, using given "*r*" (recombination frequency)	At $r = 0.5$ (independent assortment)	At $r = 0$ (complete linkage)
AABB	$1/4r^2$	1/16 (0.0625)	0
AABb	$1/2r - 1/2r^2$	2/16 (0.125)	0
AAbb	$1/4 - 1/2r + 1/4r^2$	1/16 (0.0625)	4/16 (0.25)
AaBB	$1/2r - 1/2r^2$	2/16 (0.125)	0
AaBb	$1/2 - r + r^2$	4/16 (0.25)	8/16 (0.50)
Aabb	$1/2r - 1/2r^2$	2/16 (0.125)	0
aaBB	$1/4 - 1/2r + 1/4r^2$	1/16 (0.0625)	4/16 (0.25)
aaBb	$1/2r - 1/2r^2$	2/16 (0.125)	0
aabb	$1/4r^2$	1/16 (0.0625)	0

A significant chi-square shows linkage.

The formula as given above for F_2 and test cross data are appropriate for detection of linkage in coupling phase. The same formula could be used for detection of linkage in repulsion phase, but classes are inverted as follows.

Phenotypes =	AB	Ab	aB	ab
Coupling phase =	a_1	a_2	a_3	a_4
Repulsion phase =	a_2	a_1	a_4	a_3

Linkage map

The proportion of the recombinant gamete produced in meiosis in the multiple hybrid is called recombination frequency (RF). It is computed as [(Recombinants/Total offspring) × 100]. If two genes are completely linked, RF = 0. Detection of linkage is accomplished by using the chi-square test. During the χ^2 test when null hypothesis is rejected, and presence of linkage is declared. A linear map of the genes showing the distance between two neighboring genes is directly proportional to the RF (%) between them and is known as *linkage map* or *genetic map* or *chromosome map*. Recombination fraction can vary from zero to 1/2. The distance between any two genes on a chromosome is expressed by the map units or cross-over units, which corresponds to the percentage of recombinants recovered among the progeny. For example, 1% RF is equal to one map unit or centiMorgan (cM). Map unit is considered the unit of distance in a linkage (genetic or chromosome) map. One map unit is equal to 1% recombination in the absence of double cross-over events (two separate single cross-over events between the two genes), that is, 1% RF is equivalent to 1 cM or 1 map unit. Genes 50 or more cross-over units apart behave as if they are non-linked genes and will assort independently. Fig. 3.5 presents an example of a two-point cross and calculation of RF.

Linkage map can be prepared by various ways according to the number of genes involved in estimation of linkage, that is, two-point test cross (i.e., a cross of dihybrid F_1 with its homozygous recessive parent). It provides information about RF between two genes and three-point test cross-over (i.e., a cross of a trihybrid F_1 with its homozygous recessive parent). A cross of parent with smooth, yellow seed, and purple flower phenotype with wrinkled, green seed, and white flower parent gives F_1 trihybrid with smooth and yellow seed with purple flower. The selfing of F_1 trihybrid gives F_2 phenotypes in 27:9:9:3:9:3:3:1 ratio (Fig. 3.6). A trihybrid cross is needed for determination of the order of genes in the chromosome.

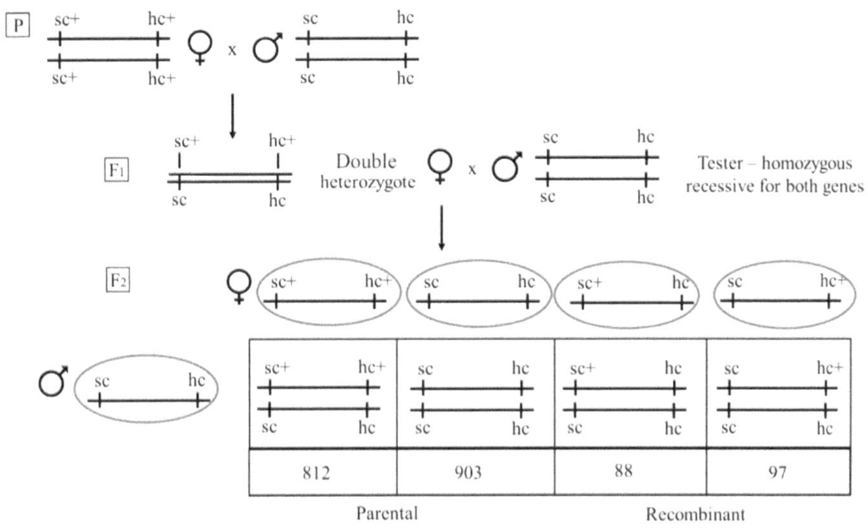

FIGURE 3.5

Hypothetical example of two-point cross in soybean for seed color (sc) and hilum color (hc) showing a cross of dihybrid F_1 with its homozygous recessive parent. Recombination frequency (RF) is $((88+97)/(812+903+88+97)) \times 100 = 9.7$. The convention to present female and male gametes are left and right, respectively. However, for ease in presentation, they are presented in the opposite manner in this Punnett diagram.

Incomplete dominance and codominance

The results of Mendel established that genes can exist in two alternate forms (i.e., one dominant and other recessive). The dominant allele expresses the same phenotypic effect in heterozygotes as in homozygotes (i.e., Aa or AA are phenotypically similar). In contrast to the complete dominance exhibited by seven traits studied by Mendel in peas, sometimes the heterozygote shows partial or incomplete dominance (also called semidominance). The phenotype of heterozygote is different from that of both the homozygotes. The inheritance of flower color in a cross of red flower (RR) and white flower (rr) in snapdragon (*Antirrhinum majus*) parents produced F_1 heterozygote with pink flower. The F_2 generation of this cross expressed both the genotypic and the phenotypic ratios which were same, that is, 1 red (AA): 2 pink (Aa): 1 white (aa). This is a modified phenotypic ratio of 3 (dominant): 1 (recessive) and is due to incomplete dominance, which produced a blended phenotype (pink type). In some case the alleles that lack a dominant and recessive relationship is called codominant alleles (i.e., similar function) and the pattern in codominance. Both the alleles are active and expressed in the F_1 heterozygote (e.g., allozymes).

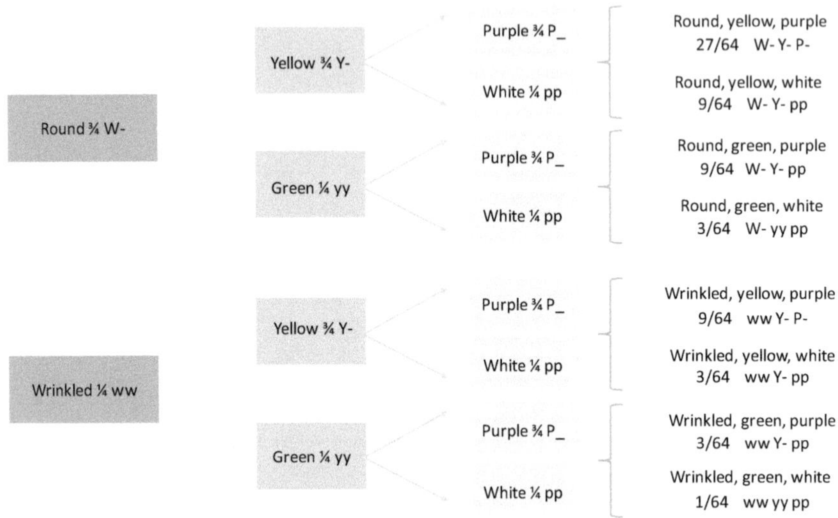

FIGURE 3.6

Forked diagram used to predict the phenotypic and genotypic ratio in the F_2 population with three dominant genes.

Multiple alleles

The Mendelian inheritance proved that genes exist in two allelic forms, one each of homozygous chromosomes. But a gene can change to alternative forms by a process of mutation, therefore, more than two alleles of a gene is theoretically possible in a population of individuals. Some genes are reported to have three, four, or more alleles in a population. Wherever more than two alleles are identified at a gene locus, it is called *multiple allelic series* or *multiple allelism.* In case of wheat as allohexaploid, red grain color is conditioned by three independent dominant alleles ($R_1R_2R_3$) acting in additive fashion, while white grain occurs when the genotype is $r_1r_2r_3$. When all three dominant alleles in one genotype is present, seed color is a very dark red.

Pleiotropy

Most of the biochemical pathways in the living organism are interconnected and often interdependent. Products of one reaction chain may be used in several other metabolic schemes, therefore, the phenotypic expression of a gene usually involves more than one trait. All of the manifold phenotypic expressions of a single gene is referred as *pleiotropic gene effects* and the phenomenon is known as *pleiotropy* or *pleiotropism.* The pleiotropic effect could be similar to that of linkage. However, the effect of linkage could be overcome by crossing over, but the true pleiotropy will never be changed. The purple pigmentation on stem, petiole, and veins of leaves is governed by a single dominant gene with pleiotropic effect in mung bean. For plant breeders, if a single gene is pleiotropic and is responsible for a positive effect on one trait but negative trait for the second trait, this relationship cannot be broken (without knocking out the gene, which will render its effect of both traits to zero). In this scenario, breeder will look to use independent genes (from a second gene) to improve the second trait. However, linked genes in repulsion (one gene responsible for a favorable effect and second gene causing a negative effect on the other trait) can be managed by creating and identifying cross-over events (recovering coupling phase linkage).

Penetrance and expressivity

The frequency with which a gene produces its phenotypic or visible effect in the number of individuals that carry it, is known as penetrance. Penetrance could be complete or incomplete and is calculated in percentage. The degree of manifestation of a character is known as expressivity, which could be uniform or variable. When the phenotypic expression of a gene is identical in all the individuals that carry the gene, it is referred as uniform expressivity. When the phenotypic expression of a gene differs in different individuals that carry the gene, it is referred as variable expressivity. Both penetrance and expressivity may be influenced by environmental conditions in some cases. The epistatic gene interactions can also affect penetrance and expressivity of genes.

Modifying genes

Sometimes the expression of major genes is influenced by the modifying genes. Such genes have small and additive effect(s). Modifying genes are primarily quantitative in nature that modify the effect of major genes either by enhancing or diminishing their expression. Therefore, modifiers could be of interest to plant breeders who concentrate on major genes in a cultivar. Modifier genes often have been reported to cause variation in traits like colors of seed, fruit aroma and taste, etc.

Threshold characters

The expression of some of the gene(s) governing a character is influenced by a environmental condition. Such characters are called threshold character (Fig. 3.7), for example, in some cases the reaction to disease pathogen in plants. Such traits

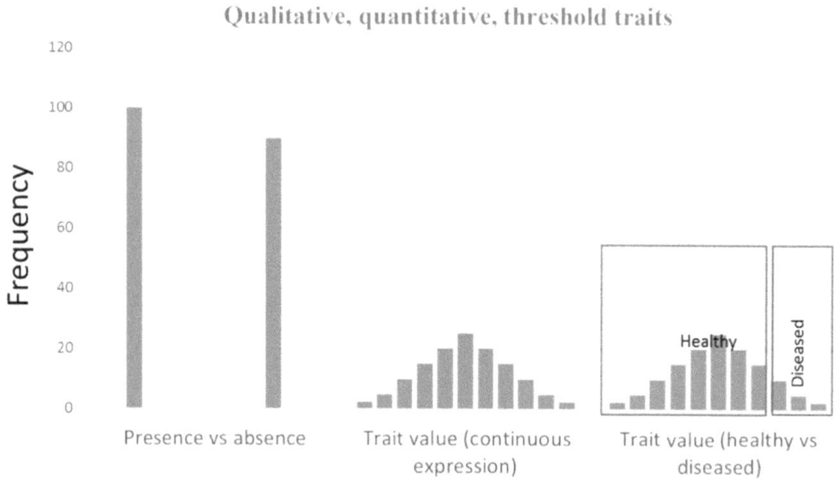

FIGURE 3.7

Frequency distribution of qualitative, quantitative, and threshold traits.

are inherited quantitatively but expressed qualitatively. For example, qualitatively inherited traits may be expressed as a presence versus absence, while a quantitatively inherited traits shows a typical normal frequency distribution for the trait expression. Examples of quantitatively inherited traits are seed yield, seed protein, and sometimes disease resistance. Although threshold traits show a quantitative inheritance frequency distribution, they express as healthy versus diseases (typical of a qualitative trait).

Quantitative inheritance

In the previous sections of this chapter, Mendelian inheritance and the various exceptions to it have been described. The qualitative characters are governed by one or few major genes located on same or different locus. Such traits are called *monogenic* or *oligogenic*. However, some characters are controlled by *multiple* genes or *polygenes* (sometimes called *minor genes*) located at different loci. Such genes have small and cumulative effects on the phenotypic expression of a character. The quantitative characters some time called *polygenic* or *multigenic* trait, show continuous variation between two extremes in the early segregating generations. Such characters are measured in terms of metrics, for example, length, weight, etc., and are readily influenced by environmental conditions. It is difficult to identify the number of genes and their effects in the segregating generations. The segregation of some five or more genes would show continuous variation similar to normal distribution.

Table 3.8 Inheritance of kernel color in wheat.

F_2 genotype	Genotypic classes	Number of R_1 and or R_2	Phenotypic classes	Kernel color
$R_1R_1R_2R_2$	1	4	1	Very dark red
$R_1r_1R_2R_2$	2	3	4	Dark red
$R_1R_1R_2r_2$	2			
$R_1r_1R_2r_2$	4	2	6	Medium red
$R_1R_1r_2r_2$	1			
$r_1r_1R_2R_2$	1			
$R_1r_1r_2r_2$	2	1	4	Light red
$r_1r_1R_2r_2$	2			
$r_1r_1r_2r_2$	1	Nil	1	White

The Swedish geneticist and plant breeder, Nilsson-Ehle in 1908 studied the inheritance of kernel color in wheat and based on the data of parental, F_1, and F_2 generations, he put forward the *multiple factor hypothesis*. He crossed two wheat varieties one with vary dark red kernel color of $R_1R_1R_2R_2$ genotype and other with white kernels of $r_1r_1r_2r_2$ genotype. The F_1 produced kernels with intermediate color (i.e., medium red, $R_1r_1R_2r_2$). In the F_2 generation, kernel color ranged from very dark red to white (Table 3.8).

He explained the results on the basis of two pairs of genes and each pair had two alleles that exhibited cumulative effects (i.e., the genes lacked dominance and their action was additive). R_1 and R_2 alleles added to redness of kernel and white contained neither of these alleles. The phenotypic classes in the F_2 showed a *continuous segregation* in the ratio of 1:4:6:4:1. This study was based only on two loci and it is now known that this trait is controlled by many genes, that is, most of the polygenic traits are controlled by many loci. If number of genes controlling a character increases, the F_2 phenotypic classes would be indistinguishable. The number of the F_2 genotypes would be 3^n (where n is the number of loci with two alleles) and number of phenotypes would be $2n + 1$.

Sometimes the progenies of a cross in the early segregating generations may fall outside the parental limits. Such segregants are called *transgressive segregants*. This is explained in a cross of wheat. Both the parents having medium red kernel color; $P_1 = R_1R_1r_2r_2$ and $P_2 = r_1r_1R_2R_2$ when crossed, F_1 produced kernel color similar to the parental lines, that is, medium red with $R_1r_1R_2r_2$ genotype. F_2 progenies showed a phenotypic segregation of 1:4:6:4:1 which is continuous segregation. The appearance of plants with extreme phenotypes beyond the parental limits is due to *transgressive segregation*.

Though the results of the above findings were clear but were less convincing as the kernel color could be classified into distinct categories (i.e., colored and

Table 3.9 Differences between qualitative and quantitative traits.

Qualitative traits	Quantitative traits
These are governed by one or few genes (oligogenes), which have discernible effect.	These are controlled by large number of genes (polygenes), whose effects are too slight to be identified individually, rather each gene has small effect.
They show discrete or distinguishable variation (i.e., distinct classes) in the segregating generation (F_2).	They show continuous variation in the F_2 generation.
These traits are least influenced by environmental conditions.	These traits are highly influenced by environmental conditions.
Such traits are concerned with numerical data.	These traits are metrical.
The data are analyzed using ratios or counts.	The data are analyzed from samples drawn from population using mean standard deviation etc.

colorless). However, subsequently East elaborated more convincingly the nature of inheritance of quantitative characters based on the findings of his studies pertaining to corolla length in tobacco (*Nicotiana longiflora*) and ear length in maize (*Zea mays*). He concluded that quantitative traits are controlled by many genes with small and cumulative effects. The greater variability in the F_2 with continuous variation resulted from the segregation and recombination of Mendelian genes. Some five genes would show continuous segregation (variation). The salient differences between qualitative and quantitative characters are given as follows (Table 3.9).

Cytoplasmic inheritance

Most of the characters of an individual are governed by nuclear genes, however, some of the traits may be controlled by *extra nuclear factors* or genes and the inheritance of such traits is known as *cytoplasmic inheritance* (also called *extra-nuclear* or *extra-chromosomal inheritance*). The factors governing the cytoplasmic inheritance are called plasmon or plasmogenes, which are present in the chloroplasts (cp-DNA) or mitochondria (mt-DNA). The plant characters which are inherited by plasmogenes are inherited in uniparental fashion by the female (egg) plant. Therefore, it shows reciprocal differences in the F_1 generation. The *cytoplasmic male sterility* (CMS) is the most extensively studied plant character of practical significance in plant breeding that is governed by plasmogenes located in the mt-DNA, which causes pollen abortion in higher plants. The CMS is observed in over 140 different plant species, and is useful in hybrid seed production of several important crop species such as maize, sorghum, pearl millet, rice, wheat, pigeonpea where it eliminates the process of hand emasculation.

Linking plant breeding with molecular biology

The classical plant breeding uses deliberate interbreeding and/or cross-breeding of closely or distantly related individuals to produce new lines or varieties with desirable traits. Plants are crossbred to introduce traits/genes from a line to an established variety and the genetic diversity is generated as a result of recombination between chromatids of homologous chromosomes, which is used to select desirable segregants/individuals, multiplied and yield evaluated and released as variety for commercial cultivation. The classical plant breeder may also use in-vitro techniques protoplast fusion, embryo rescue to make crosses successful or uses mutagenesis and polyploidy breeding to generate diversity and produce hybrid plants. However, modern plant breeding uses molecular biology techniques to select or if required to insert desirable traits (i.e., direct manipulation of DNA or by genetic engineering) into the plant. Specific gene or genes can be added to a plant (through plant transformation), or by knocking out a gene with RNA based silencing (RNAi) of plant genes, to produce a desirable phenotype. The plants resulting from adding a specific gene (interspecific, intergenus) are often called *transgenic plants*. The transgenic means genetically modified organisms that have acquired novel genes from other organisms by the process of gene transfer methods. If for genetic modification, genes of the same species or of a crossable plant (see Chapter 6: Wide Hybridization) are used with a native promoter, then the resultant plants are *cisgenic plants* (i.e., deliberate RNA based silencing of plant genes, RNAi).

The gene sequence analysis of the entire genome (i.e., complete set of chromosomes found in a nucleus of a given species) is called genomics. The genetic engineering with genomics also under pins molecular-assisted breeding [i.e., *marker-assisted selection* (*MAS*)] which is a valuable tool for accelerating plant breeding. The markers are a string or sequence of nucleic acid, which makes up a segment of DNA. The markers and DNA sequence are located on the same chromosome. They tend to stay together (i.e., they are linked) in each generation of plants that are produced. If a researcher can find the marker located within the gene also known as functional marker, it would mean that the gene itself is present. MAS can also be used to pyramid genes (i.e., to put two or more genes for a character, e.g., stress resistance genes) in a new plant variety. Scientists with major crop research organization routinely use MAS in their on-going breeding programs (DePauw et al., 2011). Further details on the use of molecular tool in plant breeding are presented in Chapter 27: Molecular Tools in Crop Improvement and Cultivar Development.

CHAPTER

Primer on population and quantitative genetics

4

Abstract

This chapter covers basic principles of population genetics and quantitative genetics that relate to plant breeding and cultivar development. The major concepts described include: types of populations in a breeding program, combining ability, qualitative and quantitatively inherited traits, gene action, phenotype, genotype and environment, genotype × environment interactions, stability analysis, heritability and its estimation, modes of selection, systems of mating, response to selection and selection theory, multiple trait selection, and generation to selection and breeding population sizes. Where applicable, practical considerations are provided for plant breeding applications.

The ways and means of genetic manipulation of crop plants depends on the mode of their reproduction, that is, self-pollinated or cross-pollinated. The self-pollinated (autogamous) plants have bisexual flowers and hence are pollinated by their own pollen grains. On the other hand, cross-pollinated (allogamous) plants have some means to enforce cross-pollination. It may involve separation of sexes into different plants (dioecy), or into different flowers of the same plant (monoecy), or pollen of the bisexual flower may be nonfunctional (male sterile), or pollen grains are functional but due to certain physiomorphological barriers (self-incompatible) are unable to effect self-pollination.

In cross-pollinated crops, each plant has equal chance to pair with every other individual plant of the population, which will only be the case with an optimum mixture of genotypes and fertilizing pollen, that is, panmixis. In a panmictic population, all individuals are potential partners and individuals in a population mate at random. There is an assumption that there are no mating restrictions (genetic or behavioral) among the individuals in the population; therefore, all recombinations are possible as each individual has an equal chance to cross with all other members (of the opposite sex) of the population. An outcome of such populations is that cross-pollinated crop cultivars are highly heterozygous and heterogeneous (due to intermating of plants). Panmictic populations are also referred as random mating population, and the smallest Mendelian populations are panmictic units. A Mendelian population is a reproductive community of sexual and cross-fertilizing individuals that share a common gene pool; where a gene pool is the total number

Plant Breeding and Cultivar Development. DOI: https://doi.org/10.1016/B978-0-12-817563-7.00020-9

and kind of genes/alleles in a sexually reproducing population that are transmitted to the next generation.

Population genetics deals with the prediction and description of the genetic structure of populations. Fundamentals of population genetics were developed for natural outcrossing species, but it is important to note that plant breeders working with cross-fertilizing (outcrossing or allogamous) species utilizes population genetics theories because the breeding methods they use are designed to increase the frequency (proportion) of desirable alleles in the population. One of the major difference between population genetics and Mendelian genetics is that population genetics principles are applied to the total products of all matings that will occur in the population, not just to one specific mating.

Genotypic frequency is the proportion of each genotype occurring in the total population. Allelic frequency is the proportion of the each allele at a locus. *Locus* designates a particular location on the chromosome, and *alleles* are the alternate forms of a locus. In popular usage, alleles are generally designated using symbols such as "A" and "a" or "B" and "b" where "A" and "B" are dominant to "a" and "b." If more than two alleles exist at a locus, or in the absence of clearly established dominance relationship, locus-alleles are listed as A_1, A_2,..., A_n or B_1, B_2,...., B_n. A locus could be a gene, that is, a coding sequence such as *Dt1* (gene controlling stem termination in soybean), or it could refer to a DNA sequence that may or may not code for something, such as a transposable element. For example, in maize (famous work of Barbara McClintock) *Ds* locus (short for *dissociation* or a locus were breakage of chromosomes occurred) along with second factor, *Ac* (short for *activator*) was necessary; and in the presence of *Ac*, *Ds* could move locations as well as cause breakages. This system is called a *two-element* system and historically has been called the *Ac/Ds* system. Other examples of locus can be *Cis-regulatory module* or cis-elements, which are a stretch of DNA where a number of transcription factors can bind and regulate gene expression. The *cis*-elements are typically located on the same chromosome (nearby gene) as the genes they control. While *trans*-elements (such as transcription factors) causes effects on genes not located on the same strand or are much farther away. Trans-acting elements recognize and bind to cis-acting elements to initiate, enhance or suppress transcription. Additionally, molecular markers such as SNP and SSR are loci not genes. These examples imply that all genes are loci, but not all loci are genes.

Hardy–Weinberg equilibrium

This law was developed independently by Hardy (1908) in England and Weinberg (1909) in Germany and it states that gene and genotypic frequencies in a large random mating population remains constant generation after generation provided there is no selection, mutation, migration or genetic drift. They showed that the frequency of genotypes in a random mating population would depend on the frequency of alleles in the preceding generation, not on the frequencies of the genotypes. Hardy–Weinberg (HW) principle is a basic concept of population genetics and describes the relationship that exists between allelic and genotypic frequencies in a randomly mating population.

The HW equilibrium is summarized as follows: p^2 AA + $2pq$ Aa + q^2 aa = 1 (or 100%). If 'p' and 'q' are allelic frequencies of A and a, respectively. If the original population has allele frequencies of p A1: q A2, random mating (random union of gametes) will produce a progeny population with genotypic frequencies in proportion p^2 AA: 2pq Aa: q^2 aa. In subsequent generations, gene and genotypic frequencies will remain unchanged if:

- random mating in a large population, which is diploid and there are only two alleles of a gene,
- selection pressure for both alleles is the same,
- no differential migration of one allele into or out of the population, and
- the mutation rate of 'A' to 'a' is the same as of 'a' to 'A' or no mutation.

The HW equilibrium requires and assumes a large random mating population; however, some inbreeding may occur even in cross-pollinated species in field. The equilibrium between loci is attained slowly over several generations, moreover under genetic linkage the rate of equilibrium between two loci is slowed down. With the example of A1A1, A1A2, and A2A2, possible matings, frequency of mating, and offspring will be:

Mating	Frequency	Frequency of offspring produced			
		A1A1	A1A2	A2A2	
A1A1 × A1A1	0.6 × 0.6 = 0.36	0.36			
A1A1 × A1A2	0.6 × 0.3 = 0.18	0.09	0.09		
A1A1 × A2A2	0.6 × 0.1 = 0.06		0.06		
A1A2 × A1A1	0.3 × 0.6 = 0.18	0.09	0.09		
A1A2 × A1A2	0.3 × 0.3 = 0.09	0.0225	0.045	0.0225	
A1A2 × A2A2	0.3 × 0.1 = 0.03		0.015	0.015	
A2A2 × A1A1	0.1 × 0.6 = 0.06		0.06		
A2A2 × A1A2	0.1 × 0.3 = 0.03		0.015	0.015	
A2A2 × A2A2	0.1 × 0.1 = 0.01			0.01	
	Σ = 1.00	Σ = 0.5625	Σ = 0.3750	Σ = 0.0625	Σ = 1.000

Genotype	Generation 1 (genotypic frequency)	Generation 2 (genotypic frequency)
A1A1	0.60	0.5625
A1A2	0.30	0.3750
A2A2	0.10	0.0625

From Generation 1 to Generation 2, genotypic frequencies have changed (with random mating), but allelic frequencies remain the same.

Allele	Generation 1 (allelic frequency)	Generation 2 (allelic frequency)
A1	0.75	= 0.5625 + 0.5(0.375) = 0.75
A2	0.25	= 0.5(0.375) + 0.0625 = 0.25

Effect of additional round of random mating:

Mating	Frequency	Frequency of offspring produced			
		A1A1	A1A2	A2A2	
A1A1 × A1A1	0.5625 × 0.5625 = 0.3164	0.3164			
A1A1 × A1A2	0.5625 × 0.3750 = 0.2109	0.1055	0.1055		
A1A1 × A2A2	0.5625 × 0.0625 = 0.0352		0.0352		
A1A2 × A1A1	0.3750 × 0.5625 = 0.2109	0.1055	0.1055		
A1A2 × A1A2	0.3750 × 0.3750 = 0.1406	0.0352	0.0703	0.0352	
A1A2 × A2A2	0.3750 × 0.0625 = 0.0234		0.0117	0.0117	
A2A2 × A1A1	0.0625 × 0.5625 = 0.0352		0.0352		
A2A2 × A1A2	0.0625 × 0.3750 = 0.0234		0.0117	0.0117	
A2A2 × A2A2	0.0625 × 0.0625 = 0.0039			0.0039	
	Σ = 1.00	Σ = 0.5625	Σ = 0.375	Σ = 0.0625	Σ = 1.000

Summary of generation to generation effect on genotypic frequencies:

Genotype	Generation 1 (genotypic frequency)	Generation 2 (genotypic frequency)	Generation 3 (genotypic frequency)
A1A1	0.60	0.5625	0.5625
A1A2	0.30	0.3750	0.3750
A2A2	0.10	0.0625	0.0625

Genotypic frequencies remain unchanged, that is, genotypic equilibrium has been reached. Genetic variability with respect to gene locus remains unchanged or constant. This is HW equilibrium (where allelic and genotypic frequencies remain constant from one generation to the next that population is in HW equilibrium). For population in HW equilibrium, genetic variability of the population does not change. This will be true in any mating system that deploys random mating. HW equilibrium assumes: large population, diploid organism, sexual reproduction, no selection, no mutation, no migration, and random mating. If these conditions are not met, population will no longer remain in equilibrium.

If "p" and "q" are allelic frequencies of A1 and A2, respectively, and if the original population has allele frequencies of p A1: q A2, random mating (random union of gametes) will produce a progeny population with genotypic frequencies in proportion p^2 A1A1: $2pq$ A1A2: q^2 A2A2.

Genotypes	A1A1	A1A2	A2A2	Total
Number	$n1$	$n2$	$n3$	N
Frequency	$x = n1/N$	$y = n2/N$	$z = n3/N$	1.0
Alleles	A1		A2	
Number	$2n1 + n2$		$n2 + 2n3$	$2N$
Frequency	$(2n1 + n2)/2N$ $= x + 0.5y$, since $x = n1/N$ and $y = n2/N$ $= p$		$(n2 + 2n3)/2N$ $= 0.5y + z$, since $y = n2/N$ and $z = n3/N$ $= q$	1.0 $p + q$

Frequencies after random mating (at equilibrium) are:

		Male gametes	
		A1 (allele frequency = p)	**A2 (allele frequency = q)**
Female gametes	A1 (allele frequency = p)	A1A1 (genotype frequency = p^2)	A1A2 (genotype frequency = pq)
	A2 (allele frequency = q)	A1A2 (genotype frequency = pq)	A2A2 (genotype frequency = q^2)

Therefore, frequencies are:

$$\text{Genotype Frequency of A1A1} = p^2$$

$$\text{Genotype Frequency of A1A2} = 2pq$$

$$\text{Genotype Frequency of A2A2} = q^2$$

$$\text{Allele frequency (A1)} = x + 0.5y = p^2 + 0.5(2pq) = p^2 + pq = p(p + q) = p, \text{ since } (p + q) = 1$$

$$\text{Allele frequency (A2)} = 0.5y + z = 0.5(2pq) + q^2 = pq + q^2 = q(p + q) = q, \text{ since } (p + q) = 1$$

In the case of two loci: if alleles A1 and A2 at Locus 1, and alleles B1 and B2 at locus 2 are present in a population.

Assume that their allelic frequencies are:

$$f(\text{A1}) = p_{\text{A}}$$

$$f(\text{A2}) = q_{\text{A}}$$

$$f(\text{B1}) = p_{\text{B}}$$

$$f(\text{A1}) = q_{\text{B}}$$

There are four gametes possible: A1B1, A1B2, A2B1, and A2B2.
Gamete frequencies will be:

$$A1B1 = r$$

$$A1B2 = s$$

$$A2B1 = t$$

$$A2B2 = u$$

where $(r + s + t + u) = 1$.

At equilibrium, gamete frequencies will become constant from generation to generation. If in the original population $D = ru - st = 0$, then the original population is in equilibrium (where $r = p_A p_B$, $s = p_A q_B$, $t = q_A p_B$, and $u = q_A q_B$). Frequencies will have to be calculated in each progeny generation to determine if it is at equilibrium. Population in disequilibrium will not reach equilibrium in one cycle of random mating when two or more genes are considered. Higher the number of genes, more cycles of random mating will be required to reach equilibrium.

Population genetics is concerned with how the frequencies of alleles in a gene pool change over time. The knowledge of population structure is important to breeding by either conventional or unconventional methods. In cross-pollinated crops, the primary emphasis is on improving populations instead of individual plants as in case of self-pollinated crops. It is, therefore, important to understand the type of variability present and its genetic control and mode of selection for changing the genetic structure, that is, the gene frequencies of breeding populations. It needs to be clarified that the discussion in this chapter is focused on improving population in cross-pollinated species; however, in several cross-pollinated species, including maize, hybrid cultivars are developed and commercially grown and the development of inbred parental lines in this case utilizes breeding methods common for self-pollinated crops.

Factors affecting equilibrium in the population

Selection is one of the most important processes for altering the gene and genotype frequencies of the random mating population because selection would allow the selected genotypes to reproduce and undesirable genotypes are eliminated. The shift in mean in the direction of selection may be observed, which could also lead to a change in variance to some extent and produce new genotypes. Selection is most effective at intermediate gene frequency and least effective at very small or very large frequencies. Selection in favor or against a rare allele may not be effective, as such allele(s) occur in heterozygote condition. However, the magnitude of these changes would depend on the heritability, number of gene (s), etc.

Migration: in plant breeding, migration involves the introgression of genes into breeding populations. The frequency of introduced gene (new alleles) into the population may be low but its effect on the host gene and genotypes could be high. In other words, migration increases variation of a population through introduction of new germplasm.

Mutation: mutations occur spontaneously in breeding populations and are generally recessive; however, dominant mutations are also observed. Mutations for easily identifiable traits could be selected for and against, depending on the desirability. Mutations for quantitative traits may go unnoticed as they are governed by genes, which have small and additive effects. Such mutations if these occur repeatedly (i.e., recurrent) may significantly affect the gene frequency of the population.

Often, dominant mutation leads to a gain of function through an increase in the activity of a gene expression or product, or new gene product(s). On the other hand, recessive mutations inactivate the affected gene leading to a loss of function. The loss of function may arise from mutation(s) that remove part of or all of the gene on a chromosome, interrupt normal expression of the gene, or change the encoded protein structure leading to an altered function. Induced mutations are used to generate additional and new variability at a much faster rate.

Genetic drift: it is also known as random drift. As population size decreases, the effect of random drift increases due to sampling error or if small sample size from original collection is taken for growing to maintain the germplasm accession/collection. Also, in smaller populations some amount of inbreeding may occur that would lead to homozygosity. As the name implies, random genetic drift is random as well as uncontrollable. For example, in a population, some individuals may leave more offspring by chance than other individuals. Let us consider a hypothetical situation in a forest where there are 50% each of two tree species. Species A is predominant in the northern part of forest and species B is predominant in the southern part. If fire destroyed 90% of the trees in the northern part of the forest, species A will be significantly reduced in number, and species B will leave more offspring, leading to a genetic drift. It is important to note that preponderance of offspring of species B is due to the chance destruction of species A and not necessarily because species B is healthier or more productive. Genetic drift is neutral to adaptation.

Types of populations in a breeding program, and mating designs

Simple populations

Simple populations consist of two to four parents. The simplest type of population is a segregating population developed from a cross of two elite lines, that is, a *two-way cross (also known as single cross) population.* An F_2 generation can be produced by self-fertilizing the F_1. F_2 and higher filial generations are developed, and selections are made in these populations in commercial breeding programs of self-fertilized crops (such as wheat, rice, and soybean). In many cross-pollinated crops, example maize, the commercially marketed products are single-cross hybrids (F_1) from inbred parents that developed after generations of selection and selfing. Creation of F_1 from two or more parents creates a breeding family (or families if more than one) or breeding populations.

Single cross

$$\begin{array}{ccc} \text{Parent A} & \times & \text{Parent B} \\ & \downarrow & \\ & F_1 & \end{array}$$

Each parent genetically contributes equally to the F_1.

Another type of segregating population is a *backcross population*, which is made from the F_1 by crossing it to one of the two parents (which will produce the BC_1F_1). The BC_1F_1 can be selfed forming a segregating population, or used in subsequent backcrossing with the same parent as used in the previous generation. The backcross population is particularly useful if one parent is superior to the other in most traits (recurrent parent), and the objective is to combine one or two genes from the unadapted (or overall undesirable parent; donor parent) into an elite line. Each backcross F_1 seed will be heterogeneous. Therefore, phenotypic testing or marker-assisted selection will be required to select the desirable plant to use in subsequent backcrossing to fix or enrich for favorable gene(s). For more details see Chapter 13: Backcross method.

Three-parent cross

Crossing the single cross F_1 to a third parent (or inbred line) and self-pollinating the resulting three-way hybrid creates a *three-way or three-parent cross population*. The generation resulting from self-fertilization is called the F_2 population. Since the third parent contributes 50% of the alleles to the final population; therefore, should generally be one of the best parents. A three-way cross is useful if one of the parents is a source of a few traits (e.g., an introduction), and one or two other parents are elite. Since the F_1 of a three parent cross are heterogeneous, marker-assisted selection can be practiced unlike in a two parent cross F_1.

$$\begin{array}{ccccc} \text{Parent A} & \times & \text{Parent B} & & \\ & \downarrow & & & \\ & F_1 & \times & \text{Parent C} & \\ & & \downarrow & & \\ & & F_1 & & \end{array}$$

Parents A and B contribute 25% each (genetically) while parent C genetically contributes 50% to the final F_1.

Double cross

A *double cross* or *four-parent cross population* is produced by crossing two initial single-cross F_1 hybrids, each formed from different parental lines. Each of the resulting individuals in the resultant F_1 will be genetically distinct. This type of cross is made when multiple traits (and genes) are being brought together in one variety.

$$
\begin{array}{ccc}
(\text{Parent A} \times \text{Parent B}) & & (\text{Parent C} \times \text{Parent D}) \\
\downarrow & & \downarrow \\
F_1 & \times & F_1 \\
 & \downarrow & \\
 & F_1 &
\end{array}
$$

Parents A, B, C, and D each genetically contribute 25% to the final F_1.

Complex populations

Some reasons to use complex populations include: development of good information for parents and full-sib families, identification of heterotic groups, estimation of general combining ability (GCA) and/or specific combining ability (SCA), and development of estimates of additive, dominant, and epistatic genetic effects and genetic correlations. Complex populations deriving from more than four parents can be constructed in several ways.

Nested or factorial designs

1. *North Carolina Design I* (nested design) (Fig. 4.1): involves mating each male parent to a different subset of female parents. Each male (N) is mated to several unrelated females (M) to produce *n* offspring per female. Offspring with common parents are full-sibs, while progenies that only share one common parent are half-sibs. This design is useful when the interest is in the variance component due to males and due to female parents within male parents. The analysis of this mating design provides information on parents and half-sibs, provides estimates on additive and dominance effects; while variance component due to males × females cannot be estimated.

♀\♂	A	B	C	D
1	x			
2	x			
3		x		
4		x		
5			x	
6			x	
7				x
8				x

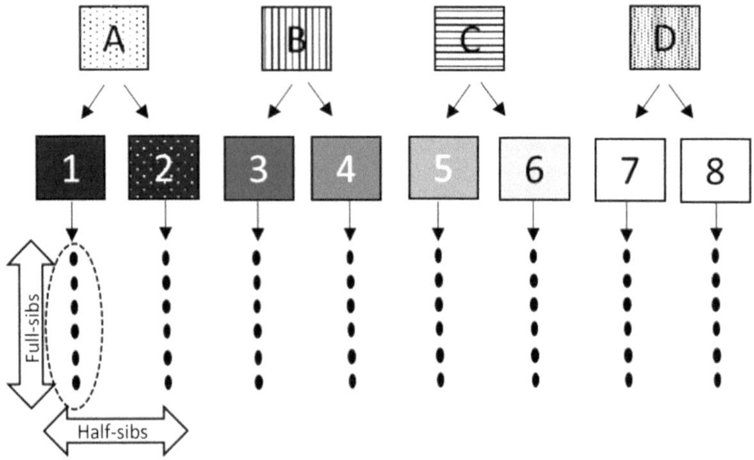

FIGURE 4.1

Example of a North Carolina Design I with eight females and four males.

2. *North Carolina Design II* (factorial design) (Fig. 4.2): this involves mating each member of a group of males to each member of the group of females. A group of males (M) is mated to an independent group of females (F) to produce ($M \times F = n$) offspring. This design is useful when the interest is in the variance component estimation due to males, females, and male × female. The analysis of this mating design provides information on parents and half-sibs, provides estimates of additive (from Var_M and Var_F) and dominance effects ($Var_{M \times F}$) enabling calculation of GCA (individual breeding value of that genotype) and SCA (individual's relative performance in specific genetic backgrounds) effects. Males and females can be from two different heterotic groups. Factorial designs allow studies to establish heterotic groups.

♀\♂	A	B	C	D
1	X	X	X	X
2	X	X	X	X
3	X	X	X	X
4	X	X	X	X
5	X	X	X	X
6	X	X	X	X
7	X	X	X	X
8	X	X	X	X

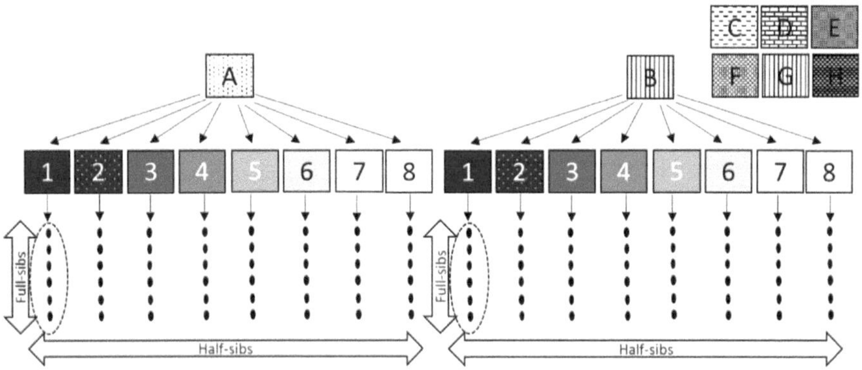

FIGURE 4.2 Example of a North Carolina Design II with eight females and eight males.

For the sake of space and page fit, only two male parents A and B are shown. Rest of the male parents (C, D, E, F, G, H) will be crossed to the same female parents (1, 2, 3, 4, 5, 6, 7, 8).

3. *Diallel cross:* in a diallel cross, all pairwise hybridizations among the parents are made in single cross combinations. Diallels can be "complete" or "full" where crosses are made in both directions, that is, including *reciprocal crosses*, as well as self-pollinations of parents. In other words, each parent is mated with every parent in the population (including selfs and reciprocals).

♀\♂	**1**	**2**	**3**	**4**	**5**	**6**	**7**	**8**
1	X	X	X	X	X	X	X	X
2	X	X	X	X	X	X	X	X
3	X	X	X	X	X	X	X	X
4	X	X	X	X	X	X	X	X
5	X	X	X	X	X	X	X	X
6	X	X	X	X	X	X	X	X
7	X	X	X	X	X	X	X	X
8	X	X	X	X	X	X	X	X

From a breeding standpoint, selfs do not have any value because recombination does not create variability if the parents are true breeding. Unless there are maternal and paternal effects and differences, the complete diallel is usually relevant for academic purposes only. In the example given above, Parent

1 will be selfed and used as a male and as a female in crosses with the other parents, thus doubling the number of crosses with each parent (e.g., 1 × 2 and 2 × 1). Diallel mating design allows the estimation of GCA (V_A) and SCA (V_{MF}) (or V_D). The advantages of NC Design II over diallel are (1) can estimate genetic variances in population(s) compared with only one population in diallel, and (2) fewer crosses need to made for generating similar information. In Fig. 4.3, it is clear that in a full diallel all cross combinations are made.

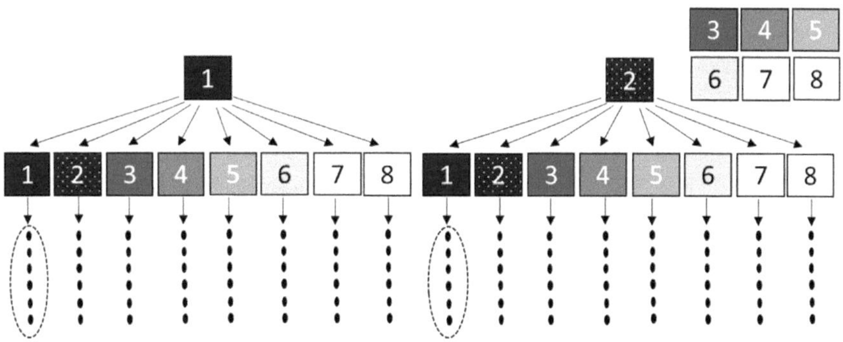

FIGURE 4.3

Example of an eight parent diallel design with eight genotypes. Due to space and page fit, only the crosses with inbred parents 1 and 2 are shown. Parents 3 to 8 will be crossed to parents 1 to 8 in a similar configuration.

- *Half diallel*: each parent is mated with every other parent in the population excluding selfs and reciprocals.

♀\♂	1	2	3	4	5	6	7	8
1								
2	x							
3	x	x						
4	x	x	x					
5	x	x	x	x				
6	x	x	x	x	x			
7	x	x	x	x	x	x		
8	x	x	x	x	x	x	x	

- *Partial diallel*: in this crossing scheme only selected subsets of full diallel crosses are made. A partial diallel can be made among any number of parents, enabling single cross pair (or a subset) of all possible pair where each parent is used as a male and a female with different parents, allowing sampling of more recombination events among favorable parents. This would allow incorporation of germplasm diversity without making a large number of crosses.

♀\♂	1	2	3	4	5	6	7	8
1		X	X	X				
2			X	X	X			
3				X	X	X		
4					X	X	X	
5	X					X	X	X
6	X	X					X	X
7	X	X	X					X
8	X	X	X	X				

All these NC and Diallel mating designs in the field requires careful field mapping and planting configuration. There are several things to consider such as flowering time, length of time pollen and stigma will be viable, and susceptibility to drought or pest. From a planting perspective in the field, considerations are (1) physical distance between parents, that is, distance between rows or plants such as side-by-side or not, (2) how much land area is available for population development, and (3) labor availability to make the planned crosses. The planting arrangements can be unstructured (e.g., unpaired parents—parents are not deliberately planted close to each other), to structured (e.g., paired parents—all parents are grown in adjacent rows or blocks). In circular crossing (also called as chain crossing) parents are sequentially crossed in pairs, for example, A × B, B × C, C × D, and so on, and loop is closed where the final parent is crossed to A. A larger area may be required for the crossing nursery when using paired parents. Parents can be arranged in subblocks representing their subgroups as will be used in a crossing nursery. Important points to consider for Diallel mating design and analysis (Griffing, 1956) are:

- Model I (fixed effect model; experimental material is to be regarded as the population on which inference is to be made)—only to estimate combining ability.
- Model II (random effect model; experimental material is random sample from some parental population. Inference is to be made on the parameters in the parent population, not on individual inbred lines)—additive and non-additive variance estimation.
- Method 1: parents, one set of F_1, reciprocal set of F_1. # of genotypes = (p^2).
- Method 2: parents, one set of F_1. # of genotypes = $(p)(p+1)/2$.
- Method 3: one set of F_1, and reciprocal set of F_1, no parents $(p)(p-1)$.
- Method 4: one set of F_1, no reciprocal set of F_1, no parents $(p)(p-1)/2$.

There are several assumptions for this mating design: no epistasis and no linkage. Neither is generally a valid assumption.

Top cross

Top cross and line × tester mating design is common in maize breeding programs where breeders are interested in developing hybrids. It is used for evaluation of SCA for a group of experimental inbreds and clones.

$$A \times \text{Tester} \rightarrow TC_A$$
$$B \times \text{Tester} \rightarrow TC_B$$
$$C \times \text{Tester} \rightarrow TC_C$$
$$D \times \text{Tester} \rightarrow TC_D$$
$$E \times \text{Tester} \rightarrow TC_E$$
$$\ldots$$
$$\ldots$$
$$\ldots$$
$$n \times \text{Tester} \rightarrow TC_n$$

The line × tester design is an extension of the top cross design. Instead of using one tester (top cross), more than one tester is used. Multiple testers are chosen to present contrasting heterotic pattern, and for their good GCA. Therefore, the design involves hybridization between lines (females) and wide based testers (m) generating f × m = fm hybrids, and provides SCA and GCA estimates.

The usefulness of these complex crosses (diallel, NC design, Line × Tester) is to estimate genetic variances, which allow us to compute combining ability. These are important to identify (select) clones (for clonally propagated species) and inbred (for cross-pollinated species) to use in the development of synthetics, determining heterotic groups and patterns to place inbred in appropriate groups, and understanding genetic variances and characteristics of important traits.

Polycross

The polycross design is used to produce synthetic cultivars, recombine selected genotypes in the recurrent selection procedure, and evaluate the GCA of the parent genotypes. Polycrosses are used to intercross a number of selected plants. Polycrosses are primarily used for cross-pollinated crop species where natural conditions (e.g., pollinating agent) allow pollinations; however, artificial pollination can also be done. In naturally pollinated nurseries, there is no control on pollen movement as it can come from a population of selected (or unselected—if no prior selection of genotype in the population is done) individuals. Moreover, there is no guarantee of successful crossing. Prolific pollen producing plants contribute pollen in more crossings than less pollen prolific producing plants. A number of aspects should be considered for successful intercrossing, and these include flowering time, wind effects, and insect pollinator activity. Flowering time is one of the most important factors, and these differences must be minimized among the parents to bias preponderance of some parents to intercross more frequently (due to overlapping flowering time) than they would by random chance. Flowering date factor can be more easily controlled in the greenhouse where polycrosses are made by hand (e.g., in alfalfa), and this is an option breeder can consider, resource permitting. If a breeder is relying on wind pollinations, the predominant direction of the prevailing wind will affect pollination and lead to a non-random pollination. For insect-pollinated crops, placing bee hives near the field is often done to ensure successful pollination. In a clonal crop, parental genotypes may be replicated in a manner to ensure that each genotype is adjacent to, or surrounded by, all other genotypes to provide equal chance and frequencies of crossing among all genotypes. If sufficient land is available, and not too many entries are included in crossing, a Latin Square arrangement (in this design, each parent is present in each row and each column) is a good way to enhance the equal chance of pollination among the genotypes. Another option is a randomized complete block design if the number of parents is large. In this design, a higher number of replications are planted (more then two); however, each parental genotype will not be surrounded by all other genotypes in equal frequency. Therefore, non-random and unequal mating occurs. To ensure that harvested seed of the polycross is representative of seed from random and equal pollinations (which is the ideal intent), three major procedures exist for harvest:

1. Bulk harvest the entire plot. this is the easiest method but will result in unequal contributions by both paternal and maternal parents to the population. The maternal or paternal parent that produces more seed will be represented in a higher proportion of seed in the lot.
2. If replications were used in the crossing design, bulk each parental genotype's seed across all the replications. This is a commonly used method, but there is an element of unequal contribution. Different clones (or inbreds or doubled haploids) of the same parental genotype may not produce the same amount of seed, so this method will skew the population towards the pollen parents that are physically closer to the highest yielding maternal parent.
3. One method that will overcome the problem listed in #2 above is to combine an equal amount of seed from each parent in each replication. This will ensure most equitable contribution of each parent. If there is a replication that did not perform to provide the minimum seed, some adjustments will be needed, leading to unequal contribution. These include adjustments among sample size per replication or pulling more seed from other replications for that genotype. Else, the lowest harvest yield plot can be used as the minimum for all other crosses.

Combining ability

Combining ability of inbred lines is of importance in determining future usefulness and commercial potential of the inbred lines for hybrid production. Combining ability can be divided into *GCA* and *SCA* (Sprague and Tatum, 1942). This concept has been very important in the commercial success of maize breeding and hybrid development. GCA is defined in terms of the average performance of a line in hybrid combinations. The GCA is calculated as the average of all F_1s having this particular line as one parent, the value being expressed as a deviation from the overall mean of crosses. From a breeding perspective, GCA is the average performance of a genotype in a cross with different tester lines, while the SCA measures the performance of a genotype in a specific cross combination in comparison with other cross combinations.

- GCA: mean hybrid performance of an inbred across a set of genetically diverse testers. GCA looks at additive gene action only.
- SCA: hybrid performance of a specific pair of genotype (i.e., two inbred lines).

SCA is a combination of non-additive (dominant gene action and epistatic interactions). SCA is defined in terms of instances in which the performance of certain hybrid combinations (between two inbred lines in a single cross) is either better or poorer than would be expected based on the average performance of the parent inbred lines. That is, each cross has an expected value that is the sum of the GCAs of its two parental lines. However, each cross may deviate from the expected value and the deviation is referred to as the SCA of the parental pair.

Estimates of GCA and SCA are applicable to the particular set of lines that a breeder has used in crossing. These crossings are generally in a factorial design (full, partial, or others such as the North Carolina designs). Sprague and Tatum (1942) reported that for unselected inbred lines, GCA was relatively more important than SCA, whereas for previously selected lines, SCA was more important than GCA. GCA is an indication of genes having largely additive effects (differences of GCA are due to the additive and additive × additive interactions in the base population) and is therefore, more important in a synthetic population. SCA is indicative of genes having dominance and epistatic effects (i.e., differences are attributable to non-additive genetic variance) and is therefore more important in a hybrid combination.

We use inbred GCA if we need to pick our testers in maize hybrid breeding (by testing progenies of HetGP1 tester with HetGp2 inbreds, or progenies of HetGp2 tester with HetGp1 inbreds; where HetGp = Heterotic Group); and we use clone GCA if we are using in synthetic cultivars (by testing progenies of different matings of all parents; therefore, we will need to measure GCA in maize from within and across heterotic patterns, and in clonal crops, all clonal parents).

Qualitative and quantitative traits

Qualitative traits are traits that are generally controlled by a single or few genes, and have phenotypes that can be classified into distinct categories. These traits are generally not influenced by environment and are recorded/scored as presence versus absence, yes versus no, or as specific color, seed shape, and so forth. Examples include the presence of awns in wheat (awned vs awnless), flower color (purple vs white), and round versus wrinkled seed (Mendel's garden pea experiment). These traits will be in the "yes versus no" classification and their expression will be the same irrespective of the environment the plants are grown, that is, genotypes with round seed will produce round seed in all environments.

Quantitative traits are controlled by several genes and produce a phenotype that cannot be classified into distinct categories, that is, there will be a continuum of phenotypes. These traits are influenced by the environment in such a way that the same genotype will produce different phenotypes in different environments. Examples of such traits are yield, protein percentage, oil percentage, and seed weight. The work of Francis Galton in 1870s and 1880s was pioneering for understanding quantitative genetics analysis of traits showing continuous distribution and lack of discrete classes.

Traits, such as plant height, are described as qualitative because they can be classified as short versus tall. However, it is important to note that plant height can occur across a range of values (cm or inches), meaning that data are not categorical. The most appropriate measurement is on a numerical scale, and this makes plant height a quantitative trait from a trait measurement perspective. The genetic definition will also define plant height as a quantitatively inherited trait due to a complex mode of inheritance governed by many genes. For example, in most crops, several plant height genes have been identified which again validates that plant height is not a truly qualitative trait. Disease resistance can be qualitative or quantitative, and the distinction between them will be driven by genetic control, influence of the environment, and phenotypic expression. Most traits that a plant breeder works to improve are quantitatively inherited necessitating more intensive phenotyping and more recently use of genomic selection (rather than marker-assisted selection, which is favored when a few genes with major effects are being selected in breeding families). Table 4.1 shows the comparison of qualitative and quantitative traits.

Types of gene action

Expression of genes can be described as additive or non-additive (dominance or epistatic). A gene acts in an additive manner when the substitution of one allele for another allele at a particular gene locus always causes the same effect. For example, $A1A1 - A1A2 = A1A2 - A2A2$. Here, the effect of substituting A1

Table 4.1 Details on qualitative and quantitative traits.

Qualitative trait	Quantitative trait
Discrete categories (distribution is bimodal)	Continuous distribution (frequency distribution does not have clearly distinguishable categories)
One or few genes control the traits	Many genes and quantitative trait loci control the trait
Large allelic effect	Small allelic effects
Inheritance studies using ratios	Genetic studies include mean and variance analyses, due to a normal frequency trait distribution
Not or minimal influence of the environment on trait phenotype	Large influence by the environment on the trait phenotype

for A2 is the same whether the substitution occurs in genotype A2A2 or in genotype A1A2 (effect on either side of the " = " sign is similar). When a gene acts additively, the maximum trait expression will occur in the genotype which possesses all the "favorable" alleles.

Non-additive gene action results from the effects of dominance (intralocus interactions, i.e., A1/A2) and/or the effects of epistasis (interlocus interactions, i.e., A1/B1 or A2/B2). Dominance effects are deviations from additivity; therefore, $A1A1 - A1A2 \neq A1A2 - A2A2$. This deviation results in a heterozygote similar to one of the parents rather than the mean of the homozygotes.

If we consider a hypothetical example of two gene loci (no linkage), the proportions of F_2 genotypes are:

Genotype	Ratio
A1A1B1B1	1/16
A1A1B1B2	2/16
A1A1B2B2	1/16
A1A2B1B1	2/16
A1A2B1B2	4/16
A1A2B2B2	2/16
A2A2B1B1	1/16
A2A2B1B2	2/16
A2A2B2B2	1/16

If we assume the genotypic value as

A1A1 = A1A2 = 4, A2A2 = 0

B1B1 = B1B2 = 3, B2B2 = 0

Assuming complete dominance at A and B loci, then -

Genotype	Genotypic value
A1A1B1B1	7
A1A1B1B2	7
A1A1B2B2	4
A1A2B1B1	7
A1A2B1B2	7
A1A2B2B2	4
A2A2B1B1	3
A2A2B1B2	3
A2A2B2B2	0

If a breeder makes selections based only on phenotype, they will select plants that have the following genotypes (A1A1B1B1, A1A1B1B2, A1A2B1B1, and A1A2B1B2) in the proportions listed below (with assumption of independent assortment at the A and B loci).

A1A1B1B1 = 1/16 of the total population,
A1A1B1B2 = 2/16 of the total population,
A1A2B1B1 = 2/16 of the total population, and
A1A2B1B2 = 4/16 of the total population.

In this case, the breeder is not able to distinguish between homozygous and heterozygous individuals of the four genotypes above because they have similar phenotypes. Therefore, if a breeder only wanted homozygous dominant (A1A1B1B1) plants and only used phenotype for selection, they will end up selecting nine out of 16 plants; however, only one out of 16 plants should have been selected if the intention is to fix the trait from further segregation. It is clear that if genotypic value is correlated with phenotypic values, selection in early generation will be effective. Additionally a breeder has to decide if they are interested in fixing allele or enriching allele, which impacts if markers (codominant) are needed or selection can be done phenotypically. A1A1B2B2 × A2A2B1B1 cross will lead to transgressive segregants where progeny will exceed the genotypic value of both parents. Additionally, codominant markers can be utilized to select homozygotes and also differentiate heterozygotes.

In a self-pollinated species where a cultivar is a pure line, non-additive gene effects can rarely be fixed and not important in the final pure line. Therefore, the selection response in earlier generation of selection for some traits is unpredictable when a trait is controlled by genes acting in a non-additive manner. In a cross-pollinated species where hybrid cultivars are used, non-additive gene effects, especially dominance effects, are important.

Epistatic interactions can be additive × additive, dominance × dominance, additive × dominance, or a higher order for three loci or more. These interactions are important for most traits, because these interactions are common.

Example below shows calculation of various genetic parameters.

Parameter	AA	Aa	aa	BB	Bb	bb
Frequency	0.25	0.50	0.25	0.25	0.50	0.25
Genotypic value (plant height in cm)	50	40	30	42	42	34
Mid-parent value (cm)		40 [(50 + 30)/2]			38 [(42 + 34)/2]	
Coded genotypic value	10	0	−10	4	4	−4
Population mean (cm)		40			40	
		A	a		B	b
Progeny mean (cm)		45	35		42	38
Average effect of allele (cm)		5	−5		2	−2
Genetic variance		50			12	
Additive variance		50			8	
Dominance variance		0			4	

$$\text{Population mean} = \Sigma(\text{Genotype frequency})^*(\text{Genotype value})$$

$$\text{Genetic Variance } (V_G) = \Sigma f_i{}^*(\text{Genotypic value} - \text{population mean})^2$$

$$\text{Additive Variance } (V_A) = 2 * \Sigma f_i{}^*(\text{Average effect of an allele})^2$$

Average effect of an allele is the effect of an allele on the mean of the progeny that inherit that allele, or in other words it is the deviation from the population mean of individuals that inherit the allele. It means that individuals that inherit "A" allele from a parent, have a trait value that is higher by 5 units (say 5 cm).

Breeding value of an individual is the portion of its genotypic value that determines the mean performance of its progeny (Falconer, 1981). *A parent's breeding value is the sum of the average effects of its alleles.* In other words, breeding value is the amount of variation in parental genotypes of breeding populations, which a breeder can use to obtain improvements that are desired. The difference between the offspring (i.e., progeny) of a particular parent and the mean of all offspring from a population is directly related to the parent's breeding value. If a parent's breeding value is high, then its offspring tend to be better than the average of the population. If the plants tend to have progeny that are about the same as the average of all progeny from the population, additive variance is low. If the plants differ in their breeding values, then additive variance is high and selection efforts will be effective (i.e., offspring of the plants that are selected will show superiority that distinguished their parents). The additive genetic variability among lines increases with increasing level of homozygosity, while additive genetic variability within line decreased at a rate of 0.5 per generation. Therefore, breeders developing pure or inbred lines should not wait too long to make within line selection.

Phenotype, genotype, and environment

Johanssen, in 1903, conducted an experiment in beans (*Phaseolus vulgaris*), a highly self-pollinated species, to study the effect of selection for seed weight using a seed lot from the cultivar, Princess. His experiments showed that: (1) selection for seed weight was effective in the original unselected population (i.e., lines selected for differences in seed weight showed consistent differences in seed weight in subsequent generations). Large-seeded parents produced large-seeded progeny, and small-seeded parents produced small-seeded progeny. (2) Selection within a line was not effective (i.e., irrespective of whether the parent was small- or large-seeded, all of the progeny of the selected seed always showed the average seed weight typical of the parent line). Johanssen concluded from this experiment that the original seed lot was composed of a mixture of different genotypes/pure lines (that were each homozygous for genes controlling seed weight), and even when the progeny of a seed lot differed phenotypically, each seed of that line possessed the same genotype for seed weight. Mathematically, the phenotypic value for an individual (i.e., a single seed in Johanssen's experiment) in a population is equal to its genotypic value plus an environmental (non-genic) deviation:

$$P = G + E$$

where P is the phenotype (eg., observed seed weight), G is the genotype (eg., genetic potential for seed weight), and E is the environment (environmental effects, i.e., factors determining the extent to which the genetic potential is reached).

In a general way for a population of seeds, the phenotypic variability is represented mathematically by the following equation:

$$V_P = V_G + V_E$$

where V_P is the phenotypic variability (total variability observed), V_G is the genotypic variability (variability due to genetic causes), and V_E is the environmental variability (variability due to environmental causes).

Genetic variability is heritable, that is, variability that can be manipulated by plant breeders and transmitted to progeny. The presence of genetic variability is essential for selection to be effective. Environmental variability is not heritable, and it can mask the true expression of a trait.

Assuming the mean value of E for all individuals across the population is zero, the mean phenotypic value would equal the mean genotypic value. Therefore, population mean is both the phenotypic and genotypic value. To prove this, consider a theoretical experiment using replicated genotypes—either as clones or as inbred lines—and measure them under regular environmental conditions. The mean E will be zero across the population, so that the mean phenotypic value would be equal to the mean genotypic value. However, in reality, plant breeders deal with segregating populations that are not genetically uniform when they are selecting, and rarely work in controlled environment for the entire breeding program and its populations.

The phenotypic value of any individual is represented as a sum of three components: (1) the mean μ of the entire population, (2) a deviation from the population mean due to the specific genotype of the individual in question, and (3) a deviation from the population mean due to the specific environment of the individual in question (environmental deviations are unique to each individual represented as E1, E2, ….).

Genotype	Phenotypic value
A1A1	μ + G1 + E1
A1A1	μ + G1 + E2
A1A1	μ + G1 + E3
A1A2	μ + G2 + E4
A1A2	μ + G2 + E5
A1A2	μ + G2 + E6
A2A2	μ + G3 + E7
A2A2	μ + G3 + E8
A2A2	μ + G3 + E9

where, μ is the population mean, G is the contribution due to genotype (different for each genotype, say $G1$, $G2$, $G3$ in example above), E is the contribution due to environment (different for each individual, say $E1$, $E2$, $E3$, …, $E9$ in example above).

$G\#$ and $E\#$ are not directly observable. Using the information above, model can be summarized as follows:

$$P = \mu + G + E$$

$G1$, $G2$, and $G3$, deviations for A1A1, A1A2, and A2A2 genotypes, respectively, are

$$G1 = G11 - \mu;\ \ G2 = G12 - \mu;\ \ G3 = G22 - \mu$$

Since $P = \mu + G + E$, $E1$–$E9$ can be calculated if the phenotypic score is known for individual, $G1$ is calculated based on the equation above (parameters known).

Since E and G are deviation from the mean, the mean of E and G is 0.

The total phenotypic variation in the population (σ^2_P) is the summation of $(P - \mu)^2$. We also know that $P = \mu + G + E$; therefore, $(P - \mu)^2 = (\mu + G + E - \mu)^2 = (G + E)^2 = G^2 + E^2 + 2GE$.

Since G and E are deviation from their means, G^2 is the variance resulting from differences in genotype and E^2 is the variance resulting from the differences in environment. $2GE$ is the term that describes the interaction between genotype and environment. If no $G \times E$, we can now write: $\sigma^2_P = \sigma^2_G + \sigma^2_E$.

Assuming no $G \times E$, this equation is most useful for clonal or highly selfing organisms, because they pass their diploid genotypes on to their offspring intact, but it is less useful for sexually outcrossing species where the genotypes are created again in each offspring by a random combination of an allele from each parent at each locus. For these species: in terms of gene action, population variability (assuming no $G \times E$) can be described as: $\sigma^2_P = \sigma^2_G + \sigma^2_E$, where $\sigma^2_G = \sigma^2_A + \sigma^2_D + \sigma^2_I$ = additive genetic variance + genetic variance due to dominance + epistatic genetic variance + environmental variance.

Genotype × environment interactions and stability analyses

Phenotype is governed by genotype (σ^2_G) and genotype × environment interactions (σ^2_{GE}). Not all variation for a phenotype is accounted for by genotype and $G \times E$ interactions, and the remaining variation is attributed to error. Any trait that you observe for a plant, a plant family, or population is a phenotype. Genotype is the genetic basis of a trait (e.g., gene or gene × gene interactions). The $G \times E$ interaction is the interaction of genotype with environment, where each genotype may perform or look different in different environments.

The environment of a single plant consists of all things other than the genotype of the individual. These factors include differences in soil, temperature, humidity, rainfall, day length, solar radiation, wind, salinity, pathogens, pests, and so forth. Environment can be a microenvironment or a macroenvironment. A microenvironment refers to a unique set of factors that alter the development of a single plant. Groups of plants growing at the same time in the same area each encountering similar microenvironments are classified as being in a macroenvironment (i.e., a class of microenvironments). For example, if a field of beans is exposed to excessive moisture stress (i.e., waterlogging), individual plants may suffer slightly different levels of waterlogging (microenvironment), but all plants would suffer some degree of waterlogging (macroenvironment). This bean field's macroenvironment will be described as waterlogged. In breeding, we are more interested in the macroenvironment, and these are classified as locations, years, a combination of location × year, or simply environments.

To describe the phenotypic value of a genotype in terms of microenvironment and macroenvironment, the equation below is useful:

$$P_{ijk} = G_i + E_k + (G \times E)_{ik} + e_{ijk}$$

where G_i is the effect of the ith genotype, E_k is the effect of the kth macroenvironment, $(G \times E)_{ik}$ is the effect of interaction between ith genotype and kth macroenvironment, and e_{ijk} is the residual composed of deviation of the jth microenvironment from the mean of such effects in the macroenvironment k and deviation of the interaction from the mean of interactions.

Genotype by environment interactions can be of non-crossover types or crossover types (Fig. 4.4A–C).

The crossover interaction is the most important type of genotype × environment interaction because different genotypes will be selected (due to their better performance) in different environments. Crossover interactions are often due to differences in the genotypic response to different environments. For example, in Fig. 4.4C, if environment 1 is disease free and that pathogen A is present in environment 2. Genotype 1 is a lower yielding, disease-resistant genotype, and genotype 2 is a high yielding, disease-susceptible genotype. Genotype 1 will have a higher yield in disease-prevalent environment 2, while genotype 2 will have a higher yield in environment 1. The goal of a breeder should be to combine the better response to both environments into a single genotype. However, a breeder needs to first determine whether yield and disease resistance are mutually exclusive. For example, it's possible that the gene for disease resistance could be linked to gene(s) that reduce yield. In this case, a breeder would need to grow a large segregating population to identify progeny with useful recombination that combines high yield and disease resistance. If, however, a single pleiotropic gene controls both disease resistance (causes resistance) and lowers yield, a breeder can only improve both traits by complementation (i.e., the building or bringing together of other useful genes to improve the responses). Additionally, breeder needs to fully understand the details of the environment—presence of weeds, disease, insect, soil drainage or flooding, management conditions, and many more to get a better handle on how to handle $G \times E$ interactions and selection procedures. These are just one of many reasons why breeders need to visit their testing sites and take useful notes.

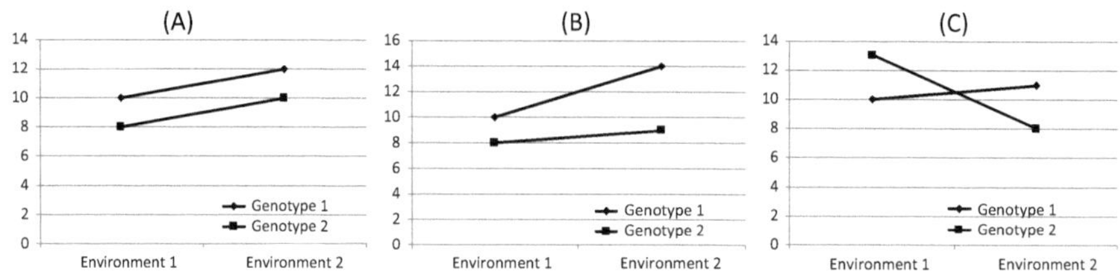

FIGURE 4.4

(A) No interaction, (B) non-crossover interaction, and (C) crossover interaction.

In literature, sometimes location and environment are interchangeably used. However, location can be defined as a physical place (with specific latitude and longitude) where genotypes are seeded and tested; and environment as a combination of physical attributes of a location and the climatic variables for that season (i.e., soil characteristics, soil fertility, topography, temperature, snowfall, rainfall, insects, diseases, and wind) that affect the plant growth and development. In a group of locations, if genotypes consistently perform the same over a number of seasons, it is considered a mega-environment. Two mega-environments have the same crossover effect that is consistently observed over a number of tests, thereby maintaining their designation as a mega-environment each consisting of multiple locations.

Stability analyses

- Type 1: a genotype is considered stable if its among-environment variance is small—with high mean and low CV in a Mean-CV analysis (Francis and Kannenberg, 1978).
- Type 2: a genotype is considered stable if its regression response over the environmental index is parallel to the mean response of all genotypes from that experiment. Example, Finlay–Wilkinson analysis (Finlay and Wilkinson, 1963).
- Type 3: a genotype is considered stable if the residual means square from regression on the environment index (type 2) is small (i.e., varieties that contribute least to the $G \times E$ interaction). Example, Eberhart–Russell analysis, Shukla's stability variance (Eberhart and Russell, 1966).
- Type 4: a genotype is considered stable if it has a low superiority measure (*Pi*) value. This ranking approach combines performance and stability into a single term. Calculates the sum of mean square differences, over environments, between the performance of a cultivar and the performance of the best cultivar within a given environment. Example, Lin and Binns superiority index.
- Type 5: a genotype is considered stable if its multiplicative variance score is small. For example, additive main effect and multiplicative interaction (AMMI). AMMI partitions the additive variance from the multiplicative variance ($G \times E$ interaction) and then uses principal component analysis (PCA) to extract the patterns from within $G \times E$ interaction component.

Mean-CV

Concurrently measures mean performance and variability of genotypes to determine optimal combination of stability and performance (Fig. 4.5).

Regression coefficient type analyses

Regression deviation types: stable varieties contribute least to the $G \times E$ interactions (linear regression and deviations from linear regression). For each entry, a linear regression of individual yield on the mean yield of all entries for each environment is computed. Cultivars are considered stable if their response to

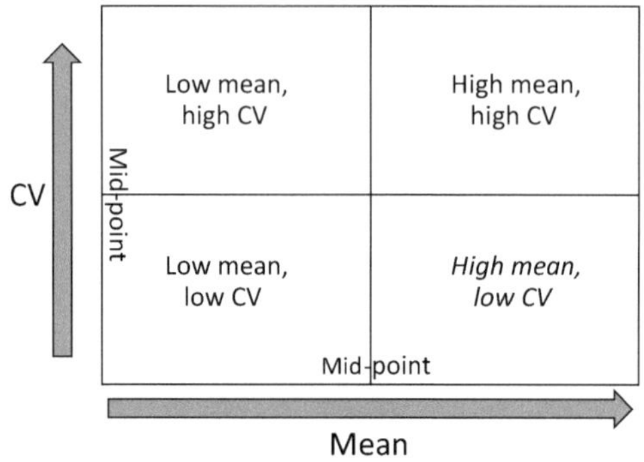

FIGURE 4.5

Mean-CV quadrants for determining genotypes with lower variability in their response variable and high mean performance.

environment is parallel to the mean response of all cultivars under test (linear regression) (Fig. 4.6). However, this method does not combine mean performance and stability into one statistic.

Superiority measure

Combines performance and stability into one term. Involves calculating the sum, over environments, of the mean square differences between the performance of a cultivar and the performance of the best cultivar within a given environment. Lower superiority value is better. This method has a tendency to be weighted toward the responsive ones in the higher yielding environments.

Multivariate approaches

Developed with an aim to combine mean performance, regression slope, and deviation from regression. Another approach is to use PCA and incorporate additive and multiplicative components (e.g., AMMI) to extract patterns from within $G \times E$. This allows to compare both genotypes and environments (can remove redundant environments). Multivariate analysis serves the following main purposes: (1) eliminate noise from the data pattern (i.e., distinguish systematic and nonsystematic variation), (2) summarize the data, and (3) reveal a structure in the data. AMMI analysis first fits the additive main effects of genotypes and environments by the usual analysis of variance and then describes the non-additive part, genotype–environment interaction by principal components analysis; therefore, AMMI summarizes patterns and relationships of genotypes and environments and gives information on $G \times E$.

Another popular analysis for the study of performance and stability is through the use of biplot analysis, for example, GGE biplot (Gabriel, 1971; Yan and Tinker, 2006). The biplot displays the G and GE of a genotype-by-environment data. In this analysis, the environment main effects (E) are removed, and the entry main effect (G) and the entry by environment interaction (GE) are retained and

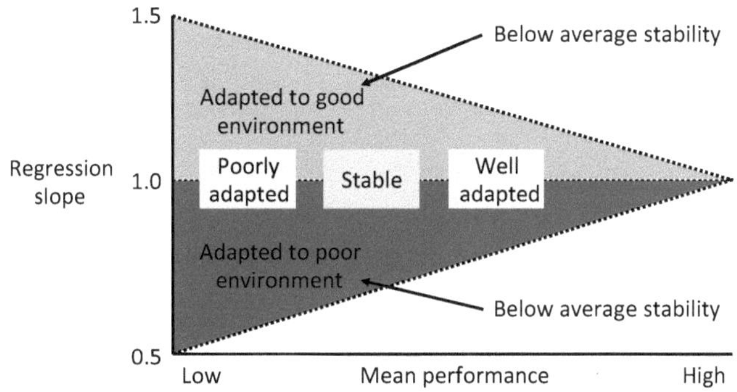

FIGURE 4.6

Graph of regression slope to study stability in performance.

combined. For the purpose of providing an overview of this analysis and its interpretation, one advanced stage plant breeding trial with 17 breeding lines and 3 checks grown in 8 locations in one year is explained (Fig. 4.7). The AMMI2 biplot plots the majority (in this case, first 2 PCs) of $G \times E$ effects (Fig. 4.7A). Projections are drawn from the (0,0) for the two PC axis, and further investigation for Location#5 (L5) shows that similar distance from the central point (0,0) but with different distances for Check#3 (C3) and Breeding line#8 (G8) from the L5 line. It means that C3 and G8 have similar $G \times E$ at L5 (the length of projection and not distance from the perpendicular line is important for determining $G \times E$). Another type of plot, AMMI1 (Fig. 4.7B), plots PC1 of the $G \times E$ effect and seed yield performance. In AMMI1 plot, genotypes closer to the horizontal line ($y = 0$) indicates less *GE* (i.e., higher stability). This information along with yield performance (x-axis) can be used to identify higher yielding and more stable lines. Similarly locations closer to the horizontal line shows less $G \times E$, that is, they have less discriminatory power to separate genotypes (in terms of $G \times E$ effects). If the percentage variation explained by the PC1 axis is lower, this information is less reliable. Environments closer to the horizontal line indicate less *GxE* interaction, and have less ability to discriminate genotypes (in terms of *GE* effect). However, the significance of above information depends on the percentage of the variance captured by PC1. GGE biplots are useful to show "What-Won-Where" to compare genotypes in specific environments. For example, in Fig. 4.7C, at L1, C2 outperforms C1 as it is closer to the L1 line through (0,0) central point. Genotypes can be compared across multiple environments, for example, C3 performs better than G12 at L4, L5, L6, and L8 while G12 is better at L1, L2, L3, and L7. Finally GGE biplot can be used to get a handle on grouping of environments (Fig. 4.7D), which shows similar environments in dashed lines.

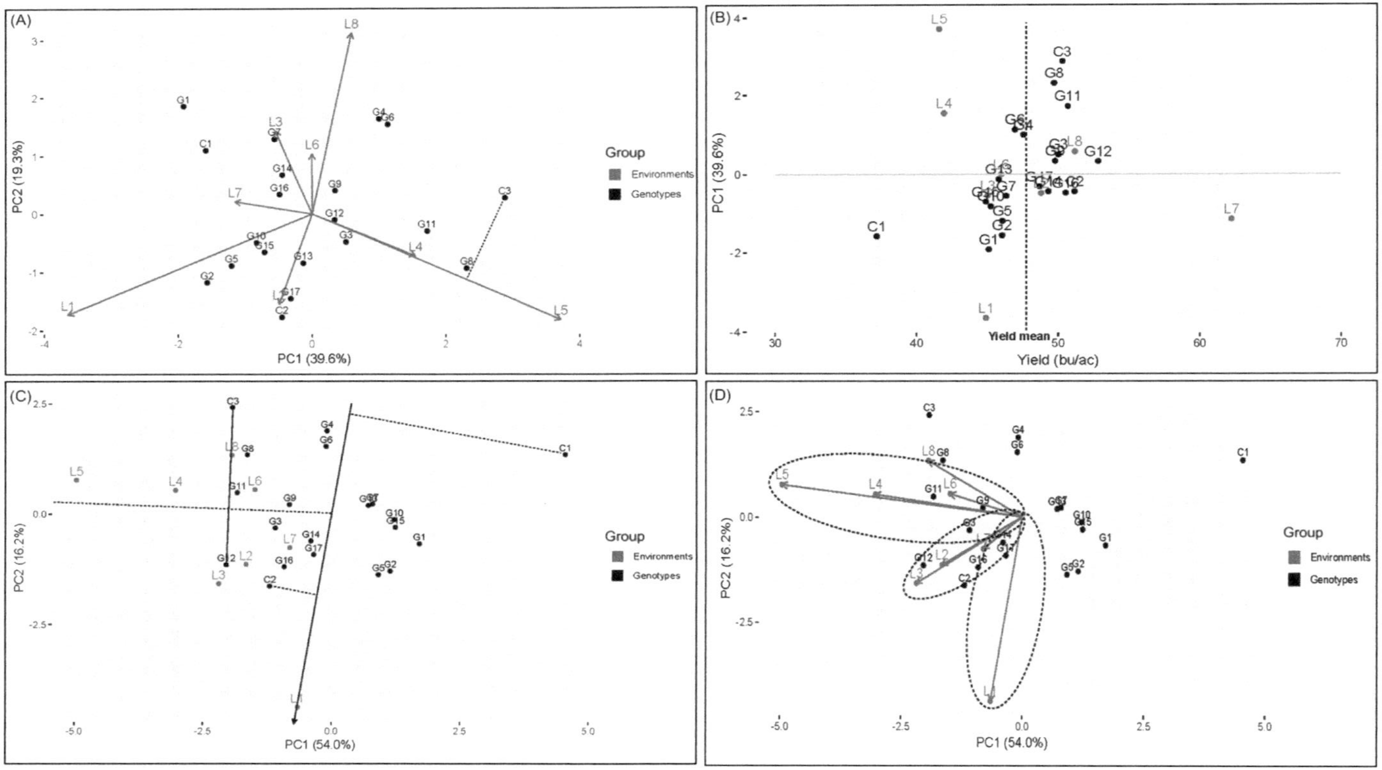

FIGURE 4.7

(A) AMMI2 biplot, (B) AMMI1 biplot, (C) What-Won-Where, and (D) environment groupings for an experiment that included 20 entries (17 experimental soybean breeding lines and 3 checks), 8 locations across Iowa, USA, in replicated testing. Data from Iowa State University's soybean breeding program.

Practical consideration for a plant breeder

There are several questions plant breeders consider when looking at multi-environment tests and its analyses: (1) Presence of mega-environments? Can the testing requirement be reduced by removing overlapping (results) locations? (2) How to pick the "winners" at each environment? (3) Strategic placement of products into given environments. (4) Which genotype or group of genotypes yield best in which environments or group of environments? (5) Which are the most stable genotypes across a range of environments? (6) Decision on putting more resources into # of years or # of locations/year. (7) Best strategy to develop broadly adapted cultivars? (8) How to select locations for testing breeding material? (9) What are the characteristics of a "good" location that should be used for testing material? These are complex questions and require a breeder to spend time on literature review and consult other breeders with more experience in the crop and institution. From a statistical viewpoint, biplot or AMMI type of analysis is a reasonable way to learn who-won-where outcomes. As previously explained, AMMI analysis first fits the additive main effects of genotypes and environments by the usual analysis of variance and then describes the non-additive part, genotype–environment interaction, by principal components analysis. AMMI summarizes patterns and relationships of genotypes and environments and gives information on $G \times E$.

A breeder should be aware of the relative importance (i.e., magnitude) of genotype × location, genotype × year, and genotype × location × year interactions to appropriately allocate resources for cultivar testing. This information will help a breeder decide the number of locations and years to be used in testing. Breeders can aim to sample one or more locations from each mega-environment (or testing zone). Each mega-environment will consist of several individual locations or sites. Within a mega-environment, the genotypes perform more similarly compared with genotypes in different mega-environments. In other words, there is little or no $G \times E$ interaction among environments within a mega-environment. In such a scenario, breeders will gain little by testing in more similar environments and should aim to test across dissimilar environments for stable performance of genotypes across a range of environments. Environmental parameters such as rainfall, soil type, physical properties, pH, and so forth, may also be a good way to cluster environments. If there are larger agroecological regions that predominantly grow a single cultivar, and if a breeder is targeting that region, the cultivar acreage map may serve as a good resource in identifying the mega-environment to use in developing a better-yielding cultivar. Another useful exercise is to perform genetic correlations (genotype means analysis) to determine whether the correlations with the environment are high or low. High correlation will mean that the predictive ability of those environments is similar; that is, breeder does not gain as much information on stability as they would gain by testing in environments with lower correlation among genotype means.

Breeders working at international institutions [members of the Consultative Group for International Agricultural Research (CGIAR) such as CIMMYT,

ICARDA, IRRI, and CIAT] have a mandate of a wider crop adaptation, while a provincial or state breeding institute's mandate will be more localized (specific area, perhaps one or two macroenvironment). CGIAR breeders often utilize many (>20–30+) diverse locations to identify cultivars with wider adaptation while breeders at a state breeding institute use fewer environments that are representative of one or two macroenvironments. Large private companies have a wider testing site network enabling a denser field tests coverage. Institutions may establish their own farms (with full infrastructural resources) at testing sites, or they may rent land from farmers.

If the mandate of the program is to develop cultivars for specific purposes (i.e., disease resistance, stress tolerance, or quality traits), then the testing sites should be selected by breeders for this objective. For example, malt barley has a very specific crop quality requirement. Stable performance of malt quality (for consistent and good quality and better taste for beer making) is a must, and the breeder will discard cultivars if they show specific adaptation for malt quality, that is, only very specific sites produce good malt, and the rest of the growing regions produce inferior malt. Similarly, if one of the main mandates of a program is to develop a disease resistant variety to Disease A, it is advisable to utilize locations where a consistent and homogenous presence of disease it noted although, breeders can equally effectively use specialized nurseries to test disease reaction while obtaining yield performance from field testing.

There may be instances, that instead of utilizing two separate locations, breeder creates separate conditions at the same field site to make selection relevant to the objective of the program. For example, breeder can set up tests in two dissimilar conditions such as dryland versus irrigated nurseries. Breeders can keep good records of seasonal patterns such as cycles of drought at the seedling stage, heat stress at the flowering stage, and years with no apparent stress for all the sites in their network. This information along with the use of long-term checks that have known and consistent responses to these stresses and weather parameters, can enable the use of a site for selection to improve tolerance for these factors. A new and more appropriate site may be identified or established, if the breeding objectives change and no current location and field site meets the need.

Factors to consider in the selection of testing sites include the following:

1. Good correlation with the performance in farmer growing conditions.
2. Ability to handle different tests (infrastructure) and ability to respond to mitigate threats (such as an ability to irrigate if needed to avoid the impact of water deficit on response to the selected treatment).
3. Low environmental error, that is, higher heritability to differentiate "keeps" (i.e., desirables) from discards.
4. Infrastructure to implement breeding decisions of timely planting, maintenance, harvest, processing, and so forth.

5. Travel distance—logistics of people travel and equipment transport to manage the site is important. It is generally difficult to maintain good sites that are long distances from breeder's central location.

Ultimately breeders need to remember that their job is to ensure that the product (cultivar) that reaches farmers helps them make a profit and that it meets the requirement of the processing industry, which buys the product from the farmers. If the breeder develops the highest yielding cultivar, but it lacks the necessary quality or protection against biotic or abiotic stress, farmers will likely not grow this variety. Therefore, breeders think of a cultivar as a "package." A package needs to have all of the ingredients that will make it readily adopted by farmers as well as the processing industry, that is, traits and characteristics to meet the needs of all stakeholders including customers.

Heritability

Heritability can be defined as the degree to which the characteristics of a plant are repeated in its progeny. Mathematically it is the proportion of total variability for a character due to genetic causes.

$$\text{Broad sense heritability}\left(h^2{}_{\text{b}}\right) = \sigma^2{}_{\text{G}}/\sigma^2{}_{\text{P}}$$

$$\text{Narrow sense heritability}\left(h^2\right) = \sigma^2{}_{\text{A}}/\sigma^2{}_{\text{P}}$$

where $\sigma^2{}_{\text{G}}$ is total genotypic variance (contains $\sigma^2{}_{\text{A}} + \sigma^2{}_{\text{D}} + \sigma^2{}_{\text{I}}$), and $\sigma^2{}_{\text{A}}$ is the additive component of the genotypic variance (variance among parents for their breeding value), and $\sigma^2{}_{\text{P}}$ is total phenotypic variance $= \sigma^2{}_{\text{G}} + \sigma^2{}_{\text{E}} + \sigma^2{}_{G \times E}$.

Narrow sense heritability is more valuable since it indicates the amount of the total observed variability due to additive gene action (which can be selected for effectively and fixed in the homozygous condition). Plant breeders want selection to be effective, that is, to identify superior individuals and use as parents to form the next generation, and selection is effective if mean of the next generation is better than the previous generation. Narrow sense heritability is the proportion of the phenotypic variance that can be effectively used to make gain from selection. If selection is effective, offspring of the selected plant will show the expression of the traits demonstrated by their parents. Broad sense heritability is less valuable since it also includes dominance and epistasis. These gene actions cannot be fixed and occur only in specific gene combinations. However, broad sense heritability is useful in clonal crops where additive, dominant and epistatic variances can be exploited. Higher broad sense heritability estimates indicates that genotypic variance is a larger component of the overall phenotypic variance, and that environmental and G x E variance components are smaller.

Heritability shows the reliability of phenotpyic value as a guide to breeding value (we measure the phenotypic values of individual but breeding value determines their influence on the next generation.

Breeding value estimate $= h^2 \times$ phenotypic value, where h^2 is regression of parent on offspring.

Environment is a component of heritability estimates, so one can understand the importance of proper and good experimentation. It helps in intelligent planning of experiments. Therefore, it is important for breeders to have a sense of the heritability of the traits for which they are making selections in their programs. This can be obtained by using data from their own experiments or, for a new program, using information available in literature or from previous experiments. The reason heritability is important is that selection response is related to heritability. The higher the heritability, the more the non-genetic reflects the genotype, and the more effective selection will be. More extensive testing (more environments, more replications) reduces the non-genetic variance and increases the heritability. Heritability can be increased more by using a higher number of locations rather than by increasing number of years. This is due to the generally smaller variance component of genotype $\times$ year relative to other components such as genotype $\times$ location. This suggests that in most cases, the breeder does not need to select on more than one year of data when selections are made in each generation as plants are achieving "true breeding" status; however, it is important to point out that this does not refer to pre-commercial testing. Pre-commercial testing is an adaptation testing to test elite experimental varieties in a range of locations across 2–3 years in large sized plots. This is necessary to ensure that variety is high performing with stable performance. The release of an improperly or deficiently tested variety can cause failures on farmer field ruining the reputation of breeder and institutions and most importantly causes economic hardships to farmers, which is unacceptable.

Heritability is used to estimate the expected response to selection and to choose the best breeding approach to improve the target trait(s). Traits with high heritability can be selected on a single-plant basis in an early generation and in fewer (even single) environments. Breeders should consider a range of heritability (rather than an absolute value) and should have some precision around their estimate (confidence interval). Higher heritability, such as 0.7, or higher narrow sense heritability means that the breeder can expect selection in early generations to be effective for that trait. High broad sense heritability only indicates that the effect of environment is smaller but does not provide insight into the relative importance of additive (which can be fixed) or non-additive (which cannot be fixed) gene effects.

Methods for estimating heritability

Each heritability estimate is unique and reflective of the method of calculation, testing environment, generation used in the estimation, and genotypes studied. While the heritability estimates will differ somewhat based on the different conditions described above, plant breeders can get a good understanding of heritability based on published literature and their, or their predecessors' experiences working with the crop on various traits.

Variance component method: comparison of segregating and homogeneous populations is applicable only to self-pollinated or clonally propagated species. This method estimates broad sense heritability and involves estimating the magnitude of various types of genetic and environmental variability.

In such experiments, $\sigma^2_{P1} = \sigma^2_{P2} = \sigma^2_{F1} = \sigma^2_{E}$

In a self-pollinated crop, the estimates of genetic variability for parent 1 and parent 2, being pure lines, will be similar to each other ($\sigma^2_{P1} = \sigma^2_{P2}$) as well as to the F_1 ($=\sigma^2_{F1}$), and these will be equal to the environmental variability (as these three are genetically uniform and therefore, any variability observed will be due to environment). The variance of the F_2, or any other segregating generation, can then be used to obtain σ^2_G ($=\sigma^2_P - \sigma^2_E$), where σ^2_P is variance estimated from the segregating generations.

Two examples for calculation of heritability are explained as follows:

Example I: below is an example of an ANOVA where a random set of genotypes was evaluated over *l* locations for *t* years, and *r* replications were used in each test. There were multiple locations, years, and replications.

Source of variation	Degree of freedom	Mean square	Expected mean square
Location (*L*)	$(l-1)$		$\sigma^2_e + r\sigma^2_{gly} + ry\sigma^2_{gl} + g\sigma^2_r + gr\sigma^2_{ly} + grt\sigma^2_l$
Year (*Y*)	$(y-1)$		$\sigma^2_e + r\sigma^2_{gly} + ry\sigma^2_{gl} + g\sigma^2_r + gr\sigma^2_{ly} + grl\sigma^2_y$
L × *Y*	$(l-1)(y-1)$		$\sigma^2_e + r\sigma^2_{gly} + ry\sigma^2_{gl} + g\sigma^2_r + gr\sigma^2_{ly}$
Rep(*L* × *Y*)	$(r-1)ly$		$\sigma^2_e + g\sigma^2_r$
Genotype (G)	$(g-1)$	MS_1	$\sigma^2_e + r\sigma^2_{gly} + ry\sigma^2_{gl} + rl\sigma^2_{gy} + rly\sigma^2_g$
G × *L*	$(g-1)(l-1)$	MS_2	$\sigma^2_e + r\sigma^2_{gly} + ry\sigma^2_{gl}$
G × *Y*	$(g-1)(y-1)$	MS_3	$\sigma^2_e + r\sigma^2_{gly} + rl\sigma^2_{gy}$
G × *L* × *Y*	$(g-1)(l-1)(y-1)$	MS_4	$\sigma^2_e + r\sigma^2_{gly}$
Pooled error	$ly(g-1)(r-1)$	MS_5	σ^2_e

All factors are considered random in ANOVA. Random effects are random levels coming from researcher's random choice of levels or from researcher's inability to control the level.

$$\sigma^2_e = MS_5$$

$$\sigma^2_{gly} = (MS_4 - MS_5)/r$$

$$\sigma^2_{gy} = (MS_3 - MS_4)/rl$$

$$\sigma^2_{gl} = (MS_2 - MS_4)/ry$$

$$\sigma^2_g = (MS_1 - MS_2 - MS_3 + MS_4)/rly$$

The phenotypic variance of the genotypic means

$\sigma^2_p = \sigma^2_g + (\sigma^2_{gl})/l + (\sigma^2_{gy})/y + (\sigma^2_{gly})/ly + (\sigma^2_e)/rly$, and if we substitute for mean squares, we will obtain $\sigma^2_p = (MS_1)/rly$.

Broad sense heritability can be calculated using the equations above.

Example II: below is an example of ANOVA where a random set of genotypes was evaluated over *e* environments (can be locations, years, or combination of years and locations), and *r* replications were used in each test.

Source of variation	Degree of freedom	Mean square	Expected mean square
Environment (i.e., *E*)	$(e-1)$		$\sigma^2_e + g\sigma^2_{r(e)} + r\sigma^2_{ge} + rg\sigma^2_e$
Rep(*E*)	$(r-1)e$		$\sigma^2_e + g\sigma^2_{r(e)}$
Genotype (i.e., *G*)	$(g-1)$	MS_1	$\sigma^2_e + r\sigma^2_{ge} + re\sigma^2_g$
$G \times E$	$(g-1)(e-1)$	MS_2	$\sigma^2_e + r\sigma^2_{ge}$
Pooled error	$ly(g-1)(r-1)$	MS_3	σ^2_e

All factors are considered random in ANOVA.

$$\sigma^2_e = MS_3$$

$$\sigma^2_{ge} = (MS_2 - MS_3)/r$$

$$\sigma^2_g = (MS_1 - MS_2)/re$$

The phenotypic variance of genotypic means $(\sigma^2_p) = \sigma^2_g + (\sigma^2_{ge})/e + (\sigma^2_e)/re$, and if we substitute for mean squares, we will obtain $\sigma^2_p = (MS_1)/re$.

The total phenotypic variance (σ^2_p) is $\sigma^2_g + \sigma^2_{ge} + \sigma^2_e$. Environmental variances are not involved in the phenotypic variance because all genotypes are evaluated in all environments and any environment is considered to have the same effect on all genotypes.

Heritability on an individual experimental unit basis is: $\sigma^2_g/(\sigma^2_g + \sigma^2_{ge} + \sigma^2_e)$.

Heritability on a genotypic mean basis is: $\sigma^2_g/[(\sigma^2_g + (\sigma^2_{ge})/e + (\sigma^2_e)/re)]$.

This estimate of heritability is obtained if the genotypes represent the population and are chosen randomly. If genotypes are not chosen randomly (i.e., selected genotypes), the ratio between genotypic and phenotypic variation is called *repeatability*. This estimate is a measure of the precision of the data and a measure of the proportion of genetic variation, which helps the breeder to detect significant difference among genotypes.

Covariance between relatives is the resemblance among relatives as measured by regression analysis. Examples are parent-offspring regression, covariance between half-sibs and full-sibs, and covariance between inbred or partially inbred families. Variability estimates can be obtained in other ways such as special mating designs (half-sib, full-sib, North Carolina designs, and so forth) or analysis of trials conducted in a range of environments. More details on its calculations can be found in a quantitative genetics textbook.

Realized heritability: for information, refer to the response to selection section in this chapter.

Modes of selection

According to Mather (1953), the action of selection upon a character in a population of living organisms may be analyzed in terms of three elementary components: stabilizing selection, directional selection, and disruptive selection (Fig. 4.8). These forms of selection are generally used to describe natural selection but could be operating during natural and artificial selection. The natural selection favors survival ability or fitness of the species. The artificial selection is directed toward specific goals in a predetermined way.

Stabilizing selection: stabilizing selection will favor toward the population mean and disfavor the more extreme variants. Therefore, in stabilizing selection, extreme phenotypes on both ends of the bell curve (population distribution for a trait) are unfavorable. In other words, it tends to narrow the variation for a character in the population. In general, stabilizing selection favors additive genetic variation in a population. This mode of selection is also called optimum selection or balancing selection. An example of stabilizing selection in a natural population could be for plant height. If we assume a natural population with a range of plant height ranging from dwarf to very tall, both dwarf and very tall may be at a disadvantage. Dwarf plants may be selected against due to competition (shading) with their neighbors, while very tall plant may be lost due to broken stalk or uprooted roots due to standing too tall in the population in high wind speeds are encountered. Selection pressures on both plant height extremes will favor selection in plants of intermediate height; and therefore, over generations, number of plants with intermediate height will increase while height extremes (on both tails of the distribution) will decrease. If distributions, before and after selection, are compared it will show a normal distribution before selection; however, postselection the distribution will be narrower and taller (leptokurtic distribution) because phenotypes at both extremes are selected against. The outcome is a reduction in variation and higher preponderance of individuals closer to the mean of the population.

Directional selection: the directional selection comes into play as the mean expression of the character departs from optimal, that is, it favors a certain genotype in a population. Selection in general has been directional in several crop plants as the plant breeders have followed directional selection to change the existing populations in a predetermined way. In natural selection, directional selection can be seen with the response of a natural population to a disease or pest. In these populations, plants that are susceptible will die; therefore, the mean of the population moves toward resistance.

Disruptive selection: the disruptive selection favors several optima for a character. Hutchinson (1965) and Doggett (1965) postulated that disruptive selection programs

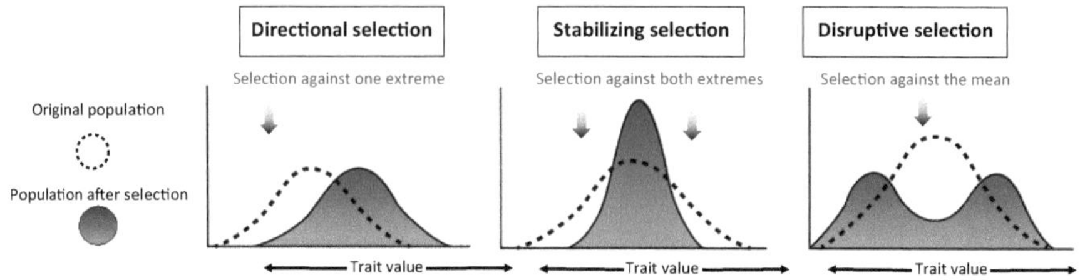

FIGURE 4.8

Directional selection can be used to move the mean in different outcomes.

can be of use in crop improvement where cultivated, wild, and semiwild forms are used. Thoday (1959) pointed out that the disruptive selection releases latent variability, which is important for breeding work and that variability can be generated for numerous traits including flowering time and seed yield. Under natural selection, if there are three different flower colors (e.g., white, purple, and pink) and there are three distinct pollinators (say, types of bees), under no selection the three types of flowers will maintain equal proportions. However, if the pollinator that pollinated purple flower was wiped out, the proportion of white and red flower will increase while purple flowered plants will decrease because they will be selected against. This will lead to the population to move toward white and red flower plants.

Systems of mating

The breeder has two basic tools to change the genetic constitution of the populations: selection and mating systems. For the improvement of any desired character, the breeder selects desirable plants, that is, more productive plant population and decides the manner in which these are to be mated. The mating systems are grouped into two broad categories - random and non-random mating.

Random mating

In this mating, the individuals are mated together at random or by chance. Each female gamete is equally likely to combine with any male gamete of the same plant or with any other plant of the population and the rate of reproduction of each genotype is equal. In other words, there is no selection. This is not possible as some form of natural selection is generally always observed in plant breeding. The true random mating does not change the gene frequencies, genetic variance and genetic correlation between close relatives in the population (we have discussed H−W equilibrium earlier). However, random mating with selection changes the mean and gene frequencies and with little effect on homozygosity, variance and genetic correlation between close relatives in the large population. Random mating in small populations lead to genetic drift, inbreeding and increase in homozygosity as discussed earlier. It is useful to preserve desirable alleles in a population like synthetics, composites and production of polycross progeny, etc.

Non-random mating

The non-random mating consists of assortative mating (genetic assortative mating and phenotypic assortative mating) and disassortative mating (genetic disassortative mating and phenotypic disassortative mating). Broadly speaking, assortative mating (also called positive assortative mating), describes situations in which individuals mates with other individuals similar to themselves, while disassortative mating (also called negative assortative mating) describes situations in which individuals mates with other individuals dissimilar to themselves. A major difference between inbreeding and assortative mating is that inbreeding affects all loci, however, in assortative mating only those loci that play a role in phenotypic characteristics which are similar are affected.

Genetic assortative mating: the genetic assortative mating or inbreeding involves mating between individuals, which are more closely related by ancestry than in random mating. The closest type of genetic assortative mating is selfing. It results in homozygosity and fixation of types, therefore, is useful to develop homozygous lines and development of inbred lines for hybrid or synthetic cultivars. As a consequence of inbreeding, the hidden variability which was otherwise protected due to heterozygosity appears and could be used (if it is favorable) in breeding.

Phenotypic assortative mating: this involves mating between the phenotypically more similar individuals than would be expected under random mating. The breeder selects individuals from the population on the basis of resemblance and these individuals are mated to each other. As a consequence of positive assortative mating, frequency of homozygous genotypes in a population is increased at the expense of heterozygous genotypes causing a decrease in the phenotypic variance. It may be used for recurrent selection schemes.

Genetic disassortative mating: this mating system involves individuals, which are less closely related to ancestry than under random mating. Organisms with disassortative mating systems tend to mate with individuals that do not share alleles (lack of identity by descent), that is, they have no common ancestors. Mating incompatibility systems (obligate outcrosser such as alfalfa) are extreme example of genetic disassortative mating. The individuals may belong to different populations like intervarietal and interspecific crosses. As a consequence of such mating, homozygosity would be reduced and heterozygosity would be increased.

Phenotypic disassortative mating: it involves mating between phenotypically dissimilar individuals. The breeder may select individuals with contrasting phenotypes. It may be used to maintain genetic diversity of germplasm, which may be useful for obtaining desirable genes in future. As a consequence of such mating system, heterozygosity would be maintained in a population and genetic correlation between relatives would be reduced. Therefore, the phenotypic variance in a population decreases as frequency of homozygous genotypes is reduced while heterozygous genotypes are increased.

Types of response to selection

Selection in a random mating population is able to change or modify the genetic structure of the population in cross-pollinated crops. Even the long term experiments could not completely exhaust open pollinated population from genetic variation. However, the following pattern of responses to selection in random mating population could be noted:

1. change in the gene and genotypic frequencies,
2. produce new genotypes,
3. shift the mean in the direction of selection, and
4. change in variance to some extent.

The magnitude of these changes would depend on the heritability of character, number of genes, etc. The responses to selection obtained in studies conducted during the past could be divided into five broad categories.

Rapid gain followed by slow responses: it has been observed for quasicontinuous character like plant height and disease resistance. The rapid initial gains are through the frequency of few genes with large effects and subsequent slow gains are due to the frequency of large number of genes with small effects.

The continued slow response for a long period: in case of corn, selection for high and low oil as well as protein content was continued for a long period without any serious loss of genetic variation. This could be due to the fact that these traits were governed by large number of genes having small and additive effects.

Slow response for a short period: in this case, response to selection will be observed for a shorter period and after it no response to selection will be observed. This could be due to undesirable correlated response for example, the germ size of corn that cannot be reduced below a certain limit in selections for oil content.

Little or no response: even in the genetically variable population, sometimes little or no response to selection is observed which could be due to inadequate selection criteria and low heritability of the character.

Recurrence of response after plateau: for polygenic characters, linkages of genes with plus and minus effect could maintain substantial variation in bound form in the population. This bound variation in certain situations could be released which may show potential unexpected response for some characteristics in a population.

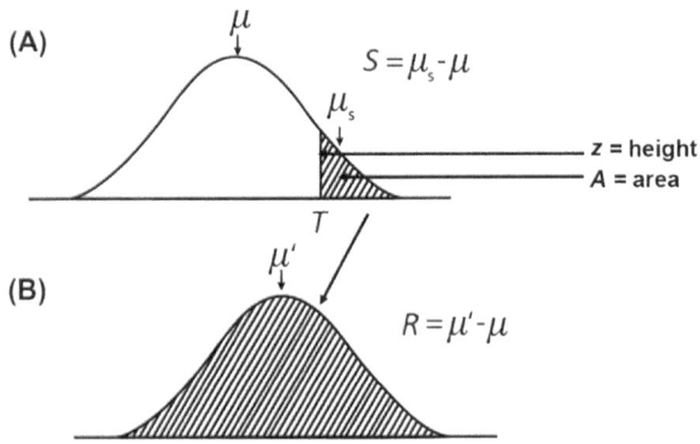

FIGURE 4.9

Example of truncation selection. (A) Phenotype distribution in parental population with mean (μ). Shaded area shows the individuals that were advanced to the next generation (individuals with phenotypes above the truncation point, T). Selected individuals (shaded area) have a mean phenotype (μ_S). (B) Distribution of phenotypes in the offspring generation derived from the selected parents. The mean phenotype is denoted with (μ'). *S* is the selection differential. *R* is the response to selection.

Adapted from Hartl, D.L., 1988. A Primer of Population Genetics, second ed. Sinauer Associates, Inc., Sunderland, MA.

Selection theory

When individuals are selected based on their phenotypic or genotypic value and practices by human it is called artificial selection (or selective breeding). Truncation is a type of individual selection and is very common in plant breeding programs. The curve in Fig. 4.9A represents the normal distribution of a quantitative trait in a population, and the shaded part, T (i.e., truncation point), represents the individuals selected for the next generation of breeding; they could be cross- or self-pollinated. The mean of the unselected population (or mean of the population in generation 1) is μ, and μ_S is the mean of selected parents. If these selected parents are mated at random, their offspring will have the phenotypic distribution in Fig. 4.9B and a mean equal to μ'. Generally $\mu_S > \mu' > \mu$.

μ' is greater than μ because some of the selected parents have favorable genes which are transferred to their offspring. μ_S is greater than μ' because some of the selected parents did not have favorable genotypes, but instead had superior phenotypes due to the favorable environment where they were tested (chance exposure to favorable environment, for example, low spot in the field that received more water, a spot in the field that received more fertilizer, or a spot in the field that was not exposed to high winds). Secondly, alleles, not genotypes, are transmitted to the offspring, and favorable genotypes may segregate, or recombination may cause breakage of favorable linkages.

The difference in mean phenotype between the selected parents (μ_s) and generation 0 (μ) is called *selection differential* (S).

$$S = \mu_s - \mu \tag{4.1}$$

The difference in mean phenotypic value between the progeny generation (generation 1) (μ') and generation 0 (μ) is called the *response to selection* (R).

$$R = \mu' - \mu \tag{4.2}$$

The prediction equation defines the relationship between S and R. For truncation selection, the prediction equation is

$$R = h^2 S \tag{4.3}$$

where h^2 is the heritability of the trait.

z/A is the frequency at the truncation point (in a normal distribution)/area under the selected portion of the curve. This is equal to selection differential/phenotypic variance. In other words, assuming that the effects of each allele are small relative to the phenotypic variation, and phenotypic values are normally distributed.

$$z/A = (\mu_s - \mu)/\sigma^2 \tag{4.4}$$

Response to selection (Breeder's equation)

It can be shown that:

$$\mu' = \mu + 2[a + (q - p)d]\Delta p \tag{4.5}$$

where Δp is the change in allele frequency of

$$A1 = (z/A)pq[a + (q - p)d] \tag{4.6}$$

where a is the genotypic effect, d is the measure of dominance, p is the allele frequency of A1, q is the allele frequency of A2, μ' is the mean phenotype of progeny generation, and μ of the progenitor generation.

Substituting $z/A = (\mu_s - \mu)/\sigma^2_p$ and Δp, the equation can be written as:

$$\mu' - \mu = \{(\mu_s - \mu) \times 2pq[a+(q - p)d]^2\}/\sigma^2_p \tag{4.7}$$

Since $S = \mu_s - \mu$, and $R = \mu' - \mu$,

$$R = S \times 2pq[a+(q - p)d]^2/\sigma^2_p \tag{4.8}$$

We already have seen that $R = h^2 S$ (Eq. 4.3); therefore, we can now define heritability in the genetic terms of a, p, q, d, and σ^2_p as

$$h^2 = 2pq[a+(q - p)d]^2\}/\sigma^2_p \tag{4.9}$$

This heritability definition is valid when the trait is under single gene control. This is hardly the case for most of the traits, and therefore, heritability (narrow sense) can be defined as:

$$h^2 = \left\{\sum 2pq[a+(q-p)d]^2\right\}/\sigma^2_{\mathrm{p}} \quad (4.10)$$

$\Sigma 2pq[a + (q-p)d]^2$ = additive genetic variation of the trait = σ^2_a.

Going back to Eq. (4.3), $R = h^2S$ can also be written as

$$R = h^2 i\sigma \quad (4.11)$$

where

$$S = i\sigma = \sigma^2 z/A \quad (4.12)$$

and i is the standardized selection differential.

The intensity of selection depends on the proportion of the total population selected. For example, below are some selection intensities based on the percentage selected (assuming a completely normal trait distribution).

% selected	**Standardized selection differential (*i*)**
0.01	3.959
0.1	3.367
0.5	2.892
1	2.665
5	2.063
10	1.755
15	1.554
20	1.400
25	1.271
30	1.159
35	1.058
40	0.966
45	0.880
50	0.798

h^2 is narrow sense heritability.

σ_p = phenotypic standard deviation (= square root of phenotypic variance).

This equation, $R = i\sigma h^2$ (Eq. 4.11), is fundamental in plant breeding. All plant breeders may not calculate an actual numerical value for selection response. However, this equation is important as it shows that selection response depends on:

- selection intensity,
- heritability, and
- phenotypic variability present in the population, for example, from a cross.

The equation $R = \mu' - \mu$ can be written as

$$\mu' = \mu + i\sigma h^2 \quad (4.13)$$

which clearly indicates that to maximize the expression of a trait in the offspring generation, the breeder needs to start with high expression of the trait, maximum heritability, and high selection intensity (although diminishing returns apply beyond a certain level).

The square root of heritability (h) is a measure of the correlation between the observed phenotypic value and the underlying genotypic value. In a breeding program, the breeder will try to maximize these factors (higher standardized selection differential, genetic variation, and heritability). One has to keep in mind that optimum balance needs to be obtained between increased expected response (which is a good thing and what a breeder is after) and increased variability of that response (undesirable characteristics of selection).

Expected genetic gain

The expected genetic gain formula is as follows:

$$\Delta G = \frac{ic}{y}\frac{\sigma_G^2}{\sqrt{\sigma_P^2}} = \frac{ic}{y}\frac{\sigma_G^2}{\sqrt{\frac{\sigma_e^2}{re} + \frac{\sigma_{GE}^2}{e} + \sigma_G^2}}$$

This formula is an extension of the response to the selection equation

$$R = h^2 S = ih^2\sigma_p = \frac{i\sigma_x^2}{\sqrt{\sigma_P^2}} = \frac{c}{y}\frac{i\sigma_x^2}{\sqrt{\sigma_P^2}}$$

i is the standardized selection differential, c is the parental control, and y is the seasons per cycle (Eberhart, 1970). The heritability equation is $\sigma^2_g/[(\sigma^2_g + (\sigma^2_{ge})/e + (\sigma^2_e)/re)]$, where r is the number of replications, and e is the number of environments.

There is sometimes confusion in the usage of terms genetic gain and response to selection. Genetic gain refers to the improvement in average genetic (or phenotypic) value in a population due to artificial selection over cycles of breeding. Therefore, genetic gain should not be referred or measured from one year to the next in the same population unless breeding cycles are involved (crossing, generation of progenies, evaluation, selection, and repeating crossing, generation of progenies, evaluation, selection, and so on). Comparison of elite variety against check and its improved performance is not genetic gain either. The response to selection equation described earlier can be used to estimate genetic gain (expected genetic gain) using population parameters, while the success of a breeding program and its genetic gain can be calculated using a long term check and including other bridging checks that overlap when a landmark check is removed. This provides continuity in comparisons. Realized genetic can be measured by regressing variety means performance (e.g., yield) across the year of release and tested in the same environment to remove confounding factors. Realized genetic gain can also be calculated as described in earlier equations.

Variability in the response to selection

CV of response to selection, that is, CV(R) (Baker, 1971) is a useful tool to investigate the influence of each parameter in the response to selection.

$$\mathrm{CV}(R) = \mathrm{SQRT}\left(\left(1 + \left(p \times h^2\right) - \left(h^2 \times h^2\right)\right) / \left(p \times n \times i \times i \times h^2 \times h^2\right)\right) \quad (4.14)$$

where n is the total number of lines evaluated, p is the proportion of lines selected, h^2 is the heritability, and i is the standardized selection differential.

For example, assume a breeder started with $n = 1000$, % selected $(p) = 0.01$, heritability $(h^2) = 0.2$, and $i = 3.959$. The CV (response to selection) would be 39% by substituting values in Eq.n (4.14).

n (total number of lines evaluated)	1000
p = proportion of lines selected	0.01
h^2 = heritability	0.2
Standardized selection differential (i)	3.959
CV(R)	39%

If a breeder started with a smaller number of lines ($n = 100$) and used the same values (% selected = 0.01, heritability = 0.2, and $i = 3.959$), the CV (response to selection) = 124%.

n (total number of lines evaluated)	100
p = proportion of lines selected	0.01
h^2 = heritability	0.2
Standardized selection differential (i)	3.959
CV(R)	124%

However, the importance of h^2 can be seen with the same calculations. If a breeder started with $n = 1000$, % selected = 0.01, heritability = 0.7, and $i = 3.959$, CV (response to selection) = 8%.

n (total number of lines evaluated)	1000
p = proportion of lines selected	0.01
h^2 = heritability	0.7
Standardized selection differential (i)	3.959
CV(R)	8%

If the breeder had started with $n = 1000$, % selected = 5, heritability = 0.7, and $i = 2.063$, the CV (response to selection) = 2%.

n (total number of lines evaluated)	1000

(Continued)

Continued

p = proportion of lines selected	5
h^2 = heritability	0.7
Standardized selection differential (i)	2.063
CV(R)	2%

These calculations show that, while the selection equation is an essential equation for any breeder to consider for trait improvement, it is also worthwhile to consider the extent of variability in relation to the mean of the population (CV). The standardized selection differential can be increased by selecting fewer lines (but this can cause increased variability of response, which is undesirable) or testing more units but selecting fewer units (this will require more resources). If a breeder makes compromises between testing more lines to advance a few, it may be done at a compromise of not doing a thorough evaluation of the lines. A less thorough evaluation will result in a lower correlation between phenotypic and genotypic values (lower h); therefore, breeder should not compromise on proper trait measurement protocols. An optimum balance needs to be achieved for each trait enabling a more thorough testing to increase the correlation between phenotypic and genotypic values (higher h) and increase the standardized selection differential. One of the ways to maximize genetic variance is to cross diverse parents. However, if crosses between diverse parents have lower unselected means than crosses between adapted (elite) parents (which will generally have lower genetic variability between them), the breeder may be reducing the mean genotypic value of the subsequent population by using these diverse parents. Therefore, crossing best × best (or elite × elite) parents is one way to maximize the genotypic mean of the starting population although it may reduce the genotypic variance, and even R. Most cultivar development programs will work with a best × best configuration, or have crosses that are at least 75% elite [e.g., (best × exotic) × best]. Selecting elite parents from different pedigree or backgrounds is a reasonable solution. However, if working with crops that have very stringent end-use quality requirements, even more care is needed with working with diverse parents.

Practical considerations for plant breeders

- Increase the numerator of the expected genetic gain formula by increasing genetic variance (larger population sizes; diverse parents, but keeping the proportion of elite parents high) and by increasing selection intensity (without getting into genetic drift problem).
- Parental control will allow for an increased response to selection. Parental control, c, can be increased by recombining genotypes where both sources of gametes originated in selected genotypes ($c = 1$), generally true in self-pollinated crops. In cross-pollinated species, $c = 0.5$ if the male gametes are

coming from unselected genotypes. Therefore, if possible, selection be conducted before pollination so that only selected genotypes contribute to the next generation. If the selected seed of selected genotypes is used for establishing the next generation, $c = 2.0$.

- Another way to increase genetic gain is to decrease the value of the denominator. This can be achieved by decreasing the number of seasons per cycle or the phenotypic variance. The phenotypic variance can be decreased by increasing the number of locations (or environments), by increasing replication, and by reducing $G \times E$ and E variances. Increasing locations is generally considered to play a more important role in reducing phenotypic variance rather than increasing replications for properly laid out experimental designs and large plots.
- Usage of proper experimental methods and field design and analysis will reduce error variance and improve confidence in the estimate of progeny performance. These methods may include augmented designs, or moving means in earlier generations where no replication is used per environment, and incomplete lattice (e.g., alpha-lattice) or RCBD in replicated tests.
- Different generations and types of progenies have different genetic variance components and therefore, affect the equation. The theoretical proportion of additive variance to total genotypic variance of half-sib is 0.25, full-sib is 0.5, and S_1 progenies is 1.
- The number of seasons required to complete a cycle can be reduced by using off-season nurseries. By using an off-season nursery with a high correlation to the home environment, selection can be facilitated for high to moderate heritability traits, and the "y" in the equation above can be reduced. If resources permit, greenhouse or a growth cabinet can be used instead of an off-season nursery and can be complemented with marker-assisted selection to increase "i" as well as "y."
- One way to obtain higher heritability is to reduce environmental effects. Higher heritability implies that selection will be more effective as "what you see is what you get."

Few recommendations to reduce the effect of environment

- If artificial selection will be performed (single plant, rows, or yield trials; early or later generations), plant breeding trials should be set on the best quality land (uniform area, less gradients in the field, highly productive). However, if the interest is to impose a stress (biotic and abiotic), systematic, and uniform stress conditions should be imposed to select for stress tolerance, although good-quality land and well placed checks and control plots are still needed.
- Use the best management practices (reflective of the recommended fertilizer; irrigation; crop rotation; and time of planting, weeding, and harvest). Some breeders advise against pest or pathogen control because natural stress occurrence will provide another trait on which to select unless you are testing the genetic

potential of a genotype in optimum conditions. However, in this case, proper and adequate placement of checks and controls are needed to ensure selection is not biased by field and pathogen presence and expression heterogeneity.

- Use check cultivars frequently. This will allow a better measure of variability.
- Use appropriate statistical designs (lattice, RCBD, partial-rep, or augmented designs as needed) and spatial adjustments, as needed.
- Use replication (improved precision, better measure of variation) and randomization (improved accuracy).

Estimated breeding value

Estimated breeding value (EBV) is expressed relative to population average, and provides an estimate of the genetic potential of the line (as a parent).

EBV can be calculated using the following formulae:

If information on both parents is available:

$$\mathrm{EBV}_{\mathrm{offspring}} = h^2 \times (P_{\mathrm{mid\text{-}parent}} - P_{\mathrm{mean}})$$

Example calculation: if parent #1 yield = 6.0 t/ha, and parent #2 yield is 5.5 t/ha, and average yield of all lines in the breeding program is 4.5 t/ha, and heritability of yield has been estimated to be 0.55,

Step 1: calculate mid-parent value = (6.0 + 5.5)/2 = 5.75 t/ha.
Step 2: calculate phenotypic superiority ($P_{\mathrm{mid\text{-}parent}} - P_{\mathrm{mean}}$) = 5.75 − 4.5 = 1.25 t/ha.
Step 3: calculate EBV(progeny) = $h^2 \times (P_{\mathrm{mid\text{-}parent}} - P_{\mathrm{mean}}) = 0.55 \times 1.25 = 0.6875$ t/ha.
Step 4: predicted yield of progeny = 4.5 + 0.6875 = 5.1875 t/ha.

If information on only one parent is available:

$$\mathrm{EBV}_{\mathrm{offspring}} = 0.5 \times h^2 \times (P_{\mathrm{single\text{-}parent}} - P_{\mathrm{mean}})$$

Example calculation of EBV: if parent #1 yield = 6.0 t/ha, average yield of all lines in the breeding program is 4.5 t/ha, and heritability of yield has been estimated to be 0.55,

Step 1: calculate phenotypic superiority ($P_{\mathrm{single\text{-}parent}} - P_{\mathrm{mean}}$) = 6.0 − 4.5 = 1.5.
Step 2: calculate EBV(parent#1) = $h^2 \times (P_{\mathrm{single\text{-}parent}} - P_{\mathrm{mean}}) = 0.55 \times 1.5 = 0.825$.
Step 3: calculate EBV(progeny of parent#1) = $0.5 \times h^2 \times (P_{\mathrm{single\text{-}parent}} - P_{\mathrm{mean}}) = 0.5 \times 0.55 \times 1.5 = 0.4125$.
Step 4: predicted yield of progeny = 4.5 + 0.4125 = 4.9125.

If information on only one parent is available, the regression coefficient equals half the heritability. This is because half represents the additive genetic relationship between the single parent and the offspring. In the second example, EBV(progeny) is lower because phenotypic information of only one parent was used. In a two parent cross, one parent was known, therefore 0.5 multiplier was used; and it is assumed they are following the average yield in the program. It is assumed that if information is available on a single parent than the regression coefficient is equal to the additive genetic relationship times the heritability. In other examples, the regression coefficients will be different and the calculation will be slightly more involved.

Multiple trait selection

Breeders will rarely ever work on a single trait. The only situation may be a germplasm breeder, who is focused on one trait only, or a program that is solely working on backcross or trait conversion. However, for genetic gain forward breeding is needed. Also multiple traits are selected in the breeding program and its pipeline.

$$\text{Genetic correlation} = [\text{Cov}(\text{Geno}_{\text{Trait1}}, \text{Geno}_{\text{Trait2}})]/[\text{Sqrt}(\text{Var}_{\text{Trait1}} \times \text{Var}_{\text{Trait2}})]$$

Therefore, Genetic correlation = (Phenotypic Corr)/($h_{\text{Trait1}} \times h_{\text{Trait2}}$), if no Env corr = 0.

There are two methods to estimate Covariance:

1. collect data on two traits, difference between Pheno(Cov) of two traits in segregating population and PhenoCov for the same two traits in a genetically homogenous population gives an estimate of GENCov
2. estimate GenoCov by measuring Trait 1 in parents and Trait 2 in progeny. If two traits have low heritability, phenotypic correlation will be low even if high genetic correlation.

Indirect selection: refers to the improvement of one trait by selection of a correlated trait. Ratio of correlated response/direct response = GenoCorr × ($i_X \times h_X$)/($i_Y \times h_Y$)

where "*X*" is secondary trait and "*Y*" is primary trait.

Indirect selection will work if genetic correlation is high, and numerator product of "*i*" and "*h*" is larger, which means secondary trait should have higher heritability that primary trait or expectation of much higher selection differential for secondary trait than primary trait.

For multiple trait selection, three main methods are suggested below.

Tandem selection: A breeder selects sequentially for each trait in successive generations. In this scenario, the population is improved first for one trait, then for the next trait, and so on. This will lead to improvement over generations. Selecting for one trait at a time is done until it is improved to a desired level (generally established a priori by the breeder). Correlated traits (+ or − ve) are important consideration, as they can be unintentionally selected. This approach can work reasonably well in germplasm improvement (adaptation traits) programs. The one disadvantage of this strategy is the long selection cycle, consequently it is generally not followed in commercial plant breeding.

Independent culling (or truncation selection): selection is practiced successively in the same generation. This is probably the most common selection strategy deployed by breeders. In this scheme, in the same generation, a breeder will generally discard all individuals that fail to meet the desired level for one trait irrespective of the value for any of the other traits. This will be followed by selection among the surviving lines for the second trait, and the process is repeated until all selections are made. In many instance independent culling may be done simultaneously for multiple traits at the first round of selection, for example, multiple agronomic traits may be available at harvest for the first round of selection. Experienced breeders will know the culling point, keeping in mind the value of the most important trait, and they may allow some relaxation for major traits when culling for traits that have less significance or importance. Breeders may also use an index to ensure that lines that fail to meet the culling value for a trait are still retained because these lines have really improved numbers for other important traits. The use of checks is very important to ensure accurate culling levels and thereby selection decisions are made. One issue with independent culling is that, with each successive trait cull, the population size and genetic variability is reduced. Multiple rounds of independent culling may be made on sets of traits within the same generation under testing in breeding populations.

Index selection: an index is developed based on the combination of the heritability and economic value of each of several traits under selection, and simultaneous selection occurs for every trait in the same generation. Each line is given an index score based on the expression and weight given to the trait. Some breeders may use a "mental" index selection. Visual selection may be done for a number of traits, an overall mental assessment may be done, and the selection is made. For example, in a space-planted nursery where single plants are growing, a breeder may make a mental assessment on the criteria for different traits such as height, seed fill, plant health, lodging, and inflorescence and using this, may either keep or discard entries. Since there are likely to be several thousand plants in a nursery, the "mental" index approach is the most feasible and should be used. There are more sophisticated methods described, such as the Pesek Baker index (Pesek and Baker, 1969), but these require estimation of variance and covariances. Using economic weights is a good compromise to remove the need to compute trait variance and covariances; that is, index can be formed based on economic weights or ranks (weighted rank-sum index). However, it is still not an easy task to develop an index.

Example of an index $= w1T1 + w2T2 + w3T3 + \cdots + w3T3$; w is the weight for a trait and T is the phenotypic value of an entry for Trait 1. Each entry gets an index, which is used to decide which entries to select or discard. However, some crops and breeding objectives will complicate the use of index selection. Also, the cost of running the program can become cost prohibitive if the number of traits are too high. This will happen because for an unbiased

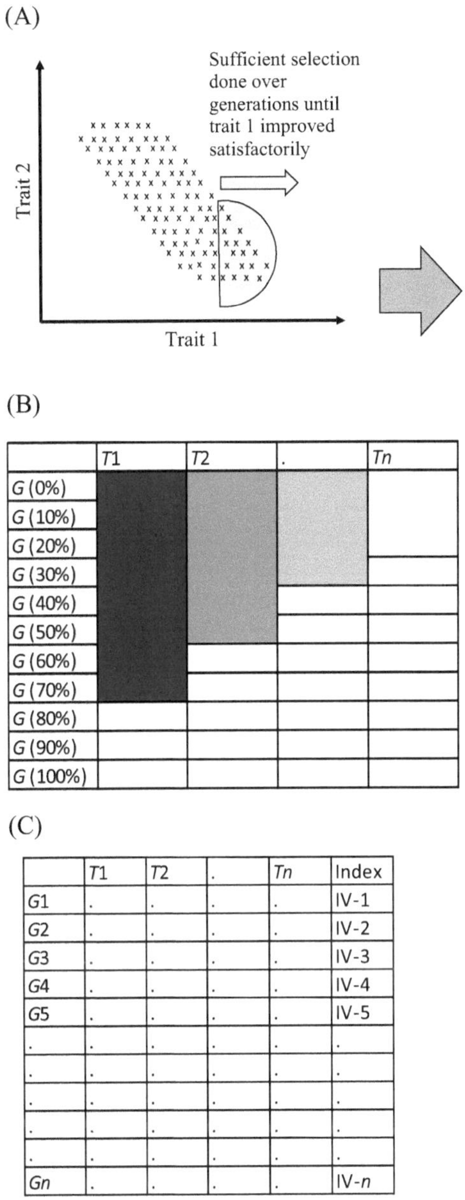

FIGURE 4.10

(A) Tandem selection. (B) Independent culling showing percentage of genotypes retained after selection for a trait. For ease, increments of 10% are shown but in practical application, culling rate will be determined by selection intensity used, or based on check data. (C) Index selection shown for multiple traits (1, 2,.., *n*) for multiple genotypes (*G*1, *G*2,. . ., *Gn*). Data for each trait used in development of index, which is then used to make selection decisions.

index, all traits on all genotypes will need to be measured. This is dissimilar from truncation selection, where in each round one or few traits are measured, and only the remaining entries undergo subsequent trait estimation or measurement. Modifications in multiple trait selection can be made to have a tiered approach; independent culling followed by index selection, each with its own or overlapping traits. Fig. 4.10 gives a representation of the three methods of multiple trait selection. These examples are generalization of each method. For example, in index selection, more than two traits may be utilized in generation of an index.

Generation to select and population sizes

In breeding programs, several contrasting breeding objectives may be targets, such as traits that have a negative correlation (e.g., yield and protein in cereals and pulses) or are linked in repulsion. Breeders may be targeting 10 traits and each trait will be under polygenic control leading to more complexities in assembling the desirable combination, and an ability to identify that combination. F_2 generation is generally the generation of maximum variability and provides the best chance to look at several genes to be selected simultaneously. Heritability is lower on a single plant basis (compared with line or plot basis), but fortunately, some of the higher heritability traits can be fixed (such as simply inherited disease resistance or plant height) in early generations including F_2. In addition, we can use markers (in marker-assisted selection or genomic selection) to have an effective selection outcome for more complex traits.

DePauw et al. (2007) provided detailed description on negative correlations and breeding strategies in wheat, essentially establishing that traits with negative correlations need to be simultaneously selected. While the progress is slower, but over longer term even grain yield and protein correlations were positively shifted. They also provide insights into population sizes and introduce a term called desirable genotype quotient (DGQ). Authors described the DGQ of a cross as that proportion of lines (P) in which a segregating locus is either heterozygous or fixed for the best allele.

For multiple loci, for example a, b, c, to z, DGQ = ($P_a \times P_b \times P_c \times \ldots P_z$). In the absence of selection, DGQ reduces to an exponential function (P^n), n being the number of desirable loci, and P will vary by generation (Table 4.2). The average size of a population that contains one such desirable genotype is represented by the reciprocal of DGQ (1/DGQ). Hence the size of the minimum population in which one desirable genotype occurs also increases exponentially.

With one gene segregation, the F_2 genotypic ratio is 1 AA: 2 Aa: 1 aa, therefore ¾ = 0.75, where 3 comes from 1 AA and 2 Aa and 4 is total number of genotypes.

Table 4.2 The effect of generation and gene differences on the desirable genotype quotient (DGQ), where a desirable genotype has the preferred allele in either the homozygous or heterozygous condition at a locus.

Gene differences	Desirable genotype quotient (DGQ)			
	F_2	F_4	F_6	DHL/RIL[a]
1	0.7500	0.5625	0.5156	0.5000
2	0.5625	0.3164	0.2659	0.2500
5	0.2373	0.0563	0.0364	0.0313
10	0.0563	0.0032	0.0013	0.0010
15	0.0134	0.0002	4.84E−05	3.05E−05
20	0.0032	1.01E−05	1.76E−06	9.54E−07
30	1.79E−04	3.19E−08	2.34E−09	9.31E−10
40	1.01E−05	1.01E−10	3.11E−12	9.09E−13

With one gene segregation, the F_4 genotypic ratio is 7 AA: 2 Aa: 7 aa, therefore 9/16 = 0.5625, where 9 comes from 7 AA and 2 Aa and 16 is total number of genotypes.

With two gene segregation, the F_2 genotypic ratio is 9 A_B_: 3 aaB_: 3 A_bb: 1 aabb, therefore 9/16 = 0.5625, where 9 comes from 9 A_B_ and 16 is total number of genotypes.

Two obvious trends are noticeable: (1) DGQ reduces as generations advances and (2) DGQ reduces as more genes are controlling the trait. If more genes are segregating, larger population sizes are needed to increase the frequency of desirable alleles in the segregating breeding populations and for a better representation of genetic variation in the populations. Also careful choice of parents should be made. While breeding programs generally strive to keep consistency in the pipeline (for standardized approaches and more efficient and cost effective resource allocation), variable breeding populations sizes can be used in some specific circumstance based on the intent of the initial crossing.

CHAPTER

Plant genetic resources

5

Abstract

Genetic diversity is a fundamental requirement for a successful plant breeding program. This chapter covers topics such as centers of origin and diversity, and gene pools. Centers of genetic diversity have played an important role in crop improvement and served as a foundational principle for germplasm expedition and collection. Similarly, gene pool concept is important to determine breeding strategies between related, closely related, and unrelated species. Examples are included in the chapter to describe the complexity as well and success in gene transfer for important crops. These two concepts lend itself to create major sources of plant materials in germplasm collection such as landraces, obsolete cultivars, currently grown cultivars, advance breeding lines, undomesticated plants, other species and genera, mutants, special genetic stocks, and plant introductions. Important considerations such as genetic vulnerability and erosion are explained. Methods deployed in germplasm conversation, and the role and activities of germplasm banks, germplasm exchanges, International Treaty on Plant Genetic Resources for Food and Agriculture, and Material Transfer Agreements are covered.

Since the early period of human civilization, humans were inseparable from their environments, resulting in the coevolution of man and crops. The population pressure during the hunting and gathering period of human development were easily relieved by moving into new territory, hence a large number of plant species were used for food on either a regular or an occasional basis. During this process, human selection influenced crop plants' morphological traits of their interest that were in a wild form (i.e., *domestication*). Domestication is bringing of wild species under human management. The wild or weedy plant species had primitive traits such as spreading/trailing growth habit, shattering of pods in legumes and spike in cereals, small seeds with undesirable color, taste, and seed dormancy, which favored their adaptation and survival in nature from generation to generation. Human selection for traits was driven by their consumption criteria (e.g., food, fiber, etc.). These traits were changed and new characters were added for selection as per human needs (e.g., large fruit size/seed size and taste requirement). Similarly, natural selection was also changed as per change in the environment (and climate), and associated migration patterns.

Plant Breeding and Cultivar Development. DOI: https://doi.org/10.1016/B978-0-12-817563-7.00009-X

Darwin described variability of plants and animals under domestication in his book *On the Origin of Species*. Genetic variability is the raw ingredient of evolution, and Darwin was struck by the range of morphological variation found in domesticated forms in contrast to their wild relatives. Genetic variation is essential for evolution in nature and is equally necessary for improvement by plant breeding. The mode of variation or pattern of evolution in crop plants is divided into three categories: the Mendelian variation, interspecific hybridization, and polyploidy. These patterns have contributed to the variation of species in both pre- and post-domestication periods, and continue to do so because these methods are currently used by plant breeders.

Genetic diversity is produced by gene mutations. Mutations may result from a change in the gene from one allele to another (i.e., spontaneous mutation). The recombination results from hybridization among individuals carrying different types of mutations. The resulting diversity is subjected to natural and artificial selection in a crop species. There are enough evidences to show that Mendelian variation has played a key role in the evolution of important crops. The single gene mutation with drastic effect (*macromutation*) or a single gene mutation with multiple effects has been found useful in the evolution of many crop species (i.e., differentiation of the domesticated species from the wild ones). Many harmful or deleterious mutations are selected against in nature in due course and useful recessive mutations such as dwarfing genes have been used in plant breeding. The *macromutations* have also been induced by the plant breeders (e.g., for plant type and nutritional quality traits), and have been used for improvement of such traits. However, mutations with small effects or with less drastic effect (micromutations) tend to be accumulated and selected in a population in nature. In plant breeding, mutations are induced through the use of mutagens (e.g., physical and chemical mutagens). Modern plant breeders have used mutation breeding in important crop species to accelerate genetic variation and to select desirable ones. Example of crops evolving through mutations (in nature) include cabbage, cauliflower, Brussel's sprouts, which all originated from a common wild species (*Brassica oleracea*) through mutations and differ from each other due to macromutations.

Interspecific hybridization

The hybridization (natural or artificial) between the two distinct species produces large variation. Depending on the distinctiveness, F_1 plant may be fertile, partially fertile, or rarely fertile, and in the F_2 generation enormous variability is produced. Of this, large part of variation may be weak, undesirable and only some of it may produce seed and can be selected for desired traits. There are examples when F_1 vigor is fixed through vegetative propagation (e.g., certain varieties of pears, plums, cherries, grapes, etc.). In few other cases, hybridization is followed by recrossing with one of the parental species (i.e., small number of genes or few genes) of one species are transferred to the other species without effecting the taxonomic integrity. This is called *introgressive hybridization*. It is believed that introgression of genes of the wild gamagrass (*Tripsacum*) to the primitive maize (teosinte) has resulted in the modern maize. Interspecific hybridization is being used by the modern day plant breeder to transfer genes for stress tolerance from wild species or distantly related species of other genera to the cultivated ones (Chapter 6: Wide Hybridization).

Polyploidy

New variability may arise in nature through reduplication of chromosome sets. The normal plant has the diploid ($2n$) number of chromosomes (i.e., two complete sets of chromosome, one derived from each of its parents). *Polyploid* plants may arise due to duplication of the chromosome sets of a single species (i.e., autopolyploidy), for example, *triploid or tetraploid*, etc. or the *duplication* of chromosome sets of two or more species (i.e., *allopolyploidy*). *Allopolyploids* results by hybridization between the two species followed by the doubling of the chromosome number. Sometimes, the hybridization between two species produces sterile hybrids, which may become reasonably fertile when their chromosome number is doubled. The doubling of chromosomes is required for smooth functioning of reproductive mechanism (i.e., presence of two chromosomes of each kind). An allopolyploid, in which the total chromosome complement of two different species is combined to form a fertile species hybrid is called as an *amphidiploid*. Sometimes, plants may be produced with multiples of only certain chromosomes or deficiencies of others (i.e., *aneuploidy*) or plants may be produced with half the number of chromosomes in the somatic cells (i.e., *haploids*).

The production of polyploids has occurred in nature and several important crop species have evolved in nature through polyploidy (Chapter 7: Haploidy and Polyploidy in Crop Improvement). It has been shown experimentally how major food crops have arisen in nature in the past. The origin of the genomes and the exact chromosome homology is studied in few important crops, for example, common wheat (*Triticum aestivum*), American uplant cotton (*Gossypium hirsutum*), various species of tobacco (*Nicotiana*), and in various *Brassica* species. The origin of naturally occurring tetraploid species of *Brassica* has been verified experimentally by crossing the diploid species and doubling the chromosome numbers of the F_1 plant produced with colchicine thereby producing an amphidiploid. The experimentally produced amphidiploid and the naturally occurring tetraploid (Table 5.1) were crossed which produced fertile progenies confirming the polyploidy origin of species.

The Japanese scientist, U. Nagaharu, described the genomic relationships among naturally occurring *Brassica* species Fig. 5.1.

Table 5.1 The chromosome number and genome relationship of three diploid and three tetraploid species of *Brassica*.

Species	Somatic chromosome number (2*n*)	Genome formula	Common name
B. campestris	20	AA	Rape, Turnip rape
B. nigra	16	BB	Black mustard
B. oleracea	18	CC	Kale, Cabbage
B. juncea	36	AABB	Indian mustard
B. napus	38	AACC	Rape
B. carinata	34	BBCC	Abyssinian cabbage

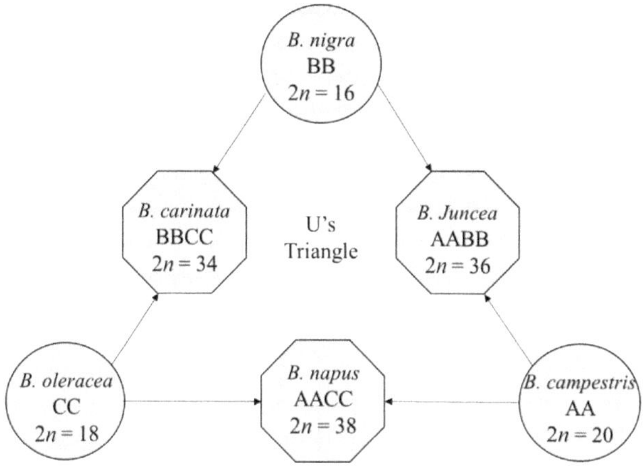

FIGURE 5.1

Origin of tetraploid (natural amphidiploid) species from the diploid species in *Brassica*.

Genetic diversity

Genetic diversity is the total amount of genetic variation in a crop species (i.e., the genes and genotypes found in a particular crop species). The crop genetic diversity is the basis of our food supply and essential to develop improved cultivars adapted to diverse and changing environments and to meet increasing demand of food supply. The genetic diversity is depleted due to genetic erosion, which is also called as loss of diversity. Nicolai Vavilov (1887–1943) was a Russian scientist and plant breeder. Vavilov (1926) published *Studies on the Origin of Cultivated Plants* which described his theories on the origins of crops concluding that each crop has a characteristic primary center of diversity that is also its center of origin. Vavilov's originally proposed six Centers of Origin and later included two more from which all our major crops were domesticated. Vavilov worked on global diversity of cultivated plants and anticipated genetic erosion; and worked with enormous determination and energy to assemble the raw materials from their center of origin.

Centers of origin

Vavilov could identify geographic regions of great diversity in contrast to other regions of uniformity or modest diversity. He believed that a limited area of high diversity or gene center must represent the center of origin of the crop in question, and that diversity was primarily a function of mutations. The mutations of over a period of time would accumulate particularly if they were preserved by deliberate selection. The *centers of origins* were separated by deserts, mountain ranges or water/sea; therefore, in these centers the development of agricultural civilizations was isolated from other civilizations. After several years of investigations, Vavilov (1951) summarized his information on the centers of origin of cultivated plants and proposed eight world centers of crop origin (Fig. 5.2). Some exceptions of these observations were noted, that is, some crops had more than one gene center, for example, barley had gene centers around the Mediterranean, in Asia Minor, Ethiopia, and China. Pattern of variation was somewhat similar in peas. There were gene centers for maize in Mexico, Colombia, and Peru. Vavilov proposed a new term "*secondary center*" for these patterns.

Primary and secondary gene centers

A crop evolved from wild species in the areas showing greatest diversity and termed them as *primary centers of origin*; and primarily due to activities of man

FIGURE 5.2

N.I. Vavilov Centers of origin of major crop plants in the world.

and developed vast arrays of genetic variants in places, called *secondary centers of origin*. The primary centers were recognized by the presence of wild forms and the prevalence of dominant alleles (Vavilov, 1951). Harlan (1975a) examined the map of the centers of crop evolution proposed by Vavilov and concluded that these gene centers coincide with centers of human civilization, which was a typically human evolutionary response to agricultural sources of food, that is, coevolution of crops and cultural man. The eight centers proposed by Vavilov (1951) are described as follows:

The Chinese Center: this includes the mountainous regions of western and central China and neighboring lowlands. It is the primary center of origin of naked oat, proso millet, soybean, radish, brinjal, opium poppy, colocasia, buckwheat, some species of *Brassica*, *Cucumis*, *Cucurbita*, *Allium*, Chinese tea, pears, peaches, apricots, plums, orange, etc. It is also reported to be the secondary center of origin of maize, common bean, cowpea, turnip, and sesame.

The Hindustan Center: this center includes India, Burma (Myanmar), Malaya, Indonesia, Thailand, the Philippines, and smaller islands. It excludes North-West India. It is the primary center of origin of rice, pigeonpea (red gram), chickpea (Bengal gram), mung bean (green gram), brinjal, cucumber, *Saccharum*, radish, okra, cotton (*Gossypium arboreum*), hemp (*Cannabis indica*), mango, orange, citrus, etc. It is a secondary center of watermelon, *Amaranthus* species, and cowpea.

The Central Asiatic Center: this includes North-West India, Afghanistan, Pakistan, the Soviet Republics of Tajikistan and Uzbekistan, and Tian-Shan. It is also known as Afghanistan center of origin. This area is the primary center of origin of wheat, club wheat (*Triticum compactum*), pea, broad bean, green gram, sesame, safflower, cotton (*Gossypium herbaceum*), onion, garlic, carrot, spinach, apricot, pear, almond, grape, etc. It is a secondary center of origin of rye (*Secale cereale*).

The Asia Minor Center: it includes the interior of Asia Minor, the whole of Transcaucasia, Iran, and Highlands of Turkmenistan. This center is also known as Near Eastern or Persian Center of origin. It is the primary center of origin of several species of *Triticum*, rye, alfalfa, Persian clover, carrot, cabbage, oat, *Allium*, lettuce, fig, pomegranate, apple, *Pyrus, Prunus*, grape, almonds, chestnut, pistachio, etc. It is considered the secondary center of origin of rape, black mustard, leaf mustard, turnip, and apricot.

The Mediterranean Center: this includes border of Mediterranean sea, Yugoslavia, Rumania, Bulgaria, and Austria. Many valuable cereals and legumes have originated in this center of origin such as durum wheat, emmer wheat, barley, oat, lentil, pea, broad bean, grass pea (*Lathyrus*), lupins, chickpea, vetch (*Vicia sativa*), clovers, rape, black mustard, cabbage, turnip, onion, garlic, beets, lettuce, asparagus, peppermint, etc.

The Abyssinian Center: this includes Ethiopia and hill country of Eritrea. It is the primary center of origin of barley, *Triticum durum*, *Triticum turgidum*, and *Triticum dicoccum*, sorghum, pearl millet, chickpea, lentil, *Dolichos lablab*, pea, grass pea, linseed, safflower, sesame, castor, coffee, onion, okra, etc. It is a secondary center of origin of broad bean.

The Central American Center: it includes South Mexico and Central America. This center is also referred as the Mexican center of origin. It is the center of origin of maize, common bean, lima bean (*Phaseolus lunatus*), melons, pumpkins, sweet potato, chillies, cotton (*G. hirsutum*), arrowroot (*Canna edulis*), papaya, guava, avocado, etc.

The South American Center: this includes the high mountainous regions of Peru, Bolivia, Ecuador, Colombia, parts of Chile, Brazil, and whole of Paraguay. It is the center of origin of potatoes, tomatoes, maize, lima bean, pea nut, Egyptian cotton (*Gossypium barbadense*), tobacco, pumpkin, cassava, pineapple, guava, rubber, etc.

Vavilov later on proposed a subcenter out of the Hindustan Center of origin and named it *Indo-Malayan* subcenter. Yam, pomelo, banana, and coconut were reported to have originated from this center. The South American Center was also divided into two: The *Chiloe* subcenter from which potato originated and the *Brazilian Paraguayan* subcenter of origin from which pea nut, rubber, pineapple originated. He also introduced a new center of origin, the *United States of America Center* of origin and two important species, sunflower (*Helianthus annuus* and *Helianthus tuberosus*) were assigned to this center.

Megagene Centers

Zhukovsky (1965) a close associate of Vavilov, proposed 12 *megagene centers* of crop plant diversity. The new areas added to Vavilov's eight centers were Australia, whole Africa, and Siberia followed by revision of boundaries to make 12 centers. The global genetic wealth including both primitive/native cultivars and their wild and weedy species is distributed to these 12 centers of diversity (Zeven and Zhukovsky, 1975). The "center of origin" of a species refers to location where that species is considered to have first appeared. The terms "center of origin" and "center of diversity" may have some similarity but there are distinctions between them and the terminology of "center of origin" is no longer preferred. Some scientists believe that it is very difficult to determine the initial geographical origin of species and prefer the term centers of diversity instead of Vavilov's centers of origin.

The *center of diversity* of a species is defined as the geographic area wherein the plant species (or genus) exhibits the highest degree of genetic variation, that is, highest number of cultivated species (or subspecies) and wild relatives, as well as gene variants (alleles) exist in that region. The center of diversity often corresponds to the area where the plant has existed the longest and will thus have had a chance to have maximum diversity. It is based on the premise that genetic variation can be accumulated only over time. It must be considered that crop plant's diversity may increase away from its center of origin due to environmental factors and human influence.

Microcenters

Within the large centers of diversity (megagene centers), small areas may exhibit much diversity (where wild species occur) than the center as a whole. These areas are referred to as *microcenters* by Harlan. According to him, plant evolution appeared to proceed in them at a more rapid rate than in other areas, particularly large geographic areas. These microcenters would be an excellent site not only to collect useful types but also to study the evolution of cultivated types.

Centers and non-centers

Harlan (1971) proposed that crop plants "originated independently in three different areas" and that, in each case, there was a system composed of a center of origin and a non-center, in which activities of domestication were dispersed over a span of 5000–10,000 km. One system includes a definable Near East center and a non-center in Africa; another system includes a North Chinese center and a non-center in Southeast Asia and the South Pacific; the third system includes a Meso-American center and a South American non-center. There are suggestions that, in each case, the center and non-center interact with each other. Crops did not necessarily originate in centers (in any conventional concept of the term), nor did agriculture necessarily develop in a geographical "center." The centers are narrow areas where crops originated in a specific time and from these areas crops spread to other places. On the other hand, the non-centers represent broad geographical areas where the crop is domesticated over a long period of time.

Harlan (1975a) observed that different crops have different geographic patterns of variation, and the interaction between space-time-genetic variation are noticeable. He attempted a classification of crops according to this interaction and most crop species could be classified into following categories.

Endemic: such crops have originated in a limited area (i.e., geographically restricted), for example, Guinea millet (*Brachiaria deflexa*) is an endemic crop in new Guinea. The wild races are wide spread in the wetter parts of Africa; however, the cultivated races are confined to the Fouta Djalon area of Guinea. Other examples of endemic crops are black fonio (*Digitaria iberia*) of West Africa, Korali (*Setaria lutescens*) and Koda (*Paspalum scorbiculatum*) of the Nilgiri hills of India, and *Panicum sonorum* of North Western Mexico.

Semiendemic: such crops have not shown very wide distribution of their cultigens from center of diversity (i.e., such crops have originated in a defined area and geographically isolated and with limited dispersal), for example, African rice (*Oryza glaberrima*) was domesticated from a wild annual species (*Oryza barthii*) in the inner delta region of the Niger river and has the maximum range of variability of African rice here and contains mostly dominant genes. From this area, African rice spread up to Senegal in the west and Lake Chad in the east. However, the entire distribution of the cultigens was not extensive.

Monocentric: for such crops, the center of origin and center of diversity are clearly discernible, for example, Arabica coffee (*Coffee arabica*) is a monocentric as its wild, semiwild, and cultivated types are found in the forests of the southern Ethiopian highlands, from here it achieved a wide distribution in recent times (i.e., restricted time), moving first to the Yemen highlands, then to Latin America by way to Europe. It is a wide spread crop with a single well-defined center and its spread is historically well documented. Other examples of monocentric pattern are rubber, cacao, and oil palm.

Oligocentric: such crops have a very wide dispersal and ancient origin, which show nodes of high variability in several areas; and therefore, are difficult to unravel the complexities of evolution of variability, for example, hexaploid wheat (common wheat; *T. aestivum*) has very wide distribution in Eurasia and shows high variability in several areas including Mediterranean, Asia Minor, North-West India, and China. Many of the crops first domesticated in the near East have achieved wide distribution and show additional centers of diversity in Mediterranean, Ethiopia, India, and China, for example, barley, pea, flax, chickpea, lentil, etc.

Noncentric: such crops are derived from wild progenitors of wide dispersals involving multiple domestications; however, a detailed analysis of the distribution of variation of the crop fails to reveal either clear center of diversity or center of origin, for example, sorghum, pearl millet, common bean, and *Brassica campestris*.

Law of homologous series

Vavilov focused on global diversity of cultivated plants. From this work he also proposed law of homologous series (Vavilov, 1920). This law suggests that closely related species and genera are characterized by similar series of variation. This makes studies of related species and genera simpler. For example, Vavilov described variability in the Fabaceae (Leguminosae) family and established its differentiation in separate genera and cultivars by monitoring characters of vegetative organs, flowers, fruits, and seeds. He also analyzed variability of characters in accessions of Vicieae, Trifolieae, Loteae, Galegeae, and Phaseoleae. This establishes the law of homologous series that despite difference between selections, the variability in characters is similar for all genera within a given family. Other examples are wheat of different ploidy level, that is, diploid, tetraploid, and hexaploid show a series of identical characters. Similar observation have been noted in the case of the diploid and tetraploid cotton.

Gene pools

The Gene Pool Concept was proposed by Harlan and de Wet (1971) as an attempt to provide a practical guide for placing existing classifications into genetic perspective (Fig. 5.3). The gene pool is also called plant genetic resources (PGR) or germplasm which could be defined as the sum total of all the alleles influencing a wide range of characters (i.e., the collective hereditary genetic wealth) that a crop

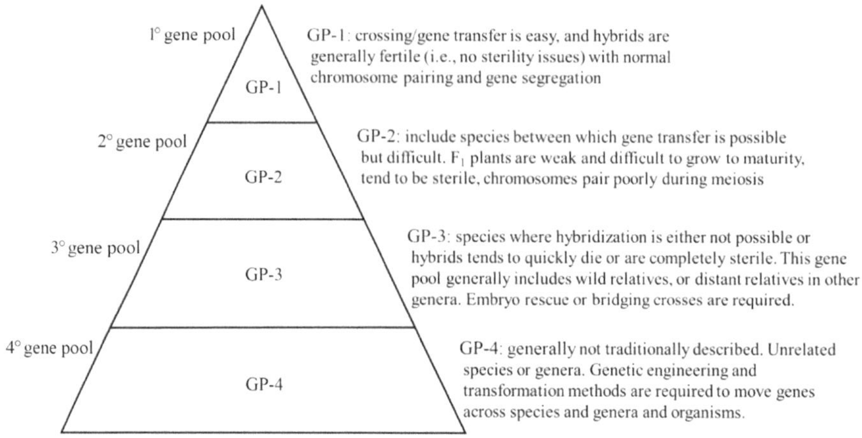

FIGURE 5.3

The primary (GP-1), secondary (GP-2), tertiary (GP-3), and quaternary (GP-4) gene pool. *Modified from Harlan and de Wet, 1971.*

has acquired over millions of years under natural conditions. It provides the basic materials or genetic stocks for further improvements of a crop in on-going breeding programs. It is used as experimental subjects by other plant scientists. However, it is not of immediate value as a cultivar. Information on the relatedness among crop plants and their relatives is useful to breeders and geneticists wishing to make crosses between related species. The purpose of the Gene Pool Concept is not to change the taxonomy; however, it serves as a guide for planning breeding activities. Various genetic resources are assigned to different gene pools of a crop species based on ease of hybridization (i.e., ability to move genes between species). The three major gene pools are: primary, secondary, and tertiary. The gene pool concept suggested that the plant breeders utilize the germplasm of gene pool 1 (GP-1) and proceed outwards.

Primary gene pool: it is also called GP-1. The primary gene pool is equivalent to a biological species itself, which includes the crop species itself and its wild and/or weedy spontaneous races as well as cultivated races. Species in a primary gene pool can be cultivars, landraces, farmer-developed or farmer-maintained populations, ecotypes, and spontaneous races (wild or weedy). Among forms of this gene pool, crossing/gene transfer is easy, and hybrids are generally fertile (i.e., no sterility issues) with normal chromosome pairing and gene segregation. It must be noted that most breeders work exclusively within this gene pool, and it is the major source of genetic variation for improvement programs. Most breeding programs, engaged in developing cultivars for commercial production, work on elite material exclusively, and spend very little direct effort on unadapted or wild relatives (because of undesirable linkage blocks, breaking of desirable linkage blocks, and epistatic interactions with undesirable genes from wild relatives).

Secondary gene pool: it is also called gene pool 2 (GP-2). A secondary gene pool will include species between which gene transfer is possible but difficult. Hybrids tend to be sterile, chromosomes pair poorly during meiosis, and where produced, F_1 plants are weak and difficult to grow to maturity. Because of some sterility in F_1s, recovery of desired types in advanced generations is generally difficult. The secondary gene pool includes related species within the same genus although not all species within a genus will be in the secondary gene pool. It is also possible that species outside the genus can be in this gene pool, for example, *Aegilops* species in *Triticum aestivum*.

Tertiary gene pool: it is also referred as gene pool 3 (GP-3). The tertiary gene pool contains those species where hybridization with the cultigen is either not possible or hybrids tends to be anomalous, lethal, or completely sterile. This gene pool generally includes distant relatives in other genera. Although hybrid sterility is common, chromosome doubling can be effective to restore fertility as it may provide chromosome homologs. The species may move out of tertiary gene pool, as new techniques are developed facilitating gene transfer. Gene transfer between a crop and a species in its tertiary gene pool is very difficult and requires embryo rescue, chromosome doubling, or bridging species to obtain hybrids. A bridging species is a third species that facilitates the exchange of germplasm between the crop species and the tertiary gene pool species by developing complex hybrids. Harlan and de Wet (1971) described a classic example using a bridging species where there was an interest to cross *Elymus* × *Triticum*. As expected, hybrid seed could not be obtained. When embryo rescue was used, very few hybrids were obtained, and even then, these were sterile. However, these researchers found that if they used an *Agropyron* × *Triticum* derivative as the female parent in crosses with *Elymus* as male parent, introgression of *Elymus* alleles was possible without the need for special technique.

Bellon et al. (1998) explained this concept in rice (Fig. 5.4) based on Harlan and de Wet (1971). The genome composition of the species in gene pool of rice is given in Table 5.2. There are two cultivated species of rice. *Oryza sativa* had its origin in south and southeast Asia and was domesticated either from a perennial wild species, *Oryza rufipogon* or from annual wild species *Oryza nivara*. The later hypothesis is widely accepted. *O. sativa* is now cultivated worldwide. *O. glaberrima*, the African rice had its origin in West Africa from *O. barthii* sometimes called *Oryza breviligulata* and remains locally important in some farming systems in those areas. These two cultivated species, primitive varieties (landraces), improved varieties, and their wild progenitors, other wild species *Oryza* with AA genome or with other genomes or species in related genera, form the gene pool of cultivated species.

The hybridization of two subspecies of *Oryza sativa* (*indica* and *japonica*) along with their introgression from wild relatives and mutation have produced a tremendously broad genetic diversity, which exists as primitive varieties and improved varieties selected by farmers and growers in the past and those improved by breeders. This makes up an important component of the primary gene pool of rice. The *O. glaberrima*, its wild progenitor *O. barthii*, and three other wild species are

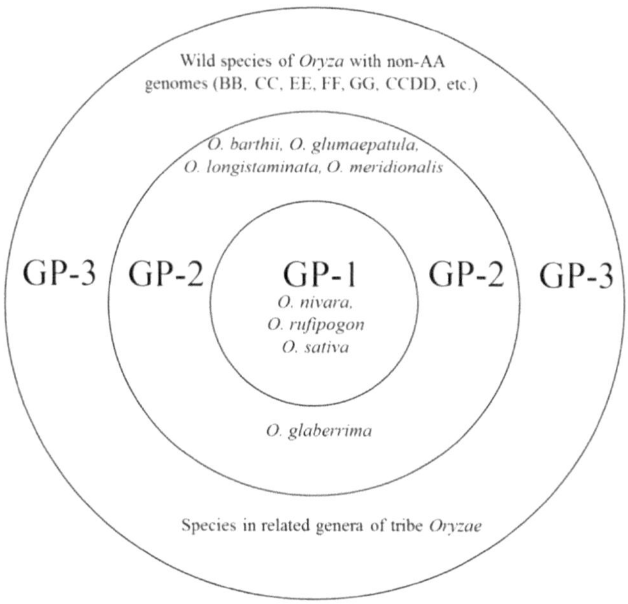

FIGURE 5.4

The gene pool of *Oryza sativa*. The GP-1, GP-2, and GP-3 are primary, secondary, and tertiary gene pool, respectively.

After Bellon, M.R., Brar, D.S., Lu, B.R., Pham, J.L., 1998. Rice genetic resources. In: Dowling, N.G., Greenfield, S.M. (Eds.), Sustainability of rice in the Global Food System. Pacific Basin Study Center, Davis, CA, USA and IRRI, Manila, Philippines, pp. 251–283.

distantly related. These form the secondary gene pool of rice. These have crossability problems with *O. sativa* with some fertility of the F_1 that leads to gene flow from GP-2 to GP-1. The wild species with non-AA genomes and the species in related genera of tribe *Oryzae*, which are distantly related to *O. sativa* form the tertiary gene pool. These species are not crossable or are impossible to cross. Hybrids with GP-1 do not occur in nature, they are inconsistent, lethal, or completely sterile. Gene transfer is not feasible without deploying drastic techniques.

Quaternary gene pool: The gene transfer techniques (plant transformation techniques) have extended the reach of plant breeding beyond the limitations imposed by sexual cross compatibility. This is called gene pool 4 (GP-4). The GP-4 should contain any synthetic strains with nucleic acid frequencies (DNA or RNA) that do not occur in nature within that species.

Gene pools are not static but change as more information becomes available or as new technologies become available for manipulating genomes. For example, soybean was reported not to have a secondary or tertiary gene pool. However, 26 perennial *Glycine* species are now considered in the tertiary gene pool, and *Glycine tomentella* has been used to transfer genes to *Glycine max* (Singh et al. 2014).

Table 5.2 The genomic composition of *Oryza* species that forms its gene pool.

Species	Genome	Somatic chromosome number (2*n*)	Distribution
Oryza sativa	AA	24	Worldwide
O. nivara	AA	24	Tropical and subtropical Asia
O. rufipogon/perennis	AA	24	Tropical and subtropical Asia, tropical Australia
O. glaberrima	A^gA^g	24	West Africa
O. breviligulata/barthii	A^gA^g	24	Africa
O. longistaminata	A^lA^l	24	Africa
O. meriodionalis	A^mA^m	24	Tropical Australia
O. glumaepatula	$A^{gp}A^{gp}$	24	South and Central America
O. punctata	BB, BBCC	24, 48	Africa
O. minuta	BBCC	48	Philippines and Papua New Guinea
O. officinalis	CC	24	Tropical Australia
O. rhizomatis	CC	24	Sri Lanka
O. eichingeri	CC	24	South Asia and East Africa
O. alta	CCDD	48	South and Central America
O. grandiglumis	CCDD	48	South and Central America
O. latifolia	CCDD	48	South and Central America
O. australiensis	EE	24	Tropical Australia
O. brachyantha	FF	24	Africa
O. granulata	GG	24	South and Southeast Asia
O. meyeriana	GG	24	Southeast Asia
O. longiglumis	HHJJ	48	Indonesia and Papua New Guinea
O. ridleyi	HHJJ	48	South Asia
O. schlechteri	–	48	Papua New Guinea

From Nanda, J.S., Sharma, S.D. (Eds.), 2003. Monograph on Genus Oryza. Sam Publishers, Plymouth, UK, pp. 283–289.

Sometimes, breeders use wide crosses to make improvement in their crop. “Wide cross” refers to a cross that involves individuals outside of the cultivated species or belonging to different species. This typically involves the secondary and/or tertiary gene pools. Even though it is difficult, it may be useful for the transfer of important traits, including pest resistance, disease resistance, or other traits not present in cultivars of that species.

Example 1: in wheat, the T1BL.1RS wheat (*T. aestivum*) – rye (*Secale cereale*) translocation has been of particular interest and was widely used in bread wheat breeding programs worldwide. A translocation results when a broken segment of one chromosome is transferred to and unites with a nonhomologous (or different) chromosome. At one point, it was estimated that several million hectares of wheat were planted to cultivars possessing this translocation (e.g., involving a tertiary gene pool; rye to wheat crop). This segment had a disease resistance cluster for leaf rust, stem rust, stripe rust, and powdery mildew, all of which are important diseases of wheat. Additionally, this segment was reported to possess genetic factors that improved grain yield and kernel weight. Resistance due to specific genes in the translocated segment have been overcome in some parts of the world, which clearly shows the continuous nature of plant breeding to develop better genetic packages (i.e., cultivars).

Example 2: in rice, the first example of the transfer of a useful gene from the wild species was the introgression of a gene for grassy stunt virus resistance from *O. nivara* to cultivated rice. Other examples are transfer of *Xa-21* for bacterial blight resistance from *O. longistaminata* to cultivated rice; cytoplasmic male sterility (CMS) sources from *O. perennis* and *O. glumaepatula* into rice for hybrid rice production. In the 1970s, grassy stunt virus epidemics, caused by transmission through the brown plant hopper (diseased rice plants produced no panicles or small panicles with deformed grains), were reported in several countries and led to severe yield losses. Of the several thousand accessions of cultivated rice and wild species of *Oryza* screened for resistance, one *O. nivara* resistant accession was identified. Plant breeders then successfully transferred grassy stunt virus resistance to improved varieties through a backcross breeding method, and resistant varieties were released for cultivation. Other examples include the transfer into elite rice genes for resistance to brown plant hopper, white backed plant hopper, and bacterial blight from *O. officinalis*.

Through regional, national, and international efforts gene banks have been established that attempt to preserve valuable agriculture biodiversity for future generations. Gene banks are genetic repositories, and contain "landraces" or local varieties of cultivated and uncultivated wild relatives. These serve to protect and preserve seed diversity as well as to provide an accessible source to provide plant breeders with seeds of interest. There are currently about 1750 institutional crop collections around the world as well as a number of community-based seed bank initiatives. The Consultative Group on International Agricultural Research (CGIAR) Program for Managing and Sustaining Crop Collections is dedicated to maintaining the 706,000 samples of crop, forage, and agroforestry resources held in gene banks at 11 CGIAR research centers around the world. Species, including those of cereals, legumes, roots, tubers, trees, and other essential staple crops, are stored in CGIAR international collections. All accessions within these collections are for international public good and are available under the terms and conditions negotiated by the International Treaty on Plant Genetic Resources for Food and Agriculture (ITPGRFA).

National Germplasm Banks

In the United States, the National Plant Germplasm System aids scientists and addresses the need for genetic diversity by acquiring, preserving, evaluating, documenting, and distributing crop germplasm. To search for genotypes of interest and place a seed order through the USDA's Germplasm Resources Information Network, the steps are first, search query, and then order form.

Breeders should determine which gene bank has the collection of material for their crop, search the gene bank, and order seed. If the requestor and gene bank are in the same country, only a Material Transfer Agreement (MTA) may need to be signed. However, this process may involves a large amount of paperwork if importing from overseas (agreements, seed importing or exporting permits, and customs documents), and planning ahead is critical to ensure that seed is received on time.

The germplasm or PGR consists of various plant materials.

1. *Landraces*: these are also called land cultivars. Landraces are primitive cultivars that were selected and cultivated under subsistence agriculture for many generations. Such cultivars have broad genetic base and possess genes for greater adaptability, durability for resistance and tolerance to adverse environments but have low productivity. They are responsive to selection for high yield but to a certain extent. When highly productive genotypes are developed through breeding the landraces undergo replacement; therefore, their conservation is essential. According to Harlan (1975b) "the landraces have a certain integrity. They are recognizable morphologically, farmers have names for them and different landraces are understood to differ in adaptation to soil type, time of seeding, date to maturity, height, nutritive value and other properties, Most important, they are genetically diverse. They are balanced populations-variable in equilibrium with both environment and pathogens, and genetically diverse."
2. *Obsolete cultivars*: the improved varieties of recent past that are rendered obsolete for commercial production because of a drawback (e.g., susceptibility to a disease), and high yielding cultivars were developed to replace them. Sometimes such cultivars are excellent source to be used as parental stock to overcome a known deficiency.
3. *Currently grown cultivars*: the modern cultivars are the most likely source of superior germplasm that is adapted to specific environment. The cultivars of this category would constitute a valuable part of working collections to be used as parents in the genetic enhancement for characters of importance.
4. *Advanced breeding lines*: the prerelease genotypes developed by plant breeders and not yet ready for release or not released are known as advanced lines or stocks. A breeder may develop and release a commercial cultivar; however, its sister lines or genotypes from other crosses that do not reach the final stage of commercialization but have genetic potential to be used as a parental strain in hybridization program, are useful for plant breeders. Some of these could have a unique trait or a combination of such traits, which are retained to be used as parent(s) in future. This is the most preferred source of parental material if the target trait for improvement is seed yield, or if the objective is to develop cultivars. This source may also include ex-Plant Variety Protection cultivars.
5. *Undomesticated plants*: in case the desired genes are not available in the cultivated types, plant breeder may like to search for useful genes from the populations of

wild types, which are source of entire genomes or specific genes that may require introgression into an adapted cultivar. The wild types have been used as source of resistance genes to biotic and abiotic stresses in several major food crops.

6. *Other species and genera*: such types are also important as sources of resistance to biotic and abiotic stresses; however, should be one of the last options in crop improvement because their use in hybridization leads to hybrid sterility, hybrid inviability, and transfer of several undesirable genes to the cultivated species along with desirable alleles.
7. *Mutants*: mutations do occur spontaneously and are also induced through the use of mutagens (physical and chemical agents) and constitute important component of gene pool. Large number of mutations are deleterious; however, some of the mutants are desirable and useful, and may be used as a source of genetic improvement of plant type, disease resistance, or quality traits (Chapter 14: Mutation Breeding).
8. *Special genetic stocks*: this category includes products of specialized genetic manipulations by geneticists and plant breeders, for example, chromosomal and genetic mutants.
9. *Plant introductions*: the plant breeders may import useful materials from outside production region (i.e., from another country) are called plant introductions. Plant introductions are likely to be successful if similarities exist in soil, climate, and photoperiodic conditions between the region of origin and the new area of production. Introductions have played a major role in plant breeding programs around the world and continue to be important. In the past, the agencies of plant introduction were invaders, settlers, traders, travelers, explorers, pilgrims, and naturalists. A number of Botanic Gardens across the world played an important role in plant introductions. Plant introduction may consist of new varieties of a crop species already grown in that area or a wild relative of the crop species or a new crop species for the area. Plant introductions are of two types.
   a. *Primary introduction*: when an introduced variety or crop as such is well suited to a new environment and it is recommended for commercial cultivation without any alternation in the original genotype. In India, several dwarf and high yielding varieties of rice and wheat were introduced during 1960s from IRRI and CIMMYT, respectively, and were released for commercial cultivation.
   b. *Secondary introductions*: the introduced variety may be subjected to selection to isolate superior variety. The introduced variety may be used to transfer superior character(s), which is called the introduction of a character(s). The secondary introductions are an important component of plant breeding program in all countries.

Acclimatization

When a crop plant is introduced into a totally new production area, it may show variable adaptation in the climatic area. They may fail to show phenotypic expression similar to that in the place of origin but may pick up later on. The ability of a crop/variety to become adapted to a new climate is referred to as *acclimatization*, which is influenced by the mode of pollination, the range of genetic variability within the crop, the longevity of the crop, and the rate of mutation. The acclimatization proceeds more rapidly in cross-pollinated crop than in self-pollinated crop because gene recombination occur more frequently in cross-pollinated crops, which adapt to the new environment. The annual crops also show frequent new gene recombinations hence are better adapted than the perennial crops. Pure lines show slight variation; therefore, are less adapted. The rate of mutation with the crop is another genetic force, which influences acclimatization.

Plant quarantine

All the introductions are subjected to quarantine. The quarantine means to keep materials in isolation to prevent spreading of diseases, *etc*. All the introduced materials are thoroughly inspected for contamination with weeds seed, diseases, and insect–pests. The materials that are suspected for contamination are fumigated or destroyed. When an introduction is received it is cataloged and is given an entry number, name, species, variety, place of origin, adaptation and characteristics are recorded. The materials are classified as exotic collection, indigenous collection, and indigenous wild.

Genetic erosion

The erosion of biodiversity is difficult to quantify in a crop species, however, some of the factors accelerating genetic erosion are described as follows:

1. The loss, fragmentation, and degradation of habitats are responsible for the depletion of valuable genetic resources in crop plants. For example, in the case of rice, specific details of population destruction and gene loss have been recorded for wild rice in the central plains of Thailand. The Bangkok area has a wide diversity of *Oryza* species including *O. nivara*, *O. officinalis*, *O. ridleyi*, and *O. rufipogon*.
2. Introduction of alien species into ecological niches has also affected genetic diversity of native wild populations of *Brassica* in Italy, as these are being introgressed into elite varieties of cole crops.
3. Change of farming systems and increased stakeholder integration into the markets have adversely affected the genetic diversity (e.g., due to consumer or farmer preferences).
4. Extension of farming and overgrazing into the natural habitats has resulted in genetic erosion.
5. Excessive harvesting beyond the levels of natural regeneration as in case of the trees harvested for timber.

6. Severe abiotic stresses like drought, flooding, air pollution, water pollution, and introduction of new pests and diseases have resulted into considerable deterioration of habitats of many wild species.
7. Modern plant breeding has resulted in the reduction of genetic diversity into two ways: (1) selection for relative uniformity from landraces, that is, the production of pure lines/clones and the use of single or double cross hybrids. This shows that the production of genotypes with narrow base has caused replacement of genetic diversity from which these genotypes were created. (2) Selection for closely defined objectives has also resulted in increased productivity with restricted adaptation. Perhaps, the biggest reason is the use of relatively few parents in crossing programs and excessive use of a very few parents in hybridizations causes a genetic bottleneck, leading to genetic diversity erosion. However, with careful planning both can be avoided by plant breeders. Another reason for reduction in genetic diversity in a plant breeding program is over-reliance of backcross based conversion program, either for a transgene or native gene, and using relatively fewer parents in this process.

Genetic vulnerability

The narrow genetic base of varieties and hybrids caused the loss of genes (i.e., genetic vulnerability). The incidence of pest and disease has been a major factor causing instability and risk in crop production. One of the most devastating epidemics caused by genetic vulnerability is the Irish potato famine of the 1840s in which more than one million Irish population starved to death due to the destruction of potato crop as a consequence of attack of late blight (*Phytophthora infestans*). It occurred from a narrow genetic base of potato plants in Ireland. The Irish mainly grew variety Lumper that was susceptible to potato blight, and since potato varieties are vegetative propagated, the Lumper potatoes were all alike genetically. The Southern corn leaf blight epidemic in the United States (1970) was caused by a newly emerged race of *Helminthosporium maydis* called race T combined with favorable weather led to major damage to the corn crop with the T cytoplasm, which was in at least 85% of the seed planted. More recently, citrus industry has been impacted by bacterial disease called Citrus Greening (*Candidatus Liberibacter asiaticus*), Huanglongbing or yellow dragon disease, and is one of the most serious citrus plant diseases in the world. The factors that promote genetic vulnerability and led to epidemics are described in detail by Singh and Singh (2005) and briefly listed as follows:

1. The commercial cultivars are generally pure lines, clones, and hybrids with narrow genetic base, hence do not have the buffering capacity to adapt itself to a change in the genetic make-up of the parasite or pathogens, which coevolves on its own.
2. Large acreages under single cultivar facilitate the spread of pathogen once infection is initiated.

3. When a new parasite or a pathogen is introduced from another locality/geographical area(s), the local varieties may not show resistance. Such parasites have caused many of the most devastating epidemics experienced by man.
4. Normally innocuous parasites to which marked susceptibility has been introduced into cultivars by breeders along with some desirable traits may be a serious threat to stability under favorable conditions, for example, Victoria oat bred for crown rust resistance at one time in 1940s was present in majority (Victoria parentage) of the US oat crop. However, Victoria and other oat cultivars with this parentage were universally susceptible to *Helminthosporium* blight a disease unknown up to that time. It was later determined that crown rust resistance and locus for *Helminthosporium* blight susceptibility were completely linked.
5. The international exchange for research, commerce, or tourism of plants and plant parts that are infected by pathogens and pests contribute toward vulnerability of the native crop species, for example, *P. infestans* mating type A1 into Europe which caused 1845 epidemic. Other examples include Chestnut blight and Dutch elm disease that originally came from the Orient, and so did Japanese beetle. (Note: international seed transfer has strict guidelines and enforcement to prevent movement of diseases and pests in new areas.)
6. Previously unrecognized strain(s) of pathogens or pests can attack otherwise resistant varieties and it often occurs so rapidly as to curtail the commercial use of otherwise satisfactory varieties. Wheat stem rust, Ug99 strain is an example.
7. Pathogen may overcome major gene resistance (i.e., failure of vertical resistance) soon after a variety is released because of the intense selection pressure on the pathogen. This dependence on a single source of resistance is a major problem. Breeders can strengthen and lengthen the plant resistance against disease and pest epidemics by utilizing diverse resistance genes and "stacking" or "pyramiding" multiple genes in commercial lines. If there is a lack of diverse genes to protect against pathogen or insects, crop refuge can be effectively used to extend the longevity of gene, for example, the use of *Sm1* gene to protect against orange blossom wheat midge [*Sitodiplosis mosellana* (Géhin)] in Canada.

The role of germplasm breeder or prebreeding is to mitigate the risk of genetic vulnerability, and requires the integration of genes from nonelite sources (including; landraces, and wild/weedy relatives).

Prebreeding

The wild species are a reservoir of useful genes for biotic and abiotic stresses and diversification of cytoplasmic genetic male sterility; however, identifying the desired traits in the wild type of germplasm has often been ignored due to the confounding effects of many undesirable traits under natural habitats and poor expression of plant growth under domestication. In recent time, plant breeders and geneticists are increasingly interested to identify useful traits and transfer them to cultivars. The process of incorporating desired genes from wild species into an adapted variety or genotype is termed as *prebreeding* or *parental line breeding* or *germplasm enhancement*. The desirable traits from wild species can be accomplished by using cultivated types as female and wild species as male parent. The progenies after evaluation are backcrossed to cultivated type and resulting progenies are again subjected to screening and evaluation to help reduce the presence of undesirable genes of wild species. Such an approach has been used in tomato, rice, grain legumes to transfer useful traits. Marker assisted back crossing or marker assisted selection can be used to more effectively transfer genes from wild species to cultivated types (Chapter 27: Molecular Tools in Crop Improvement and Cultivar Development).

Germplasm exploration and collection

One of the main purposes of germplasm exploration and collection is to reduce the risk of genetic erosion of important genes and genetic stocks as well as to utilize the useful genes in crop breeding to increase the stability of crop cultivars and hence reduce the risk of yield loss. *Exploration* refers to collection trip organized in the centers of diversity (as discussed earlier in this chapter). The *collection* is taping of genetic diversity from various sources including national and international centers. N. I. Vavilov was ahead of his time, and was among the first to recognize the need for intensive plant collecting, research, and preservation. He organized and took part in more than 100 collecting mission in major agricultural areas of the world. He insisted on systematic classification, evaluation, maintenance, and utilization of the vast collections assemble at the Institute of Plant Industry, Leningrad. The US Department of Agriculture (USDA) began efforts to plant germplasm collection in 1898 under the leadership of David Fairchild. The first collection PI 1 was a cabbage accession from Moscow, collected in 1898. In the period from 1946 to 1971 the USDA sponsored some 65 plant exploration expeditions. These covered a wide array of plant species including oilseeds, forages, vegetable crops, potatoes, sorghum, beans, peas, citrus, and medicinal plants. Later on additional collections were made for several crop species by National centers in different countries (Table 5.3), which are spread out across the globe where collections of one or more important species of that country have been collected and preserved. Additionally the international centers were established in areas of genetic diversity. The CGIAR has now 15 international centers. Of these, eight concentrate on specific crops, for example, CIAT, CIP, CIMMYT, ICARDA, ICRISAT, IITA, IRRI. These centers have strong crop breeding activities including collection, conservation utilization, and supply of germplasm resources of their mandated crops. These centers work in close cooperation with national centers.

The CGIAR established the International Board of Plant Genetic Resources (IBPGR) in 1973 at Food and Agricultural Organization (FAO) headquarter in Rome. The basic function of IBPGR is to promote an international network of genetic resource

Table 5.3 Examples of few international agricultural research institutes related to field crops, their mandate, and germplasm holdings.

International center	Established in the year	Mandate crop(s)
International Rice Research Institute (IRRI), Los Banos, Philippines	1960	Rice
Centro International de Mejormiento de Maiz Y Trigo (International Center for the Improvement of Maize and Wheat) (CIMMYT), El Baton, Mexico	1966	Wheat Maize Triticale
International Center for Tropical Agriculture (CIAT), Palmira Cali, Colombia	1967	Cassava Beans Maize Rice
International Institute of Tropical Agriculture (IITA), Ibadan, Nigeria	1968	Cassava Bambara Groundnut Cowpea Soybean Wild *Vigna* Yam Rice
International Center for Agricultural Research in Dry Areas (ICARDA), Morocco	1976	Barley Wheat Lentil Chickpea Faba bean Forage
International Crops Research Institute for the Semi-arid Tropics (ICRISAT), Hyderabad, India	1972	Sorghum Pearl millet Minor millets Pigeonpea Chickpea Groundnut
Centro Internacional de Papa (CIP), Lima, Peru	1971	Potato Sweet potato Andean roots/tubers
West African Rice Development Association (WARDA), Monrovia, Liberia	1971	Rice
International Center for Research in Agroforestry (ICRAF), Nairobi, Kenya		Sesbania
International Network for the Improvement of Banana and Plantain (INIBAP), Monterriersur-Lez, France		Musa
International Livestock Research Institute (ILRI), Kenya	1974	Forages
Asian Vegetable Research and Development Center (AVRDC), Taiwan		Mung bean Brinjal Tomato Pepper

activities to further the collection, conservation, documentation, evaluation, and utilization of plant genetic germplasm and thereby contribute to raising the standard of living and welfare of people throughout the world. The close cooperation of FAO of the United Nation and IBPGR and other agencies has established priorities for collections. The letter "E" has been used by FAO to designate emergencies, whereas at lower levels of priority categories I, II, and III are used species by species and region by region. Earlier, priorities were first endangered crop species and later for endangered areas/countries; however, presently the priorities are often set out by areas first and crop species second. The IBPGR was updated to include element of research and training and was transformed to a new autonomous organization, International Plant Genetic Resources Institute (IPGRI) in 1992. The IPGRI has the following four major objectives:

- To assist countries particularly in the developing world in needs of PGR conservation and strengthening their links to users.
- To build international collaboration in crop and geographic area.
- To develop strategies and technologies for the conservation of PGR.
- To provide an information service.

The collection should not limit only to the gene centers but should also include peripheral areas/regions of the species distribution where species is under environmental stress and useful mutations could have been selected in this region.

The genetic resources collector samples population rather than individuals, as it is aimed to capture maximum amount of diversity for minimum number and size of samples. The sample size should represent 95% of the total genetic diversity from an area. The diversity could be sampled randomly (i.e., non-selectively). For this 50–100 individuals may be collected per site with some 50 seeds taken from each plant [except in species with large seeds, for example, coconut (*Cocos nucifera*), Marshall and Brown, 1975]. The collecting sites should be scattered throughout an area. However, mixed sampling (i.e., selectively and randomly) is also suggested by some scientists so that morphological variation and entire variation are sampled. Selective sampling could be best for vegetatively propagated species because of their nature and strong artificial selection that has been applied. The nonselective sampling would capture the greatest range of diversity. Some of it may be of immediate use and others may be of use in future. Adequate field records must be maintained.

Germplasm conservation

The PGR conservation involves managing and using resources in a manner that does not deplete them. It is essential for maintaining the continued availability of biodiversity in a crop species for its sustainable use in crop improvement. There are two basic approaches to germplasm conservation.

Ex-situ conservation: in this approach, genetic resources (seeds, organisms, or individuals) are actually removed from their original habitat or natural environment (i.e., off-site). The *ex-situ* samples are preserved in a gene bank. The condition of storage differs depending on mode of reproduction of species. For seed propagated species, seeds are dried to bring it to an appropriate moisture level and then usually placed in hermatically sealed moisture proof containers before storage. The seed stores are relatively safe, easy to maintain because they use minimum space. The handling of germplasm is easy. This is a cheap method of conservation of germplasm. For vegetatively propagated species, the materials may be maintained as full plants in field banks or botanical gardens. The cuttings or other vegetative parts may be stored for short-term period of time at moderately low temperature and humidity. The clonal species could be maintained through cryopreservation for long duration.

In-situ conservation: it consists of preserving the integrity of genetic resources by conserving them within the evolutionary dynamic ecosystems of their original habitat (i.e., "on-site") where environmental changes are constantly taking place, and is sometimes also called "in agro" or inter situ conservation. It requires establishment of natural park or biosphere reserves, national parks or special legislation to protect endangered areas or threatened species, which are also called *gene sanctuaries*. In India, the gene sanctuary for wild relatives of citrus is established in the Garo Hills of Assam. In Ethiopia, gene sanctuary for conservation of wild relatives of coffee was set up in 1984. *In-situ* conservation would help to preserve together the wild species under the complete natural or seminatural ecosystem, thereby allowing the evolutionary process to continue in their natural habitats. This is a long-term and dynamic conservation approach, which has great value; however, the natural reserves can cover only a very small portion of the total distribution area of a major species. The management of such areas is problematic and also a costly method.

The germplasm collections could be divided into three categories based on their duration of preservation and their use in crop improvement.

1. *Base collections:* it includes maximum number of samples (accessions) of a crop species, which are conserved for long-term (i.e., up to 50 years or more). These are also called principal or whole collections. The seed samples are dried to a low humidity (5% ± 1%) and stored at a subfreezing temperature of −18°C or −20°C in sealed containers. The plant cells, tissue or other vegetative materials could be stored at extremely low temperatures between −150°C to −190°C in liquid nitrogen. This is called *cryopreservation or freeze-preservation*. The base collections are maintained for long duration, therefore, are maintained only for the purpose of regeneration and are used only when germplasm from other sources is not available. A subset of base collection (a limited set of accessions derived from base collection) representing the spectrum of whole collection is called *core collection*. Core collection can be assembled based on combination of molecular marker profile, pedigree information, geographical origin, etc. Sometimes, the core collection is too large and for this reason, a minicore collection can be created for their use in single studies.

2. *Active collections*: active collections comprise the same materials as in base collection; however, these are available for distribution to plant breeders. These are conserved for medium term (i.e., 10–15 years or more). The samples are dried to a moisture content of above 5% and are stored at 0°C.
3. *Working collections*: such collections include elite germplasm that is adapted and frequently used by breeders in the on-going breeding programs. Seed samples are dried to a level of 8%–10% moisture content and are stored at 5°C–10°C for short-term (i.e., 3–5 years).

In any gene bank, base and active collections should be kept separately. Routine germination tests are carried out after every 5–10 years to ascertain the reduction in seed viability. For regeneration, sufficiently large samples may be taken (50–100) to prevent genetic drift and must be allowed to self or sib-pollinate. Some scientists, however, considers that smaller number would be adequate.

Seed materials differ in viability characteristics and based on viability, seed may be divided into two groups, that is, orthodox and recalcitrant seed.

1. *Orthodox seeds*: the viability of such seeds can be prolonged under conditions of low moisture and low temperature storage, for example, cereals, pulses, oilseeds (except soybean, which has poor viability under reduced moisture cold storage), etc.
2. *Recalcitrant seeds:* some seeds show a very drastic loss of viability with a decreased moisture content below 12%–30% (e.g., coconut, coffee, cocoa, rubber, and oil palm). Such species could be maintained for long duration using in vitro techniques.

It is difficult to maintain live samples of vegetatively propagated crops as vegetative propagules. Therefore, in vitro storage of germplasm using tissue culture (suspension cell, callus, and meristematic tissues) is useful for the regeneration of full plants from these systems. Meristem cultures are favored for in vitro storage. The meristem cultures would be advantageous because of several reasons.

- Exact genotypes can be conserved indefinitely free from pathogens and without loss of genetic integrity.
- Meristem cultures of vegetatively propagating cultigens, such as potatoes (*Solanum tuberosum*), yams (*Dioscorea* spp.), cassava (*Manihot utilissima*), sweet potatoes (*Ipomoea batatas*), and other crops would be useful because their seed production is very poor.
- A reasonably long regeneration cycle can be envisaged for meristem culture. Potato meristems can be stored at the temperature of −196°C in liquid nitrogen with successful regeneration of plants from these cultures.
- Regeneration of forest and fruit trees germplasm through seed requires very long time (10–20 years) to reach sexuality, but meristems can be subcultured in a few minutes.

Molecular conservation

With the advent of molecular marker technology and with lowering costs and ease of its utilization, PGR agencies are using them for:

- Data acquisition: determine genetic diversity of existing collections to plan exchange, collection and research strategies (e.g., core collection assembly, usage for genome-wide studies and other genetic studies for useful gene identification, and strategies for overcoming underrepresented genetic holdings).
- Maintenance and identity: use of marker data to identify duplicate or redundant accessions and ensure that the collection is meaningfully diverse.
- Characterization: molecular data can supplement passport and phenotypic data to obtain more accurate records of accessions.
- Community resource: marker characterization provides invaluable resource to the scientific community to use accessions for research applications and for the identification of previously unknown genes, for example, the USDA soybean germplasm collection (of more than 20,000 accession has been genotyped with the 50K SNP chip as a community resource).

Evaluation of germplasm

It consists of screening the germplasm for morphological, physiological, biochemical, plant pathological, and other characteristics. However, for better use by the plant breeders such evaluations are not enough. It is desirable that the accession are classified into various categories along with the significance of each of these, like resistance to biotic and abiotic stresses, earliness, dwarfness, productivity, and quality traits. The germplasm lines could be screened in the field for agronomic traits and biotic stresses, while physiological and quality traits can be assessed through laboratory techniques. It is desirable to use multilocation trials and collaborative approaches for screening of biotic and abiotic stresses, and to phenotype other important traits. Some of the accessions may not appear to be of immediate value to the plant breeders but could be needed in future, hence, should not be discarded. The evaluation is a continuous process as new genetic stocks are flowing into the gene banks.

Documentation of germplasm

It refers to compilation, classification, analysis, storage, and retrieval of data. The documentation is also called as information system. Passport data include source and origin data of the accessions as well as the best fitting taxonomic identification (according to the curator). More specifically passport information consists of (1) accession data, and (2) collection data.

Accession data consists of gene bank designation, accession number, donor name and country, donor identification number, other numbers associated with the accessions, scientific name (genus, species, subspecies, variety, and cultivar group), cultivar name/pedigree, acquisition date, date of last regeneration or multiplication, accession size, number of times accessions regenerated, originator, breeding method (relating to cultivars: unselected population, mass selection, individual selection, crossing, mutation, backcross, polyploidization, and other methods), and number of plants grown during last regeneration.

Collection data consists of collector's name, number, date of collection of original sample, country of origin, province/state where sample has been collected, location of collection site, latitude of collection site, longitude of collection site, altitude of collection site, collection source, status of sample, local/vernacular name, number of plants sampled, photograph, herbarium specimen, growth habit for wild material, if under cultivation: cropping, topography, pests and diseases of collection sample, and other notes from collector.

It is a prerequisite to transfer the evaluation data in a systematic manner to the computer so that it is used as and when required. This could be accomplished with the help of descriptors developed for each crop in the past. The documentation would provide information of various activities of PGR and would help explorers, evaluators, and curators as well plant breeders.

Distribution of germplasm

It is one of the important activities of gene banks/centers. The seed of specific germplasm lines in small quantity are supplied to researchers on request, generally free of cost. Proper records are maintained about the distribution of material. The user is required to send report of its merit to the distributor. Generally a MTA is needed between the provider and recipient.

Material Transfer Agreement

A seed or plant part MTA is an agreement that allows for transfer of seed or plant part without transfer of title. An agreement is a contract between a seed provider and seed recipient. The provider maintains ownership of the seed transferred at all times during the agreement dates and beyond. Transferred seed is received and used by the recipient according to the terms listed in the legally binding contract. In the agreement, the provider may impose conditions of audit or term-bound reports on the usage of material, and the recipient has to abide by these conditions. When a plant breeder makes a request for seed, s/he will work with their organizational designate in the Office of Intellectual Property and

Commercialization or the office that manages intellectual property (IP) and technology transfer. Similarly the plant breeder should notify the same office when s/he receives a request for seed/plant part and should never give out material without going through proper steps and approvals. MTAs are signed by the sending breeder and their organizational representative, as well as the recipient breeder and his/her organizational representative. Plant breeders or researchers utilizing the material (i.e., seed/plant part) are ultimately responsible for fulfilling the obligations of the MTA and must follow the regulatory terms. Plant breeders need to remember that an MTA is a legal and binding document. Broadly speaking, an MTA will have the following sections:

- *Introduction*: short text on the type of material or purpose.
- *Parties*: describes the sender and recipient and their organizational affiliations.
- *Definitions:* describes scientific terms such as material (seed, genotype, etc.).
- *Description of use of the materials*: conditions on what can and cannot be done using the material.
- *Confidential information*: lists any specific confidentiality clauses.
- *IP rights*: this is where licensing, royalties, inventions, and conditions are listed.
- *Warranties*: these are to protect the sender or provider and to stipulate that material does not come with any warranties.
- *Liability and/or indemnification:* recipient assumes all liability for damages that may arise from how/what recipient does to the material after transfer, and sender is not liable.
- *Publication:* provides a description on publication rights for recipient.
- *Governing law:* describes which jurisdiction laws will apply (state and country). In most cases where provider and recipient are from separate countries, the two parties may not define this section.
- *Termination:* date of termination of the agreement. It may also describe what the recipient has to do with leftover material from the agreement. Most likely the original material sent by the provider is expected to be destroyed.
- *Signatures:* the agreement is not considered to be executed until all necessary signatures are obtained, and material should not be sent until the MTA is officially signed. Signatories are the official organizational representatives and generally include the provider breeder and recipient breeder. Note, however, that the provider and/or recipient may or may not be a breeder but may be a researcher in some other discipline such as genetics.
- *Exhibits or appendices:* list of material or data accompanying the material.

It is important that breeders realize that if an MTA was signed to send or receive seed, it is a legal document and breeders are legally bound to follow the conditions. While plant variety protection allows breeder's exemption, MTAs are a useful legal tool for establishing all rights on the use of the seed provided by originator to recipient. MTAs are also useful for advanced breeding lines, lines under patent protection, germplasm releases, etc.

The International Treaty on Plant Genetic Resources for Food and Agriculture

The Standard Material Transfer Agreement is a mandatory model for parties wishing to provide and receive material under the Multilateral System of the ITPGRFA. It is the result of lengthy negotiations among the contracting parties to the treaty and may not be varied or abbreviated in any way. However, as a template, it contains some paragraphs and sections that need to be completed for each use. The MTAs that use the standard template are private agreements between the particular providers and recipients, but the Governing Body, through FAO as the Third Party Beneficiary, is recognized as having an interest in the agreements. The standard template has been developed to ensure that the provisions of the Treaty regarding the transfer of Plant Genetic Resources for Food and Agriculture under the Multilateral System are enforceable on users.

Text for Farmer's Rights from The International Treaty on Plant Genetic Resources for Food and Agriculture

Farmers' Rights: in Article 9, the TPGRFA recognizes "the enormous contribution that the local and indigenous communities and farmers of all regions of the world, particularly those in the centers of origin and crop diversity, have made and will continue to make for the conservation and development of plant genetic resources which constitute the basis of food and agriculture production throughout the world." It gives governments the responsibility for implementing Farmers' Rights and lists measures that could be taken to protect and promote these rights:

- The "protection of traditional knowledge relevant to plant genetic resources for food and agriculture;"
- "The right to equitably participate in sharing benefits arising from the utilization of plant genetic resources for food and agriculture; and"
- "The right to participate in making decisions, at the national level, on matters related to the conservation and sustainable use of plant genetic resources for food and agriculture."

The ITPGRFA also recognizes the importance of supporting the efforts of farmers and local and indigenous communities in the conservation and sustainable use of PGRs for food and agriculture and includes a funding strategy. In this strategy, "priority will be given to the implementation of agreed plans and programs for farmers in developing countries, especially in least developed countries, and in countries with economies in transition, who conserve and sustainably utilize plant genetic resources for food and agriculture."

Utilization of germplasm

Uses of germplasm: methods to use germplasm in breeding programs include direct release as cultivars (less likely in species with breeding efforts, more likely in orphan crops where some selection may be done on a plant introduction or landrace prior to release as a cultivar). The second and more appropriate use of germplasm is for introgression of single gene traits from the wild species or unadapted germplasm into the elite cultivars. As previously explained, germplasm accessions particularly breeding lines from other institutions and/or ex-PVP lines are very useful for certain traits and genetic background especially, if a breeder is starting a new breeding program. One disadvantage of the use of ex-PVP lines is that genetic gain starting baseline will lower as these lines are more than two decades old. Both cultivated and wild type of PGRs have been used in different crops. The cultivated types have been used either as such, directly to develop a variety through selection; or as a donor of important traits in cross breeding and as a variant for genetic studies. The wild and/or weedy types have often been to transfer gene(s) for tolerance to biotic and abiotic stresses and sometimes to develop new sources of cytoplasmic–gene male sterility. Careful use of germplasm materials can also be very beneficial to bring in "useful" genetic diversity using traditional or molecular marker technologies.

As the related, wild and/or weedy species are reservoirs of genes for biotic and abiotic stresses, these would continue to be of great use in future as new virulent biotypes/races emerge and new environmental challenges are confronted. The wild types will be useful in the improvement of agronomic, physiological, disease, pest and quality traits in cereals, grain legumes, oilseeds, forages, fiber plants, *etc*.

CHAPTER

Wide hybridization

6

Abstract

Wide hybridization or distant hybridization is crossing between two different species or genera, and has been used successfully to move genes and to create new crop species. It has been used effectively in row crops and vegetables. In ornamental crops it has been used due to its ability to create novelty. The barrier to wide hybridization includes external (e.g., spatial isolation, ecological isolation, mechanical isolation, and cross incompatibility) and internal (hybrid inviability, hybrid sterility, and hybrid breakdown). In this chapter, some of the major examples of wide hybridization in multiple crops, such as wheat, rice, cotton, sugarcane, tomato, pigeon pea, chickpea, mung bean, and urd bean, are presented. For practical applications, advantages and limitation of wide hybridization are included.

The crosses involving two different species, or two different genera are called wide hybridization or distant hybridization. Wide hybridization was attempted for the first time by Thomas Fairchild in 1717. He successfully produced interspecific hybrid between the Carnation (*Dianthus caryophyllus*) and Sweet William (*Dianthus barbatus*). Wide hybridization has also been used for synthesizing crop species. Triticale (*X Triticosecale* Wittmack) was produced by Rimpau in 1891 in Sweden from a cross of wheat and rye (*Secale cereale*). This cross was produced initially in 1876 but the progeny was sterile. However, he produced fertile amphidiploid ($2n = 56$) through colchicine treatment and named the plant as Triticale.

The production of octaploid triticale (AABBDDRR, $2n = 56$) from a cross of hexaploid wheat (AABBDD, $2n = 42$) × rye (RR, $2n = 14$), and of hexaploid triticale (AABBRR, $2n = 42$) from a cross of durum wheat (AABB, $2n = 28$) × rye (RR, $2n = 14$) has provided encouragement to plant breeders to attempt other distant crosses. A Russian scientist Karpechenko in 1928 produced an intergeneric amphidiploid, *Raphanobrassica* from a cross of radish (*Raphanus sativus*) and cabbage (*Brassica oleracea*). The aim in producing *Raphanobrassica* was to synthesize a crop species that could combine the root of radish and leaves of cabbage. The allotetraploid did combine these characteristics, but in the opposite direction. It had leaves like radish and roots like cabbage, which was not a desirable combination; however, this work leads to other attempts resulting in hybrid "Raparadish" (*Raphanus sativus* × *Brassica rapa*) and is used as fodder.

Plant Breeding and Cultivar Development. DOI: https://doi.org/10.1016/B978-0-12-817563-7.00028-3

The development of Triticale and *Raphanobrassica* are examples of intergeneric hybridization and synthesis of new crop species. Triticale is grown commercially in Europe and other countries globally. Improved triticale varieties have been released for commercial cultivation by breeding institutions including International Maize and Wheat Improvement Center (CIMMYT), and Agriculture and Agri-Food Canada (AAFC).

Wide crosses have played a major role in the origin of modern types in some plant groups, although its role in other groups has been trivial. Intergeneric hybrids were frequently used in ornamentals especially among roses, orchids, and lilies (see Emsweller et al., 1937). In orchid, hybrids between two genera are always joining and shortening of the two generic names, for example, Brassocattleya (*Brassavola* × *Cattleya*), Odontonia (*Odontoglossum* × *Miltonia*), and Ascocenda (*Ascocentrum* × *Vanda*). Usually, hybrids with three genera in the background are written similarly. For example, Brassolaeliocattleya (*Brassavola* × *Laelia* × *Cattleya*), and Vascostylis (*Vanda* × *Ascocentrum* × *Rhynchostylis*). The success of intergeneric hybrids in ornamentals could be due to vegetative propagation in contrast to seed propagation in field and vegetable crops. In the asexually propagated species, once superior hybrid is obtained, it can be perpetuated indefinitely, which is usually not possible in field or vegetable crops. In addition, the demand for quality of attributes (novelty) itself is often desirable in ornamentals. This is likely to be achieved by interspecific or intergeneric hybridization. Generally speaking, in field and vegetable crops, wide hybridization has been used to transfer specific gene(s), for example, genes for resistance to diseases from one species to another by backcross breeding. The wild relatives/ancestors possess genes for resistance to diseases, insects, and nematodes in almost every crop. These species have acquired a wealth of resistance genes to biotic agents over long periods of host–pathogen interactions. Often genes from wild species/genera have resistance against a wide range of races, and these genes are sometimes termed "super genes." For example, the 1AL.1RS translocation segment from "Insave" rye cultivar conditions resistance to greenbug (genes *Gb2* and *Gb6*), wheat curl mite (*Cmc3*), rust ($Sr1R^{Amigo}$) and powdery mildew (*Pm17*), while 1BL.1RS from "Imperial" and "Petkus" rye cultivars provide resistance to leaf rust (*Lr26*), stem rust (*Sr31*), stripe rust (*Yr9*), and powdery mildew (*Pm8*) in wheat.

Barriers to wide hybridization

Several problems are associated with the process of making crosses between distantly related species/genera. The problems associated with barriers to wide hybridization could be external (prior to hybridization or premating) or internal (after hybridization or post-mating), which are described along with their remedial measures. The knowledge and understanding to the barriers of wide crosses are essential to increase the chances of success.

External factors

Such factors are due to the environment in which species or genera are grown. These factors cause prefertilization barriers in making crosses.

1. *Spatial isolation*: isolation associated with geographic distances is the physical separation of time, distance, and environment that may result into sensitivity to photoperiod (this can be manipulated by modifying the length of day or night).
2. *Ecological isolation*: cereals have spring and winter types belonging to the same species and may be grown side by side are ecologically isolated. In northern latitude of the great western plains of Canada and the United States, spring wheat is seeded in April/May and harvested in September/October, while winter wheat is planted in August/September and harvested next year in July/August. In semi-tropical environments, spring wheat is grown in their winter season (planted in November/December and harvested in April/May). For winter habit cereals, vernalization is required to initiate flowering. At CIMMYT, spring and winter flowering are achieved using elevation to provide the cold requirement for vernalization by November sowing of winter cereals and a May planting of spring cereals at the same site at the Toluca (elevation of 2640 m). Seeds after hybridization may be grown at specific sites to screen for winter or spring habits and for other agronomic desirable features.
3. *Mechanical isolation*: such isolation was observed in species whose flowers get adapted to insect pollination that too by only specific insect species. In this situation, two compatible species, which are pollinated exclusively by insect may not hybridize in nature leading to conclusion that two species are sexually incompatible. For example, dicot plant Sage tree-shrub in California, southwestern United States, and Northwestern Mexico belonging to genus *Salvia* has a mechanical isolation between two genera. The two-lipped flowers of *Salvia mellifera* have stamens and style in the upper lip and is pollinated by small- or medium-sized bees that carry pollen on their backs from flower to flower, whereas *S. apiana*, with long stamens and style, is pollinated by pollinators (large carpenter bees and bumble bees), which carry pollen from plant to plant. These pollinators of one species cannot pollinate the other *Salvia* species, even if they visit flowers of the other species. This happens due to a lack of contact of pollen with the style of the other species.
4. *Cross incompatibility*: the inability of functional males gametes of a species/ genera to effect fertilization with the female gametes of another species/genera is called as cross incompatibility, which leads to the failure of zygote formation. It could be due to lack of pollen germination, or insufficient growth of pollen tube to search ovule or inability of male gamete to unite with egg cell. Pollen tube fails to reach the embryo sac as in some species crosses of *Datura*. Style of the female species is longer than pollen tube would normally grow as in crosses of maize and *Tripsacum*. Sexual reproduction can often be successful by making reciprocal crosses using species with short style as female or by

shortening or splicing longer styles. Screening for compatible parents, use of immunosuppressants, mechanical manipulation of styles, and bud pollination would help in overcoming these problems.

Internal factors

Such barriers are genetic in nature and they cause post-fertilization or post-mating problems in making crosses.

1. *Hybrid inviability*: in certain wide crosses, fertilization occurs, and zygote formation takes place; however, hybrid zygote does not grow into a normal embryo under usual condition and is referred to as hybrid inviability. This could be due to: (1) unfavorable interaction between the parental genomes; (2) unfavorable interaction between cytoplasm of one species and nuclear genes (genome) of another species; and (3) unfavorable interaction between embryo, endosperm, and maternal tissues. Proper choice of parents, making reciprocal crosses and application of growth hormones would increase chances for the development of zygote into viable seed. In such situation, embryo can be dissected and transferred to the culture medium. The plants can be regenerated from the embryos in culture medium. The protocols for embryo culture are developed for various plant species.
2. *Hybrid sterility*: hybrid sterility is the inability of a hybrid to produce viable offspring. The problem is more serious in intergeneric than in interspecific crosses. In intergeneric crosses, hybrid sterility may vary from complete sterility to complete fertility. The hybrid sterility is primarily due to a lack of structural homology between the chromosomes of two species. The meiotic abnormalities result in no pairing or reduced pairing of chromosomes. Sometimes sterility is due to small structural changes and is not detectable during meiosis which is called cryptic structural hybridity coined by Stebbins (1945). This type of sterility could be overcome by doubling of the chromosomes of the F_1 hybrid through colchicine. The chromosome doubling would lead to normal chromosome pairing and production of viable gametes.
3. *Hybrid breakdown*: it is a major problem in interspecific crosses. In some wide crosses involving different species, F_1 plants show normal development and reasonable fertility but their F_2 progenies die in seed or early in development, or they develop into weak plants. This is referred as hybrid breakdown. This could be due to the segregation and recombination of favorable combination of dominant and recessive genes. In the interspecific cross of closely related *Gossypium hirsutum* and *G. barbadense*, the F_1 plants are vigorous with normal fertility; however, the plants in the F_2 generation are weak, sterile, or agronomically unsatisfactory. The backcrossing of F_1 with the parents, followed by selection could be useful in this situation.

Although there are obstacles and barriers in the production of wide crosses, they have utility; therefore, interspecific and intergeneric crosses have been produced and used not only in ornamental plants or fruit trees and various berries, but have also been practiced in an array of field crops used for food, feed, fiber, and fodder. The wide cross has been made with the following considerations.

- To increase genetic variation in the crops where limited genetic variation is available.
- To enhance and exploit heterosis.
- To understand the phylogenetic relationship among species.
- To produce new character expression and new genotypes beyond the range of either parent for quantitative traits.
- To improve cultivars by transferring specific genes for traits such as disease, pest and stress resistance, and also for some quality traits.

Some specific cases of practical significance in plant breeding are presented in this chapter.

Triticum spp. (wheat)

The genus *Triticum* includes a polyploidy series of species and hybrids ranging from diploids ($2n = 14$) to hexaploid ($2n = 42$). *Triticum aestivum* and *T. durum* are widely cultivated species. Another species, *T. dicoccum* is sparsely cultivated in certain parts of India. There is enough evidence that the bread wheat has evolved in nature from three diploid species, *T. monococcum* (AA), *Aegilops speltoides* (BB), and *Aegilops squarrosa* (DD) through intergeneric hybridization and genome building. The natural crossing between *T. monococcum* ($2n = 14$) with *A. speltoides* ($2n = 14$) by chromosome doubling has resulted in evolution of tetraploid emmer wheat (AABB, $2n = 28$). The hybridization in nature between the tetraploid wheat with diploid *A. squarrosa* ($2n = 14$) following by doubling of chromosome has resulted in the evolution of hexaploid wheat.

Different methods have been used for transferring genetic materials to bread wheat from other cultivated species or from wild relatives, which are both diploid and tetraploid. Where alien/donor species are closely related to wheat and carries one or more of the A, B, or D genomes, it is possible to directly cross these species to wheat using special techniques of embryo culture to transfer genes into wheat. Some of the diploids could be crossed with hexaploid or first crossed with tetraploid and then with the hexaploid. However, tetraploids can also be crossed directly with the hexaploid wheat. Homologous chromosome pairing and crossing over would occur in the hybrids. The selection among resulting segregants would provide an opportunity to isolate desirable segregants. Success was achieved by hybridization of durum wheat (*T. durum* = *T. turgidum*) a tetraploid species with AB genomes to common wheat varieties which is hexaploid with ABD genomes. For example, high protein content (*Gpc-B1* gene) and disease resistance genes from *T. dicoccoides* to *T. durum*

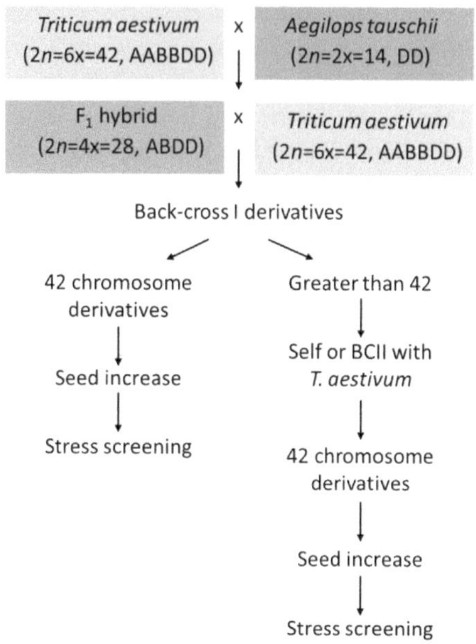

FIGURE 6.1

Transfer of genes from *Aegilops tauschii* to *Triticum aestivum*.
From Mujeeb-Kazi A, S Rajaram. 2002.

and *T. durum* to *T. aestivum* have been transferred. Synthetic wheat (hexaploids) are artificially created by crossing tetraploid (durum wheat, AABB genome) and wild goatgrass (*Aegilops tauschii*, DD genome). Synthetic wheat [Synthetic hexaploid wheat (AABBD'D') is developed by creating a fertile hybrid between tetraploid durum wheat (*T. turgidum*, AABB) and diploid wild goat grass (*Aegilops tauschii*, D'D')]. It provides broader genetic base to hexaploid wheat, and is a useful strategy to introduce agronomically useful traits into hexaploid common wheat from wild genetic resources. Merezhko (1998) analyzed the impact of plant genetic resources on wheat breeding and inferred that the most significant results have been achieved through hybridization with *T. durum* and *T. dicoccum*. Only a few bread wheat cultivars have been developed with the participation of *T. turgidum, T. persicum, T. timopheevi, T. spelta, T. compactum*, and *T. sphaerococcum*. The richest gene pool of species belongs to *Aegilops* species. The transfer of valuable genes from the *D* genome of *Aegilops* is now much easier (Fig. 6.1). An amphidiploid with AD genome is synthesized which is cytogenetically stable and can be crossed readily with *A. tauschii* and bread wheat, and serves as intermediate bridge for introgressive hybridization.

Embryo rescue may be needed in development of synthetic wheat, and regenerated plants are triploid, necessitating the use of colchicine to double the chromosomes. In cross involving differing ploidy levels, geneticists recommend the

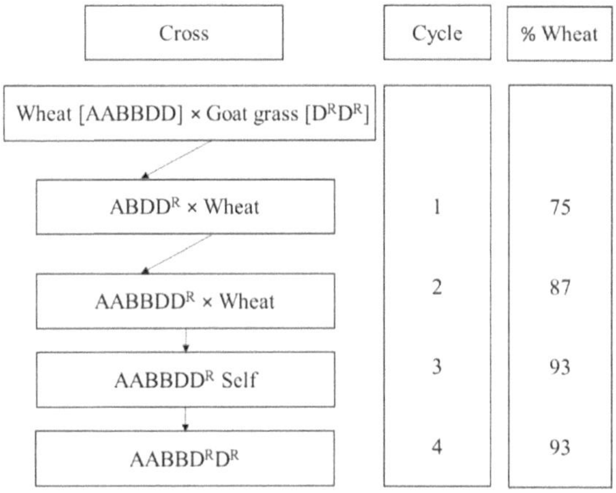

FIGURE 6.2

Gene transfer from *A. squarrosa* to hexaploid wheat. In this figure, D^R = goatgrass genome with a resistance gene transferred to wheat.

Adapted from Gill, B.S., Raupp, W.J., 1987. Direct genetic transfer from Aegilops squarrosa *L. to hexaploid wheat. Crop Sci. 27, 445–450.*

use of higher ploidy level as a female parent. Gill and Raupp (1987) suggested that the use of goatgrass (*A. tauschii or A. squarrosa*) may be suitable for direct gene transfer. They observed complete homology with little genetic interaction between D genome of wheat and D genome of goatgrass. However, interspecific hybridization between the two species is difficult and only few tetraploid hybrids were reported. Researchers used *A. squarrosa* as female and noticed difficulty in successful crossing, particularly F_1 abortion, hybrid lethality, and also high male and female sterility of F_1 hybrids. However, if wheat is taken as female parent as suggested in Fig. 6.2, better results would be achieved. Genotypes, polyploidy level, crossing procedure, and cross direction are important to generate viable hybrids in *Triticum*, and previous research has shown that generally higher ploidy species should be used as a female parent.

Some of the donor species are not closely related to bread wheat and do not have a common genome with wheat. The pairing and recombination would not occur in such crosses. The use of bridging cross and production of amphiploid would be helpful. While newer examples are available, examples presented below are historically important to demonstrate efforts prior to the use of modern tools.

Sears (1956) used this technique for the first time successfully to incorporate leaf rust resistance from *Aegilops umbellulata* to *T. aestivum. A. umbellulata* is a diploid ($2n = 14$) wild grass from Mediterranean region having UU genome and has no genome common with *T. aestivum*. The cross between *T. aestivum* and *A. umbellulata* does not produce viable seed. Therefore, he made a bridging cross between *A. umbellulata* and *T. dicoccoides* ($2n = 28$) and produced F_1 amphidiploid, which was backcrossed twice with *T. aestivum* and selection for leaf resistance was made. He successfully developed addition line carrying the rust resistance gene on an additional *A. umbellulata* chromosome. X-ray irradiation of this addition line resulted into translocation of a small segment of *A. umbellulata* carrying resistance inserted to 6B chromosome of wheat. The resistance line was named "Transfer." The gene was transferred from Transfer to Thatcher wheat through backcrossing. Knott (1961) used the procedure of Sears to transfer stem rust resistance from *Agropyron elongatum* (EE, $2n = 70$) to *T. aestivum*. One of the translocations was transmitted normally through the gametes and was used as source of stem rust resistance in Australian cultivar Eagle and Kite.

A similar technique was used by Driscoll and Jensen (1964). They successfully transferred a leaf rust resistance and a powdery mildew resistance gene from rye (*Secale cereale*) to bread wheat (*T. aestivum*). They irradiated a leaf rust resistant wheat line carrying a rye chromosome followed by screening the resultant progenies to select a line carrying an induced translocation of the desired type called "Transec." This line had two resistance genes from rye inserted into wheat chromosome 4A.

Oryza spp. (rice)

The cultivated rice (*O. sativa*, $2n = 24$ AA) is crossable with its closely related wild species with AA genome. The genes of interest have been transferred through conventional crossing and backcrossing procedures. The cross between *O. sativa* and *O. rufipogon* ($2n = 24$ AA) produces partially fertile F_1. The crosses of *O. sativa* × *O. glaberrima* and *O. sativa* × *O. longistaminata* were highly sterile in F_1. Introgression lines have been produced at International Rice Research Institute (IRRI) through embryo rescue following hybridization between elite breeding lines of rice and several distantly related species of *Oryza*. An example of transfer of a useful gene from wild species is the introgression of a gene for grassy stunt virus by G.S. Khush in 1977 at IRRI, Philippines. The grassy stunt virus is transmitted by brown plant hopper and the disease can cause heavy losses in yield in epidemic areas. Crosses were made between improved varieties IR 8, IR 20, IR 24, and *O. nivara* (Accession 101508). Following three backcrosses with improved varieties, the gene for resistance to grassy stunt virus was transferred. IR 28, IR 29, and IR 30, all resistant varieties to this disease were released in 1974. Subsequently many other varieties, for example, IR 34 and IR 36 at IRRI and other varieties by national programs were released.

Lin and Yuan (1980) in China were the first to transfer a cytoplasmic male sterile (CMS) source from wild rice, *O. sativa* f. *spontanea* to develop CMS lines for commercial hybrid rice production. The cytoplasmic source was designated wild abortive (WA), which refers to a male sterile wild rice plant having abortive pollen. The WA type of cytoplasm has been used extensively in hybrid rice development in China and in other countries. Other sources of CMS lines with the cytoplasm of *O. nivara* (RPMS1) and *O. rufipogon* (RPMS2) were also later developed. These lines showed a gamephytic male sterility with a restorer reaction different from the WA source of cytoplasm. In another example, *Xa21* gene for bacterial leaf blight resistance was transferred to cultivated rice from *O. longistaminata*. Genes for rice tungro disease tolerance and tolerance to moderate level of acidity have been transferred from *O. rufipogon* into rice cultivar IR 64.

Gossypium spp. (cotton)

There are four cultivated species of genus *Gossypium*. *G. hirsutum*, the American upland cotton, and *G. barbadense*, the Egyptian cotton, are also referred to as the new world species, and are tetraploid. On the other hand, *G. arboreum* and *G. herbaceum* are referred to as old world species or Asian species and are diploid. The crosses between the cultivated tetraploid ($2n = 52$) species as well as between the cultivated diploid ($2n = 26$) species are fully fertile. There are several old world diploid species and new world diploid species. All the tetraploid species with AD genomes have one genome of 13 chromosomes similar to cultivated old world diploids, and another set of 13 chromosomes similar to new world diploid species. In India, interspecific hybridization has been used for the production of improved cotton varieties. Interspecific crosses between *G. hirsutum* and *G. barbadense* were made in Tamil Naidu, and MCU-2 variety was developed with longer staple length. Crosses between tetraploid and diploid species were used to develop superior varieties. In Gujarat, 170-Co.2 was developed from a cross of *G. hirsutum* × *G. arboreum* and 134-Co.2M from a cross of *G. hirsutum* × *G. herbaceum*. Hybrids (e.g., DDH-2, Pha-46, DH-7, and DH-9) have also been developed from crosses involving Asian diploid species; *G. arboreum* and *G. herbaceum*. The incorporation of genes from wild species to cultivated cotton may not be always preferred because such crosses may affect the fiber strength, texture, and other positive combination of traits. However, if necessary, wild species of cotton may be crossed with cultivated diploid species followed by colchicine treatment of F_1 to double the chromosomes. The amphidiploid of such a cross can be crossed with cultivated tetraploid species and backcrossed to recover desirable traits. Such a procedure in the United States from a tri-species hybrid [(*G. arboreum* × *G. thurberi*) × *G. hirsutum*] has exceptionally strong fiber. The F_1 from cross between Asian cultivated diploid cotton (*G. arboreum*, $2n = 26$) and wild American diploid ($2n = 26$) species subjected to colchicine treatment resulted in the production of amphidiploid ($2n = 52$), which was similar to upland cotton *G. hirsutum*.

Saccharum spp. (sugarcane)

Interspecific hybridization has played an important role in the development of present day varieties of sugarcane. It has been a general practice to cross noble cane, *S. officinarum* ($2n = 80$) with other species to combine high yield of sugar of the *S. officinarum* clones with hardiness and disease resistance of the other species. This process is called "nobilization". Two to three backcrosses may be made with *S. officinarum* to recover satisfactory sugar content and other desirable qualities of the noble cane parent. Many of modern varieties of sugarcane have been developed from crosses of *S. officinarum* ($2n = 80$) $\times$ wild species *S. spontaneum* ($2n = 40$ to 148) and/or *S. barberi* ($2n = 82$ to 142). These varieties have inherited high sugar content of *S. officinarum*, disease resistance of *S. spontaneum*, and cold tolerance and vigor of *S. barberi*.

Sugarcane is vegetatively propagated; therefore, fertility in the F_1 is not important. There is little gene exchange in the interspecific hybrids, rather there is an increase in chromosome number as a result of such crosses. Chromosome number of sugarcane varieties may vary from $2n = 100$ to 125. The wide cross of *S. officinarum* ($2n = 80$) as female and *S. spotaneum* ($2n = 40$ to 148) or *S. barberi* ($2n = 82$ to 124) or *S. sinense* ($2n = 116$ to 118) or *S. robustum* ($2n = 60$ to 148) as male would produce F_1 with chromosome number $2n + n$. Due to abnormality in the process of fertilization and embryo formation, the somatic number ($2n$) of chromosomes are transmitted rather the gametic number (n). A cross of *S. officinarum* $\times$ *S. spontaneum* would produce F_1 with $2n = 136$ chromosome ($40 + 40 + 56$). If this hybrid ($n = 68$) is backcrossed with *S. officinarum* as the maternal parent, the BC_1 plant will have $40 + 40 + 68$ (or $2n = 148$) chromosomes. Additional backcrosses do not result in further increase in chromosome number. However, this phenomenon does not occur in reciprocal crosses, when *S. officinarum* is used as the pollen parent and *S. barberi* or *S. sinense* or *S. robustum* are used as female parent. The F_1 hybrid has $n + n$ chromosome contributed from both parents.

Lycopersicon spp. (tomato)

The genus *Lycopersicon* (now Genus *Solanum* section *Lycopersicum*) is divided into two subgenera on the basis of fruit color, position of stigma, etc. Fruits of plants of *Eulycopersicon* are usually red or yellow in color when ripe and this subgenus includes the cultivated tomato; while fruits of *Eriopersicon* remain green or purple green throughout development. The red fruited subgenus *Eulycopersicon* has two species: *L. esculentum* (=*Solanum lycopersicum* L.) and *L. pimpinellifolium* (*S. pimpinellifolium* L.). The green fruited subgenus *Eriopersicon* included *L. hirsutum* (*Solanum habrochaites*), *L. peruvianum* (=*Solanum peruvianum*), and *L. chilense* (=*Solanum chilense*) and also few other species of less importance. No barrier of any consequence exists to the hybridization between *S. lycopersicum* and *S. pimpinellifolium* to gene recombination in the hybrid. *S. pimpinellifolium* has contributed more genes for disease resistance than other tomato species, with numerous varieties having this species genome in its parentage.

Most of the species of subgenus *Eriopersicon* are self-incompatible. The *L. hirsutum* (=*S. habrochaites*) is a source of resistance to leaf mold, *Septoria* blight, bacterial canker, tomato foot and stem rot, and tobacco mosaic virus. Crossing some of the *Eriopersicon* species (e.g., *L. chilense*) with the cultivated tomato, *L. esculentum,* is unsuccessful as the stigma of *L. chilense* does not accept pollen from *L. esculentum* leading to flower abortion. However, if *L. chilense* is used as a male parent (pollen source), fruit formation can happen albeit with few viable seeds, but with embryo rescue plants can be obtained. The F_1 hybrids are successfully mated to *L. esculentum* only as male and to the wild parent only as the female. Similar patterns exist for *L. esculentum* and *L. pimpinellifolium* with wild relatives, where fertile plants can be obtained by embryo rescue when these species are crossed with *Eriopersicon* parent as a pollen source. The wild species have been reported to possess genes for resistance to bacterial wilt, bacterial canker, *Fusarium* wilt, gray leaf spot, leaf mold, leaf spot, *Verticillium* wilt, curly top, mosaic, and more. The resistance to these diseases has been transferred to commercial cultivars. A dominant gene providing resistance to *Septoria* blight and a gene that causes sharp change in the content of carotenoid pigment have been transferred from *S. habrochaites* to *S. esculentum. S. peruvianum* is resistant to many diseases of tomato including curly top virus, spotted wilt caused by virus, and resistance to root knot nematode, as well as a source of genes for high vitamin C content. For making crosses between *S. esculentum* and *S. peruvianum* the former is used as female and embryo culture is resorted to establish seedlings. Genes conferring resistance to *Septoria* blight and root knot nematode have been transferred to *L. esculentum* from *L. peruvianum*. The wild species have been used as parent more frequently in genus *Lycopersicon* compared with other crops.

Note: Linnaeus in 1753 placed tomato in the genus Solanum as *Solanum lycopersicum* L. Soon afterwards, Philip Miller (1768) placed it in its own genus naming it *Lycopersicon esculentum*. While the later nomenclature has been widely used until early 2000s, recently cytological and molecular data have prompted several scientists to propose reverting tomato's scientific name to *Solanum lycopersicum* L. Currently, both names are reported in the literature but gravitating toward the Linnaeus nomenclature.

Cajanus spp. (pigeon pea)

The genus *Cajanus* includes some 32 species, and of these 17 are reported from the Indian subcontinent, 13 are reported from Australia, and 1 from New Guinea.

C. scarabaeoides is common to all areas. The significance of wild species in an on-going pigeon pea breeding program is highlighted by Singh (2014). The wild relatives possess genes for resistance to diseases, insect-pests, high protein content, and source of cytoplasmic genic male sterility systems.

Cajanus scarabaeoides ($2n = 22$) is considered as progenitor of cultivated pigeon pea, *C. cajan* ($2n = 22$). *C. scarabaeoides* is reportedly resistant to all the major diseases of pigeon pea, and resistant to pod borers. An interspecific cross between *C. cajan* (cv. Pant A-3) as female and *C. scarabaeoides* (accession JM 4147) as male was made by Verulkar et al. (1997). Pant A-3 is determinate type and JM 4147 is prostrate type. The F_1 hybrid plants were intermediate in growth habit and were fertile. In the F_2 generation an array of plant types was observed. The inheritance of resistance to pod-borer (*Helicoverpa armigera*) and pod-fly (*Melanagromyza obtusa*) was studied in this cross. Individual plants of parents, F_1 and F_2 were screened for pod-borer using excised leaf squares technique; while, for pod-fly screening, the damage to pods of individual plants were examined. The antibiosis to pod-borer was governed by a single dominant gene. The resistance to pod-fly in this accession of *C. scarabaeoides* was governed by two recessive genes. Elite lines from this cross were selected that possessed resistance to pod borers and gave good yield. The crosses of *C. cajan* with *C. lineata, C. albicans, C. cajanifolius*, and *C. sericea* were made successfully at ICRISAT, India. *C. platycarpus* is incompatible with *C. cajan* as pollen tube fails to reach ovary and only empty pods are formed, possibly parthenogenically. In a reciprocal cross using *C. platycarpus* as the female, fertilization was observed; however, embryos aborted after 20 days. The embryo rescue was used to produce hybrid plants, which were morphologically intermediate but pollen sterile. The wild progenitor of pigeon pea, *C. scarabaeoides*, has great potential in improving pigeon pea. The species also has higher number of clusters and pods per plants, which could also be useful for increasing yield per plant. More research is needed on other wild species for the exploitation in pigeon pea improvement. At ICRISAT, seven cytoplasmic male sterility (CMS) system have been developed by integrating the cytoplasm of wild species with genome of cultivars of cultivated pigeon pea through interspecific hybridization followed by selection and backcrossing.

Cicer spp. (chickpea)

Chickpea, *Cicer arietinum* L. ssp. *arietinum* (domesticated chickpea), is one of the first grain legumes to be domesticated. It contains 9 annual (including cultivated chickpea) and 34 perennial species (Ladizinsky and Abbo, 2015). Three annual wild species: *C. arietinum* L. ssp. *reticulatum* (progenitor of cultivated chickpea), *C. echinospermum*, and *C. bijugum* are closely related to chickpea. Both *C. reticulatum* and *C. echinospermum* have been used as sources to increase yield and to improve stress tolerance of *C. arietinum*. The importance of wild species in chickpea improvement has been highlighted by Singh (2014) and is presented in Table 6.1.

Table 6.1 The usefulness of wild species in chickpea breeding.

Wild species	Desirable traits based on published literature
C. reticulatum (A)	Cold tolerance and AB resistance
C. echinospermum (A)	Cold tolerance
C. judaicum (A)	Fusarium wilt resistance
C. pinnatifidum (A)	BGM and AB resistance
C. yamashitae (A)	AB resistance
C. cuneatum (A)	AB resistance
C. montbretti (P)	AB resistance
C. anatolicum (P)	AB resistance
C. bijugam (A)	BGM resistance

A, *Annual species;* AB, *ascochyta blight;* BGM, *botrytis gray mold;* P, *perennial species.*

Of these species, *C. reticulatum* and *C. echinospermum* are crossable routinely, although there may be partial fertility with the later; whereas, *C. pinnatifidum* is crossable using growth regulators, and ovules were collected to dissect embryos to obtain hybrid plants. The cross of *C. judaicum* × *C. pinnatifidum*, *C. judaicum* × *C. bijugam*, and *C. pinnatifidum* × *C. bijugam* are successful but F_1 hybrids were partially sterile. The cross of *C. judaicum* × *C. cuneatum* produced seeds but F_1 hybrids were completely sterile.

C. reticulatum, the wild progenitor of chickpea is of special significance as it grows vigorously under field experimentation and possesses acceptable plant traits. It is cross compatible with chickpea. Earlier crosses of desi chickpea were attempted with *C. reticulatum*. In Table 6.2, an example of successful cross between these species is described that led to the development of higher yielding line PG-063 (Fig. 6.3A and B).

At Pantnagar University (India), an elite line of kabuli chickpea, PG 92–97, which is an erect type with cream colored seeds of medium size, and tolerant of BGM was crossed with the accession (JM 2100) of wild progenitor that is prostrate type with tolerance to abiotic stresses; however, is prone to shattering and have undesirable seed characteristics (Fig. 6.3A). Five F_1 hybrid plants were obtained which were fully fertile with intermediate growth habit to parents and prone to shattering like wild species. The plant to progenies were grown F_2 through F_5 and individual plants were selected (Table 6.2). During selection emphasis was laid on plants with erect/semierect plant growth habit with higher pods/plants, acceptable characteristics. All plants with pod shattering characteristic were discarded. Enormous variability for growth habit, seed traits, and yield components was observed in the segregating generations. Seven F_6 promising progenies (PG-057 to PG-063) were selected and seed were multiplied, while replicated trials were conducted with inclusion of checks and elite entries from other crosses in test. PG-063 was also tested in All India multi-location replicated trials. PG-063 was registered in 2009 with NBPGR, India, (registration No. of PG-063 is

Table 6.2 Stepwise description of a wide cross of *C. arietinum* × *C. reticulatum* to develop elite line PG-063 (Singh DP, unpublished data).

Year	Generation	Number of plants/progenies grown/ selected/evaluated
1997–98	(PG92–97 × JM 2100)	Cross attempted.
1998–99	F_1	Five plants were grown successfully in field. Plants were fully fertile.
1999–2000	F_2	All the seed of 5 plants were grown. In all 320 plants were grown and 36 desirable plants were selected using visual selection.
2000–01	F_3	36 plants to progenies were grown and from these progenies 44 desirable plants were selected using visual selection.
2001–02	F_4	44 plants to progenies were grown and from these only 14 desirable plants were selected using visual selection.
2002–03	F_5	14 plants to progenies were grown and from these 10 superior progenies based on visual selection were selected
2003–04	F_6	10 selected progenies were grown. Of these, finally 7 progenies (PG-057 – PG-063) were selected. The seed of each progeny was harvested separately.
2004–05	F_7	Seed multiplication of seven progenies was taken up.
2005–06	F_8	Station trial of these 7 progenies + 18 other progenies from varietal cross + 5 check varieties (4 desi + 1 kabuli) was conducted.
2006–07	F_9	Multi-location replicated trial of PG-063 in All India Coordinated Trials were conducted.

INGR 09108) as a superior germplasm for high yield per plant, bushy, and dwarf plant type (Fig. 6.3B).

Vigna spp. (mung bean and urd bean)

Mung bean or Green gram (*Vigna radiata* (L.) Wilczek) has many desirable traits, such as erect growth habit, early maturity, and long pods, with large number of seeds. Black gram or Urd bean (*V. mungo* (L.) Hepper) possess comparatively durable resistance to yellow mosaic virus, *Cercospora* leaf spot, synchronous maturity, non-shattering of pods, more bunches (cluster) per plant, and higher methionine content. The cytogenetic studies showed that these two cultivated species are closely related with chromosome differentiation having occurred through one major reciprocal translocation. Attempts made to combine the desirable

(A) (B)

FIGURE 6.3

(A) The F_1 (*middle*) between *Cicer arietinum* cv PG 92–97 (kabuli type, *right*), as female and *C. reticulatum* germplasm accession JM 2100 from Turkey, as male (*left*). (B) PG-063 is a derivative of above wide cross and registered as a superior germplasm (INGR 09108) with NBPGR, New Delhi.

(A) (B)

FIGURE 6.4

(A) The F_1 plant (*middle*) along with parents, mung bean cv. BDYR-1 as female (*right*) and black gram cv DPU88-31 as male (*left*). (B) The mature pods of F_1 (*middle*) of mung bean cv. BDYR-1 as female (*left*) and black gram cv. DPU88-31 as male (*right*).

features of these two cultivated species of Asiatic *Vigna* genus have been successful. The crosses of *V. radiata* as female and *V. mungo* as male were attempted by several workers since 1960 (see Singh, 1990). The reciprocal cross of *V. mungo* × *V. radiata* was rarely attempted and was successful through embryo culture. The usefulness of interspecific crosses in genus *Vigna* in the on-going breeding program has been emphasized by Singh (2014). In the experience of authors, the crossed pod of *V. radiata* × *V. mungo* has one seed and rarely two seeds. The F_1 plants have partially filled pods (Fig. 6.4). In the F_2 and F_3 generations, few plants akin to black gram were observed which were sterile and finally dead, and the remaining populations were akin to mung bean with few characters like, yellow mosaic virus and *Cercospora* leaf resistance and more clusters per plant and more pods per cluster. Large number of plants in the segregating generation showed mottling of seeds (green background with dark gray spot on seed coat). Therefore, such crosses have been used to transfer resistance to foliar diseases and synchronous maturity from black gram to mung bean. However, some of the elite lines from such crosses have added advantage of more clusters per plant and more pods/cluster. At Pantnagar University, the studies on interspecific hybridization of *V. radiata* × *V. mungo* were continued for about three decades by author (D.P. Singh). As a result of this work, Pant Mung-4 was developed and released in 1997, and it is the first variety developed directly from an interspecific cross of mung bean (cultivar T-44) and black gram (cultivar UPU-2). Pant Mung-4 has more pods per cluster and more clusters per plant and has multiple resistance to foliar diseases in contrast to T-44, which was highly susceptible to foliar diseases. Resistance to foliar diseases and clustering pattern from black gram were transferred from cv. UPU-2 by using pedigree selection. From the above findings it can be inferred that with routine techniques *V. radiata* × *V. mungo* crosses could be made and superior varieties of mung bean could be developed through recombination breeding. However, rarely progenies, similar to mung bean and black gram through routine procedure have also been selected from interspecific crosses. The cross of *V. radiata* cv. BDYR-1 as female and *V. mungo* cv. DPU 88-31 as male has produced progenies similar to mung bean and urd bean genotypes (Fig. 6.4). From this cross seven elite lines akin to mung bean and six akin to black gram were selected for unique characteristics and registered with NBPGR, India (Table 6.3). These elite lines had new combinations of desirable traits, not seen in parental lines or other available released cultivars. These traits included higher seed yield, protein, iron and/or zinc, and disease resistance (Singh, 2014).

Success has also been documented in transferring disease resistance traits and diversify yield genes from wild progenitor of mung bean (*V. radiata* var. *sublobata*) into cultivated mung bean, and wild progenitor of black gram (*V. mungo* var. *sylvestris*) into cultivated black gram. In both the cases, cultivated species was used as female and their respective wild progenitor as a male parent (Singh, 1990). The wild species are trailing type, with undesirable traits like pod shattering and small black seeds but with disease resistance traits. The F_1 generation between these interspecific crosses are fully fertile with trailing growth habit. In the segregating generations, tremendous variability for plant type, maturity duration, yield traits, disease resistance, and seed characteristics were observed. Large number of progenies continued to show undesirable traits such as shattering of pods and mottling of seed coat for several generations. Some desirable transgressive segregants for yield traits in both mung bean and urd bean were observed. Resultant progenies were used as a valuable parental source in a forward breeding program, although none of these could be released as cultivar as these were not significantly superior when compared with released cultivars. At other places, for example, at TNAU, India, a variety, cultivars from an interspecific cross of *V. mungo* cv. Vamban 3 × *V. mungo* var. *sylvestris* has been released for

Table 6.3 Stepwise description of an interspecific cross of *Vigna radiata* (cv. BDYR-1) × *V. mungo* (cv. DPU 88-31) (Singh DP, unpublished data).

Year	Generation	Number of plants/progenies grown/selected/ evaluated
2001	BDYR-1 × DPU 88-31	Cross attempted.
2002	F_1	Two plants, partially fertile with intermediate morphological traits.
2003	F_2	2 plants gave rise to 26 F_2 plants. All the plants were harvested separately.
2004	F_3	26 progenies (103 plants) were grown and seed of all the 100 surviving plants was harvested separately.
2005	F_4	100 plants to progenies (87 akin to mung bean and 13 akin to black gram) were evaluated in augmented design. 32 desirable progenies were selected.
2006	F_5	32 progenies were evaluated for yield components and protein, zinc, and iron content. One mung bean and one black gram varieties and parental cultivars were used as checks.
2007	F_6	Retested in two on-station replicated trials. 1. 21 progenies akin to mung bean + 4 checks as in F_5 2. 11 progenies akin to black gram + 4 checks as in F_5
2008	F_7	Seven lines akin to mung bean and six lines akin to black gram were registered with NBPGR as superior germplasm.

cultivation in the year 2011 (Vamban-7). Other successful examples include the transfer of resistance to bruchid beetles (*Callosobruchus chinensis* and *C. maculatus*) from *V. radiata* var. *sublobata* to mung bean in Thailand.

Advantages of wide hybridization

Compared with intraspecific crosses, interspecific or wide crosses have been less frequently used. However, wherever these have been used, they have shown tremendous advantage to plant breeding programs. Some of these situations are described below:

1. *Creation of new crop species*: wide hybridization can lead to a creation of new crop species. Triticale is an example of new crop evolved from an intergeneric cross of *T. aestivum* and *Secale cereale*. Triticale combines yield and grain quality of wheat and winter hardiness of rye. *Nicotiana digluta* has been synthesized from a cross of *N. tabacum* and *N. glutinosa*.
2. *Simultaneous improvement of two species*: the wide crosses between mung bean (*Vigna radiata*) and black gram (*Vigna mungo*) have been attempted to

simultaneously improve both the species leading to cultivar registrations. These mung bean cultivars have plant type and seed characteristics of mung bean and disease resistance traits of black gram. Similarly, elite lines of black gram have inherited erect growth habit, longer pods with more seed per pod from mung bean.

3. *Transfer of genes for biotic stresses*: the wild species have genes for resistance to diseases, insect-pests, and nematodes. The genes for disease resistance have been transferred more frequently from wild species to cultivated species through crossing and back crossing/through selfing. This is due to the fact that simple and reliable screening techniques are available to pathogens whereas the techniques to maintain colonies (i.e., rearing) and controlled infestation of insect pest are comparatively tedious. Disease resistance from wild species have been transferred in an array of food crops: wheat, rice, sugarcane, soybean (e.g., yellow mosaic virus from *Glycine soja* to *G. max*), okra (yellow virus mosaic virus from *Abelmoschus manihot* to *A. esculentus*), and potato (late blight resistance from *Solanum demissum* to *S. tuberosum*). Resistance to insect-pests have also been transferred from wild species to cultivated species. Some of the examples where resistance from wild species/wild progenitor have been transferred to cultivated species are resistance to brown plant hopper and white backed plant hopper in rice (from *Oryza officinalis* to *O. sativa*), resistance to pod borers in pigeon pea (from *Cajanus scarabaeoides* to *C. cajan*) and resistance to weevils in mung bean (from *Vigna radiata* var. *sublobata* to *V. radiata*). These examples are not exhaustive as numerous other examples (crops, traits) exist.
4. *Transfer of genes for abiotic stresses*: the wild species have been successfully used to transfer genes for abiotic stresses as a result of this the cultivated species show improvement in their adaptation. The winter hardiness in wheat has been transferred from *Agropyron*. In rice, tolerance to aluminum toxicity has been transferred from *O. rufipogon* to cultivated rice *O. sativa*.
5. *Transfer of genes for quality traits*: in some crops, wild species have been used to improve quality traits. For example, lint strength from *G. thurberi* has been transferred to *G. hirsutum*, and carotenoid content in tomato (B gene from *L. hirsutum* to *L. esculentum*). The starch content in potato tubers has been improved by utilizing wild species germplasm. In tobacco, leaf quality was improved by utilizing *N. debenyi*. Grain protein content gene (*GPC-B1*) from tetraploid wheat (*Triticum turgidum* ssp. *dicoccoides*) has been transferred to hexaploid wheat to increase seed protein content.
6. *Transfer of yield genes*: the wild species have genes for higher yield and its components. In case of grain legumes of *Vigna, Cajanus*, and *Cicer* genera it is observed that as such their wild progenitor(s) have long pod bearing length, that is, they produce more pods per plant. Pod number is the most important yield component. The wild progenitor of pigeon pea has higher number of bunches per plant. Therefore, the wild progenitors of these pulse crops have been used to introgress these genes to cultivated varieties. High yielding transgressive segregants of *V. radiata* × *V. radiata* var *sublobata* in the selfed progenies and

V. mungo × *V. mungo* var. *sylvestris* in the BC_1F_2 and in selfed progenies of *C. cajan* × *C. scarabaeoides* and *C. arietinum* × *C. reticulatum* of wide crosses have been observed by breeders. The introgression of yield genes from wild oat, *Avena sterilis* to cultivated oat (*A. sativa)* was achieved by Frey and coworkers at Iowa State University in 1970s (see Lawrence and Frey, 1975). They crossed *A. sativa* with *A. sterilis* followed by backcrossing of F_1 four times to *A. sativa* and selected pure lines (transgressive segregant) with more than 20% higher yield than recurrent parent.

7. *Alteration of mode of reproduction*: CMS systems have been developed in Pigeon pea by integrating the cytoplasm of wild species and nuclear genes of cultivated species, and hybrid varieties have been released. In several other self-pollinated field crops (e.g., rice, wheat, barley, cotton, etc.), CMS have been developed using wild species and used for hybrid seed production. Genes for apomixis have been transferred from *Tripsacum* to maize for research and potential breeding applications.
   Self-incompatibility alleles from *Brassica campestris* have been transferred to the self-compatible *B. napus* for the production of hybrid seed. Diploid *Brassica* species, for example, *B. rapa* (syn. *B. campestris*), *B. oleracea*, *B. nigra* are largely self-incompatible, while tetraploid *Brassica* species such as *B. napus*, *B. juncea*, and *B. carinata* are generally self-compatible, although exceptions (variety specific) have been noted in both diploids and tetraploids.
8. *Other uses*: wild species have been used to transfer dark green color and excellent leaf texture in lettuce and bright red thin flesh in red peppers.

Limitations of wide hybridization

It is true that the wild species are reservoir of several desirable genes for biotic and abiotic stress resistance and have not been fully exploited by the plant breeders. There are certain inherent problems in making wide crosses using wild relatives/species successful. Some of these limitations are described as follows.

1. *Hybrid sterility*: the crosses between cultivated and wild species may result in a F_1 hybrid which may be varying in its fertility (from sterile to fertile). The hybrid sterility could be due to chromosomal or genic factors, which could be overcome with the use of techniques described earlier in this chapter.
2. *Linkage between gene of interest and undesirable traits*: however, care may be taken so that undesirable traits of wild species (such as poor plant type, pod shattering, and undesirable seed characteristics) are not transferred along with desirable trait(s). The seed shattering trait of wild species remains associated for several segregating generations. The wild progenitor has been used as male parent to transfer the traits like disease and insect resistance through backcrossing with recurrent parent, and has not been used as female except when CMS lines are to be developed as in case of pigeon pea. Linkage drag is more commonly

associated with the transfer of favorable genes from wild species to elite lines, and as a result several generations of large population sizes (for sampling sufficient recombinant events) are required to break the undesirable linked genes from the gene of interest. It is advisable to invest resources to manage large family sizes in such breeding crosses to identify desirable recombinants.

3. *Gene transfer take longer time*: it may take about 5-10 years to incorporate even the Mendelian genes from wild species to cultivated species using either backcrossing with recurrent parent or using selfed progenies of F_1 hybrid with large segregating population to recover the genotypes of cultivated types. Use of molecular markers have significantly reduced the time to gene transfer, along with background selection to reduce the proportion of wild parents (and only move the gene of interest); however, research efforts are needed to develop such molecular solutions (e.g., development of useful markers, and cost effectiveness to make them routinely available for all breeding program).
4. *Segregation in F_2 onward does not represent entire variation*: as a result of more pronounced heterozygosity (due to more segregating alleles at each locus) of F_1 hybrids, the F_2 and later generations of an interspecific cross show a very wide range of segregation (from very poor or sterile type to very vigorous type). Large number of segregants would have entirely new characteristics, different from those of either parent. However, morphologically the recombinants may look fascinating but not represent total range of variation, which is expected from such crosses. The segregation for different characteristics do not fit in the Mendelian ratio in the F_2 generation, due to irregular meiosis leading to reduced survival or elimination of some of the gametes and zygotes.

However, if plant breeder has patience they can achieve good success. There is need to grow large segregating populations. Both pedigree based and backcross method could be used. In pedigree breeding method, plant to progenies may be grown from F_2 onwards and selection of individual plants can be delayed to F_3 generations onward to get reasonable population sizes. The plants with undesirable traits like: trailing (spreading) growth habit, shattering of pods and unacceptable seed traits should be discarded and plants with desirable traits may be selected. Due to large family size in breeding population (i.e., cross), high culling rate can be deployed.

There is a need to involve a greater number of wild species for their use in improving tolerance to biotic and abiotic stresses and wider adaptation in ongoing breeding program of field crops including food, feed, fiber, and vegetables. Favorable genes are also present to increase yield, and more efforts are needed to harness these genetic resources to increase the rate of genetic gain in major and minor crops. Investment of public sector resources in germplasm development can benefit the human population to meet the need of a challenging climate and dynamic stress agents. The challenges and a tedious path of working with wide hybridization should not deter breeding programs to utilize natural variation in related species. Plant breeding program need to be forward thinking with a long term vision rather than short term gains.

CHAPTER

Haploidy and polyploidy in crop improvement

7

Abstract

Changes in chromosome number, gene mutation, and genetic recombination are important sources of genetic variation, and resulted in evolution of numerous crop species. Doubled haploid (DH) production has been considered as a major factor in reducing the amount of time needed to develop pure lines and inbred lines in crops that are amenable to tissue culture. The major methods of DH production are anther culture, microspore culture, and interspecies or intergeneric crosses. In this chapter, DH production methods and its success in wheat, maize, and barley are described, and advantages/disadvantages are presented. Polyploidy and its usefulness in plant breeding are also covered.

In addition to mutation and genetic recombination, the change in the number of chromosomes is an important source of genetic variation that resulted in the evolution of a number of crop species. The nuclear genes, which produce enzymes and finally a trait, are located on chromosomes; therefore, a change in the chromosome number leads to observable change in the trait of an individual. Each chromosome exists as a member of the homologous pair in diploid species, and number of such pairs of chromosomes is specific to a particular species. For instance, wheat has 21 pairs, rice 12 pairs, maize 10 pairs, and *Arabidopsis* has only 5 pairs of chromosomes. A basic set of chromosomes constitute a genome containing one member of each pair of chromosomes and represents the minimum number of chromosomes that has survived in a species.

Haploids

For plant breeders it is important to note that both sperm and egg are haploid, that is, they contain a single set of chromosomes from the male and female parent, respectively. Fertilization unites the single set of chromosomes in the sperm nucleus with the single set of chromosomes in the egg nucleus to have a complete pair of chromosomes (diploid) in the zygote. Several crop species, for example, banana, sweet potato, and grasses have three or more sets of chromosomes, and are called polyploids. Grass species are generally polyploids.

Plant Breeding and Cultivar Development. DOI: https://doi.org/10.1016/B978-0-12-817563-7.00008-8

Haploid terminology is used in the context of an individual that has a single set of chromosomes. In crop improvement, these individuals will not be viable and reach maturity and seed set; therefore, this term 'haploid' is used to describe haploid seed of diploid species. If the individual is a haploid but derives chromosome sets from a polyploid species, it is referred to as polyhaploid. Meanwhile, a haploid arising from a diploid species is referred as monohaploid, and now more generally a referred as haploid.

Conventionally, the ploidy level of an organism is denoted by 'x', which refers to the number(s) of sets of chromosomes or number of chromosomes in a genome complement (i.e., *genomic chromosome number*, also called as monoploid number). Therefore, diploids are written as $2x$, whereas tetraploids are written as $4x$ as they have four sets of chromosomes. Terminology for listing the gametic number of chromosomes (i.e., number of chromosomes in gametic cells, or the number of chromosomes in a single complete set of chromosomes, or simply *gametic chromosome number*) is 'n', therefore, $2n =$ number of chromosomes in somatic cells (i.e., the somatic chromosome number), while $n =$ reduced number following meiosis in egg (female gamete) and sperm (male gamete). For example, a tetraploid wheat (i.e., durum wheat) would be represented as $2n = 4x = 28$. In common wheat (hexaploid wheat), chromosomes are derived from three different ancestral species (see Chapter 6: Wide Hybridization), where each had seven chromosomes in its haploid gametes (i.e., A, B, and D genomes each had seven chromosomes). Therefore, the monoploid number (x) is 7 and the haploid number is 21 (3 genomes $\times$ 7 chromosomes in each of the three genomes). The somatic cell in common wheat has 42 chromosomes [six sets of 7 chromosomes—three sets from the female (egg cell) and three sets from the male (sperm cell), which together form the plant genome]. For common wheat, $2n = 6x = 42$; which means $x = 7$, and $n = 21$ ($n = 3x$). While common wheat is a polyploid, it behaves as a diploid due to the *Ph1* gene, which prevents pairing between related chromosomes.

In maize, spontaneous haploid has been found. Spontaneous haploids have been reported in cotton as a result of semigamy, and in pepper as a result of twinning (twin-embryo haploids) of an embryo from a gametophytic cell other than the ovum (apogamy). These are major causes of spontaneously occurring haploid in angiosperms. Spontaneous haploids occur *in vitro* from: (1) polyembryony (one synergid develops without fertilization—apogamy, at the same time as the fertilized egg leading to twin embryos: one is haploid and the other is diploid), (2) pseudogamy (development of an unfertilized female gamete after stimulation by male nucleus), (3) semigamy (reduced male and female gametes participate in embryogenesis but nucleus fusion does not occur, leading to sectoral chimeric plants), (4) androgenesis (maternal nucleus is eliminated or inactivated before fertilization of egg cell and haploid contains chromosome set of males only in its cell), and (5) preferential elimination of chromosomes (observed in wheat, barley in intergeneric or interspecific crosses).

With the reporting of haploid plants (Blakeslee et al., 1922), interest grew in developing techniques for inducing haploidy. Soon after this discovery, utilization in breeding programs was suggested but remained a theoretical consideration until the early 1970s. It was discovered that haploids can be produced *in vitro* by anther or pollen culture, unpollinated ovary culture, chromosome elimination using the *bulbosum* method in barley, and wheat × maize hybrid method (wide hybridization) to develop wheat haploid plants (Fig. 7.1A and B). Wide

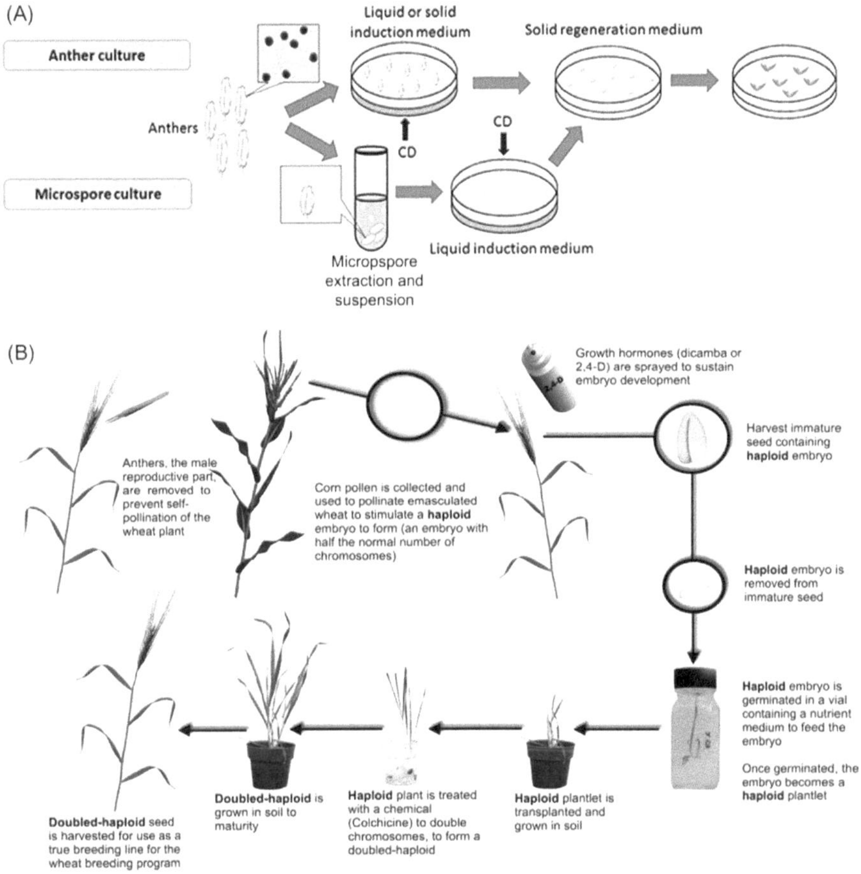

FIGURE 7.1

(A) Anther and microspore culture methods for doubled haploid production and (B) doubled haploid production in wheat using the wheat–maize pollen method. *CD*, Chromosome doubling.

(A) Yan, G., Liu, H., Wang, H., Lu, Z., Wang, Y., Mullan D., Hamblin, J., Liu, C., 2017. Accelerated Generation of Selfed Pure Line Plants for Gene Identification and Crop Breeding. Frontiers in Plant Science, v8: 1786. (B) Humphreys, D.G., Knox, R.E., 2015. Doubled haploid breeding in cereals. In: Al-Khayri, J.M., Jain, S.M., Johnson, D.V. (Eds.), Advances in Plant Breeding Strategies: Breeding, Biotechnology and Molecular Tools. Springer, New York, pp. 241–290.

hybridization (interspecific or intergeneric) results in the production of haploids due to chromosome elimination, and colchicine is then used to double the chromosome to obtain $2n$ and a completely homozygous plant.

For plant breeding purposes, haploids have potential only if large numbers can be produced efficiently in many diverse parents that are used to create breeding populations. Efficient use of the haploid technique in crop breeding program requires: (1) a dependable method of producing haploids, (2) a method in which haploid production represents a random sample of gametes, (3) a reliable method of doubling haploid chromosome number, and (4) adequate technical competence and suitable physical faculties.

Common wheat

Bread wheat (*Triticum aestivum* L.) has seen tremendous success due to the development of doubled haploid (DH) cultivar using the wheat × maize (Fig. 7.1B), anther or microspore culture system. The DH technology was rapidly adopted by wheat breeders in their wheat breeding programs starting in early 1990s as it leads to the production of true breeding lines in one generation significantly reducing the time to develop cultivar and helps increase the rate of genetic gain. In Canada, from 1997 to 2011, 25 DH cultivars were released. The predominant method to generate DHs has been wheat × maize pollination (use wheat as female and maize as male) to generate haploid embryos followed by chromosome doubling, while anther, pollen and microspore culture has also been used. In Canada, from 2006 to 2009, more than 30% of all acreage was on DH wheat cultivars for the premium wheat class (Canada Hard Red Spring). More information can be found in Depauw et al. (2011). Winter wheat breeders have been able to reduce time to develop cultivars in half, due to DH methods, which is a significant advantage compared to traditional methods that could take more than 15 years.

Maize

DHs in maize is a great success story. S.S. Chase is known for his work in the development of the 'monoploid method' for producing homozygous diploid lines of maize, which sparked interest in DHs in maize breeding. Chase conducted his work in late 1940s at Iowa State College and later as a research geneticist at DeKalb Agriculture Association, Inc., and further developed his methods leading to its availability as a plant breeding tool. In his reflections on the monoploid (i.e., haploid) method (Chase, 1952), he listed that the two major problems in the development of homozygous diploid lines of maize from monoploid sporophytes: (1) the production and early recognition of large numbers of monoploids, and (2) the derivation of self (and homozygous) diploid progeny from these monoploids. Through a combination of favorable seed and pollen parent as well as genetical, morphological and cytological sorting techniques, reasonable production and early recognition of monoploid were made possible despite a low rate of occurrence of monoploid (1 in 1000). The occasional spontaneous doubling phenomenon of monoploid was a solution to the second problem (where about 10% of untreated monoploid lead to diploidization). From the modest start of Chase at Iowa State College developing about 100 homozygous diploid lines in early 1950s to current time, maize DH have come a long way as it is estimated that majority of private company maize breeding programs in North America and Europe generate their inbred lines primarily through DH methods.

Haploid inducers have the ability of generate seeds with haploid embryos (i.e., haploid seed) when they are crossed to normal genotypes (inbreds). Haploid inducers can be paternal and maternal haploid inducers. Maternal haploid inducers - when inducers are used as the pollen-source parent leading to transmission of only the maternal nuclear genome to haploid embryo, while paternal haploid inducers are used as seed-bearing parent because only the paternal nuclear genome is passed to haploid embryos. Paternal inducers transmit both nuclear and cytoplasmic genome to the haploid embryo. Paternal inducers transmit both nuclear and cytoplasmic genome to the haploid embryo (Trentin et al., 2020). Paternal inducers have been used to introduce cytoplasmic male sterility (CMS) to inbred lines. The steps involve the usage of a haploid inducer line with CMS. This haploid inducer is pollinated with inbred lines that need to be converted to CMS (also called as maternal inbred). The cross will generate paternal haploids that carry CMS. These paternal haploids are then pollinated with pollen from the inbred line (maternal inbred) to produce a new diploid inbred line with CMS. Compared with backcrossing, this method is quicker to introduce CMS, and no residual genome of the CMS donor parent is left. Also compared with other *in vivo* methods, chromosome doubling is not needed. In maize, due to the advancement in haploid induction, induction rates have improved from 0.1% to more than 10%.

Haploid induction rates (HIR) = number of kernels with haploid embryo divided by all kernels studied.

For DH production in maize, maternal haploid inducers are used as these have high HIR. The haploid inducers are specialized genetic stocks (e.g., Stock 6, WS14, KEMS, RWS, MHI, PK6, UH400, and many others) used as a pollen source for an F_1 female (produced by a cross of two inbred parent lines) to generate haploids. The female maize plant (F_1) is a normal diploid, but the use of inducer line as pollen parent leads to progenies with kernel segregation (in the ear that was pollinated with inducer line pollen) diploid ($2n$) kernels (that are self's and discarded), and certain fraction of haploid (n) kernels due to anomalous fertilization. Since the kernels with a haploid embryo have a regular triploid ($3n$) endosperm, they have normal germination. The inducer lines have been bred to contain indicator traits to help sort haploid kernels from diploid kernels. These include a dominant anthocyanin (purple) color marker [generally, *R1-Navajo* (*R1-nj*)] that expresses in the seed aleurone (below the pericarp, aleurone is a one cell thick outermost layer of the maize endosperm) and in the embryo (that developed into corn plant) in the haploid inducer stock line. Therefore, it can be used as a phenotypic marker to differentiate regular diploid and haploid seed. Haploid seed will have colorless embryo and purple aleurone, while regular hybrid diploid seed will have purple embryo and aleurone. No coloration in either embryo or aleurone indicates self-seed (or outcrossing contamination), and are discarded. While the sorting is generally done manually and is tedious, modern imaging techniques, and automation will likely make the sorting process faster and more efficient. The expression of the *R1-nj* color marker can vary depending on the genetic backgrounds and environmental conditions, but so far this gene has been used extensively for maize DH production (Fig. 7.2). False positives can be identified by morphological markers such as 'glossy' or anthocyanin coloration of the stalk as the seedling emerges (after planting of seed identified as haploid).

From the selected haploid kernels, spontaneous chromosome doubling occurs. To increase the doubling (of chromosomes) rates, colchicine treatment is applied. Another method of differentiating haploids and diploids is built on high oil content; as diploid seed have higher oil content than haploid seed due to Xenia effect. China Agricultural University series of haploid induction lines were developed that combined HIR with high oil content for sorting by near nuclear magnetic resonance enabling faster and automated throughput for haploid seed selection. In Germany, University of Hohenheim series haploid induction lines with high HIR and oil content have been developed.

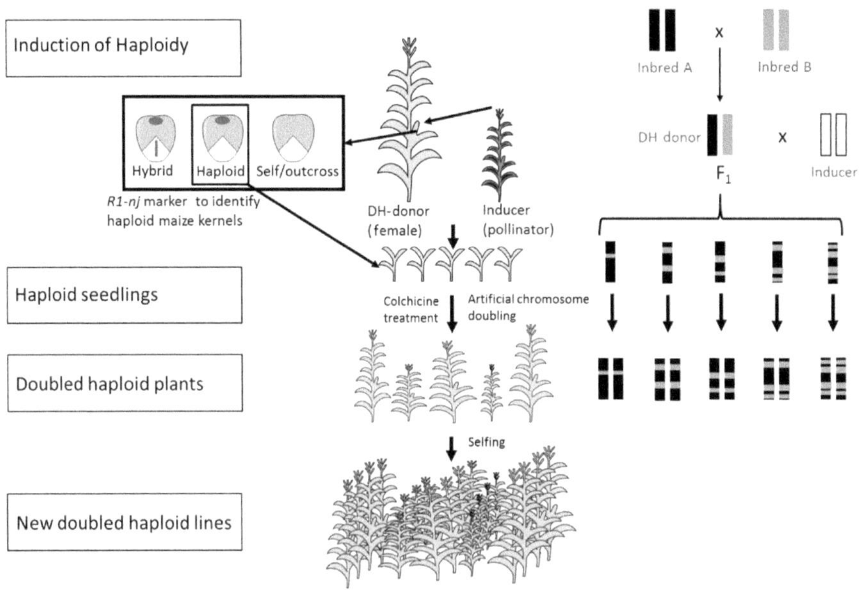

FIGURE 7.2

A schematic diagram showing maternal doubled haploid production in maize.

Compared with tissue (e.g., anther) or cell (e.g., microspore) culture, haploid induction is less technical and more economical. The simplicity of this method allows its easier integration within maize breeding programs, and this method can be used in the breeding company or through a third-party service provider (for a 'fee for service' agreement). The advantage of this method is an ability to consistently produce sufficient number of DH lines in a short time and lower costs.

Advantages of DH methods in maize and pure line crops:

1. Shortened breeding time compared with traditional way to develop maize inbred lines (or pure lines in self-pollinating crops such as wheat), which significantly positively impacts the rate of genetic gain.
2. Cost efficient, less labor intensive, no need for specialized labs to develop DH lines.
3. More extensive testing of hybrids (from DH inbred line parent crosses) to determine combining ability of DH lines. For pure lines, more extensive yield testing of DH lines can be performed.
4. For plant variety protection requirements of distinctness, uniformity, and stability, DH lines are advantageous as they are completely homozygous.
5. There is less effort to maintain lines and no inbreeding depression.
6. Marker technology (for MAS and genome wide prediction and selection program) can be rapidly applied even at haploid stage leading to reduced overall costs and better resource management.
7. Easier to implement backcrossing scheme to transfer genes, or convert lines to a new biotech trait.
8. Multiple nursery screening(s) can be initiated as soon as the DH line is developed (due to complete homozygosity).

Barley

The frequency of haploid production is high in barley with the *bulbosum* technique. Barley cultivars have been released as a result of barley haploid breeding, and the *bulbosum* method is used in select program laboratories that have tissue culture laboratory set up or accessible. In 1979, the first reported DH barley cultivar 'Mingo' was released in Canada using the *bulbosum* method.

Some of the earlier work in barley includes Kao and Kasha (1969) who reported that interspecific crosses (done reciprocally) between 28-chromosome autotetraploid cultivated barley (*Hordeum vulgare* L.) and tetraploid *Hordeum bulbosum* produced 14-chromosome all 'diploid' barley plants resembling cultivated *H. vulgare* barley. Subsequently, Kasha and Kao (1970) reported on a technique to obtain haploids in barley using an interspecific cross between cultivated barley (*H. vulgare*) and *H. bulbosum*. Their major findings were that (1) a high frequency of haploid embryo can be obtained, (2) only haploid forms were obtained from a cross between diploid parents, (3) haploids can be obtained from any given line of cultivated barley (i.e., no genotype effect was noticed), and (4) the haploids always have the cultivated barley genome, which means that the doubling of chromosomes can lead to breeding applicable DH lines.

The formation of gametes in *H. vulgare* following pollination by *H. bulbosum* is a result of fertilization and not parthenogenesis. Although a hybrid zygote is formed following this interspecific cross, subsequent embryogenesis and endosperm development undergo chromosome elimination. Although fertilization occurs, the *H. bulbosum* chromosomes are lost during the first few mitotic divisions of the embryo. The elimination process is preferential for the loss of *H. bulbosum* chromosomes. The chromosome instability of the hybrid cells of the developing embryo probably causes both physiological and developmental instability, which in turn causes the embryo to abort. Therefore, the embryo must be removed 8–10 days after hybridization. Following excision, *in vitro* culturing of the embryo removed from the endosperm provides better conditions for survival; many embryos will differentiate slowly into plants, most of which are haploid. The chromosomes of the haploid plant can be doubled by colchicine treatment. The resulting DH or dihaploid plant is homozygous for all gene loci in the haploid gamete of the *H. vulgare* parent used in the initial interspecific cross. The steps in the development of DH lines in barley using *bulbosum* method are:

Step I. Define objective(s), select appropriate parents, and hybridize parents to produce 15–20 F_1 seeds. This procedure may be completed under controlled environmental conditions over a 3- or 4-month period.

Step II. Grow F_1 progeny in environmentally control facilities. Hybridize F_1 plants with the wild barley *H. bulbosum* pollen to produce haploid embryos, which are then placed in *in vitro* culture. This procedure may require 3–6 months.

Step III. Grow the embryos into seedlings. At three to five leaf stage, treat plants with colchicine to cause chromosome doubling, and grow plants to maturity. This requires 3–6 months. Haploid DNA can be collected to run markers of choice for marker assisted selection or genome wide prediction (based on results, undesirable haploids can be culled at this step).

Step IV. Increase seed from doubled homozygous haploids (i.e., DH lines) for replicated field trial evaluation. This requires a growing season.

Step V. Begin replicated yield trials for performance testing. Additional testing can be done in specialized disease or abiotic stress nurseries. Proceed to multi-location yield trials to determine genotype × environment interaction (stability analysis) for the release of commercial cultivar.

The method described above provides a broader guideline for methods that rely on preferential chromosomal elimination (including wheat × maize system).

Anther, Pollen and Microspore Culture

The *H. bulbosum* method in barley, wheat × maize system in wheat, and haploid inducer system in maize have become well-defined procedure but other methods also exist. Haploid plants can also be produced directly from pollen or anther (Guha and Maheswari, 1964), and this played a major role in developing interest in haploidy. However, anther or pollen culture can be used only if specialized tissue culture labs are available. In anther culture, the developing anthers are excised at the most desired stage (generally targeted at the uninucleate stage) aseptically from unopened flower and then cultured on nutrient medium in sterile conditions. Microspores in cultured anther develop into callus tissue or embryoids that give rise to haploid plantlets. In pollen culture (also called as microspore culture), pollen grains (targeted at the uninucleate stage) are obtained from the anther and then cultured on nutrient medium. With this procedure, microspores without producing male gametes develop into haploid embryoids or callus tissue giving rise to haploid plantlets.

Broadly in pollen culture, the steps are (1) growing healthy donor plant, (2) remove and collect floral organs (see anther, and pollen as examples above) from the donor plants, (3) pretreatments (e.g., heat, cold, osmotic stresses, or starvation) to help switch gametophytic pathway to sporophytic development of microspore, (4) isolation of microspores, (5) transfer of microspore to media, (6) regenerate haploids, and (7) artificial chromosome doubling (e.g., through dipping cut haploid plants with colchicine). In most application chromosome doubling is involved except in ornamental species where haploids may be multiplied repetitively and used directly. The main difference between the procedures of anther and pollen (microspore) culture is the isolation of microspores from the anther prior to culture in pollen culture, whereas in anther culture the whole anther is cultured. Pollen culture is preferred over anther culture, because plants may arise from various parts of the anther, with the result that a mixed population of plants with various ploidy level might be obtained in a given culture.

Anther culture has been applied successfully in tobacco and rice, which has stimulated intensive research on other species of economic importance. Haploids can be obtained in many economically important crops, but the frequency of production is important to decide which method to use. Anther culture has been successful in both wheat and rice, but the frequency of haploids from cultured anthers has been either low (about 0.1% of cultured wheat anthers) or, in the case of rice associated with albino seedling production. Albinism is one of the major issues in anther or pollen culture in rice. However, improvements have been reported with the addition of zinc sulfate or copper sulfate in pretreatment media. Additionally, several genotypes have been reported to have lower DH production efficiency, for example, *indica* rice type has lower DH production than *japonica* type. A similar albino problem in anther induced barley haploids necessitates that problems of poor regeneration and high numbers of albinos need to overcome. The *bulbosum* technique for haploid production in barley does not suffer from this problem, hence remain the chosen method in many labs.

Other crops

In addition to wheat and maize, cultivars were produced by the haploid technique in oilseed rape (*Brassica napus*) and in tobacco (*Nicotiana tabacum*). In the dioecious asparagus (*Asparagns officinalis*), haploids have been developed from polyembryonic seeds and by anther culture. In asparagus, male plants (YY) outyield female plants (XX). Haploids produced by anther culture resulted in 50% homozygous YY male plants, which when crossed with any female plant produced entirely male progenies.

Techniques for chromosome doubling

Colchicine treatment is the most common procedure used for chromosome doubling by applying colchicine, an antimicrotubule agent. This treatment has proved especially successful with barley, maize, and wheat where careful consideration has been given to the length of colchicine treatment and to additional compounds that improve and facilitate the doubling effect. Colchicine can also be used to haploid plants obtained from the anther culture technique. Main techniques to separate haploid (self) and DH plants are: flow cytometry, root tip squash, and morphological measurements. Due to the complexity of the process (application time, rate, other chemicals, artificial stress, etc.) and the genotypic dependency, it is difficult to develop a single protocol for colchicine application that is applicable in all crops. Therefore, for illustration purposes only one example is given using barley. The apical meristems are treated for 5 hours with a 0.1% colchicine solution mixed with a penetrating agent, such as 2% dimethylsulfate, and a surfactant. Colchicine can also be applied by capping the plant, by the immersion method or by first splitting the base of each stem with a razor blade before immersion. The addition of gibberellic acid and benzyladenine to the colchicine mixture can increase plant survival and the number of doubled lines in *vulgare* × *bulbosum* method. In the first generation of DHs, seed set is reduced by colchicine treatment. But subsequent generations recover from the effects, and the use of DHs in breeding programs is not hampered. However, colchicine has been reported for adverse health hazards. Therefore, should be carefully handled with proper safety precautions and safe experimental procedures and handling. Safer alternatives include: amiprophosmethyl and pronamid, as well as antimitotic herbicides that are being explored for being less toxic without reducing DH generation rate.

Another method of chromosome doubling is to take stem and root sections of haploid plants and culture them *in vitro* on a medium that initiates shoot differentiation. Because some cells in the stem and root segments are normally diploidized, diploid plants are obtained when such cells give rise to shoots. A success rate of nearly 10% or higher is necessary for the haploid technique to be a useful plant breeding tool.

Advantages and uses of haploids

1. The haploids are useful source for the development of pure homozygous lines in context to breeding. The main advantage is the reduction in time to develop

new varieties in contrast, to conventional plant breeding program which takes about 8–10 years to develop a pure homozygous variety with sufficient performance data, whereas by the use of anther culture the period can be reduced to few months or 1–2 years (to obtain sufficient seed quantity). Homozygosity is attained in the quickest possible way; therefore, it makes research and breeding easier. It is important to note that for cultivar release, precommercial testing requirement and time remains the same for both DH and non-DH Lines. The only saving of time is in the line development phase.

2. In case of self-pollinated crops, it takes four to six generations of selfing before a breeding line is selected for precommercial testing and release as a cultivar. The presence of dominance and heterozygosity in earlier filial generations adversely affects the selection efficiency of individual plants (prior to achieving pure line status), and as a result the genetic correlation between selected plants in one generation and the next (i.e., progeny performance) is also lower. With the availability of DH lines, these problems are circumvented.
3. In cross-pollinated species, a DH line is used as a parental inbred line or a tester inbred line rather than as a cultivar; therefore, tests for combining ability can be initiated earlier than in a comparable non-DH system.
4. Pure homozygous DH lines can be used for the production of pure F_1 hybrids (homogenous hybrid).
5. Haploid cell cultures are useful material for studying somatic cell genetics. In case of haploids, a mutation can be detected easily as majority of mutations are recessive; therefore, not masked by the presence of dominant allele.
6. The haploid production using anther culture results in induction of genetic variability. By anther culture not only haploids but also plants of various ploidy levels and mutants are obtained and can be used in the breeding programs.
7. The haploids can be used for the generation of male plants exclusively. In case of *Asparagus*, male plants have a higher productivity and yield earlier in the season than female plants and haploids are produced from anther of male *Asparagus* plants (XY) these are either X or Y. Chromosome doubling of Y results in super male plants YY, which can subsequently be raised by vegetative propagation.
8. In winter wheat, where only one generation of production is possible per year, DHs make a dramatic positive effect to the genetic gain as it essentially reduces the time to develop cultivar in half.
9. Haploids have provided a major contribution to basic cytogenetic research through the development of trisomics and monosomics, which was achieved with the use of Chinese spring wheat. In addition to this, the complete series of nullisomics and tetrasomics were produced. These aneuploids have provided basic information on wheat evolution, cytogenetics, and plant breeding. The work done on *Ph1* gene on chromosome 5B, revolutionized cytogenetic thinking about chromosome pairing in polyploid.
10. Haploid plants are very useful materials to be used for the production of genetic stocks like aneuploid lines, which are very useful in genetic studies

and gene mapping. DH lines are preferred for genomic studies as the other methods of mapping (F_2, $F_{2:3}$ families, recombinant inbred lines) can either not be repeated or they still have residual heterozygosity, while DH are immortal and completely homozygous, therefore, every plant in the DH line is identical. This prevents problem in genome sequencing if RILs have minor differences due to heterozygosity.

11. The DHs are used as mapping population in genome mapping and lead to more accurate genetic maps and QTL/gene localization.
12. The genotype × environment interaction can be studied more precisely in double haploid lines due to complete homozygous nature in comparison to, other inbred lines obtained in conventional methods of breeding.
13. Haploids have been used for the production of disease and insect resistance and stress tolerant breeding lines in different crop species, and marker assisted selection can be done at the haploid plant stage to reduce the population of undesirables without using further resources to develop the DH line and increase the seed. Genomic selection can be applied at haploid stage, and only selected plants are doubled to create lines.
14. Haploids are an attractive resource for genetic transformation for rapid recovery of transgenic line with fixed homozygous state.

Disadvantages of haploids

1. The production of haploids requires high level of management and expertise in tissue culture if anther or pollen culture is being used.
2. The relatively high incidence of albinism has been found in anther and pollen cultured plants.
3. The success rate of plant regenerated from the callus produced from anther is sometimes low.
4. There can be genotype specificity as well as differential rate of haploid production due to genotype, chemicals, rates, stages, labs, etc. Parents that have genotype specificity generate fewer DH line progenies irrespective of the other parent used in the cross. This implies that smaller sized breeding populations will be created from such parents.

Polyploidy

Two types of polyploids were described by Kihara and Ono (1926): autopolyploid and allopolyploid. In autopolyploid plant species, same set of chromosome is doubled (e.g., AAAA, or BBBB types of autotetraploid), whereas allopolyploidy is the product of interspecific hybridization (e.g., AABB or AACC or BBCC are case types of allotetraploid); therefore, contains separate set of non-homologous chromosomes. An autopolyploid is formed by duplicating a genome within the same species, such as potato (*Solanum tuberosum*), alfalfa (*Medicago sativa*), and sugarcane (*Saccharum*), whereas an allopolyploid is derived from hybridization between different species followed by chromosome doubling or from fusion of unreduced gametes between species. Allopolyploids are generally more diverse than autopolyploids. Polyploids are a product of rare mitotic or meiotic event (e.g., non-disjunction), which causes the formation of gametes with a complete set of duplicate chromosomes (i.e., diploid gametes in diploid organisms). If a diploid gamete fuses with a haploid gamete, a triploid zygote will forms, while a fusion of diploid gamete with another diploid gamete creates a tetraploid zygote. Triploids are generally unstable and sterile. Examples of natural autopolyploids includes alfafa, coffee, peanut, potato, and sugarcane. These species arose through spontaneous chromosome doubling. Whereas examples of allopolyploid include blueberry, oat, oilseed rape, strawberry, upland cotton, and wheat.

The convention to present genomes of different species is to use unique letter of the alphabet to distinguish each genome. In allopolyploid species, the term 'homeologous chromosomes' is used to differentiate chromosomes that are similar but not homologous (e.g., chromosomes 1A, 1B, and 1C in wheat).

Polyploids are classified as below (Grant, 1982):

Autopolyploids (the genomes that come together in the polyploid are identical, hence is called 'Auto')

- strict autopolyploid *AAAA* and
- interracial autopolyploid *AA**AA***.

Autopolyploids are also called polysomicpolyploids and the ploidy level starts at triploid or above.

Amphiploids (Clausen et al., 1945) includes segmental allopolyploids, genomic allopolyploids, autoallopolyploids, and aneuploids.

- Segmental allopolyploid: for example, $A_1A_1 \times A_2A_2$ leads to $A_1A_1A_2A_2$ (the genomes in the species are partially homologous to each other; therefore, exhibit partial multivalent formation).
- Genomic allopolyploid: for example, *AA* × *BB* leads to *AABB* [these are also known as true allopolyploid or a disomic polyploid and are derivatives of parental hybridization with structural dissimilarity between their basic genomes ('x'). The ploidy level is tetraploid and beyond].
- Autoallopolyploid: for example, *AAAA* × *BB* leads to *AAAABB* (these species combine the characteristics of both autopolyploids and allopolyploids).
- Aneuploids: involve the gain or loss of a single chromosome. For example, a diploid plant ($2n = 2x = 10$) can have the following aneuploidy conditions with somatic chromosome numbers presented in brackets:
  - trisomic [2n + 1 (extra)] will have 11 chromosomes,
  - monosomic [2n − (deficient)] will have 9 chromosomes, and
  - double-trisomic [2n + 1 + 1 (2 extra)] will have 12 chromosomes.

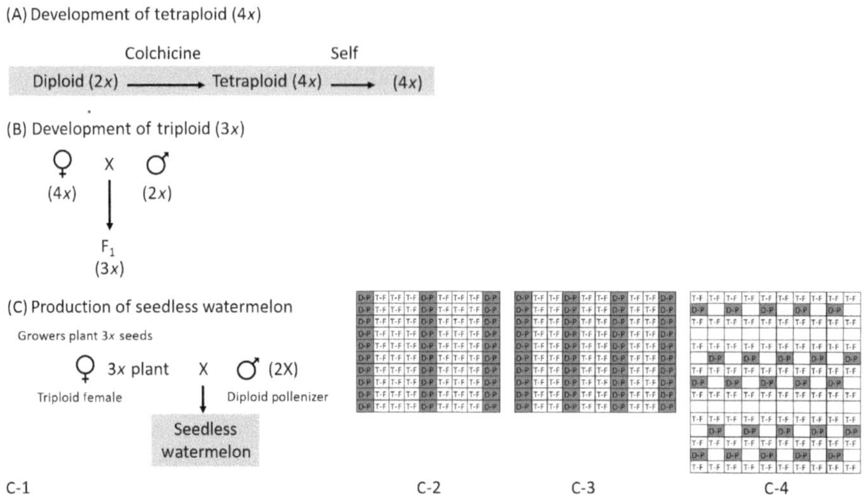

FIGURE 7.3

(A) Development of tetraploid plant, (B) development of triploid plant, and (C-1) scheme for the development of seedless watermelon, and (C-2, C-3, and C-4) three examples of field arrangements of seedless watermelon production. (C-2) One pollenizer row alternating with three triploid female rows, (C-3) one pollenizer row alternating with two triploid female rows, and (C-4) arrangements of pollenizer rows every third triploid female rows in alternating rows to give a 3:1 ratio of female to male plots. *D-P*, Diploid Pollenizer; *T-F*, Triploid Female.

Increased ploidy level has been reported for an increased vigor, for example, in ornamental and horticulture species. Seedless watermelons are one example of the role of plant breeder to harness polyploidy and develop modern economic species (Fig. 7.3).

Polyploidy has been suggested to provide an advantage of heterosis, where polyploid progeny of two diploid progenitors is more vigorous and healthier than either of the two diploid progenitors. It also creates heterozygosity preventing accumulation of recessive mutations. Additionally, polyploidy provides a 'buffering' effect, where polyploid offspring has more copies at any given locus; therefore, shielded from the deleterious effects of recessive mutations. Polyploidy can also lead to new gene functions due to multiple copies of the genes. Self-incompatibility can be overcome in polyploid organisms due to genome interactions. Despite the advantages, polyploidy can also lead to problems in normal completion of cell cycles (mitosis and meiosis) due to sheer size of the genome and associated single irregularities. Polyploid species may also be less amenable to genetic modification and stability of expression of gene (e.g., alfalfa genetic modification). Development and utilization of molecular tools including markers are also a challenge in polyploid species. Due to a larger genome, balance of cell volume and genome size is disrupted.

CHAPTER

Hybridization and selection in self-pollinated crops

8

Abstract

Hybridization is the first step in the development of pure line varieties in self-pollinating crops. Hybridization creates recombinants, bringing favorable features from parents in the cross, and are advanced and selected in segregating populations while reaching homozygosity. After performance testing in multiple locations over years, pure lines are released as cultivars. This chapter describes the main steps in the creation of pure lines, setting plant breeding objectives, consideration in selection of parents, procedures in hybridization, handling segregating generations, and strategies to bring multiple traits in a variety.

Self-pollinated (self-fertilizing) species are also called autogamous species or inbreeders. Various plants characters such as homogamy, cleistogamy, chasmogamy, etc., favor self-fertilization. Hybridization is the first step in the development of pure line for self-pollinating species and refers to the mating between genetically dissimilar varieties. The plants with combined desirable features of the dissimilar parents are selected from the segregating populations and are advanced for increased homozygosity at each loci. The seed of selected progenies are increased for testing, eventually leading to the development and release of pure line cultivar. Hybridization is of three types:

- Intervarietal or intraspecific (involves varieties of a species, i.e., crossing of varieties with the same species),
- Interspecific (involves two species of a genera), and
- Intergeneric (involves two genera of a family).

Normally, intervarietal (i.e., intraspecific) hybridization is used for developing superior varieties, while the other two types of hybridization (interspecific and intergeneric) have been used primarily for transferring specific character(s) or genes; and therefore, have been explained in Chapter 6: Wide hybridization.

Early history of hybridization

In 1694, Camerarius discovered sex in plant. He was a professor of natural philosophy at the University of Tubingen. Prior to Camerarius, in ancient times sex in plants was

Plant Breeding and Cultivar Development. DOI: https://doi.org/10.1016/B978-0-12-817563-7.00003-9

known (especially, in date palm); however, no empirical evidence was generated to determine the facts. Camerarius, through experimentation, established that pollen is indispensable for fertilization and classified the male and female flowers as pollen producing and seed bearing plants, respectively. He discovered that if male flowers from monoecious (maize) and dioecious plants (*Mercurialis*, spinach, hemp) were removed, no seed was produced. He also gave an idea of producing new types of plants through crossing. Cotton Mather in a letter to James Petiver in 1716 observed that ears from yellow grain maize planted next to red and blue maize have red and blue kernels scattered among them; and therefore, appears to be the first recorded observation of natural varietal hybridization and wind pollination (which was officially described by Paul Dudley in 1724 with wind pollinations for variety crosses in maize). Though not explicitly explained, these observations also give investigation on the resemblance of the offspring to the male parent in plants, which is officially credited to Fairchild. Thomas Fairchild in 1717 produced the first artificial plant hybrid by crossing Sweet William (*Dianthus barbatus*) with carnation (*Dianthus caryophyllus*), an interspecific cross. The hybrid plant obtained was sterile and called Fairchild's Mule. This was the first recorded hybrid in plant, as until that point only hybrids in animals were known.

As early as 1727, a seed store "Coq de la Bonne Foy" in France was run by Claude Geoffroy and Pierre d'Andrieux. Their daughter married Philippe-Victoire de Vilmorin and the couple inherited the seed store and established the first documented plant breeding centered company "Andrieux-Vilmorin" now called Vilmorin SA. The Vilmorin's Company initiated plant breeding work and produced many improved varieties of crops through hybridization. The first systematic investigations into plant hybridization in a number of species were conducted by Joseph Koelreuter (1760–66) in Germany. He made many crosses in tobacco. He reported that hybrids between *Nicotiana rustica* and *N. paniculata* were sterile. He also emphasized hybrid vigor in F_1.

Louis de Vilmorin (1856) developed the progeny test, that is, evaluation of the worth of plants on the basis of performance of their progenies. This principle is also known as the Vilmorin Isolation Principle or Vilmorin Method, which involved selection of single plants followed by separate testing of their progeny (i.e., progeny test). Thomas Andrew Knight (1759–1835) developed several varieties of fruits and crop plants and reported that male and female parents in pea contributed to F_1 offspring equally and segregation occurred in next generation. Knight observed dominance of phenotype as well as segregation of seed coat color; however, he did not provide evidence through ratios as later done by Mendel. While Knight's work predated Mendel's discoveries, Knight was involved in practical breeding. Knight is also credited for making first wheat cross in Europe. Patrick Shirreff started a small grain cereal breeding program in 1819 in Scotland. He is also credited to be one of the first significant practitioner of selecting pure lines but more importantly he was one of the first wheat breeder to obtain superior varieties. He proposed that new varieties can only be obtained through crossing landraces and introductions. His work gives systematic experimental plot planning for comparative trials of different strains. While he considered the first hybrid (cross of two parents) and its subsequent generation as "hybrids" as law of segregation was yet unknown, that is, he was likely developing varieties that was in segregating forms. Shirreff released several wheat and oat varieties in his career and was perhaps one of the first plant breeders to emphasize the necessity of the final test of the product (not just variety performance) to ensure needs of wheat millers and bakers were met. In 1886, William J. Farrer made complex crosses in wheat and developed several Australian varieties of wheat. The most striking example was the variety "Federation".

Sagaret (1826) was one of the first researcher that used the word "dominate" in reference to characters in progeny from crossing of parents with contrasting characters and described distribution of characters, which became widely known through Mendel's systematic experiments. Gartner (1849) used hand pollination to perform 10,000 experiments in crossing using 700 species from 80 genera leading to 350 hybrids. He observed no dominance, dominance, and partial dominance of one parent in some hybrids; and more importantly upon observing high vigor, noted that it will be useful for agriculture applications.

One of the greatest landmark in the biological science was the elucidation of laws of inheritance: the law of segregation and law of independent assortment by Gregor Mendel (1865, published in 1866) in garden pea. These discoveries have given hybridization a scientific basis. The laws of inheritance were independently rediscovered in 1900 by de Vries, Correns, and Tschermak.

W.M. Hays born near Eldora (Iowa) was a plant breeder at the University of Minnesota, and is credited for using progeny test for the first time in America (1890) and initiated pure line breeding from older heterogeneous varieties. The pure line concept was further developed and systematically explained by Johannsen (1903) for breeding self-fertilized crops. Hays also originated and practiced centgener method (growing 100 plants per generation in a head row from a single selected plant from the previous generation and making single plant selection among the 100 plants; repeating the process in each generation) that is based on principle of uniformly spacing plants and planting at uniform depth so allow individual (single) plant comparison and selection within and among rows.

Shull's (1908, 1909) biggest contribution was publishing his work describing experiments on inbreeding and crossing in maize, which led to the establishment of basis of hybrid maize breeding. Shull was studying inheritance of number of rows of grains per ear. Using self-fertilization, he developed inbred (he called them pure lines at that time) lines that differed in the number of rows of grains per ear (but each line was uniform and true breeding), and reported a decline in vigor and productivity with inbreeding. However, crossing of these inbred lines lead to uniform and highly vigorous and productive hybrids and were superior to the open-pollinated population from which the inbred lines were developed. The problem (in early 20th century) with Shull's inbred lines were that they were weak and unproductive; and therefore, did not make for good female inbred parent line (Note: in modern maize breeding, female inbred lines are bred to be highly productive—this allows them to be economical in seed increase and hybrid seed production). Therefore, hybrid maize breeding while recognized as a potential new breeding scheme was neither practical nor economical at that time. Jones (1917) gave the idea of double cross hybrid (a cross between two single crosses) in corn to circumvent the problem associated with single cross hybrids at that time and laid the foundation of US hybrid maize breeding revolution.

Steps in the development of pure line cultivars

During the 20th century hybridization was widely used for crop improvement. It continues to provide the most certain basis for continuing improvement in adaptation, yield, stress tolerance, and quality traits. Improved self-pollinated cultivars are obtained through three basic approaches.

1. Introduction: a breeder assembles cultivated varieties currently grown in other regions (domestically or internationally) and identifies pure lines that exhibit desirable characteristics and adapts to the new target area. A breeder must be aware of licensing and material transfer agreement (MTA) issues (see Chapter 23: Intellectual Property Rights and Protection) before acquiring lines from different sources. In self-pollinating crop species, introductions are generally received as pure lines, but they may contain mixtures of different genotypes or offtypes and may require roguing out these plants to obtain clean and uniform seed for commercial release.
2. Selection: a breeder acquires and assembles landraces, identifies the best genotypes from the landrace variety, and releases one or more for commercial production. This approach has applicability for orphan crops but is not generally used for major world crops. Introductions can also be used as a starting population source to select varieties, especially if they are mixtures of genotypes. See Chapter 9: Mass and Pure Line Selection.
3. Hybridization: a breeder makes hybridization between parents with desirable characteristics, evaluates, and selects superior genotypes for commercial production. This is the most common approach for developing new cultivars, and the first step is the establishment of objectives and identification of parents for hybridization.

Pure line cultivars are developed in self-pollinating species. Pure line cultivars are homogeneous and homozygous and, once created, can be maintained indefinitely by selfing. *Inbred lines are different from pure lines* although sometimes people use the terms interchangeably. Inbred lines are developed in a cross-pollinated species through inbreeding, and these lines are used as parents in the production of hybrid and synthetic cultivars. Inbred lines are not meant for release to the farmer for commercial production because inbred lines suffer from inbreeding depression. Their yield will be lower than that of a hybrid and even an open-pollinated variety. Inbred lines are homogenous and homozygous, similar to pure lines; however, artificial selfing must be done at each generation to maintain or increase seed. Sib-mating can be used to avoid severe inbreeding depression.

The main steps in the development of pure line cultivars using hybridization are as follows:

Step 1: define objectives.
Step 2: form the genetic base by creating segregating population(s).
Step 3: perform selection and develop pure lines.
Step 4: conduct yield trials (testing of experimental lines) and seed multiplication.

Step 1: Breeder clearly defines biologically possible and stakeholder needs driven objectives before any other activity happens. Breeders may consider the following when defining objectives:

a. Will it meet the needs of the producers, processors, and consumers? The best way to accomplish this is by having direct interactions with these three groups. Reading newsprint and other informational sources will also give a breeder an indication of the requirements by these groups. Breeders attend farm shows and farm group meetings, meet and visit processing companies, marketing companies, and colleagues in other disciplines. In crops with export potential, in addition to national stakeholders it is beneficial to learn export market requirements from overseas clients and partners. These interactions allow breeders to be current and relevant in setting objectives for their program.
b. Available resources. Breeder conducts a survey and study to determine if s/he has the necessary resources to achieve breeding objectives? These resources include germplasm, infrastructure (field and lab), expertise, equipment, team strength, financial etc. For example, if a breeder would like to select for resistance to a disease using molecular markers, do they have access to a marker screening lab or a disease nursery. Without required resources, it will be almost impossible to meet the objective of developing cultivars with resistance to that disease.

Without clear and logical objectives, the breeder risks working aimlessly, analogous to driving a car without knowing the destination. Setting objectives allow the breeder to make strategic decisions such as:

1. Parents that have the necessary complementation of traits to develop progeny that possess desirable traits from both parents.
2. Breeding methodology for generation advancement.
3. Selection strategies and plan for any specialized nursery or tools.
4. Traits and generation of selection for each.
5. Seed production and distribution.

Step 2 : Create breeding families, i.e., segregating populations. Based on the objectives, the breeder can pick the parental material from:

a. Advanced lines from their own breeding program.

b. Advanced lines from another breeding program.
c. Released cultivar (e.g., Ex-PVP).
d. Germplasm line from gene bank or a prebreeding program.
e. Introductions (from other countries) from colleagues or gene bank.
f. Mutant line populations (unselected or selected).
g. Wild relatives that are crossable or resources available to do embryo rescue, if needed.

Crops that have a long history of breeding effort will rely on cultivars and advanced breeding lines and, in special cases, introductions as the choice of parent material. Crops with less breeding effort will rely on populations and introductions. The majority of parents in a breeding program will be the best lines (advanced lines or newest cultivars) derived from the continuous breeding cycle of the program. The use of elite experimental line(s) as a parent as early as possible, makes one of the most significant long-term impact on genetic gain. Most successful breeders, start using advanced (and elite) experimental lines as a parent after 2 years of multi-location data. If using genomic prediction, a line can be used even earlier. However, a balanced approach is likely the best—where parents are used as soon as breeder feel comfortable that enough data is available in a mixed portfolio of several parents that are well known for their performance and few that have less data. Sometimes, breeder make anticipatory crosses (with insufficient data); and as new information becomes available, entire breeding populations can be discarded if the parent(s) are deemed unsatisfactory and majority of other high-potential breeding populations are available. Unlike animal breeding, phenotypic performance of a line is generally used as a guide to its use as a parent in hybridization and not its breeding value. Breeding value of varieties in self-pollinating crop species can only be computed once its progenies have been tested and characterized; therefore, in some breeding programs certain parents are used more extensively as they overwhelmingly generate superior progenies. A breeder has to be aware that for higher genetic gain, newly developed breeding lines (from forward breeding) should be used as a parent at a higher rate than only using older but proven parents all the time.

All of the traits desired in the cultivar must be present in the parents. Parental selection has to be based on reliable and complete data, for example, yield testing, adaptation testing, stress assessment, and end-use quality. If breeder does not have specialized nurseries, they can collaborate with other breeders in the same or different organizations to send material for testing and characterization. Material transferred across organizations may need a MTA, which is a legal agreement between breeders and/or organizations. MTAs will define the scope of agreement and rights of provider and recipient.

The number of crosses made by a breeding program depends on various factors such as objectives, available resources, and breeding method (determines number of breeding lines generated). In self-pollinated crops with pure line cultivars, it is assumed that the parent seed being used is a pure line (homogenous and homozygous). But if the breeder decides to accelerate the incorporation of a trait (more rapid genetic gain) and uses a line that is still segregating visually for a trait, a larger number of F_1 plants will be needed to increase the probability of recovering the desired recombinants.

The cross configuration will also dictate how many F_1s to create. If a cross is made between two pure lines, all F_1s will be heterozygous and homogenous. If a three-way cross is made using three different pure lines, the F_1s of the three-way cross will be heterozygous and heterogeneous, necessitating a large number of F_1s to be created to develop large sized F_2 population and subsequent early generation populations. There is considerable debate about the relative importance of the number of crosses versus the population size per cross. In most scenarios, there will be an inverse relationship between the number of crosses and population size, and this is primarily due to resources available in a breeding program. Irrespective, careful decisions should be made in parental selection based on prior information and scientific principles should be used.

Step 3: Selection and develop pure lines. once the crossing scheme is determined, and crosses made, the next stage is to choose an appropriate breeding method to develop and advance the breeding populations. The populations are advanced through generations, they will be composed of an array of different pure lines where genetic variability exists among but not within lines. This happens because all methods in pure line breeding lead to an increase in homozygosity, a reduction in the genetic variance within families, and an increase in the genetic variance between families. Cultivar development aims at identifying the best homozygous lines.

Selection should commence in early generations and, preferably, as early as in F_2; as it is the generation of maximum variability and the stage at which the minimum population size required to observe the desirable type is lowest. It is also important that a breeder makes selections in each generation, if possible, to continue with the development of pure lines and to eliminate undesirable types. This may be accomplished through phenotypic or genotypic selection with molecular markers. If a high value marker or set of markers is/are available and linked to the trait of interest, it will be beneficial for a breeder to use the marker(s) to enrich and advance the desirable types while eliminating the undesirable types.

Consider, for example, a trait under recessive gene control which, in the fixed homozygous recessive state, will never segregate to give the desirable dominant allele. Evaluating lines with the undesirable homozygous recessive genotype in this case will be a drain on resources. Therefore, the use of molecular markers (as they are not influenced by environment) lends a breeder confidence in the selections made, provided the molecular marker is robust, tightly linked to or on the gene, and not background dependent.

For selections in early generations, a breeder handles several thousand plants or hills or rows. Therefore, it is very important to handle this material in a selection environment that is very similar to the target region or, if breeding for a subregion, is representative of the target environments of different agro-ecological regions. In other words, ensure that the environment is ideal for selecting for the targeted traits. For example, choose a dryland environment if breeding for drought tolerance, or use irrigated or sufficient rainfed field nurseries if breeding for high-input environments. It is best to grow these early generation trials at a location to which the breeder has easy and quick access. This enables observation of the material to facilitate making breeding decisions, and also from a logistical viewpoint gives better management of the trial location. It is extremely important that a breeder eliminates controllable sources of variation such as weeds, non-uniform land, animal damage, and non-uniform application of chemicals and irrigation etc.

In early generations (F_2 to F_4), selection is restricted to traits of high to moderate heritability and repeatability, whereas in later generations (F_4 to F_6 or F_8 depending on the complexity of the crop genome), evaluation is more detailed and involves multi-location testing and replication to select for lower heritability traits. In situations of relatively large number of entries, but limited resources, as in early generation trials, single replication yield plots may be used to identify material for advancement. Statistical approaches and spatial adjustments, such as running mean, nearest neighbor adjustments, partial replication, or augmented designs, can be used to identify promising lines with reduced resources. As generations advance, more seed is available, and population sizes are sufficiently reduced to allow for increased replications and locations. Selection is then done for traits of lower heritability such as yield (in larger sized plots) and for end-use quality. Techniques such near-infrared spectroscopy can be effectively utilized in earlier generation testing of end-use quality traits. Selection is based on cut-offs developed in comparison with checks or industry requirements. Note that cut-offs for traits are variable in every test and require appropriate checks for comparison.

Step 4: Yield testing, pre-commercial testing. Final stages in the breeding cycle will involve lines that are considered pure lines (non-segregating) and have been retained after all selection activity up to this point. At this stage, more extensive testing of the few most promising pure lines from a cross is done for agronomic performance and other important traits, including end-use quality if applicable; this implies that none to several progenies from breeding populations may be left. These progenies likely represent the same year of crossing, if the program follows a yearly structure of activities consistent for each population created per year. Multi-environment testing is done for adaptation and stability, and environments may be locations or a combination of locations and years. At this stage, trials will be grown using the lattice design, or other incomplete block if the number of entries is large, or randomized complete block design if fewer entries are remaining; detailed observations are made, and data are collected. Since a fewer number of lines are tested, a more detailed assessment is feasible for an increased number of traits. For optimal use of resources and to ensure the timely adoption of pure lines, breeder needs to proactively initiate seed multiplication alongside advanced yield testing for the production of sufficient quantities of certified seed for the launch of pure line(s) as cultivar(s).

In the case of doubled-haploid (DH) lines, steps 3 and 4 are very closely aligned. They can even be considered as one because once the DH is generated, lines can go into multi-location replicated testing where seed quantities permit.

Setting plant breeding objectives

Plant breeding objectives for a program will depend on geographical adaptation trait requirement, prevalent biotic and abiotic factors that influence production, uses of the cultivar, and factors that are important to farmers and end-users. Plant breeder should be aware of the requirements of all stakeholders such as farmers, seed growers, processors, and consumers. Without the knowledge of what is desired in a cultivar, plant breeders cannot effectively develop solutions for the stakeholders that meet their needs. Objectives need to be clearly defined and the outcomes should be measurable. For example, "increasing grain yield" is a loosely defined objective as it lacks a measurable outcome. One way of establishing an objective for grain yield is to list the objective as "to increase grain yield in a new cultivar that is 5% higher than mean of predominant cultivars used as checks (assume, check A, check B, check C) after three years and multi-location agronomic and performance testing." In this objective statement, measurable outcomes are clearly stated (eg. 5% increase over clearly defined check or checks, length of testing, benchmark checks). The listing of 5% is for educational purposes only, and a breeder will set the desired program targets and testing requirements based on their program set up and expected deliverables. It is important to remember that in the above example, grain yield is likely just one of the several traits a breeding program aims to improve, so clear objectives (with measurable outcomes) for all required traits need to be established prior to making crosses. In addition, objectives may change due to market or production system issues; therefore, plant breeding programs need to have sufficient flexibility to update objectives and continue the breeding pipeline with minimal disruption. Finally, one needs to consider "must have" traits compared to "nice to have" traits. Must have traits include traits that farmers/processors/consumers absolutely require to grow/process/consume the economical part of the new cultivar. This defers from nice to have traits that are non-essential and likely of minor significance and value to the stakeholders, but do give some advantage in the current situation or projected to be useful in future.

Selection of parents

The selection of parents is important as it determines the nature and extent of variability in the segregating and advanced generations and the potential limit for selection. The objective could be the development of cultivar, or introgression of exotic genes for purposes of population improvement or for the development of cultivars. In self-pollinated crop species, two or three parent crosses are common, with the condition that targets for traits of interests are/can be achieved with the two or more parents. This implies that selection of parents is a crucial step in cultivar development, and this choice is governed by the breeding objectives and availability of genetic material to enable achieving these objectives. As the number of traits increase, the probability of finding all the desired characteristics in the two parents decreases necessitating an additional parent(s) in the crossing scheme for the initiation of the development of breeding population. Breeders can approach their crossing scheme plans similar to a financial investment plan, which consists of stocks, bonds, cash, and government securities among others. The idea is to balance the risks and rewards. In a breeding program, it is not uncommon for breeders to make varying combinations of safe and speculative crosses. An example of a safe cross will be the use of two well characterized elite parents, while an example of a speculative cross will be to use one elite well characterized parent and the other parent is a pure line with less extensive data (eg., fewer environment performance data) but suggested exceptional performance in one of more "must have" traits. In a breeding program, it is useful to include more parents (to increase the genetic diversity) and to make some safe crosses to ensure that deliverables are timely met. Irrespective of a public or privately controlled and operated program, documentable success is essential so that healthy funding can be received to effectively run the breeding program. The number of crosses attempted may not just depend on the availability of suitable parents, but also on resources available for making crosses and for handling the segregating generations and testing.

Procedures of hybridization

In hybridization method of breeding self-pollinated crops, the parent varieties/lines/cultivars are artificially cross-pollinated. Manual hybridization is most commonly used and it is relatively easy with field crops, which have large flowers. However, it is tedious in crops that have small and fragile flowers, such as forage grasses and some of the grain legumes. The knowledge of flowering habit of the crop with which breeder is working is necessary. The hybridization can be done in field or in controlled conditions (greenhouse, growth cabinet chamber, and glasshouse). If parents differ in heading dates then staggered planting dates will be necessary to ensure that female and male flowers are available at the same time to make artificial hybridization. In day length sensitive crops, artificial lighting is used to vary day lengths in order to bring diverse parents to flowering stage at similar dates, and in some cases allow the plant to grow vegetatively before flowering, giving sufficient number of flowers on a healthy plant.

Generally, the technique of making manual crosses consists of removing anthers of female parent with tweezers before pollen shed (a process known as emasculation) and collecting pollen from male parent and transferring it to the stigma of the emasculated plant. The procedures for emasculation and pollination vary with the crop. However, care is taken that the style or the very delicate stigma is not damaged. In most crops, as soon as the anthers are removed, the emasculated ear/flower must be covered with a small glassine/cellophane bag to exclude any foreign pollen. The base of the bag should be folded over and held with a paper clip. A small card tag/price tag should be loosely threaded around the base of the ear/flower bud(s) and on this the cross (name of the female and male parents) or code, date of emasculation/pollination, and the worker's initials are recorded.

In few crops, for example, sorghum (*Sorghum bicolor* L.), methods have been developed to control anther dehiscence using high humidity, which is created by covering the panicle with a plastic bag prior to flowering. This method, known as plastic bag emasculation, circumvents the need to manually emasculate each floret to remove anthers allowing higher numbers of crosses to be made. Emasculation in forage grasses may be done successfully with a hot water treatment if there is a temperature at which pollen is destroyed but ovary remains viable. It is especially useful when large number of plants or panicles are needed for crossing. For example, hot water treatment for emasculation has been shown in smooth brome grass, *Bromus inermis*. The temperature and time differ for most species and genotypic differences may exist within a species. After the inflorescence is emasculated, it is covered with a bag. Pollen must be collected and brought to the emasculated plants for hybridization. If pollen is collected on paper within a petri dish or in a glass container, the female flowers can be bent and placed in direct contact with the pollen or pollen can be applied with a brush.

In wheat, manual emasculation and pollination is generally followed. However, modifications have been shown to be effective. For example, Ethrel (2-chloro ethyl phosphoric acid) has been used as male gametocide by M.A. Beek at the University of Wageningen. The lines/varieties were mixed and sown, and the resulting mixture was divided into male and female strips and the female strips were sprayed with Ethrel. It gave best results when applied as an aqueous solution of 2000 ppm Ethrel a.i. (1000 L water/ha) at growth stage 41; with 0.05% (v/v). Agral added as wetting agent. Gibberellic acid 3 (GA 3) was applied at 150 ppm 3 days after the application on all plots to promote ear emergence. Ethrel performed reasonably well (60%–80% cross pollination depending on dose) in the polycross procedure. These are just very few of numerous examples that are available in the literature and plant breeders in specific crops can find most widely used methods in their crops through literature search as well as with discussions with other breeders in that crop.

Growing the F_1 hybrid generation

The F_1 (F = Filial) seeds from the crosses may be sown either in a field nursery in wider spacing or in pots in controlled conditions. It is always desirable to sow the parents for reference purposes, and several times parental trait information can also be used to determine true crosses from selfs (although this does not require growing F_1 and parents together) by looking at female trait expression for maternally inherited traits as well as for traits where female/male have recessive/

dominant trait expressions. Expression of flower color trait in F_1 can be used to differentiate self and crossed seed—in a cross of white female color flower soybean and purple male color flower soybean, a white colored flower in F_1 is a self; because purple is dominant over white flower color in soybean. Similarly, other morphological traits can be used to determine true crossed F_1 seed in the absence of molecular markers. F_1 plants provide the seed for F_2 generation; therefore, these may be grown under optimum conditions so as to produce maximum quantity of F_2 seed. The F_1, although highly heterozygous, will be all homogeneous for each breeding population in a two parent cross, that is, will have similar genotypes and phenotype (will also look exactly alike or identical) for each F_1 plant. Observations on hybrid vigor and character dominance may be recorded. Any obvious self's of the female parent may be discarded. In doubtful cases F_1 plants may be harvested separately and grown as progeny rows. The F_2 progenies that do not segregate can be rouged out. Molecular markers can also be utilized to confirm crossed seed with self's by using the seed chip (of the F_1 plant) or plant tissue. A set of molecular markers can be assembled by the breeding program that can help to differentiate the parents and crossed seed in each annual hybridization cycles.

Genetic basis of combination breeding

Planned hybridization in self-pollinated crops is taken up to recombine the desirable traits from different parents. The basic principles of heredity: (1) law of segregation, that is, members of a pair of alleles separate clearly without blending with each other during formation of germ cells or gametes; and (2) law of independent assortment, that is, separation of alleles at a locus is not influenced by separation of alleles at other locus are operative.

When parents differing in number of genes are hybridized and F_1 is advanced to produce F_2 generation (artificial selfing is not required in self-pollinated crops), in the F_2 generation, parental types are obtained. Additionally, segregation and recombination of genes produces new genotypes. Maximum amount of variation is expected in the F_2 generation. This is because genes of both the parents are brought together in the F_1 generation and the alleles of the genes for which the parents differ are in heterozygous condition. The heterozygosity is reduced on self-fertilization by one-half in each generation, irrespective of gene pair and homozygosity is increased due to natural self-fertilization (Table 8.1). In self-fertilized crops, we denote $S_1 = F_2$ for filial generation designation and so on.

1. *Number of loci*: as the number of different genes distinguishing the parent's increases, the number of possible genotypes expected in the F_2 generation increases markedly assuming no linkage (Table 8.2; also see Table 3.2). For example, in a crop species having 10 chromosome pairs and if only one gene is different on each chromosome for a cross of two parents (varieties), the F_1

Table 8.1 Percentage of completely homozygous plants in the segregating generations of self-fertilization.

	Number of genes segregating					
Generation	**1**	**2**	**3**	**4**	**10**	***n***
S_1	50	25	12.5	6.25	0.098	$(1/2)^n \times 100$
S_2	75	56.25	42.188	31.641	5.631	$(3/4)^n \times 100$
S_3	87.5	75.563	66.992	58.618	26.308	$(7/8)^n \times 100$
S_4	93.75	87.891	82.397	77.248	52.446	$(15/16)^n \times 100$
S_5	96.875	93.848	90.915	88.074	72.798	$(31/32)^n \times 100$
S_6	98.438	96.899	95.385	93.895	84.429	$(63/64)^n \times 100$
S_7	99.219	98.444	97.675	96.911	92.457	$(127/128)^n \times 100$
S_8	99.609	99.220	98.833	98.447	96.162	$(255/256)^n \times 100$
S_9	99.805	99.610	99.415	99.221	98.064	$(511/512)^n \times 100$
S_{10}	99.902	99.805	99.707	99.610	99.028	$(1023/1024)^n \times 100$
S_m	$[(2^m - 1)/2^m]^1 \times 100$	$[(2^m - 1)/2^m]^2 \times 100$	$[(2^m - 1)/2^m]^3 \times 100$	$[(2^m - 1)/2^m]^4 \times 100$	$[(2^m - 1)/2^m]^{10} \times 100$	$[(2^m - 1)/2^m]^n \times 100$

Where m = number of generations of self-pollination, n = number of genes segregating, and number of generations of self-pollination will be $S_1 = F_2$ *and* $S_{10} = F_{11}$.

Tab 8.2 Different types of phenotypes produced in the F_2 generation from parents different for number of allelic pairs.

Number of allelic pairs (or heterozygons loci)	Kinds of phenotypes in F_2 assuming	
	Full dominance	**No epistasis and dominance**
1	2	3
2	4	9
3	8	27
4	16	81
10	1024	59,049
N	2^n	3^n

After Briggs, F.N., Knowles, P.F., 1967. Introduction to Plant Breeding. Reinhold Publishing Corporation, Davis, CA, pp. 124, 426; Shebeski, L,H., 1967. Wheat and wheat breeding. In: Nielsen, K.F. (Ed.) Proc. Canadian Centennial Wheat Symp. Modern Press, Saskatoon, Saskatchewan, pp. 249–272.

would produce 1074 types of gametes and 59,049 different genotypes in the F_2 generation. The minimum F_2 population for chance in presence of each genotype (one plant of each genotype) would be 108,457. Large number of these genotypes will be heterozygous at one or more loci; and due to continued selfing, heterozygotes would disappear and homozygotes produced due to selfing are of significance to the plant breeder as these are selected to form a variety (in self-pollinated crops where the variety is a pure line).

As indicated in Table 8.2 (also see Table 3.2), with increase in number of generations of self-pollination, the frequency of completely homozygous plants increases. As the heterozygosity decreases considerably in F_5 and F_6 generations, the selection of superior plants/progenies may be more effective. The F_2 population size usually plant breeders grow is 1000–5000 plants per breeding population. A population of 5000–10,000 plants is more appropriate in small grain cereal crops such as wheat and rice where breeders are working on higher than average number of traits (compared to other self-pollinating crops where fewer quality traits are required). The complex characters such as yield and adaptation that are quantitative traits are controlled by many genes.

The F_2 generation for such traits show continuous variation. Such variation is also controlled by Mendelian genes similar to discontinuous variation as observed by Nilsson Ehle, a Swedish plant breeder who studied the inheritance of kernel color in wheat (*Triticum aestivum*). The parents had different genetic constitution, that is, $R_1R_2r_2r_2$ and $r_1r_1R_2R_2$ and F_1 had $R_1r_1R_2r_2$ genetic makeup. He observed a phenotypic segregation of 15 (red): 1 (white) in the F_2 generation with 5 phenotypic classes; 1 (white): 4 (light red): 6 (medium red): 4 (red): 1 (dark red). Some of the segregants in the F_2

generation transgresses the limits of both the parents, that is, these transgress the degree of intensity of expression of the parental features in the positive or negative direction and are called transgressive segregants. The red color becomes stronger with an increasing number of dominant alleles, and weaker when they are missing. Thus, the F_2 distribution of kernel color showed quantitative features with continuous variation. Genes affecting formation of a quantitative feature are called polygenes. The effect of a single polygene is small and environment has modifying effects on such gene(s). These ratios and gene actions have been explained in detail in Chapter 3, Genetics in Relation to Plant Breeding.

2. *Number of alleles at each locus*: the number of alleles segregating for each gene will also determine the amount of variability in the F_2 generation. Multiple alleles may not be a serious consideration in self-pollinating diploid crops so long the crosses are attempted between two pure lines. The maximum number of alleles in F_1 and F_2 generation is two for a diploid inbred parent or pure lines—one allele contributed by each parent; therefore, will have two homozygotes and one heterozygote in the F_2 generation. However, for non-inbred parents or heterozygous parents or complex crosses multiple alleles may be involved, and with more than two alleles at each locus the number of possible genotypes further increases. Two loci each with two alleles segregating would result in nine different genotypes in the F_2 generation. Similarly, for three loci each with two segregating alleles, 27 kinds of possible genotypes in F_2 generation will be present.
3. *Linkage among segregating gene loci*: due to linkage certain genes situated on the same chromosome are inherited as a block from parents to offspring. The linkage has no effect on gene frequency and to the homozygosity with each generation of selfing; however, it has an effect on the occurrence of specific gene combinations. This means the genes which are more closely linked in parental combinations (non-recombinant types) will be appearing in higher frequencies to recombinant in succeeding generations than expected with random segregation. The magnitude of this effect would depend upon the strength of linkage (Table 8.3).

   If two desirable genes are linked, then linkage is an advantage to the plant breeder. For example, if two desirable genes are closely linked (say, 0.10 map units apart), 20.25% of the plants in the F_2 would have both homozygous desirable genes, whereas only 6.25% plants would have both homozygous desirable genes if the two genes were independent. However, if a desirable gene is closely linked to an undesirable gene the linkage is serious disadvantage to the plant breeder. In order to get separated the desirable and undesirable gene combinations large population size must be grown to identify those rare recombinants that need to be selected and advanced.

Table 8.3 The effect of linkage on the proportion of AB/AB (homozygous genotypes) expected in the F_2 generation from a double heterozygous AB/ab (coupling) or Ab/aB (repulsion).

	Percent of AB/AB or ab/ab in the F_2, if the F_1 is	
Cross over value (*p*)	**AB/ab (coupling)**	**Ab/aB (repulsion)**
0.50 (no linkage)	6.25	6.25
0.30	12.25	2.25
0.10	20.25	0.25
0.05	22.56	0.06
0.01	24.50	0.0025
P	$(1/4)\ (1-p)^2 \times 100$	$(1/4)\ p^2 \times 100$

Handling of segregating generations

As previously explained, genetic segregation will begin with the F_2 generation (in a two parent cross, where each parent is homozygous and homogenous) and heterozygosity will be reduced by one-half with each succeeding selfed generation. The larger the number of genes in which the crossed parents differ from one another, the smaller is the prospect of obtaining the desired combinations in the F_2 especially if they are caused by recessive genes. In every population of self-fertilizing crops, the homozygous plants will only have homozygous offspring and the heterozygotes will always segregate to homozygous forms leading to an increase of homozygotes and decrease of heterozygote (Table 8.1). The homozygous forms can be calculated for each generation with freely segregating genes as, $[(2^m - 1)/2^m]^n$, where m is the number of generation of self-pollination, and n is the number of genes segregating (Table 8.1). A problem often faced by a plant breeder is how many plants should be raised to be 95% or 99% sure of recovering one or more plants exhibiting the trait, that is, the number of plants necessary to recover a trait (Table 8.4). Size of F_2 population will depend upon breeding objective, that is, number of genes and characters to be combined, facilities available, and the crop. Usually 1000–10,000 plants may be grown as described previously.

Reduction of inter-row competition

One major problem in selection nurseries is early segregating generations inter-row/plot competition, that is, the effect on the performance of a particular progeny row of adjacent or neighboring rows. One of the factors responsible for inter-row competition is variation in plant height in cereals and growth habit in legumes in the early generation nurseries (bulk or short rows, e.g., F_2, F_3) as well as in early generation yield testing plots. Some of the ways to minimize such inter-row effects in breeding trials are: (1) to select and grow morphologically similar materials, for example, semidwarf and

Table 8.4 Number of plants (*n*) necessary to recover a required number of plants with trait (From Sedcole, 1977).

		r = number of plants to be recovered								
p	***q***	**1**	**2**	**3**	**4**	**5**	**6**	**8**	**10**	**15**
0.95	1/2	5	8	11	13	16	18	23	28	40
	1/3	8	13	17	21	25	29	37	44	62
	1/4	11	18	23	29	34	40	50	60	84
	1/8	23	27	49	60	71	82	103	123	172
	1/16	47	75	99	122	144	166	208	248	347
	1/32	95	150	200	246	291	334	418	500	697
	1/64	191	302	401	494	584	671	839	1002	1397
0.99	1/2	7	11	14	17	19	22	27	32	45
	1/3	12	17	22	27	31	35	44	52	71
	1/4	17	24	31	37	43	49	60	70	96
	1/8	35	51	64	77	89	101	124	146	198
	1/16	72	104	132	158	182	206	252	296	402
	1/32	146	210	266	218	268	416	508	597	809
	1/64	293	423	535	640	739	835	1020	1198	1623

Where n *is the number of plants necessary,* q *is the probability of the occurrence of the trait, and* p *is the probability of recovering* r *plants with the trait.*

tall lines, in adjacent but in separate trials, (2) plant wider plots, (3) discard edge rows, (4) harvest only center rows of each plot, and (5) plant guard row on either side of the test plots. Another factor to consider is to grow breeding populations that are more similar (genetically, agronomically, morphologically) in progeny rows and early yield trials to avoid inter-varietal and inter-row competition. Chapter 26 on field plots designs in plant breeding describes these aspects and solutions in more detail.

CHAPTER

Mass and pure line selection

9

Abstract

Artificial selection has been practiced for thousands of years by humans to make improvements in plant species. Mass selection is one of the earliest methods of artificial selection that enabled domestication of crop plants. Another important type of selection was performed in landraces that are heterogeneous consisting of multiple genotypes and selection of individual plant progenies gives rise to pure lines. The pure line and mass selection theory, procedures, genetic basis, their merits, and limitations are described. The information on the application of mass selection and pure line selection for variety purity and breeder seed production are included.

Natural selection has resulted in the evolution and domestication of new species. Along with natural selection humans have knowingly or unknowingly practiced some selection as per their needs (i.e., artificial selection, or generally referred as "selection"). One of the key difference between "selection" and "natural selection" is that in artificial selection there is non-random variance in survival or reproduction of entities, which may be at an individual or category levels and is not imposed by nature alone (as is the case in natural selection). Artificial selection describes the deliberate choice of individuals for breeding in each generation and the advancement of select individuals. Artificial selection has been practiced for thousands of years by humans to make improvements in plant species. For example, artificial selection led to the rise of modern maize from its progenitor, teosinte. Mass selection is one of the earliest methods of artificial selection. The success of selection depends on the presence of genetic variability.

Mass selection was used in ancient times by farmers to select heaviest or largest seed by wind/hand/sieves or by selecting plants in field/cobs/panicle from the standing crop(s) that met human needs. The populations of self-fertilized crops contain a mixture of different groups of individuals. These groups are uniform within themselves; however, they are different from each other and serve as raw materials for selection. During the process of selection, a few individuals having certain superior characteristics are selected and are used as parents for the next generation or progeny. These are examples of directional selection, which is a form of artificial selection in which phenotypically superior plants are chosen for

Plant Breeding and Cultivar Development. DOI: https://doi.org/10.1016/B978-0-12-817563-7.00018-0

breeding. As a result of selection, the hidden genetic variability of populations are shifted. Therefore, selection can be defined as a process either natural or artificial through which individual plants or group of plants are sorted out from the mixed population. (Note: in plant breeding context, where a breeder makes artificial selection there may be element of natural selection due to prevailing weather, soil, and air variables).

Selection was practiced in ancient times by farmers. Early crop breeders during the 18th and 19th century, through selection, developed numerous varieties of many important self-pollinated crops. In general, the progenies from superior single plants were more uniform than the remaining population. Different progenies had different agricultural values. Selection was effective only when differences existed in the population of original variety (land variety). Selection was widely used for the improvement of major self-pollinated food crops such as wheat, oat, barley, and grain legumes.

The important features of selection in self-pollinating crops are as follows:

- Selection is effective only if heritable variation is present.
- Selection directly does not act on genes but acts on genotypes and as a result, the frequency of genotypes is altered.
- Selection cannot produce new genes; however, new genetic combinations may be produced through natural and/or artificial cross-pollination (i.e., through recombination).

The land varieties or landraces are heterogeneous and are the original source materials for selection. Such varieties consist of population mixtures containing a number of genotypes, largely homozygous types. Natural selection in land varieties/races has resulted in cultivars well adapted to the local environment. The individual plant progenies of homozygous types give rise to "pure lines". Pure line is the progeny of the single self-fertilized homozygous plant. Natural cross-pollinations between pure lines produce heterozygous plants and homozygous types in the following generations.

The pure line theory

The Svalof Experiment Station established by Swedish Seed Association in 1866 developed pure line selection. The pure line theory was proposed by W.L. Johannsen in 1903. He obtained commercial seed of Princess variety of bean (*Phaseolus vulgaris*), a highly self-fertilized crop and studied variation for seed weight. The seed lot Johannsen studied had seeds of different sizes. He selected seeds (of different sizes) and grew them separately. He chose 19 progenies as distinctly different lines (pure lines) originating from original commercial seed lot and determined seed weight of each. The plant to progenies were planted and seeds were harvested from each of the growing plants. The line No.1 had

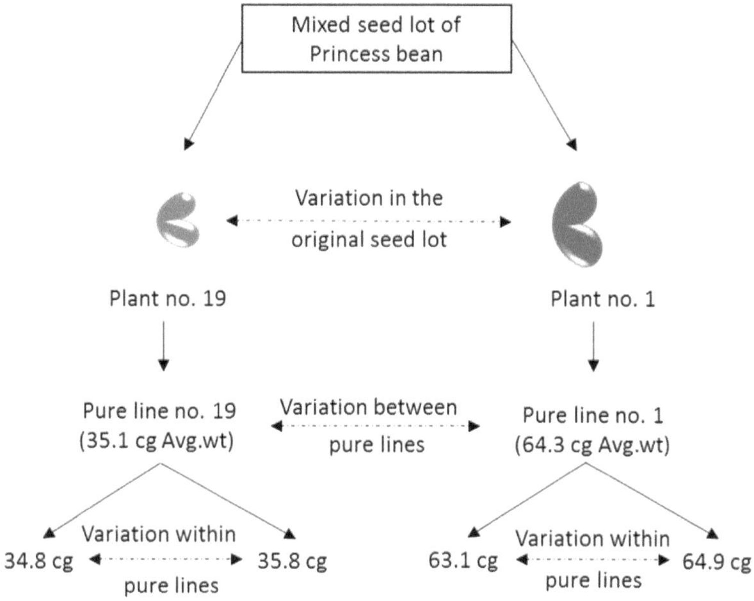

FIGURE 9.1

Between and within progeny variation in a commercial seed lot of Princess bean variety by W.L. Johannsen.

heaviest seed weight of 64 centigrams (cg) and line No.19 was lightest with seed weight of 35 cg covering a continuous range of seed size in between (Fig. 9.1). He first separated seeds in each pure line into classes of 10 cg size and grew them separately. He observed that classes of different sizes within a pure line produced progeny with the same average seed weight. The categories of 20-30, 30-40, and 40-50 were noted in line No.13; however, the average weight of progenies were 47.5, 45.0, 45.1, and 45.8 cg, respectively. This suggested that within a pure line seeds of all sizes produce progeny with the weight that is characteristic of the line. He further selected large and small beans in each line for six successive generations (Table 9.1), and in each generation the light line was selected for light seeds and the heavy line for heavy seeds. After six generations of selection in line No.1, the mean weight (progeny) of light line was 69.07 cg and of heavy line was 67.66 cg. This suggested that the average weight of each line was remarkably constant generation after generation. Parent–offspring correlations within pure lines and between 19 pure lines were worked out. Within pure line No.13 no relation between seed weight of parent–offspring was recorded ($r = -0.18 \pm 0.038$), whereas significant parent–offspring correlation (0.336 ± 0.08) was observed for the mixed population.

Table 9.1 Effect of selection for 6 generations in line No. 1 of the Princess bean variety.

Year	Mean weight of parent seed (cg)			Mean weight of progeny seeds (cg)		
	Light line	Heavy line	Difference	Light line	Heavy line	Difference
1902	60	70	+ 10	63.15	64.85	+ 1.70
1903	55	80	+ 25	75.19	70.88	− 4.31
1904	50	87	+ 37	54.59	56.68	+ 2.09
1905	43	73	+ 30	63.55	63.64	+ 0.09
1906	46	84	+ 38	74.38	73.00	− 1.38
1907	56	81	+ 25	69.07	67.66	− 1.41

After Johannsen, W.L., 1926. Elements der exacten Erblichkeitslehre. Gustav Fischer, Jena.

From the experiment of W.L. Johannsen following conclusions could be drawn:

- The original seed lot of commercial variety of bean had a mixture of several pure lines and the variation (in seed lot) was both genetic and environmental.
- Selection was effective in separating bean seeds.
- After a pure line was developed, further selection within the line was ineffective because within line variation was only due to environment and non-heritable. A pure line was defined as the progeny of a single self-fertilized homozygous individual.
- The progeny test is important in plant breeding to differentiate the effect of heredity (genes) and environment toward phenotype. The phenotype reflects the effects of genotype, that is, genetic worth of selected plants and environment.

Genetic basis of pure line selection

Johannsen through his experiments proved that the variation for seed weight in the original commercial seed lot of Princess bean variety was due to both heredity and environment and the variation within pure line(s) was due to differences in micro-environment of the plants of these lines. The selection of 19 pure lines in a self-pollinated crop was the result of homozygosity attained through continuous self-fertilization for several generations. Mendel demonstrated the effect of inbreeding, and showed that continued self-fertilization caused a decrease in heterozygote Aa by one half per generation (Table 9.2).

This showed that in a population after four or five generations of selfing the heterozygotes were reduced to a small number and homozygotes AA and aa are increased to an equal proportion. Reduction in heterozygosity at each locus occurs at the above rate, irrespective of the number of other heterozygous loci.

Table 9.2 Percentage of heterozygous and homozygous plants after self-fertilization of a heterozygous individual at a single locus (Aa).

Generation of self-fertilization	Heterozygous (%)	Homozygous (%)
S_0[a]	100	Nil
S_1	50	50
S_2	25	75
S_3	12.5	87.5
S_4	6.25	93.75
S_5	3.125	96.875
S_{10}	0.098	99.902
S_m	$(1/2)^m \times 100$	$[1 - (1/2)^m] \times 100$

[a]$S_0 = F_1$.

For "n" heterozygous gene pairs the proportion of completely homozygous plants after "m" generation of self-fertilization is computed as: $[(2^m - 1)/2^m]^n$

All heterozygous loci approach homozygosity at the same rate, that is, homozygosity at a given locus is not affected by the number of gene pairs (n). It is computed as: $(2^m - 1)/2^m$.

The proportion of completely homozygous individuals increases with slow rates as the number of gene pairs (n) increases. However, increase in the rate of homozygosity is independent of the number of genes (n). As the number of genes increases, the proportion of completely homozygous plants after a given number of generations decreases. The effect of selfing is so strong that 100 independent genes after 10 generations of selfing would render more than 90% of population completely homozygous at all loci. The above explanations hold good if there is equal survival of all gametes and genotypes and all the genes are inherited independently. Differential survival of genotypes would alter the rate of homozygosity. Genetic linkage would increase the proportion of homozygous individuals in any generation but not affect the percentage of homozygosity. As a result of long term or continuous selfing, the population includes different kinds of 2^n homozygous families/types for a character, where n represents the number of heterozygous gene pairs. A very large number of families is possible even when a small number of gene pairs are segregating.

Pure line selection

It is one of the oldest and simplest method of breeding. John Le Couteur and Patrick Shirreff initiated the practice of selecting single heads or plants. Johannsen through his experiments with Princess bean cultivar and inheritance of seed size laid the scientific foundation to pure line selection. Pure line cultivars are developed in self-pollinating species. Pure line cultivars are homogeneous and homozygous and, once created, can be maintained indefinitely by selfing. However, pure lines can develop variability and sometimes have some residual heterozygosity; therefore, breeders have been able to exploit it to develop improved cultivars.

Inbred lines are different from pure lines although in literature, at times, these two terms are used interchangeably. It is important to distinguish them: (1) inbred lines are developed in a cross-pollinated species through inbreeding, and these newly developed inbred lines are used as parents in the production of hybrid and synthetic cultivars; (2) inbred lines are not meant for commercial release to the farmer for commercial production because inbred lines suffer from inbreeding depression. Their yield will be lower than that of a hybrid and even an open-pollinated variety; and (3) inbred lines are homogenous and homozygous, similar to pure lines; however, manual selfing or sib-mating must be done artificially at each generation to maintain or increase seed.

Pure line selection procedures are based on selection of individual plants and their progeny testing. The selections are made from a genetically variable population such as landrace cultivar. Contrasting with pure lines, landraces are highly heterogeneous generally consisting of a broad genetic base providing a larger genetic variation. The additional value comes from accumulated favorable genes for adaptation and disease traits, as well as at times represent alleles that have been selected by humans for sensory nuances. The variation between plants may be large and can occur for only one or two or many distinct characters such as resistance to a disease. Many of the improved cultivars of the self-pollinated crops, particularly of cereals and grain legumes during the first half of the 20th century, were developed worldwide by pure line selection, for example, Croscat and L67 pure line cultivars of common bean.

General procedure of pure line selection

Step I: Selection of large number of plants or heads from the original heterogeneous population. The number of individual plants is not a fixed amount but varies according to the resources available. However, 500–1000 plants selections are made based on desired phenotypic traits such as plant height, maturity, disease reaction, seed color, straw strength, or head type, etc., from the genetically variable original population (landrace or land variety) in the field. The base population may be space planted at a low seed rate so that superior individual plants can be clearly identified. Selection could be made on more than one occasion: (1) if there is wide range of ripening or maturity dates within the base population, and (2) select plants that stand well (i.e., show better reaction against biotic or abiotic stresses) after adverse weather conditions or disease and/or pest epidemic or in artificially created stress nurseries. This is most important step of pure line selection because nearly all of the genetic diversity exists between plants and the lines derived from them (and very little variation is observed within lines). The desirable types omitted at this stage cannot be recovered later on.

Step II: The seed of selections (mother plants) in step I are sown in unreplicated progeny rows. The parameters of step I are better observed in the progeny rows, that is, between progeny selection is practiced. The progeny rows with uniform phenotype and the desired characteristics are saved for further testing. The undesirable progeny rows are discarded. The visual evaluation would enable the rapid elimination of lines with obvious defects or with no particular merits. Natural epidemic or artificial screening may further assist in this process. Superior lines could be sampled and grown in the same way in the following year for reassessment (Step III or directly to Step IV).

Step III (Optional, if genetic variation within progeny rows in step II is noticed): Seed from selected progenies of step II may be sampled to grow progeny rows. This is to eliminate undesirable selections and to reduce large number of selections to a few promising selections. Selections that have the potential of producing commercially acceptable cultivars are maintained.

Step IV: Replicated field trials of selected progeny rows are conducted for agronomic evaluation. The existing popular cultivar(s) should be used as check(s). The trials may be conducted at more than one location for 2–3 years along with appropriate agronomic, disease, and quality checks in appropriate tests.

Step V: Selection of the best progeny row (pure line) for seed multiplication (this can happen concurrently with testing in Step IV) and release.

How pure line become impure?

The varieties of self-pollinated crops do not remain pure or homogeneous over the years for all the characters, that is, during the process of cultivation after their release, pure line cultivars show variation for some trait(s) which could be qualitative or quantitative. The degeneration or deterioration in pure line could arise due to the variation from the following sources.

Mutation: In evolution and development of new plants, specific gene mutation(s) and both auto- and allo-polyploidy have provided the great majority of raw materials in the origin of genetic variation and have been useful to plant breeders. The chromosomal aberrations have been of limited use to plant breeders. Specific gene mutations (point mutations) have been studied in detail; however, mutations in quantitative traits or characters have not been studied thoroughly due to the difficulty in distinguishing their phenotypic effects. In general, mutations occur spontaneously and are random, recurrent and have specific effect on phenotype. The frequency of mutations varies from one gene to another.

In the past, farmers and plant breeders have selected favorable spontaneous mutations for a character and released superior cultivars of self-pollinated crops directly or used these in cross breeding. For example, semidwarf wheat ("Norin 10") and rice ("Dee-geo-woo-gen") cultivars credited for the "Green Revolution". These spontaneous mutations were used to develop short, strong strawed varieties that produced high grain yields in response to fertilizer inputs. There are several examples of useful spontaneous or induced mutations having been used in the on-going breeding program for the genetic improvement of field crops such as mutations for male sterility (e.g., for hybrid production), quality traits (e.g., brown mid-rib mutant in maize for better digestibility), and disease and pest resistance. Not all mutations have a positive impact. The unfavorable mutations have resulted in deterioration of pure line cultivars, and at times necessitated going back to breeders rows to produce cleaner version of breeder seed for seed production and certification. See chapter 14: Mutation Breeding for more details.

Natural crossing and recombination: In self-pollinated crops, there is some amount of cross-pollination. The natural cross-pollination between two diverse pure lines for character(s) may produce genetic variation. The spontaneous mutations also occur regularly in a cultivar (as described above) and the natural cross-pollination of these with pure lines would also produce genetic variation (Note: the selfing of heterozygotes, e.g., a later filial generation seed or progeny row, through segregation and recombination produces new phenotypes). The new phenotype(s) with improved agricultural value could be used for development of new pure line(s); however, generally speaking this is rare. The harmful or undesirable genotype (based on phenotypic traits or genetic screening) if not identified in time and rouged out, leads to the deterioration of pure line cultivar(s) in due course of time. Due to these considerations, care is taken to produce breeder seed of cultivars of self-pollinated crops in isolation from other cultivars.

Mechanical and field-based mixture: Mechanical contamination with other cultivars is possible during sowing, harvesting and threshing, and storage. Some varieties/crops have seed dormancy, as in several grain legumes, which could lead to varietal admixture. This can happen if same plot is used year after year for the seed production of same crop. These mechanical admixtures could be avoided if care is taken during the cultivation processes, including not planting the same crop species in seed increase fields in successive years based on reported length of seed dormancy in their crop species, and follow proper guidelines to avoid varietal admixtures.

How long does a pure line remain pure?

This depends upon the particular crop, its genetic stability (accumulation of gene mutations over a long time and the amount of natural cross-pollination). The stability is influenced by the care with which seed is produced. If the presence of off-types are carefully monitored and eliminated, purity can be maintained for a longer time.

Merits of pure line selection

1. Pure line varieties have high degree of uniformity of the end product, which is desired by consumers.
2. Pure line is a useful method in self-pollinating crops, which has been neglected by breeder for genetic improvement in crops with insufficient breeding efforts.

Limitations of pure line selection

1. Improvement is limited to the isolation of the best genotype(s) already present in the mixed population.
2. The varieties developed through pure line selection have high degree of uniformity, which makes them vulnerable to pests and diseases.
3. Development of an outstanding pure line would lead to fast erosion of genetic variation.
4. Pure line selection is only applicable in crops with limited prior breeding or in orphan crops.

Mass selection

It is one of the earliest methods and has been used for the improvement of landraces of self-fertilized crops. However, in present time consumers require uniformity in the end product of crop species. This shows that there is limited utility of mass selection in self-fertilized crops compared to pure line selection in breeding. Mass selection or phenotypic recurrent selection is described in Chapter 17, Recurrent Selection in Self-Pollinated Crops. In breeding of self-fertilized crops, mass selection has two uses:

a. Improvement of landraces.

Landraces/varieties that are grown in rainfed and marginal land, can benefit by mass selection method for rapidly improving these varieties through phenotypic mass selection and discarding obviously unproductive and defective plants/individuals (too early or too late, susceptible to diseases and varying in plant height etc.). During phenotypic mass selection, the agronomic and horticultural features of variety for adaptation and yield are retained. The size of population to be selected for mass selection may vary from several hundred to a few thousand individual plants. The selected plants are harvested and seed is composited without progeny testing. However, it will be better if the seed of each selected plants is harvested separately and plant-to-progeny testing is carried for the purity and resemblance with the original variety (see pure line selection above; the progenies showing variation with the parent variety are discarded and the rest of lines are harvested in bulk). The mass selection may be repeated for one more year so that remaining heterozygous off-type plants are removed. Such cultivars after a few generations of selfing would consist of population of phenotypically similar but otherwise different pure lines. It would provide stable yield against pests and diseases and adverse environmental conditions.

b. Maintenance of pure lines cultivars and breeder seed production.

Mass selection is largely being used for maintenance breeding, that is, to maintain the genetic purity of the existing pure line varieties of self-fertilized crops. The other use is to produce breeder seed. The original increase plot of the highest purity seed source is grown. The off-type progenies are observed regularly at different stages of development and mutants, natural hybrids, and varietal admixtures are removed before maturity to avoid cross-pollination. The remaining plants are bulk harvested. These are grown in a bulk plot again next season (or year) and the cycle can be repeated, if further purity is desired.

While mass selection is an easy and quick method, genetic purity and phenotypic uniformity can be an issue. The other method for breeder seed increase in row grain crops is progeny testing. In the progeny testing method, candidate variety is grown in an increase plot, and some 100–1000 representative individual plants from the candidate variety are selected. The number can vary according to crop and desired seed amount for seed certification and production. These individual plants are seeded into progeny rows (called breeder rows). Selection is practiced within and among progeny rows to remove the non-conforming plants within a row, and non-conforming rows among its sib-rows. The remaining progenies are each separately harvested in bulk, remnant seed per row is kept in seed reserve and the rest of seed from each pure row are bulked to constitute breeder seed. The breeder seed can be recreated from the reserved sample from each breeder row, by repeating the process as often as necessary to maintain purity. The biggest advantage of progeny testing over mass selection for breeder seed production is genetic purity and phenotypic uniformity. Disadvantage is that it takes more time and seed volume may be less (depending on the crop).

The main classes of certified seed are:

- (a) breeder seed (or prebasic seed)—directly produced by plant breeder that developed that cultivar,
- (b) foundation seed (or basic seed)—directly produced from breeder seed by qualified seed growers, experimental stations, or approved agencies. Needs very experienced staff to ensure criteria are met,
- (c) registered seed—if needed, and
- (d) certified seed—acceptable genetic identity and purity to be approved and certified by certifying agency.

There are standard color tags that identify each certified seed class: (1) white—breeder or foundation seed, (2) purple—registered seed, and (3) blue—certified seed. In the United States, each state has its own certifying agency, and the role of certifying agencies is to verify the genetic identity and purity of seed through required field and seed inspections. Private seed companies that do not use the seed certification system maintain their own high standards of genetic identity and purity as farmers will show trust in the information provided by the seed company. However, each country has their own regulations for certification, and breeder should contact the national certification agency to learn the procedure applicable in the country.

The general steps of the two important uses of mass selection, that is, improvement of landraces and the maintenance of purity of pure line cultivars/breeder seed production are different and these are described as follows (Table 9.3):

Table 9.3 General procedures of mass selection

Improvement of land races/varieties	Maintenance of genetic purity of pure lines
Step I. Phenotypic mass selection of large number (few to several thousand) of plants (or ears) from a variable population of land variety and discard the remaining population at maturity. The selected plants are harvested and seed are bulked together without testing. Normally too early or too late types and susceptible plants are discarded during the selection.	**Step I**. Some 100–1000 individual plants representing the parental pure line variety are selected from a large field of increase (preferred source) or yield testing plot (unharvested rows so as not to bias yield test results).
Step II. The population from the bulked seed of step I grown next year, and the process of phenotypic mass selection is repeated.	**Step II**. The progeny (seed) of the selected plants in step I is grown in next year. The off-types progenies; including spontaneous mutants, recombinants from out-crosses, and admixtures with other varieties are observed at different growth stages of the crop and are roughed out before maturity. The seed from remaining progenies is harvested in bulk to form pure seed. Reserve bulk seed is kept for future increase and stored properly.
Step III. The yield testing to compare original and selected population may be conducted, and land variety is released for cultivation.	**Step III**. The steps I and II may be repeated every year or as required, primarily to obtain sufficient and required seed amounts.

Merits of mass selection

1. It is a useful method of rapidly improving widely adapted landrace(s) without making drastic changes and with minimal efforts.
2. Mass selection is an important method for the maintenance of purity of released pure line cultivars of self-fertilized crops.

Limitations of mass selection

1. Plants are selected based on their phenotype alone and not genotypic testing. The phenotypic selection need to be repeated as it is not possible to know whether the selected plants are superior due to hereditary or environment.
2. Improved varieties are less uniform as compared to pure line varieties for the end product and the quality.

CHAPTER

Bulk method

10

Abstract

The bulk or population method is considered one of the easiest and cost- and resource-efficient method of handling segregating populations. This method is useful when the parents used in creating breeding families (i.e., breeding populations) are of adapted types, and characters under consideration are polygenic in nature. The main steps are to grow the segregating progenies as breeding populations year after year (or season after season) until a satisfactory degree of homozygosity is reached. In early segregating generations, populations are planted in tight plant-to-plant spacing (creating interplant competition) in bulk plots, generally from random sample of seed from the previous season total harvest. Bulk method can be modified as per the need, facilities available, and in program objectives. In this chapter, features and procedures of bulk method are described, along with the genetic basis, applications, merits, and limitations.

The Swedish plant breeder Nilsson-Ehle in 1908 proposed bulk method for the first time to combine winter hardiness of the "Squarehead" variety with high yield of the "Stand-up" variety of winter wheat. He emphasized that growing large populations would increase the chance of high yielding types appearing among the winter hardy selections (Newman, 1912). Nilsson-Ehle enabled natural selection by discarding plants that had suffered winter damage but produced some seed, increasing the rate of shift toward hardy types.

Natural selection operates during bulk method of breeding and the frequency of unadapted types is decreased in the population due to interplant competition and adverse environmental factors. At times, natural selection can have an adverse and unintended effect on the frequency of desirable types as a result of interplant competition and random sampling of seed for sowing the segregating generations. In general, during bulk breeding, artificial selection is not practiced; however, there is always an opportunity to remove plants with obvious defects like susceptible types, that is, some negative type selection may be followed. Due to the excessively simplistic nature of this method, modifications have been proposed to increase the likelihood of selecting favorable genotypes. Florell (1929) advocated the use of artificial selection to eliminate undesirable and weak plants in bulk method of breeding. Allard (1960) suggested that there is no need to depend entirely only on natural selection. The

Plant Breeding and Cultivar Development. DOI: https://doi.org/10.1016/B978-0-12-817563-7.00026-X

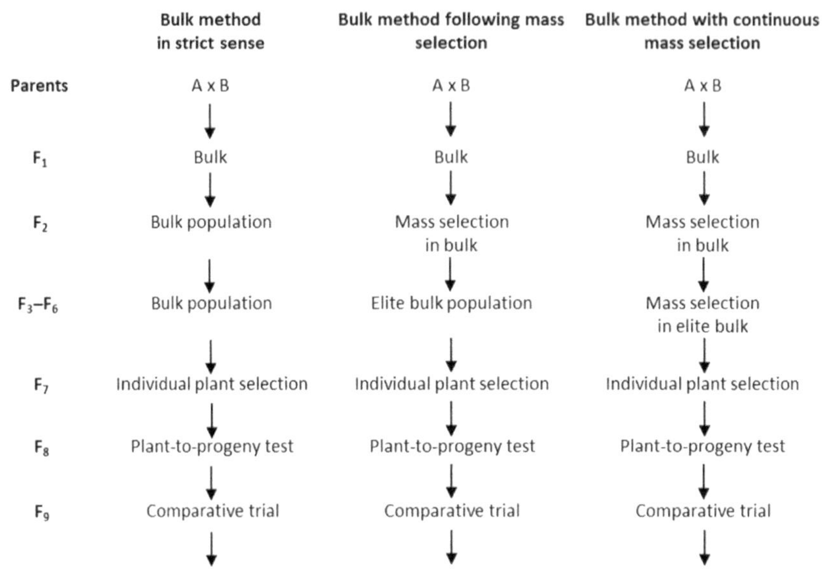

FIGURE 10.1

Different versions of the elite bulk and mass selection in elite bulk method of breeding.
After Mac Key (1962).

selection for character of interest can be practiced at any time during the period of bulk propagation. Mac Key (1962) has suggested some kind of positive/negative mass selection, which lowers the proportion of undesirable types (Fig. 10.1) and size of population during bulk breeding can be more or less restricted. He suggested that the differences of morphology, maturity and disease resistance, etc., may be useful criteria for mass selection, and natural selection is sometimes useful for characters such as winter hardiness, preharvest sprouting, and seed shattering. Other suggestion for visual selection includes plant height, resistance to diseases and maturity duration in early generation bulks during bulk breeding (Jensen, 1988). Once the breeding population(s) reach desired level of homozygosity, for example, F_5 to F_7 generations, population(s) are space planted, desirable plants are pulled, grown in progeny rows, followed by yield testing (similar to pedigree method of breeding; see chapter 11 - Pedigree method).

General procedure of bulk method

For the success of bulk breeding method, it is desirable that larger population of segregating generations are grown under appropriate conditions for the variety to be bred. This would help in the expression of all gene combinations with the expected probability. The natural selection would proceed as expected. The general outline of bulk method in different generations in a small grain crop is described as follows and are also presented in Fig. 10.2.

Selection of parents: Desirable parents are selected and desired number of crosses are made between them. Each cross creates a breeding family (or breeding population). The following scheme is presented per breeding population (i.e., each breeding family) basis.

F_1 generation: About 50–100 F_1 plants are space planted along with parents and after removing the selfed F_1 plants, harvested seed is bulk harvested.

F_2 to F_4 generations: Two to three thousand plants are grown and entire plot at maturity are harvested in bulk. A random sample of seed from the total harvest is used for sowing a similar F_3 plot in the following season. The process is repeated until F_4 generation or until plants reach the desired level of homozygosity. The F_2 to F_4 plots are planted in close spacing and natural selection to abiotic

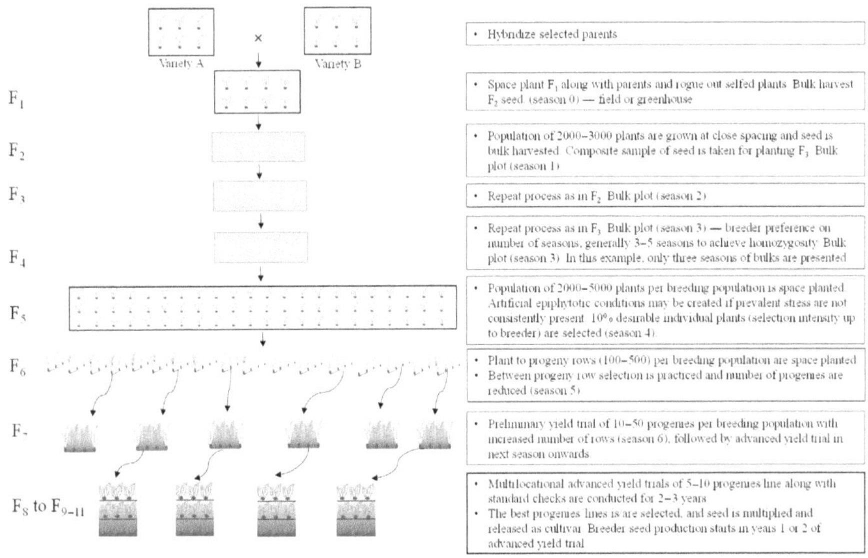

FIGURE 10.2

General outline of bulk method of breeding. This is a hypothetical example for a small grain cereal, and the number of plants, progeny rows, and preliminary and advanced yield trials vary between different breeding programs. *Arrows* shows progeny rows, or entries that were selected for the next generation of testing. In general scheme of bulk method, artificial epiphytotic conditions are not created; however, breeders can utilize such nurseries for important traits, if appropriate. Breeders include relevant checks interspersed in earlier generations (in field where bulk plots are grown), and as a part of experimental design (replicated and randomized) in yield trials. Breeders make adjustments for expected germination percentages to ensure optimum population size before deriving progeny rows.

stresses may reduce the frequency of unadapted types and as a result of self-fertilization high level of homozygosity is achieved.

F_5 generation: The F_5 bulk seed is space planted and the population size could be increased (2000–5000 plants). This would help in the better expression of individual plants as well as for expression of genetic variation. Some 5-10% of the desirable single plant selections are made. Selection intensity can be higher or lower as per breeder preference.

F_6 generation: The superior individual plants (100–500) selected in F_5 are harvested separately and the seed is grown in plant-to-progeny rows, like those in pedigree method for their evaluation. The selection in F_6 is practiced between progeny rows. The less productive (inferior) progeny rows are rejected. Selection can be based on seed yield, agronomic traits, adaptation traits, disease and quality traits. Some 10–50 superior progeny rows are bulk harvested (per progeny row) separately.

F_7 generation: The desirable progenies selected in F_6 are put in a preliminary yield trial in larger plots and after selection (by the end of season) number of progenies/lines are reduced to 5–10. The superior progenies are bulk harvested.

Note: The population size F_5 to F_7 may be reduced considerably. Artificial selection pertaining to pests and diseases may be exerted during this period in specialized nurseries in addition to yield tests.

F_8 to F_{10} generations: Multi-location advanced yield trials are conducted for 2–3 years. During this period emphasis would be on evaluation of agronomic and quality traits. The number of lines/progenies are reduced further. The best progenies (or strains) or progenies is/are selected (from all breeding families that are undergoing selection in the same generation that year) and recommended for release as cultivar.

In each generation of yield tests, appropriate experimental designs are needed including relevant checks. These checks can also serve as benchmark to determine selection culls levels.

Modifications of bulk method

The schemes of bulk method can be modified by the breeder as per the need, facilities available, and the program objectives. The selection could be imposed during any generation but selection should not be so restrictive that the population size of the bulk is severely reduced. Harlan et al. (1940) in United States suggested that the bulks should be put in comparative trials and evaluated for agronomic characteristics and the populations may be selected for further use in breeding. The modifications in bulk method following mass selection in F_2 and bulk method with continuous mass selection have been suggested by Mac Key (1962). Early Generation Testing (EGT) describes the procedure for selecting superior lines of families before they are homozygous. It also refers to a specific use where the genetic worth of a breeding population is determined by analyzing yield data from a segregating (early generation) bulk plot, and entire populations deemed inferior after testing, are removed. Higher yielding populations are advanced for further assessment and segregating populations can be handled with any of the methods (pedigree, single seed descent, and bulk). In a modified method, for example, yield testing is started in an early generation (e.g., F_2, F_3, or F_4) to make selections. The early generation bulks are grown in yield plots (two or four row plots); therefore, more resources are required to handle EGT to assess the genetic worth of a bulk population(s) for advancement in the breeding pipeline. Nonetheless, EGT allows elimination of breeding populations that are inferior through the use of replicated and multi-environment testing. In addition, selection for traits of lower heritability can be practiced to discard inferior populations. Thus, EGT testing in this scenario can be done for one or two generations followed by selection of superior plants and yield testing of these lines.

This method and its variations have been popular for a long time. Harrington (1952) used the mass-pedigree method for developing superior varieties of cereals. He recommended growing 10,000 F_2 plants per cross. Harvesting one ear per plant, bulking these to form the F_3, repeating the procedure under suitable selection pressures, in the F_3 through F_5 as in the F_2 generation. About 5000 plants may be grown per cross in the F_5 and after selection in the field and in the laboratory, approximately 500 plants may be retained to grow F_6 as ear rows and about 15% are selected, advanced to F_7 and put into trials. Jensen (1988) has suggested that the desirable segregants for characters like suitable maturity duration, short plant height, and resistance to diseases may be selected based on visual rating in the F_2 or yield testing of F_3 progenies followed by one bulk of all selected lines remaining for F_4 and F_5 from which superior lines may be selected. This may be a useful scheme as it could reduce the problem of numbers and also interplant competition as well. Some small grains breeders, use a modification of the bulk breeding method, such as a combination of bulk method with the pedigree method or other methods. Numerous modification can be utilized to piece together the advantages of different methods, such as growing early segregating generations (F_2 and F_3) under the bulk method, and subsequent generations are advanced using a pedigree method approach (see Brown and Forsberg, 1987). Allan (1987) suggested that when the bulk population contain contrasting alleles for morphological characters, the plants at F_6 could be subdivided into several phenotypic groups to prevent the loss of desirable but less competitive genotypes.

The bulk method allows natural selection to act and remove undesirable genotypes from the population (i.e., per cross). The choice of growing environment will dictate the type of traits to be selected for or against; therefore, care needs to be exercised in selecting environments that are suitable for realizing the objectives of the program. While natural selection was the primary selection force when this method was developed, most cultivar development breeders use a modified approach to integrate stress (abiotic, biotic) regime by setting up specialized plots to make selections, on a per plant basis, from each breeding population. In all breeding methods, natural selection is always operating but the artificial imposition of selection criteria allows to favorably change the desirable gene and allele frequencies compared to natural selection alone. It is important to note that most current agriculture production happens in human modulated conditions (e.g., fertilizers, sometimes irrigation too); therefore, plant and genotype selections need to happen in experiment conditions mimicking agriculture production field practices.

Salient features

- Segregating generations of each breeding population are advanced to homozygosity through bulking material within populations.
- This is a low cost, less technical method of breeding.
- Natural selection is used to remove undesirable plants, which may or may not imply suitability to agriculture production system and fields.
- In modified bulk method, an artificial selection environment can be used to select for a trait of interest. Bulks can be grown in a disease or other stress nursery to select for that trait. Markers can also be utilized to select for desirable traits to constitute the bulks. Another modified bulk method may include, single plants or inflorescences per plant are selected at each generation; while in the bulk method, plants from the entire population are harvested and seeded (full or subsample) in the next generation. Another modification can be if each breeding population is subjected to multiple methods for leading to homozygosity and selection, for example, combining features of bulk and pedigree methods in the same population.
- EGT of bulk populations may be done for yield testing to make a decision on population retention based on ranking among populations.
- It is not necessary that a genotype with better adaptation and survival ability is also agronomically the best. It is ideal to develop a variety that combines both adaptation and superior agronomic performance.

Genetic basis of bulk method

The continued bulking of crosses developed from wide range of germplasm for several generations (30–40) under the influence of natural selection would cause evolutionary changes as a result of change in the gene frequencies in a population. Therefore, it has also been referred to as evolutionary breeding. It refers to overall worth of a group or the population as a whole through the selective advantage of natural selection. For varietal development, 4–6 generations of bulking would be sufficient. In general, the desirable effect of short-term bulking would be on the number of those genotypes that have better adaptation and survival ability to a particular location and ultimately would reduce the number of genotypes with poor adaptation and extremely sensitive types (these could be eliminated if natural selection is severe). The positive effects are more pronounced on traits related to abiotic stresses like soil stresses and temperature stresses, and biotic stress such as disease resistance etc. The susceptible plants in each bulk may not survive up to maturity and are eliminated altogether or are reduced in number drastically as these produce less number of seeds. Therefore, in each generation the size of bulk may be reduced proportional to the genetic segregation for the stress-related trait.

In the bulk breeding method, the individual plant selection is deferred to a stage, when heterosis and dominance does not fool the breeder to select the best heterozygous instead of best homozygous types when sufficient level of homozygosity is attained. At this stage, each available genotype is represented by several individuals with less chance to escape visual selection by the breeder. Bulk breeding may be a useful approach in small grain cereal (e.g., wheat and barley) to breed for cold tolerance or preharvest sprouting in areas where these problems can be severe. In this situation, breeding families can be moved forward in a bulk method approach and the season those stress conditions are experienced by the crop, natural selection should eliminate susceptible plants in the breeding family and the surviving plants will have a genetic constitution for better stress tolerance to that factor. This method was successfully used in soybean breeding, where population of soybean grown on soil infested with soybean cyst nematode increased the frequency of segregates with resistance to the pest after several generations of inbreeding (Hartwig et al., 1982).

Three historical studies are reported to emphasize the relevant features and effects of natural selection during bulk method of breeding:

1. Harlan and Martini (1938) in their study used seeds of 11 varieties of barley representing a wide range of adaptation types from wide adaptation to specific and even few poorly adapted to all locations. The equal proportion of seeds were mixed and the mixture was grown for 4–12 years under very different environmental conditions at 10 experiment stations in northern and western parts of the United States. Each year, harvest of seed was used to plant two plots the following year. One plot was again harvested in bulk and the other used to take count of plants of each of the 11 varieties in remaining population. At the end of 12 years, at Ithaca (NY, the United States), the variety "Manchuria" was clearly a dominant genotype. This dominance was evident by the second year. None of the cultivars were eliminated until sixth year. "Deficiens" and "Meloy" were lost. This showed that natural selection in barley is an effective tool for the selection of desirable genotypes. One or two varieties were predominant at some of the locations and other varieties at these locations were rapidly eliminated and the change in composition of varietal mixture was slow to moderate depending on location. This could be due to difference in magnitude of environmental stress and the genetic make-up of the varieties.
2. Singh and Johnson (1969) studied the effect of natural selection on morphological and physiological characters under conditions of strong selection in a composite

Table 10.1 Barley varieties involved in the composite cross (From Singh and Johnson, 1969).

Varieties	Genotypes
Beecher	vvbbRR
Fjola	vvbbRR
Hannchen	VVbbRR
Herta	VVbbRR
Husky	vvbbrr
Jet	VVBBRR
O.A.C.21	vvbbRR
Peatland	vvbbRR
Plains	vvbbrr
Proctor	VVbbRR
Sanalta	VVbbrr
Titan	vvbbrr
Trebi	vvbbRR
Vantage	vvbbrr
Velvon II	vvbbrr

Where; vv = 6-rowed, VV = 2-rowed, bb = non-black seed, BB = black seed, rr = smooth awn, and RR = rough awn.

cross of 105 crosses involving 15 varieties of barley (Table 10.1) The composite was grown for 5 years at nine experiment stations of widely different latitudes in Canada, the United States, and Norway. The F_1 to F_3 generations were grown at Edmonton. After the harvest of F_3 seed, samples of 105 crosses were bulked in equal parts by weight to form the composite and the sample of seed of this composite was supplied to each station. The composite was grown for five generations (F_4 to F_8) under conditions of strong competition, that is, high seeding rate which was uniform for each year at all stations. After 5 years, a 10 lbs. sample of resulting seed was recovered from each station. Comparative tests of these naturally selected materials, with the unselected original F_3 seed stock held in storage as control were grown in replicated drilled and spaced plant plots at Edmonton in 1964 and 1965.

Significant differences were observed among plants from different stations for heading date, maturity period, yield, spike type, and awn type. Little difference was observed for plant height and seed size. There was little or no effect on collar type, neck shape, neck length, leaf width, or shape of basal rachis internode. In general, acclimatization of the composite cross at different locations was quite obvious. Disruptive selection observed for heading, maturity, and yield appears to be due to natural selection, which sorted out types most suited for various locations. The differences between the original stock and the final generation samples obtained from different stations showed that under natural conditions a population may become readily adapted to widely different climates. This adaptation was brought about by natural selection operating on the original genetic variability present in the population. Natural selection acted strongly against black kernel, the elimination of which was probably due to genetic linkage with a character of low competitive ability.

3. Three composite crosses (CC), CCII (generations 13, 23, and 45), CCV (generations 5, 10, 21, and 30) and CCXXI (generations 5, 9, 14, and 16), and five cultivars were studied to evaluate yield stability of composite cross populations of barley and to determine whether natural selection has resulted in the production of genotypes that are both high yielding and stable in yield (Soliman and Allard, 1991). They concluded that the composite cross populations offer an opportunity to produce cultivars that show small or no genotype–environment interaction. Several CC or gene pools of spring barley have been developed from a cooperative breeding program of the ARS, USDA, and University of California (Table 10.2).

Table 10.2 Different gene pools of spring barley developed in the United States and registered by the Crop Science Society of America.

Composite cross (C.C.) or gene pool	Registration number	No. of parents and crosses involved	Important features
C.C. II, Calif. F_{40}	Reg. No. GP 1	Complete intercrossing of 28 diverse parents.	From F_2 to F_{40} approximately a 50% yield gain resulted from natural selection of the "fittest" genes and associations. It was used to study evolutionary breeding method, population dynamics, etc.
C.C.V., Calif. F_{25}	Reg. No. GP 2	Developed by intercrossing 31 parents (all 6 row types).	It was used in studies of population dynamics.
C.C.XII. Calif. F_{26}	Reg. No. GP 3	Developed by intercrossing 26 parents and backcrossing their progenies to the F_1 of bi-parental cross.	It embodies more heterozygosity for smooth awns and white aleurone and is more productive than C.C.II.
CC.XIV, Calif. F_{23}	Reg. No. GP 4	Developed by intercrossing 9 California cultivars and also involving genetic male sterility.	It has unique advantage of fostering more persistent heterozygosity in concert with evolutionary breeding.

(Continued)

Table 10.2 Different gene pools of spring barley developed in the United States and registered by the Crop Science Society of America. *Continued*

Composite cross (C.C.) or gene pool	Registration number	No. of parents and crosses involved	Important features
C.C. XIV Calif. O_{15} F_7	Reg. No. GP5	Derived from random manual selection of 1500 or more male sterile heads, with naturally crossed seeds in 15 successive completely heterozygous generations, followed by 7 generations of mixed selfing and outcrossing as occurs in a population having some msms genotypes.	It may be interesting material in many genetics oriented disciplines.
C.C.XV, Calif. F_{15}	Reg. No. G.P. 6	It is related to C.C. XIV, Calif. F_{23} but seed generation was lost. The lineage was then deliberately adjusted back for 8 generations to acquire a tandem relationship with O_7F_{15} (except for the O_7 generations).	This is a rich resource population in study and exploitive potential particularly if the related populations in O_7F_{15} and $O_{15}F_7$ are also used.
C.C.XV, Calif. O_7F_{15}	Reg. No. GP 7	It involved the same crop history as C.C. XV, Calif. F_{15} except that 7 generations of complete heterozygosity were added.	This gave greater variability and may be useful to imaginative researchers.
C.C.XV, Calif. $O_{15}F_7$	Reg. No. GP 8	It involved complete outcrossing (msms x Msms) and manual selection of the crossed seeds on msms for 15 successive generations, followed by 7 generations of mixed selfing and outcrossing. There were some 700 parents in this gene pod.	This population truly represents the most massive and persistent genetic recombination in the history of barley breeding.

Modified after Suneson (1969).

These three and other historical studies suggest that the concept of composite breeding and bulk breeding could be combined and superior lines can be isolated from short-term bulks. The CC can be developed from single crosses among many cultivars/pure lines, bulking their F_1 seed, and growing the composite as a bulk for several generations. The concept of long-term bulk or extended bulk could be useful for augmenting new allelic resources in an on-going breeding program.

Application of bulk method

This method has been used for the genetic improvement of self-fertilized crops like cereals including small grain cereals, grain legumes and oil seeds; however, it is not suited for horticultural and vegetable crops, or crop species where the ultimate goal is to commercialize hybrid seed. In the past, bulk breeding for four to six generations have been found appropriate followed by pedigree selection. Long-term bulking was found to be most appropriate for evolutionary breeding and population improvement (Suneson, 1969).

Merits of bulk method

1. The method is simple, easy, and less expensive.
2. It is more adapted to breeding for genetically complex characters which are difficult to evaluate on a single plant basis.
3. Natural selection reduces the frequency of unadapted types in the population.
4. Self-fertilization in bulk from F_2 to F_4/F_5 increases homozygosity. The preselected F_5/F_6 provides homozygous phenotypes in an otherwise heterogeneous population.
5. The selected F_5/F_6 plant-to-progeny rows would have negligible negative variants than if selection had already started in the F_2.
6. Breeder is relieved of keeping pedigree records, which saves time and labor.

Limitations of bulk method

1. The pure lines that perform best in pure culture may not survive best in competitive heterogeneous population from F_2 to F_4 as the characteristics that make them most productive in agriculture production system may not overlap with natural selection.
2. The inheritance of characters cannot be studied in this method.
3. Inadequate seed sampling for planting next generation(s) in bulk method may reduce genetic variability and also create a bias in the genetic composition.
4. Natural selection operating in the population may modify gene frequencies in an undesirable direction, that is, against agronomically desirable types, for example, individuals with short stature and/or large seeds are poor competitors in mixture, hence these may be reduced in number or may be eliminated in due course.

CHAPTER

Pedigree method

11

Abstract

In this chapter, features and procedures of pedigree method are described, along with the genetic basis, merits, and limitations. In pedigree method, plants with the desired combination of characters of the parent varieties are selected starting in the F_2 generation, and the progeny of each selected plant is reselected in succeeding generations until genetic purity (homozygosity) is reached. Pedigree method can be modified as per the need, facilities available, and the program objectives. In the pedigree method, records of selected individual-plants and their families are kept accurately so that the pedigree of every selected line is known at all times enabling tracing of ancestral relationship among plants. The major modifications of the pedigree method are described using examples in legume and cereal crops. This chapter also includes details on writing conventions of pedigrees, selection history and identity.

The pedigree method was first described by H.H. Love in 1927; however, Louis de Vilmorin (1816–1860) (see Gayon and Zallen, 1998) and Svalof breeding program have been using pedigree type breeding since the mid to late 1800s. Love (1927) proposed pedigree-based approach as a crop breeding method due to its ability to comprehend inheritance control of traits and plant selection. The pedigree selection was most commonly used method of breeding in self-fertilized crops, such as wheat and rice. More recently, variants of the method have been developed that combine pedigree method of generation advancement and selection with another method to gain programmatic efficiency at a reduced time and cost.

Similar to other methods of generation advancement and selection, the process starts with the hybridization of parents that bring the complementary suite of traits for the desired cultivar outcome. The F_1 of a two-parent cross (where both parents are homozygous and homogenous), although likely highly heterozygous (for each pair of segregating alleles; if sufficiently genetically diverse parents were crossed to create the breeding family), will have similar genotypes and will look identical. The selection procedure in each two parent breeding population begins in the F_2 generation, as F_2 plants are genetically distinct and genetic diversity is at a maximum in the F_2 population. At this stage, at each locus, 50% of the plants will be heterozygous, and in each successive generation

Plant Breeding and Cultivar Development. DOI: https://doi.org/10.1016/B978-0-12-817563-7.00023-4

heterozygosity is halved. By the F_5 generation, homozygosity is theoretically up to 94% and by F_8 it is 99%.

Before starting to select plants, it is necessary for the breeder to acquaint with the major characteristics of the crop and their genetic control and should have an ability to distinguish between desirable and undesirable plants in segregating generations. To discard entire breeding population early in the program, the F_2 generation's mean and variance can be tested in replicated trials. Crosses with high mean and high variance, and perhaps average mean and high variance, expected to show transgressive segregants for yield and its components can be retained while other types of breeding population (e.g., low mean and low variance, high mean and low variance) are discarded. If F_2 populations are unreplicated, a breeding population can be omitted based on poor vigor (in comparison with established checks and other populations as control), as these are unlikely to provide superior segregants. For making effective selections, breeder needs to regularly observe breeding populations during the crop season, identify and tag plants that look promising with desired combinations of characters, make final selections and record numbers selected at maturity. Selected plants/ears are harvested, dried, and threshed individually and another round of selection can be made on the basis of grain characters.

Depending on the resources for handling the F_3 generation, appropriate number of F_2 plants are selected. Other considerations include trait: biology, genetic control, heritability, etc. Allard (1960) suggested that at least 50 F_2 plants may be taken but up to 10% of the F_2 plants could be selected and usually the ratio of F_2 individuals to F_3 families is higher for wide crosses than for varietal crosses involving genetically similar parents. However, according to Harrington (1952), 5%–20% F_2 plants can be selected. The F_2 and F_3 populations are highly genetically variable; therefore, superior plants for characters of high heritability having obvious visible differences such as plant height, ear type, disease, and pest resistance and sometimes also grain appearance can easily be selected. Selection for yield and yield traits is deferred to later generations. It is helpful and generally necessary to include check cultivars (released variety of the area), and spreader/infector in the F_2 and subsequent segregating generations. The usage of checks varieties helps in calibrating visual selection index, while spreader/infector rows help create a disease pressure to allow plant selection for such disease traits. Due to the record keeping, all parent-offspring and sib-relationship are collected and kept, every progeny in each breeding population from every generation can be traced back to the original F_2 plant. The availability of pedigree information is also useful to make decision on retaining more or less sibs. For example, if pedigree and selection history are kept, breeder can determine higher genetic worth sib-families and can advance the entire family for next generation of field testing. The motivation for this stems from the requirement in some crops, where several traits are under selection and keeping higher potential and worth families increases the likelihood of finding pure lines that meet or exceed majority if not all the trait thresholds.

General procedures for pedigree selection

As previously explained, there are numerous variation of pedigree methods. However, if the breeder is using more than one generation of selecting plants within a progeny row and keeping detailed records of familial and sib-relationship during generation advancement, it is considered "pedigree" method. A general outline of pedigree method is provided in Fig. 11.1.

Selection of parents: in pedigree method of breeding, selection of parents for hybridization is made judiciously. While two-parent crosses are more common, three or higher number of parents may be used to develop the breeding population (or also referred as breeding family) depending on the breeding objectives. Parents chosen to create the breeding populations are generally cultivated types, that is, cultivars and elite lines. For example, one of these parents can be a released variety that needs improvement for more than one characteristics. The second parent complements the first one. However, if the objective of the cross is to improve one characteristic controlled by one to three genes or QTLs, marker-assisted backcrossing method will be most effective. It is important to consider that pedigree method attempts to bring together several desirable genetic factors from parents into the new variety. Breeding program will make a large number of crosses to create new breeding populations (eg., 50 – 200 or more) each year, and ensure genetic diversity among parents to maximize genetic gain (see Chapter 3: Genetics in Relation to Plant Breeding and Chapter 4: Primer on

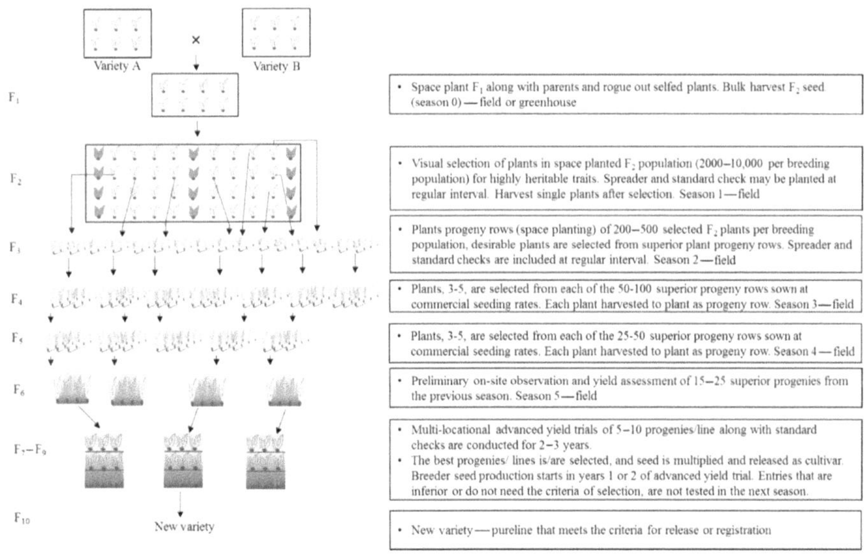

FIGURE 11.1

General outline of pedigree method.

Population and Quantitative Genetics). To obtain sufficient quantity of F_2 seed, sufficient number of cross attempts are made; and ideal number of crosses will depend on the crop species and breeding objectives in conjunction with available resources. The information described below is generalization from a single breeding family and uses cereal grain as example. Every breeding program and each crop species will have differences from the scheme below. It is important to note that if a breeding program makes 100 new breeding families (i.e., populations), this scheme will be running concurrently on each of the 100 breeding families. New breeding populations are created each year for continuity in a breeding program.

F_1 generation: some 50–100 F_1 plants are space planted. Parental cultivars are planted along with F_1 plants for comparison and to identify hybridity. Possible selfed plants, which resemble female parent, are removed.

Hybrid combination can be set up such that recessive allele for a visually distinguishable trait such as flower color is present in the mother, while the dominant allele is present in the male parent. For example, in soybean cross attempt can be made using the female (e.g., white flower, recessive) and male (e.g., purple flower, dominant) that will allow selfed seed to be easily distinguished, because F_1 cannot be white flowered. All while flowered F_1 plants are removed so that only true crossed F_1 plants produce F_2 seed. The knowledge of cytoplasmic and nuclear genetic inheritance of morphological traits is required to pick genes that can guide the choice of parent to use as male and female and to distinguish selfed versus crossed plants. If no such genetic traits are available, breeders can utilize molecular markers to establish heterozygosity among random loci, or a breeder can obtain plant to progeny from each F_1 and grow F_2 in a progeny row to confirm segregation.

F_2 generation: selection begins in the F_2 generation, which has the maximum genetic variation. The highest frequency of genotypes possessing desirable alleles on all segregating loci are found in the F_2 generation and the frequency of such genotypes declines in subsequent generations. Generally large population (2000–10,000 plants per breeding population) is space planted. The reason of space planted is to ensure an ability to select on a per plant basis (due to maximizing the individual-plant expression) without confounding neighboring plants, as well as to obtain sufficient quantity of seed for next generation progeny row. The local or standard varieties and susceptible checks (spreader/infector rows) sown at regular intervals in F_2 nursery is useful. Checks provide the benchmark to provide comparisons and establish selection criteria, while spreader/infector rows can help create disease pressure to provide an environment to make selection for necessary biotic traits. Superior plants for characters of high heritability are selected and each selected plant is documented. Breeder or staff will put a tag or spray paint selected plants for ease in plant identification at the end of season when seed are mature and ready for picking. Breeder may start selecting during the vegetative stage but definitely at reproductive and later stages depending on traits under selection (disease reaction, plant architecture, flowering date, etc.). If tags or another method of "select" identification is not done at the time of trait expression, it becomes impossible to pick plants at maturity. Several round of "selection walk" may be done, and in each successive round if a previously

selected plant is not retained because it fails to meet the selection criteria at that time, the tag can be removed. If spray paint is done, stem or tiller of such later discarded plants can be bent, so at the time of seed collection those later unselected plants are not used. Plant parts (spikes, panicle) or whole plant are taken depending on breeding scheme and seed requirement for $F_{2:3}$ progeny row. This decision is taken after due consideration on breeding scheme, labor, space, and time requirement. Whole plant pulls require more labor, time, and space (for storage before threshing) but provides plenty of seed for next generation plots, while single spike or panicle harvest is quick requiring less time and space but limited seed is available. Generally breeder should attempt to establish these breeding tests at a location that represents the target environment. It is also beneficial to utilize a location that works with program's logistical needs for proximity, storage, etc.

F_3 generation: each plant selected in the F_2 is space planted in the F_3 generation as plant-to-progeny rows along with check varieties and if feasible, spreader rows (spreader rows assist with spread of disease enabling selection for diseases) at regular intervals to help in selecting superior plants. Each progeny row should contain 20–30 plants, or a sufficient number to do within-row selection. It is important to note that a breeder should increase the number of progeny rows as genetic variation among line is larger than within line at this stage. This implies that genetic gain will be more by growing more progeny rows and selecting efficiently rather than growing few progeny rows with large number of seed in each progeny row. Selection in the F_3 is on a progeny or line basis, that is, decisions are first made on the basis of entire progeny performance, selection of plants to go on to the F_4 generation is still on a single plant basis. Some of the progenies may be discarded entirely and in other progenies single plants (3–10) may be selected. Selection of single plant (whole plant or its inflorescence) is selected at each generation from each row; however, in some visibly inferior rows, the breeder may not make any within-row selections. The selection in the F_3 generation should be more rigorous than in the F_2 as appearance of an F_3 progeny row is a more reliable indicator of potential than that of a single F_2 plant, and partly because the volume of material must be reduced to manageable proportions in later generations. The number of plants selected is generally less than the total number of F_3 progenies grown. For example, if 200 progeny rows per breeding family (i.e., breeding population) are grown, selection will happen at two levels: (1) 20–40 progeny rows may be selected and (2) 3–5 plants per progeny row are selected. These principles are used as the breeding funnel needs to reduce the numbers in each generation to enable adequate handling of plant material in trials. Generally, these selections are made on traits that have moderate to high heritability and can be visually selected.

F_4 generation: the F_3 plants are also grown as space planted plant-to-progeny rows as in previous generations. The progenies become more homogeneous (homozygosity is 87.5%) but not sufficiently homozygous to be mass propagated. Differences between them will now be greater than differences within progenies; however, the final selection will have to be again on a single plant basis within

selected progenies. The number of progenies going forward to the F_5 generation should again be drastically reduced. Appropriate checks are planted so that selection for traits under selection can be made. These checks may include standard checks for plant height, days to maturity (although it is a complex trait and interacts with environment, it can assist a breeder to narrow the range of selection to somewhat match the target environment), disease checks, etc.

F_5 generation: the single plants selected in the F_4 from superior progenies are planted together (in blocks) in F_5 generation. From F_3 to F_5, sibs (plant to progeny rows from the plants picked in the same rows) are grown next to each other to make selections. In these generations, breeders may follow ideotype based selection to tag plants (See Chapter 25: Breeding of Crop Ideotypes). In the F_5 generation, the emphasis from plants will be shifted to overall features of the progenies as plants will be sufficiently homozygous (93.8%); therefore, progenies may be planted at commercial seeding. Within line genetic variation is reducing at each generation; therefore, selection starts to move towards among lines. The F_5 plant-to-progeny rows of a F_4 selected line may be harvested in bulk. Any doubtful plant-to-progeny row with off-type plants may be rouged out. Number of selected progenies should be considerably reduced (25–50) per breeding population. Appropriate checks are planted so that selection for traits under selection can be made. Checks can be included for comparisons, as described in the F_4 generation.

F_6 generation: in general, segregation at any given locus will be minimal; therefore, preliminary yield assessment for agronomic evaluation can be conducted in replicated or unreplicated single or multiple locations. Emphasis may be on selection for observable and measurable traits with low heritability such as grain yield, grain quality, etc. Appropriate check varieties are included. Only entries deemed superior for yield and other breeding objective targets are retained for further testing. However, entries that are comparable to yield checks but with improved performance in other high importance traits (e.g., specific disease, insect pest, and quality trait) can be retained, if these lines are considered useful for use as a parent, as a check, or targeted commercialization. These decisions are made relative to the checks; therefore, the choice of check is very important.

F_7–F_9 generations: replicated multi-locational yield trials (advanced yield trials) of superior selections/progenies along with two to three (or more) checks are conducted in different years and seasons.

Each breeding program may follow a different population size and number of plants per row. These numbers will depend on the objective of the program and cross created to develop the breeding populations (eg., diverse or more parent crosses generally require a larger population size), and available resources (technical, infrastructure, and human resources). While in each generation, specific traits are selected, selection of multiple traits happens simultaneously in specialized nurseries (on plant traits) and harvested seed (on seed traits). If the correlation between home and off-season environments is high, selections can be made in each generation using more than one growing season per year significantly advancing the genetic gain.

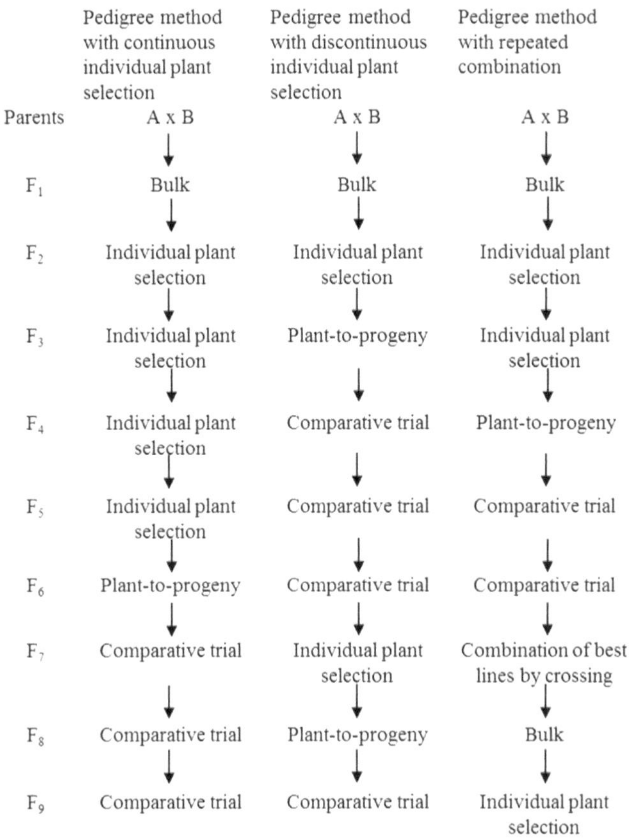

FIGURE 11.2

Different versions of the pedigree method of breeding.

After MacKey, J. 1962. The 75 years development of Swedish plant breeding. Hodowla Roslin Aklim. i Nasiennictwo 6, 437–467.

Modified pedigree methods

Depending on the breeder, the crop, facilities available, and the need of program, the pedigree method has been modified (Fig. 11.2).

Early modifications of pedigree method

Mac Key (1962) reviewed the 75 years' development of Swedish Plant Breeding and referred that Nilsson-Ehle when treating his cross populations first used the pedigree method with continuous individual-plant selection. The pedigree method became successively improved at Svalof (Akerman and Mac Key, 1948) and tremendous improvement was made by combining continuous selection with a progeny test including yield trials (Fig. 11.2). This was possible by completing individual-plant progeny plots based on the selected plants with a yield test based on the harvest from the selection plot of the preceding year. The procedure helped for the selection of the most promising lines for further reselection until sufficient homozygosity was reached. This method was laborious, expensive, and time consuming; therefore, it was modified. The individual-plant selection instead of being a continuous process for several generations was restricted to the F_2 and in some cases to the F_3 generation only.

The pure form of pedigree method is rarely used by breeders anymore, due to the length of time it takes (one filial generation per year, in latitudes which only permit one crop season in a year), necessity to grow the segregating material at home environments, the need to keep track of pedigree (because both among and within progeny rows are made), and costs associated with managing the numbers. Majority of programs that use pedigree method, use a modification that may combine aspects of two or more methods (including pedigree) for filial generation advancement. These modified methods are built to reduce the time to development, make more strategic selections (with the use of genomics and phenomics tools), and reduce the complexity of handling breeding materials. Most breeding programs aim to streamline their breeding pipeline with the choice of breeding method as it helps avoid errors in record collection, storage and curation, better utilization of available resources, cost reduction (personnel and equipment), faster cycle time (parents to progeny and their usage as a parent in next cycle of crossing), and infusion of technology among other factors.

In one modification of pedigree method, single plants are selected in the F_2, F_3 and F_4 (Lupton and Whitehouse, 1957). After the selection of single F_4 plants, the remaining plants from that row were bulked. This bulk was used for yield testing in F_5 generation, while plant to progeny rows were also concurrently grown in that season. The procedure was continued in F_6 and F_7. A notable feature of this scheme is the concurrent testing in the F_5 and subsequent generations of bulk rows in various trials for yield, disease resistance and quality, while single plant, ear row and family selection is proceeding for homozygosity. A modified pedigree-based on the above principles was used at the Plant Breeding Institute, Cambridge, to develop a series of high-yielding disease-resistant quality winter wheat varieties.

Early generation yield testing

Shebeski (1967) proposed this method to assist wheat breeder to select lines for improved yield potential and bread making quality in early generations. This method consist of growing large F_2 populations (10,000 plants) and a large number of F_2 single plants are selected based on general appearance, vigor and ability to produce a minimum of 750 seeds. However, the difficulty of identifying the best F_3 lines requires much greater attention. For this, each F_3 lines was derived from a single F_2 plant and was compared with a control plot. The use of large F_3 plot of 750 plants in

3 row plots 5 m in length with a contiguous bulk control plots to correct soil variability was advocated in wheat crop by Shebeski (1970). The row spacing was 15 cm within plots and 60 cm between plots to minimize inter plot competition. DePauw and Shebeski (1973) further studied the effectiveness of assessing unreplicated F_3 lines for yield in a wheat cross. They studied correlation between F_3 lines and related F_4 bulk mean yield; and expressed as a percentage of adjacent control, it was 0.59**. Similar values were observed between F_3 lines and related F_5 family mean yield. This suggested that by growing a large sample of the progeny of each F_2 genotype adjacent to a control, it was possible to discriminate among F_3 lines for heritable quantitative differences. However, this method is not routinely used due to its inherent large resource requirement.

Several breeding programs, however, follow a modified method of early generation testing (EGT) (e.g., yield testing at F_3 or F_4 generation) to make selection decisions. In this approach, early generation lines are grown in yield plots. Although, more resources are required to handle EGT, this method allows elimination of inferior genetic materials (lines). That is, selection for traits of lower heritability is practiced to discard inferior lines. This yield testing can be done in replicated and/or using multi-environment trials. This approach has been successful because genetic correlation (r_G) between genotypic value of an F_2 plant and the genotypic value of a random recombinant inbred line derived (without selection) from the same F_2 plant is high. With moderate to high heritability traits, correlation (r_P) between the phenotypic value of a line in early generation and the genotypic value of a descendant recombinant inbred line is also high (increasing value with increasing heritability) (Bernardo, 2014). Since $r_P = r_G h$, this method will show effectiveness for discarding undesirable genotypes for further testing for even low to moderate heritability traits.

Genetic basis of pedigree method

Pedigree method is helpful in developing a cultivar as a replacement of a well established variety in self-fertilized crops. Keeping this in view, one parent is usually selected based on its proven performance in the area of use, and the other parent is picked to complement the first one. Selection of individual plants is continued until the population is near homozygosity (i.e., from F_2 to F_4/F_5). At this stage, selection is practiced among families. This could be ascribed due to the fact that hybrids are expected to segregate for large number of genes and every F_2 individual would be different from other individual. Many loci will become homozygous in F_3 and F_4 generation; however, within families segregation would be still observed. That is why selection is based on best plants in best families from F_2 through F_4/F_5. By F_5/F_6 generation, most families are expected to be homozygous at most loci, i.e., minimal within family genetic variation; therefore, as generations advance, within families selection is not useful and among families selection is more effective.

Pedigree method has also been useful to correct some specific weaknesses or defects of established varieties. It is also expected that during pedigree breeding transgressive segregants would be recovered, hence yield and quality will also be improved. This is also referred to as transgressive breeding. While the term "selection" is generally used, it is important to mention that numerous breeders consider "discarding" rather than "selection" *per se* for progeny row and individual plant visual rating. It is due to lower accuracy and precision in measurement and assessment; therefore, while it may be more difficult to determine the "best" plants, it is easier to identify "worse" plants. This is more applicable for progeny row selection.

Application of pedigree method

The pedigree method is most suitable for improving specific characters that are easy to identify and can be readily visually selected, like qualitative disease resistance and morphological traits. Many of such characteristics are governed by Mendelian genes. For traits that are complex and time consuming, such as seed yield and end-use quality, direct selection can be deferred until F_6/F_7. At this stage, population is homozygous and the number of progenies is fewer enabling selection for traits with a quantitative mode of inheritance. The cultivars developed through pedigree method have traditionally been relatively narrow genetic base. However, this is not due to the modality of pedigree method of breeding, but more reflective of the crops where this method was primarily used (e.g., rice and wheat). In these crops, due to quality requirements, parents are chosen to meet the quality requirement creating a narrower genetic base. Pedigree method can be used effectively for a number of row crops, where the objective is to develop a pure line variety.

Examples of pedigree method and its modifications in legume cultivars

Mung bean yellow mosaic virus (MYMV) is very devastating disease of mung bean (*Vigna radiata*) and black gram (*Vigna mungo*) in south-east Asia. At G.B. Pant University of Agriculture and Technology, Pantnagar, India, modified pedigree method of selection was used to develop high-yielding and MYMV resistant varieties of these crops (Singh, 1988). The early segregating generations (F_2–F_4) and (M_2–M_4) developed using intra- and inter-specific hybridization and gamma-ray breeding, respectively, were subjected to screening for yellow mosaic disease.

Highly susceptible lines to MYMV were planted after every five or six rows of the test materials. No insecticide was sprayed, to facilitate the natural build-up of vector whitefly (*Bemisia tabaci* Genn.). Individual plants with moderate to high degree of resistance to MYMV, along with erect growth habit, profuse podding, good pod length, having 10–11 seeds per pod, average seed size (4 g/100 seeds) were selected in F_2. For seed appearance, selection was made after harvesting by inspecting the packets of seed lot of each F_2 plant. Plant to progeny rows were grown in F_3 and three to five plants from each of the superior progeny rows were selected. The selected individual three to five plants were planted in bulk in F_4 generation. The basis of selection in F_3 and F_4 was similar to F_2; however, inferior progenies were removed. The remaining superior progenies (selections) were included in on-station preliminary yield evaluation trials and for quality parameters testing. These superior selections were also multiplied simultaneously and were evaluated at three sites in F_6 generation for yield stability and uniformity at the regional stations of the University. The most promising candidates were tested in multi-locational yield trials for 2–3 years. Using this scheme, high-yielding and MYMV resistant varieties were developed and released. These included mung bean varieties - Pant Mung 2, Pant Mung 3, Pant Mung 4, and Pant Mung 6; and urd bean varieties - Pant Urd 31, Pant Urd 35, and Pant Urd 40. These are the leading varieties of mung bean and urd bean in India. Pant Urd 31 is photo-thermo-sensitive, hence is cultivated in all states in different agro-climatic niches and cropping systems.

Of the above varieties Pant Mung 2 is a product of gamma-ray breeding, Pant Mung 3 was developed using intraspecific hybridization, whereas Pant Mung 4 and Pant Mung 6 were developed through interspecific hybridization of mung bean crossed with urd bean. The urd bean varieties were the product of intraspecific hybridization.

Visual selection for yield components, plant type and resistance to diseases in early segregating generations (F_2–F_4) in chickpea and lentil was also used at Pantnagar for developing higher yielding varieties (Singh, 1997). Pantnagar is a hot spot for botrytis gray mold (BGM) (*Botrytis cinerea*) of chickpea, and rust (*Uromyces fabae*) of lentil. The promising F_5 progenies (selections) of chickpea and lentil were evaluated in on-station replicated trials along with released cultivars (checks) in timely planted (after harvest of maize) and in late planted (after harvest of rice) conditions. The progenies with early seedling vigor and superior yield traits in both conditions were selected for testing in multi-locational replicated yield trials for yield *per se* and resistance to *Fusarium* wilt and other foliar diseases, and were released. Chickpea variety, Pant Gram 186 and lentil varieties, Pant lentil 5, Pant lentil 6, Pant lentil 7, and Pant lentil 8 have been released for cultivation. All of these varieties are higher yielding. Chickpea variety Pant Gram 186 is resistant to *Fusarium* wilt and BGM and above listed varieties of lentil are resistant to *Fusarium* wilt and rust diseases. Pant Gram 186 is a popular variety in rice-based cropping system of Indo-Gangetic plains in northern India. The lentil varieties are the leading varieties in the seed chain and are widely grown in northern India. Reaction to *Fusarium* wilt and *Ascochyta* blight was tested in multi-location yield trial.

Examples of modified pedigree method in the development of wheat cultivars

In wheat, AK Singh with his colleagues at Swift Current Research and Development Center of Agriculture and Agri-Food Canada developed numerous durum wheat cultivars using a modified pedigree method, and he also participated in the development of several common wheat cultivars using a modified pedigree method (Ruan et al., 2016, 2019; Cuthbert et al., 2016, 2017; Singh et al., 2015a, 2017a,b; Clarke et al., 2009a). These varieties were generally developed with alternating generations of within line single plant selection (at home locations) and among progeny row selection (in winter nurseries) to minimize the time to developing varieties, with better cost effectiveness and integrating multiple traits in the finished variety in a forward selection approach. Some of these varieties have been grown on over million acres attesting to the success of this breeding approach, especially when the breeding objectives encompass numerous agronomic, disease–pest, and quality traits. These cultivars meet or exceed the checks performance for more than 20 measured traits further demonstrating the effectiveness of a modified pedigree method in small grain cereal such as wheat in situations where multiple trait selection and breeding are needed.

Merits of pedigree method

1. The method is especially useful for characters with simple genetic patterns. This method is suitable for the improvement of easily identifiable and simply inherited character(s); however, transgressive segregation for yield and other quantitative characters can also recovered.
2. Record keeping of plants and progenies would provide genetic information for all time to come, and helps exploit sib-relationship within and among family genetic variation.
3. Inheritance of character(s) from pedigree records could be used in research studies, and help understand the genetic control for application in future breeding methods.
4. The selection is based on both phenotype and genotype; therefore, it is reliable and effective method of selection for superior types in the segregating generations.
5. Maximum opportunity is available for selection of superior plants and progenies during the early segregating generations. This allows for an automatic elimination of weak and inferior types, and focus can be directed on genetic material with higher worth.

Limitations of pedigree method

1. Maintenance of accurate pedigree is tedious and time consuming and relatively expensive. Sometimes documentation is difficult. If a breeder does not adequately utilize the familial relationship information, the merit of pedigree method (through detailed record keeping) is lost.
2. The method is more suited for qualitative traits but not optimal for traits such as horizontal resistance where backcross method is more suitable.
3. Selection for yield is not effective in early segregating generation, hence some of the valuable genotypes may be lost.
4. The method requires at least 10–12 years to complete, if only one growing crop season is possible and large number of crosses are difficult to be handled. However, as previously described, pedigree method is rarely implemented this way (one growing season per year in Northern latitudes), and modifications have been made to shorten the breeding cycle including the use of contra-season nurseries. Modified pedigree methods are most common, and rarely a pure pedigree method is practiced anymore.

Writing pedigree and selection history

In this section, pedigree and selection history writing is described. Although, this section is included in the pedigree method chapter, the concepts are applicable to all breeding methods (bulk, pedigree, doubled haploid, single seed descent, single pod descent, etc.).

$F_{\#}$ and $S_{\#}$ symbols

In plant breeding, "$F_{\#}$" and "$S_{\#}$" symbols are used to designate the number of generations of selfing or sib-mating. The "F" symbol is generally used in self-pollinating species to denote the filial (i.e., family) generation of offspring following a cross between two or more parents, while the "S" symbol stands for selfing and is used in cross-pollinated species to denote generation of self-pollination. The subscript (#) represents the specific generation ($F_{\#}$), that is, F_1 is the first generation following a cross and subsequent generations are designated F_2, F_3, F_4, etc. based on the number of generations the offspring are self-pollinated or sib-pollinated (i.e., pollinated by a sibling plant in the same progeny row). In other words, $F_{\#}$ refers to the subscript to the generation in the sequence of generations following a cross (F_1, the first generation after the cross; F_2, the second generation after the cross, F_3, the third generation after the cross, and so on). If both of the parents of a cross are homozygous then the F_1 offspring will be homogeneous (i.e., all plants will be uniform and genetically the same) and heterozygous at individual loci. If either or both of the parents are heterozygous then the F_1 offspring will be heterogeneous (non-uniform) and primarily heterozygous at individual loci except for non-segregating loci. In a complex cross involving more than two parents (even if the parents are homogeneous), the F_1 generation will be heterogeneous and heterozygous.

Single gene example with two homozygous parents

Parent 1 (AA) x Parent 2 (aa)

↓

F_1

(Aa)

In the example above, both parents were developed by breeder after undergoing several generations of selfing so that they are homozygous at all loci. Female parent is always listed on the left most side. The cross between these two unrelated parents produces F_1 progeny that are all uniformly heterozygous (Aa), and F_1 progeny population (all F_1 from this cross) will be homogeneous since each F_1 will be "Aa" type at this locus. When an F_1 plant is self-pollinated or when two F_1 plants are crossed with each other, F_2 seed is produced. If parents were homozygous then the F_2 generation is the first generation when the offspring are heterogeneous (i.e., segregating for different parental alleles). The F_2 generation is typically the generation when selection for simple traits begins. Self-pollination of F_2 plants produces F_3 plants, self-pollination of F_3 plants produces F_4 plants, and so on.

Parent 1 (AA) x Parent 2 (aa)

↓

F_1 all Aa (a homogeneous/uniform population of heterozygous plants)

⊗ (⊗is the symbol that indicates selfing)

F_2 AA, Aa, or aa (heterogeneous population of genetically different plants)

⊗

F_3

In cross-pollinated species, "$S_{\#}$" is used instead of "$F_{\#}$." The symbol S_0 can be used to describe the progeny from a single cross between two homozygous parents as either:

1. Similar to F_1 (in self-pollinated) which indicates that the plant was not derived from self-pollination. That is, if $S_0 = F_1$, the argument is that no systematic self-pollination has occurred.

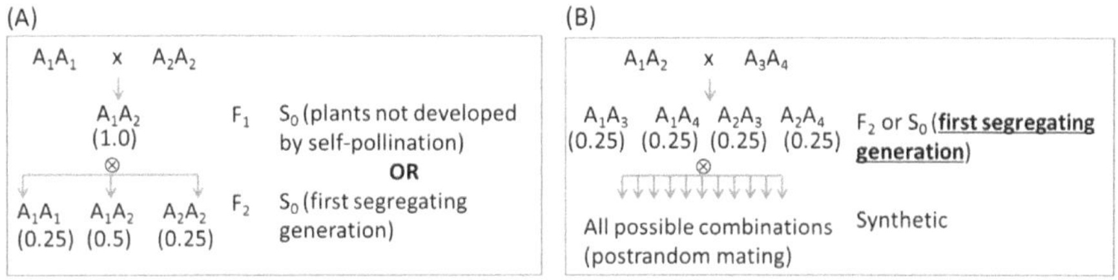

FIGURE 11.3

Representation of S_0 generation and equivalency to F generation terminology in: (A) when crossing two inbred parents and (B) when crossing two non-inbred parents.

2. Similar to F_2 (in self-pollinated) where population is formed by random mating in which the population is heterogeneous and heterozygous. That is, S_0 or F_2 refers to the base population, where the base population is heterozygous and heterogeneous and contain two alleles per locus at frequencies of 0.5.

Therefore, it is important that the breeder clearly describes what they are referring to in a particular situation, and then be consistent in the usage of S terminology. In Fig. 11.3A, the numbering convention S_0 is explained. If non-inbred parents are used, the production of genotypes can be S_0 (similar to F_2, as first segregating generation), while postrandom mating will create all possible combinations and result in a synthetic not a S generation (Fig. 11.3B).

Using "$F_{x:y}$" or "$S_{x:y}$" to describe breeding lines according to the generation they were derived

$F_{x:y}$ or $S_{x:y}$, describes "x" as the generation where single plant was harvested separately to give rise to the derived line, and y represents the current generation of inbreeding of the plants within this derived line.

Breeding lines (or genotypes) are derived from individual plants at various generations. An $F_{2:4}$ line refers to an F_4 line that was derived from a single F_2 plant. The F_2 plant was selfed to produce F_3 seeds, which were then grown in a single F_3 progeny row, self-pollinated, and then harvested as a F_4 bulk of many or all of the F_3 plants in this row. Based on this scheme, each individual F_2 plant gives rise to a genetically distinct $F_{2:4}$ line. These lines are also described as "F_2-derived lines in the F_4 generation" or simply "F_2-derived F_4 lines" (Fig. 11.4A).

An $F_{3:4}$ line refers to an F_4 line (or progeny row) created from a single F_3 plant growing in an F_3 progeny row that was produced from the seed of a self-pollinated single F_2 plant. An individual-plant was selected from an F_3 row to produce F_4 seed and this seed when grown represents an $F_{3:4}$ line. These are also described as "F_3-derived lines in the F_4 generation" or simply "F_3-derived F_4 lines" (Fig. 11.4B).

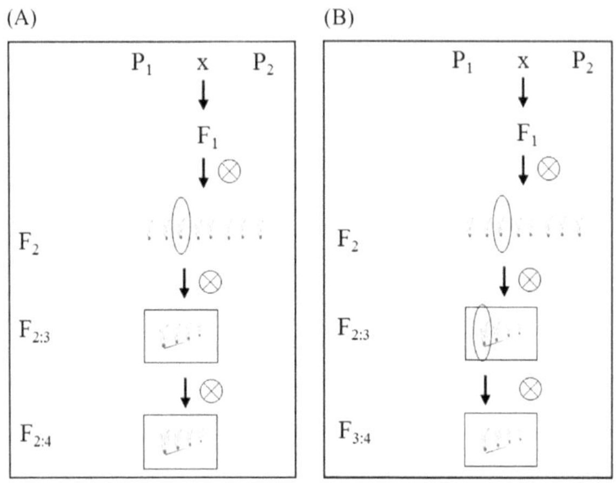

FIGURE 11.4

(A) Development of $F_{2:4}$ line and (B) development of $F_{3:4}$ line.

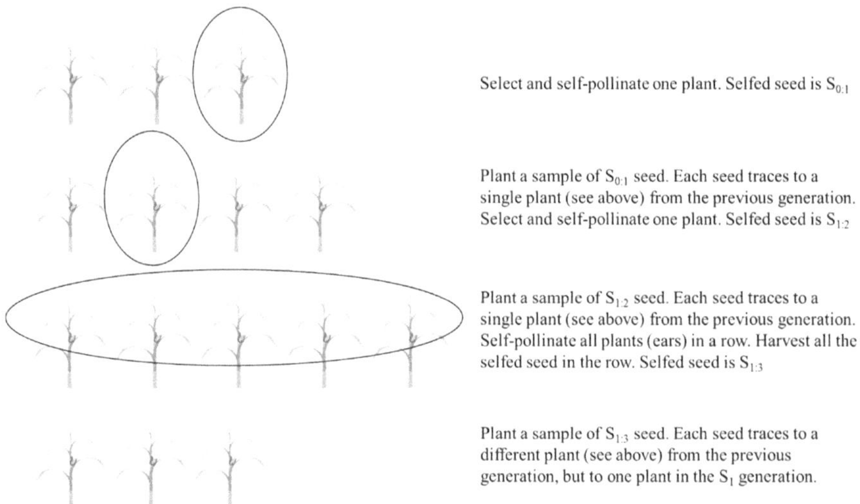

FIGURE 11.5

Development of $S_{1:3}$ line in maize.

The difference from $F_{2:4}$ lines is because the first subscript designates the generation of the last individual-plant selection. For proper generation information, plant breeder need to identify the recent most generation in which a single plant was harvested to give rise to the current line of descent. In cross-pollinated crops, such as maize, generation of selfed line is described in Fig. 11.5. In this figure, S_0 is considered equivalent to F_1.

Writing a standard pedigree

Each organization perhaps follow a different standardized system for recording pedigrees. Common nomenclature is described in the text below for educational purposes.

The female parent is always designated by listing it first (starting from the left) followed by the male parent (on the right). For example, A is the female parent and B is the male parent in an (A × B) cross. An (A × B) cross can also be written as A/B. In more than two parent cross, '/' symbol to write pedigree is advised over 'x' symbol usage.

If an F_1 (A/B) plant is pollinated with parent C, and the F_1 is used as the female and C as the male, the resulting three-parent cross would be designated as A/B//C.

Subsequent crosses with parental materials D, E, F, and G used sequentially (all as males) are indicated using a number to record the cross order in the following way: A/B//C/3/D/4/E/5/F/6/G.

If the example above is changed to use D and F as female parents, with E and G remaining as males, the cross would be recorded as follows:

Step 1: A/B is the first cross.
Step 2: A/B//C is the second cross, where A/B is the female.
Step 3: **D/3/**A/B//C is the third cross, with D as female, and A/B//C as male.
Step 4: D/3/A/B//C**/4/E**, with E as male, and the 4-parent cross as the female. Note: bold and underline text is for information and instructional purpose only. In writing a pedigree, you will not have to bold text. One will simply write the pedigree as D/3/A/B//C/4/E.

The inclusion of "F/5/" as the female and "/6/G" as a male completes the pattern.

Writing a backcross pedigree

In multiple backcrosses, the sequence of these letters from left to right corresponds to the sequence in which the backcrosses are made. Backcross pedigrees include an asterisk (*) and a number indicating the dosage of the recurrent parent. The asterisk and the number are placed next to the crossing symbol (/) that divides the recurrent and donor parents. The following are examples of pedigree formats involving backcrosses:

A is the recurrent parent: A*2/B of the initial cross and has been used as a parent two times. Therefore, A*2/B indicates one backcross or a BC_1 cross.
B is the recurrent parent: A/3*B, and has been used as a parent three times. Therefore, A/3*B indicates a BC_2 cross.
A*2/B is therefore, A//A/B; and indicates that A was used as a female in both F_1 and BC_1.
A/3*B could be B/3/A/B//B and indicates that A was used as a female, F_1 was then used as female, and BC_1F_1 was used as male.
A/3*B could be A/B//B/3/B and indicates that A was used as a female, F_1 was then used as female, and BC_1F_1 was used as a female.
The $F_{\#}$: derived symbols as previously described for regular crosses will follow the $BC_{\#}$ designation. For example, $BC_1F_{2:4}$ or $BC_2F_{2:4}$.

Assigning an identity number to each cross or backcross

Every cross, i.e., breeding family or breeding population, should receive a unique ID number that will allow everyone in the breeding group to recognize the year the cross was made (e.g., 2014), a cross number (e.g., 1001), and perhaps the target purpose of the cross (e.g., HO for high-oil, or abbreviation for another specific trait or market segment). Breeding programs can use numbering codes for purpose of crosses, for example, elite crosses, diversity crosses, specific quality traits, specific or generic stress tolerance etc.

Recording selection history using a Breeder's cross identification designation

Each plant population or breeding family (i.e., population generated from the initial cross), segregating lines and advanced lines are assigned a Breeder's Cross IDentification (BCID) and a selection history. The BCID is important because it gives each breeding family (or population) its own unique distinction, and can explain several aspects of the cross, for example, the year cross was made, and with selection history there is a record on the process of selection including where and how the initial cross was made, and where and how subsequent selection steps occurred for each generation of selection. Selection history can identify the number of individual plant(s) selected and the letter indicates the location where selection took place and/or under what specific conditions selection was conducted. Each breeding program follows their own naming system, so the examples presented here are hypothetical, not exhaustive, and should be used for learning purposes only.

Example 1 ISU20S110-51A-05C-010A-010C-8A-0C

BCID = ISU20S110, a single cross was made at Iowa State University (code: ISU) in 2020 (code: 20), in soybean (code: S) and the cross (i.e., breeding population) was 110th cross in that breeding cycle (code: 110). The selection history reflects that a modified pedigree selection method was used. In the F_2 the "51A" indicates that this genotype was the 51st individual-plant among those selected at location "A" (code A is the home location) in this population. The F_3 designation of "05C" indicates that 5 plants were selected and harvested in bulk from the F_3 progeny row grown at location "C" (code C is winter nursery location). Seed from the bulked F_3 progeny row was planted at location "A" in the F_4 and 10 plants were selected and harvested in bulk. Similar scheme was used in F_5 at location "C." In the F_6, a single plant was selected from this genotype (8th plant) at location "A" and constituted the seed for the next generation. In the F_7 (or more appropriately, $F_{6:7}$) all plants in the progeny row at location "C" were harvested in bulk, as shown by the designation "0C." This created the genotype ISU20S110-51A-05C-010A-

010C-8A-0C. This progeny hereon is named with this convention, if no further selection is practices within the line.

Example 2 ISU19S052-010A-010C-010A-05C-0A

BCID = ISU19A052, a single cross was made at Iowa State University (code: ISU) in 2019 (code: 19), in soybean (code: S), and the cross (i.e., breeding population) was 52th cross in that breeding cycle (code: 052). The selection history indicates three generations (F_2–F_4) of selection in which 10 plants were bulked from the progeny row (or plot) for each season at either the "A" or "C" locations. In the F_5 generation the genotype selected was the fifth plant from the bulk plot at the "C" location. In the F_6 a complete plot bulk was harvested at location "A." This lead to the creation of the genotype ISU19S052-010A-010C-010A-05C-0A.

Example 3 ISU19S102-A-C-A-002C-0A

BCID = ISU19A102, a single cross was made at Iowa State University (code: ISU) in 2019 (code: 19), in soybean (code: S), and the cross (i.e., breeding population) was 102th cross in that breeding cycle (code: 102). The selection history indicates three generations (F_2–F_4) of single pod descent and selection in which pods were selected from the plot and bulked for next generation seed source. This happened in each season at either the "A" or "C" location. In the F_5 generation the genotype selected was the second plant from the bulk plot at the "C" location. In the F_6 generation, the seed from the previous season single plant selection was planted as a progeny row and this plot was harvested in bulk at location "A." This leads to the creation of the genotype ISU19S102-A-C-A-002C-0A. If the same format is followed for all populations, this line could also be written as ISU19A102-002 for this unique line if the remaining steps are standardized and pipeline is identical.

These three examples show some variation in naming convention, however, it is important to note that each program uses their own codes to write BCID and selection history. Programs can use different series numbers to align with the breeding objectives, for example, 100 series for elite yielding crosses targeted for cultivar development, 200 series for germplasm development crosses and so on. Separation that is more refined can be used for specific trait targeted breeding, such as fatty acid, amino acid, carbohydrates, end-use quality, specific diseases, abiotic stresses, target areas etc. It is very important to have clear guidelines on writing convention and strict adherence to follow it.

CHAPTER

Single seed descent method 12

Abstract

The single seed descent (SSD) and its modifications are valuable methods that can generate progeny rows representing the genetic variability of the F_2 generation. This method or its variants have been successfully used in a number of crops, particularly small grain cereals and legume crops. The SSD based rapid generation advance method coupled with genomic selection can help in rapid cycles, positively impacting genetic gain. In this chapter, generalized SSD method, important variants of this method, genetic basis, applications, advantages, and limitations are described.

The single seed descent (SSD) method is motivated by the need to maintain genetic variability of F_2 generation (generation of maximum genetic variance) in a breeding population for later generation selection in yield plots. The maximum genetic variability is expressed in the F_2 population, except in clonal crops with self-incompatibility, where highly heterozygous clones are used as parents, and F_1 generation represents maximum genetic variance. The SSD method contrasts itself from the bulk method that also maintains the genetic variation of F_2 or other early generation. In bulk method, there is a risk of losing valuable types in the population by interplant competition (natural selection), and genetic variation may diminish as the breeder is not able to use the entire harvest for sowing the next generation (genetic drift).

Since SSD does not suffer from these factors, the integrity of genetic variability is preserved until individual plant selection is initiated, and subsequent yield testing stage is done. This contrasts with other breeding methods, where some variability is lost in early generations through natural or artificial selection. The SSD method overcomes these weaknesses, and also speeds up the breeding cycle by rapidly inbreeding a population prior to initiating individual plant selection. The method consists of advancing large number of F_2 plants through F_4 or F_5 generations by taking one randomly selected seed per plant. The focus in early stages of the scheme is on attaining homozygosity as rapidly as possible without selection. The selection starts after attainment of homozygosity in field and controlled-environment growing of SSD populations. Once the desired level of homozygosity is reached, seed of single plants are bulk harvested (each plant constituting a pure line variety). Seed of each line is then grown in progeny rows

Plant Breeding and Cultivar Development. DOI: https://doi.org/10.1016/B978-0-12-817563-7.00024-6

(single or paired short rows, depending on species and seed availability per harvested plant) or head rows. Selection can be made on agronomic, morphological, disease and pest, and quality traits. The method capitalizes on growing several generations per year in a combination of home field, greenhouse/growth chamber, and winter nurseries.

The origin of SSD method is traced back to the work of plant scientist C.H. Goulden (1941), who suggested the use of growth chamber for rapid plant cycle turnover of breeding populations in small grain crop species. Goulden's approach included planting of F_2 seeds, then harvest one seed from each plant for the F_3 generation, and continue this process until F_5 or F_6 generation when pure line variety is developed. Goulden attained F_6 generation in 2 years by conducting multiple plantings per year using greenhouse and off-season field nurseries.

Grafius (1965) during a visit to Europe was motivated to further improve Goulden's method. This motivation stemmed from two observations that could enhance the rapid cycling procedure: (1) *Arabidopsis* genetic work and short cycling, and (2) lack of nutrient can still take barley plants to maturity. In Grafius words "Why not make an Arabidopsis out of barley and oats?" In his experiments, he reduced nutrient and water application and grew plants in higher densities showing that it was possible to grow 8000 plants in 3.5 feet × 26 feet bench. This can represent an entire breeding program's populations equivalent of 8000 plant rows. In addition to saving time, financial savings with this method was also noted. While bulk method will likely be cheaper, the efficiency gain in SSD through reduced cycle time was enormous. It also allows populations to be handled in relatively easier growing conditions and away from threats of environmental factors. Grafious also highlighted the importance of harvesting seed earlier than physiological maturity saving time, and thereby enabling even more rapid cycling.

Johnson and Bernard (1962) used the term SSD for the first time while working with soybean. Brim (1966) built on SSD method and proposed a modified pedigree method (now called as the modified SSD). Currently, small grains programs worldwide extensively use SSD, and three to four generations can be easily completed in one year for these crops, for example, barley and oats. Also, with an efficient usage of home locations and off-season nurseries, soybean breeding programs routinely advance three generations per year, and with proper planning even four generations per year. For example, soybean breeders do not need to wait until physiological maturity stage and pods can be picked earlier (up to 3 weeks earlier) to shorten the crop cycle. These pods are gently dried to prevent fertility loss and also inhibit growth of mold.

Recently, Watson et al. (2018) presented a method called "speed breeding" that was proposed to shorten generation time (seed to seed). Authors showed completing up to 6 generations per year for wheat (*Triticum aestivum* L. and *Triticum durum*), barley, chickpea, and pea (*Pisum sativum*). Four generations per year were completed for canola (*Brassica napus*). Speed breeding is conducted in fully enclosed controlled-environment chambers and allows research and breeding applications. This is achieved due to manipulating the photoperiod (e.g., extending photoperiod to 22 hours light/2 hours dark for wheat and barley) and the use of cost saving light-emitting diode supplemental lighting. These changes lead to shortening the life cycle by approximately half of the control conditions used in their study. Speed breeding is an incremental gain to previously reported approaches, such as the use of greenhouse or screen house facilities to grow breeding material in flat beds that allows the completion of life cycles more rapidly as explained earlier (this method is called rapid generation advance (RGA) in small grain cereals, and has been practiced by breeders for decades). In the RGA method, four or more generations per year are completed. The speed breeding approach can be integrated in a SSD method to achieve pure line in one year or less; however, it requires specialized infrastructure. The most attractive use of RGA or similar methods such as speed breeding is when it is integrated with a genomic selection pipeline. The combination of rapid generation cycling with genomic prediction is a valuable tool for programs that have the required infrastructural and financial resources.

One additional facet of SSD that has been exploited by breeders is to create specialized plots for disease screening based selection, and/or use integration of molecular markers in the generation advancement and selection. This can be achieved on leaf tissue sample or chipped seed, as per the availability of resources. These approaches are still within the realm of SSD because the genetic variability of a breeding population is preserved, but undesirable plants can be discarded reducing the resource footprint. However, it means that sufficient population sizes will need to be grown in earlier generation. For example, consider a scenario where a two parent cross is created. Parent 1 has favorable allele of gene A (AA), and Parent 2 has unfavorable allele of gene A (aa). In the F_2 generation, population will segregate in 1AA: 2Aa: 1aa ratio. Breeder can select 25% plants that are fixed for the favorable allele, but this causes a more severe reduction in population size. Breeder can keep 75% of the population (keep AA and Aa) genetic configuration plants while accepting that "Aa" types will further segregate. These decisions need to be made by considering how important is that trait or gene. If it is an absolute must have gene, breeder may discard all Aa and aa genotypes to fix the genetic classes early so resources are better utilized. SSD method works well in self-fertilizing crop species where additive and additive x additive genetic variances are predominant for traits of interest.

General outline of single seed descent method

The early segregating generations of SSD from F_2 through F_4/F_5 are advanced in greenhouse or in off-season nurseries. Each F_2 plant is represented in the homozygous form in F_6/F_7 generation. Three to four generations per year of spring wheat, oat, and barley can be completed. The general outline of SSD method (random SSD method) for a breeding population is described as follows and is also given in Fig. 12.1.

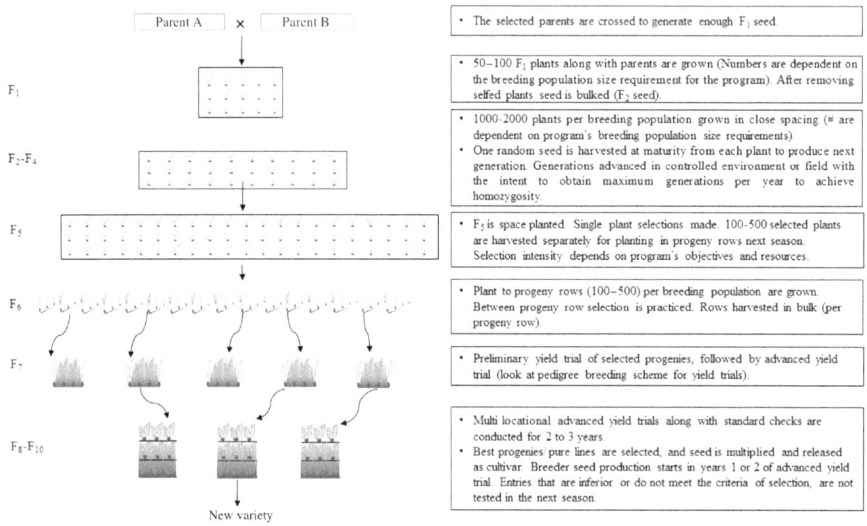

FIGURE 12.1

Flowchart of the single seed descent method of breeding.

Selection of parents: the carefully selected parents are crossed to generate enough F_1 seed to produce large F_2 population (i.e., breeding family).

F_1 generation: depending on the crop and reproductive biology, (generally 50 to 100) F_1 plants along with parents are grown. Careful assay to identify selfed plants is needed to remove non-true F_1 plants (i.e., selfed plants). After removing selfed plants, seeds from the remaining F_1 plants are bulked (F_2 seed). Selfed seed can be removed by observing morphological characters or using molecular markers. A large F_1 population may be grown to ensure adequate recombination among parental chromosomes to produce F_2 seed. The F_1 generation may be grown in greenhouse in trays or in field.

F_2 generation: some 1000–2000 F_2 plants are grown in close spacing (these numbers vary from crop to crop and are influenced by the size and resources of the program). From each F_2 plant, at maturity, one random seed is harvested. Harvested seed is bulked within that breeding family. It is important to consider mortality and germination percentage to determine the population size in the F_2 and later generations. This ensures that adequate population size is available at the final stage of SSD (i.e., at single plant harvest to create progeny rows). For example, if 80% germination rate is observed for a crop and harvested seed in the breeding program, and the final desired breeding family population size is ~200, the following planning can be done.

F_2 = 500 planted seed. Harvest one seed from each of the 400 emerged plants.
F_3 = 400 planted seed. Harvest one seed from each of the 320 emerged plants.

$F_4 =$ 320 planted seed. Harvest one seed from each of the 256 emerged plants.
$F_5 =$ 256 planted seed. Harvest seed of each ~ 205 emerged plants to form progeny rows.

This will give 205 progeny rows. If crop species produces 50 seed per plant in these field or controlled environment generation advancement, 205 progeny rows (each with 50 seed) will be available from this breeding family. If the program makes 100 crosses per year, 20,500 (100 breeding families × 205 F_5 lines per breeding family) progeny rows will be created per year. If artificial selection is practiced, these numbers can be reduced to only move the meritorious progeny rows for field testing. These selections can be made using phenotypic or marker based approaches.

F_3 and F_4 generations: the F_3 and F_4 seeds are grown to produce 1000–2000 plants in close spacing. At maturity, one random seed is harvested from each plant to produce next generation.

F_5 generation: the F_5 generation is space planted, and entire seed of selected individual plants (100–500; depending on selection criteria imposed by the breeder) is harvested separately.

F_6 generation: the F_6 plant-to-progeny rows (100–500) are grown in field conditions. Standard checks are included in this generation. Visual selection is made among progenies. The desired number of superior progenies are selected, and inferior progenies with visible defects are rejected.

For all breeding methods where progeny rows are grown, depending on programmatic setup, progeny rows can be grown in a single row or paired rows at one location, or in single or paired rows at more than one location. Planting progeny rows at one location is one of the biggest risk taken by a plant breeder, due to the problems associated with weather events where an entire field is severely impacted (damaging flooding, hail, or another significant and adverse natural or artificial event etc.). Therefore, planting progeny rows at two locations mitigates loss and also provides genetic and genotype x environment interaction estimates. However to accomplish this, careful planning needs to be done to ensure adequate quantity of seed are available to plant sufficient sized plots at more than one location, and enough resources are available to manage planting, maintenance, data collection, and harvest at both locations. It is also possible that breeder may plant one location as an observation nursery (and a back-up seed source) where only notes are taken with no harvest, and the main location is harvested and serves as a seed source.

F_7 generation: the preliminary yield trials at few locations of selected progenies along with standard checks are conducted, and low yielding plant-to-progeny rows (lines) are rejected.

F_8 and F_9 generations: the multi-locational adaptive trials (at more locations than preliminary yield trials) of superior lines selected in F_7 along with standard checks for two to three years are conducted. The very best line(s) is/are selected for release. Breeder seed production commences, in parallel with yield testing to ensure adequate seed is available for commercial launch.

F_{10} generation: seed multiplication and release as variety.

The scheme described in Fig. 12.1 is a generalized representation that can be modified from crop to crop based on the objectives of the program and facilities that are available to the breeder. The example given above is presented for an overview, and the numbers described per breeding population are presented as a broader example, which vary among programs. Breeders plant relevant checks interspersed in earlier generations and as a part of the experimental design (replicated and randomized) in yield trials. In the single pod descent (SPD) or single pod selection (e.g., in soybean), the scheme will be similar except one random pod per plant is picked instead of a single seed (F_2–F_4). In modified single seed or single pod selection, breeder may impose artificial selection on phenotype or use marker data to make selections per generation reducing population sizes in each successive filial generation. Breeders also make adjustments for expected germination percentages to ensure optimum population size before deriving progeny rows. Also, breeders may pick two seed per plant in SSD (or two pods per plant in SPD), where one set of seed or pod (one per plant) goes to the next generation, while the second set is kept in reserve, in case of crop failure or other issues.

Briefly, in SSD or SPD, generations are advanced to homozygosity rapidly. In the case of small grain crops such as wheat, barley, and oats, three seasons can be easily completed in artificial growing conditions (greenhouse, etc.), that is, RGA. Limited space is needed to keep a population size of 250–300 seeds per breeding family. SSD plots (field or controlled-environment conditions) can be grown in a disease or other stress nursery to select for that stress trait while the population advances to homozygosity. SSD is a cheaper, less technical method of breeding. Rapid inbreeding and homozygosity are achieved. There is no need for record keeping of individual plants while advancing through SSD. The SSD method saves in field space, time, and effort in harvest and book-keeping.

Examples of single seed descent method and its variations

Single random seed method of SSD: Kaufmann (1961, 1971) observed that early generation selection may not be effective due to linkage among important characters. He used a single random seed to advance large number of F_2 plants to a stage of near homozygosity. This procedure is also called random method and has been described in detail earlier in this chapter (step by step). The method has been found useful to develop rapidly large number of homozygous lines, which trace back to different F_2 plants. No selection is practiced in the early segregating populations.

The F_2 generation is grown under normal spacing to have enough seed to take one random seed from each plant and remaining seed is kept in reserve. Same procedure is followed from F_3 through F_6. In F_6 entire seed of each plant is harvested separately to grow plant-to-progeny rows in F_7. The progeny rows (lines) with desirable characters are harvested and sown in multirows in F_8. After agronomic evaluation, superior lines for different characters are retained for advanced replicated yield trials.

The excellent lines from yield trials (at least for one feature) are planted from reserve F_3 seed (harvested from F_2 plants). These could be selected for their specific features in F_3 and F_4 for further use in breeding. The method is suitable for identifying most promising crosses for their further exploitation. This method was described by Kaufmann (1971) as "essentially the same as the "modified pedigree method" outlined by Brim (1966) for soybean breeding."

Single-spike or single hill method of SSD: to overcome the issues associated with genetic drift (introduced by poor crop stand during SSD cycle) and tediousness of collecting single kernel from each plant, some breeders use single-spike descent method or single hill method of SSD. The method consists of harvesting one spike from each F_2 plant, which is usually maintained as a separate line. Some 5–25 seeds from each F_2 plant are sown in pots in greenhouse or in hills in the field, and a single-spike is harvested from the line to be used for the next generation of selfing. The scheme is fast and allows the breeder to keep a seed reserve but requires more space than strict SSD. While this method can significantly expand in numbers and scope because of low rates of attrition due to emergence losses, the advent of genomic prediction can provide advantages wherein genetic potential of individual early generation plants can be predicted and only select plants are chosen to pick single-spike and repeating the process. These methods will reduce genetic variability, and therefore, adequate attention needs to be paid considering the overall program rather than individual breeding families.

Modified pedigree method: the method described by Brim (1966) is a modification from the pure form of SSD, as he proposed to harvest one pod per plant at each generation as parental material although only one seed from the pod serves as parental material. Brim (1966) devised a scheme for advancing generations to the desired level of inbreeding in soybean as that proposed by Goulden (1941) in small grain cereals. Each F_2 plant in the breeding family is advanced by single seed descent. In the F_2 and succeeding generations, only one seed is used from each plant and serves as a parent for the next generation. For practical considerations, a single pod (with two or more seed) is taken from each plant for bulking. When the desired level of homozygosity is achieved, each progeny row (that traces to a different F_2 parent is maintained in bulk). Additionally, he described selection for characters of high heritability (such as plant height, disease and seed shattering) that can be successfully performed on single plants. In the original SSD method, genetic variability of a breeding family is preserved (i.e., no artificial selection until plants reach high level of homozygosity).

Modified pod descent: in this method, instead of single seed descent a variation is followed: one pod per plant is picked, and in succeeding generation this can be handled in two ways: (1) Single pod descent: seeds of each pod are grown in a single hill, and again one pod is picked per hill and (2) Bulk pod method: seeds of each pod are bulked and grown, and in each successive generation, one pod is picked, and seed of all picked pods are bulked. The bulk can be divided into two, one for planting the next generation and the other half is held as reserve. Both methods ensure maintain the breeding family population size. Artificial selection can also be utilized at each generation to move the trait means for simply inherited characters.

RGA method of SSD: a scheme of SSD using RGA was suggested by Goulden (1941). This method has been routinely used in North American small grains program to move from initial cross to yield trials in 2–3 years, depending on the use of winter nursery for progeny rows while earlier generation are handled in controlled-environment conditions.

Genetic basis of single seed descent method

In general, in case of self-fertilized crops only additive genetic variance (V_A) along with additive x additive type of epistasis can be exploited through selection. Large proportion of the genetic variance is additive for economic traits. Only one-half of additive genetic variation is expressed in the F_2 generation, which must be inbred to complete homozygosity for manifestation of total additive genetic variance ($2V_A$). Brim (1966) observed that when genetic variance for a character is all additive, the means of generations do not change with inbreeding. However, as a result of inbreeding, variance among progeny means will increase and variance within progenies will decrease. Brim and Cockerham (1961) showed that in soybean expected progress from selecting the upper 5% of progenies increased with inbreeding but at a decreasing rate. This suggested that additional gain from inbreeding could be seen with time and effort, but there would be an advantage to inbred to the F_3 and/ or F_4 generation. When gene interactions of the additive type are important, testing at higher levels of inbreeding than the F_3 or F_4 is likely to be more efficient since the selection of number of superior lines is practiced when population has attained fairly good levels of homozygosity (F_5/ F_6). The selection would be useful only among progenies rather than within progenies. The SSD is similar to the bulk method, which also ends up with population consisting of mixture of pure lines as a result continuous bulking for four to five generations of inbreeding. During this period, natural selection has also been operating in both the methods, if populations are grown in field. Artificial selection in SSD could be practiced in the F_2 generation if the favorable conditions are available as per the objectives of the program, which is similar to pedigree method of breeding. Selection could be deferred to F_6/F_7 generation under space planted conditions. As previously described, marker technology provides attractive alternatives to harness the potential of SSD.

Application of single seed descent method

SSD has been widely used for the development of superior varieties of crops like, wheat, oat, barley, soybean and tomato. Alternatively, SSD is a potential method for predicting the potential hybrids for further exploitation in self-fertilized crops. Through SSD, superior lines are developed in the shortest possible time, and completely homozygous lines derived from a cross involving diverse parents can be used for QTL analysis. The SSD method and its variants are most effective when they are combined with phenotypic or molecular selection tools, as there is limited use to carry forward "undesirable" genotypes (which is not known until progeny row or later testing). This causes a drain on a breeding program, and resources can be channeled to more accurate and precise characterization of genotypes that show merit.

Merits of single seed descent method

1. It is an easy and rapid method to attain homozygosity, and three to four generations can be completed per year in greenhouse/off-season nursery. Consequently, less space is required to grow early generations.
2. Detailed book-keeping and note-taking are considerably reduced.
3. SSD can be applied at any stage of the program, and it can be combined with pedigree and bulk selection methods.
4. Large numbers of random pure lines are produced, permitting better exploitation of the potential of a cross.
5. Through SSD, the entire range of variability is retained as every plant originates from a different F_2 plants. The variability is available for selection at F_6 where homozygosity is near complete. At this stage, heritability is higher and hence evaluation and selection can be concentrated between lines. Selection is not complicated by heterozygosity as SSD emphasizes the additive genetic variance and reduces the influence of non-additive component of the total genetic variance.
6. SSD also provides an opportunity for recombination to occur for many generations before selection is practiced. When favorable genes are dispersed between the parents of a cross or in close linkage with unfavorable genes (either in coupling or repulsion), such recombination would occur equally in the bulk method without selection; however in the bulk method with natural selection, total genetic variability will be more limited than in SSD.
7. Seed chipping and marker technology can be effectively used to combine marker assisted selection and genomic selection at any and all filial generations, F_2 onwards.

Limitations of single seed descent method

1. Since no pedigree records are maintained, the identity of superior F_2 plant is lost. Selection is based on individual phenotype and not progeny performance.

2. The segregating populations may not be subjected for field evaluation until F_6 or F_7; therefore, the segregating populations are not subjected to stresses faced in the field. If the parents are adapted type, then this limitation may not be as important.
3. The number of plants from F_2 to F_4 are equal as from each plant, only one seed is harvested to take up the next generation. Some of these seeds may not germinate due to inviability, or some of the plants may not set seeds due to natural factors resulting in the loss of some valuable plants leading to genetic drift. A sample of one pod per plant or one spike/hill could avoid this situation.
4. The assumption that single seed represents the genetic base of a F_2 plant does not necessarily hold true unless extremely large and infeasible population sizes are maintained.
5. The SSD method may not be the most appropriate way to select populations originating from parents, which differ for many gene pairs.
6. If doubled haploid technology is available and not cost limiting, SSD may be less preferred.

CHAPTER

Backcross method

13

Abstract

The backcross method involves the crossing of hybrid (F_1 or its progenies) of two parents to either parent, with an aim to improve an elite line by adding one or few genes from the donor parent. This method does not operate in isolation of previous forward breeding approaches (pedigree, bulk, single seed, and doubled haploid), but it complements them. A breeding program will run an active forward breeding program (that will develop elite varieties) and use the backcross method to add one or few genes (disease resistance, herbicide resistance, and end-use quality characters) to enhance the value of variety or help meet market requirements. This ensures that one or few genes from a donor parent added to an established variety (recipient parent) without disturbing the basic genotype of the elite variety. In this chapter, features and procedures of backcross method along with the genetic basis, its modifications, merits, and limitations are described. Several scenarios in the application of backcross method are explained.

Harlan and Pope (1922) described a basic idea to shift variation towards one of the parents as a method of small grains breeding. They introduced a very strict version of backcross, which was extensively used by Briggs and coworkers in wheat breeding program at Davis, California. In greenhouse and growth chambers, this strict backcross became extremely rapid and reliable, as long as the character to be transferred is sufficiently independent of environment for expression and not inherited too complexly. The backcross method is now routinely used in commercial seed companies for the transfer of genetically modified gene(s), for example, herbicide tolerance, insect resistance, to convert elite lines and pure lines in self-pollinated species and inbred lines in cross-pollinated species, prior to commercialization and/or production of hybrid seed. Due to challenges in hybridization (lower frequency, manual dexterity, and labor intensive) in self-pollinated crops, commercial companies may not necessarily use the backcross method to convert (non-GM to GM) lines in each breeding cycle. Instead, once a new GM trait has been incorporated in the germplasm pool, breeders may work with this material removing the need to constantly use backcross to convert lines. However, when a new GM trait is introduced, breeders will have to convert elite lines and cultivars by removing the previous GM gene and inserting the new one; overall, a very cumbersome task. On the other hand, in maize hybrid development programs, female inbred lines are bred and developed in a non-GM background, and depending on

Plant Breeding and Cultivar Development. DOI: https://doi.org/10.1016/B978-0-12-817563-7.00016-7

market placement decision, the backcross method is used to convert them to GM lines prior to crossing with the male inbred line for hybrid production. The male parent in a hybrid themselves can be created through the backcross breeding method to provide an array of trait options, and this will prevent the need to convert each newly developed female inbred line.

According to Briggs and Allard (1953), for a backcross scheme to be successful, the following three requirements must be satisfied: (1) a satisfactory recurrent parent must exist that meets the requirement of the breeding objectives, (2) the expression of character from donor parents is retained through several backcrosses, and (3) the genotype of the recurrent parent must be reconstituted after the method is completed with a reasonable number of backcrosses executed with populations of manageable size.

The recurrent parent

The desirable cultivar or parent that receives the desired genetic trait possesses a high preponderance of genes for other traits and is called recurrent parent. It is also called as the recipient parent and is an elite line. A satisfactory recurrent parent is available in any established breeding program because these recurrent parents are elite varieties that have been developed using forward breeding approaches. In some situations, a breeder may use variety developed using the backcross method as a recurrent parent in a second cycle in order to add more genes. Most of such cultivars have excellent combination of genes and are cultivated widely in a particular area or region; therefore quality as recurrent parents in backcross breeding is an important factor. Keeping in mind the need to increase genetic gain, breeder will maximize their return on investment to start using the backcross method on most promising candidate varieties, which may be still undergoing final stages of precommercial testing. This will ensure that a new improved (i.e., converted) variety or line is available without any time delay.

Maintenance of the character under transfer

The cultivar or parent that contributes the simply inherited characteristic(s) is called non-recurrent parent or donor parent. The non-recurrent parent is used only once in the hybridization in a backcross scheme. The breeder need to be concerned only with selection for the character being transferred during the backcrossing. At the same time, all other characters are taken care of automatically by the backcrossing procedure, and the breeder need not to pay attention to them or to pay little attention (Allard, 1960). However, it is observed that some of the intensity of character may be lost even when its genetic control is predominantly monogenic (Briggs and Allard, 1953). None of the bunt-resistant, backcross-derived wheat varieties were equal to "Martin" the donor parent in resistance.

Upon artificial inoculation, "Martin" was nearly always completely free from infection, whereas the derived lines showed as much as 3% or 4% infection. Several other similar examples have been cited by the breeders; however, such a loss of resistance has not limited the utility of resistant varieties if they were still fairly resistant and better than the recurrent parent for the specific characteristic. This loss of resistance can be attributed to epistatic gene interactions. Therefore, before commercialization of backcross-derived variety, breeders should appropriately phenotype to ensure that trait expression is within the breeder established threshold, and meets the requirements for targeted market segment and farmer requirements.

General out-line of the backcross method

The method involves crossing the recurrent parent with the donor parent and F_1 hybrid, and/or its progenies are backcrossed to one of the parents (recurrent parent). The second step is repeated two to six times and at the end recovering the recurrent parent genotypes with added attribute(s). Through each successive backcross generation, genes from non-recurrent parent are reduced to half (heterozygosity is reduced to half) as with selfing, and by the sixth backcross the genotype has recovered over 99% of the recurrent parent. For attempting backcross, recurrent parent is generally used as a female parent because more seed of the recurrent parent is available, and elite line's non-nuclear DNA is retained. Also, generally pollen supply is not as constrained as female flowers. However, as described in the previous section, sufficient number of backcrossed seed need to be used. The number of backcross generation would depend on the expression of gene, that is, nature of inheritance, level of germplasm from recurrent parent desired, and objectives of breeding. If only phenotypic assessment is used at each generation, a selection pressure or artificial selection is imposed after each backcross generation to identify desirable types (for next generation of crossing) and discard undesirable plants. If marker-assisted backcrossing is deployed, phenotypic selection may not be needed. The size of population in the backcross generation would depend upon the number of genes to be transferred. Extensive yield testing of the end product of the backcross method is not required as the new lines are similar to recurrent parent, except for newly introduced characters. However, if fewer generations of backcrossing has been imposed, yield testing is desirable to ensure that donor parent genome is not causing poorer performance in the new line.

Gene transfer from exotic germplasm, wild and weedy relatives of cultivated species is possible through backcrossing. It would take longer time to remove the undesirable agronomic traits (such as, seed color, shattering habit etc.) of the donor parent. Two or more different but simply inherited characters or from different sources in a variety, two or more series of backcrosses, one for each character, can be attempted and on completion crossed together to combine different characters. This is called *convergent improvement* or *double backcross.*

Dominant gene transfer

In case a single dominant gene (RR) with high expressivity governing resistance to a disease, phenotypic selection is adequate for selecting desirable individual plants, and recessive genotypes are discarded (Fig. 13.1).

Selection of the parents: the recurrent parent (A) is widely grown cultivar but is susceptible to a disease. The donor parent (B) is selected, and it possesses a dominant gene (RR) for resistance.

Making crosses: 10–20 hybridizations between A and B parents are attempted, F_1 harvested and bulked. These 10–20 number of attempts are governed by crop's reproductive biology and anticipated success rate.

F_1 generation: the F_1 plants of A × B are grown. The recurrent parent "A" is susceptible (rr), and donor parent "B" (RR) is resistant. All the plants of F_1 generation will be resistant (Rr).

BC_1 to BC_5/BC_6 generations: the F_1 hybrid plants (Rr) are crossed to recurrent parent (A) to obtain the first backcross (BC_1) generation. The BC_1F_1 generation would segregate in the ratio of 1 (rr): 1 (Rr) for susceptibility and resistance, respectively.

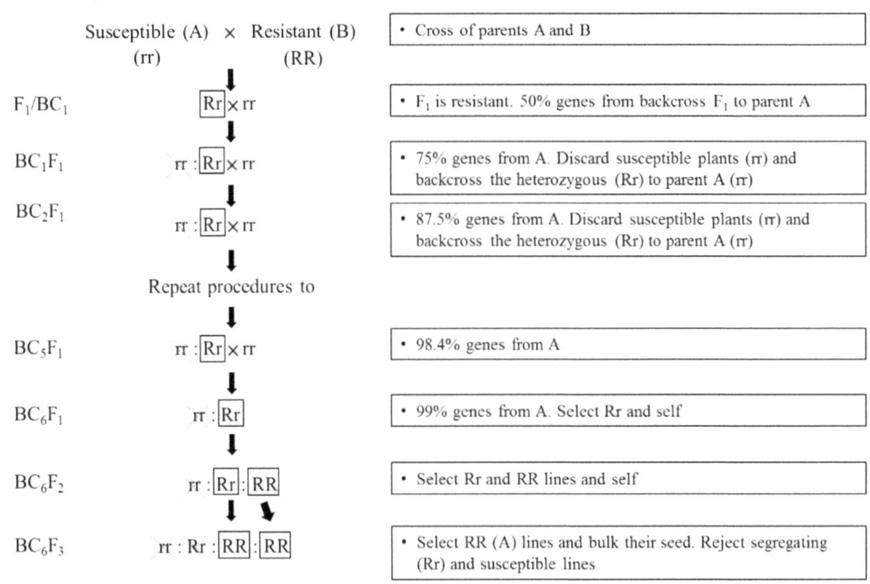

FIGURE 13.1

General outline of backcross breeding to introduce a dominant gene (RR) into susceptible variety, where × = undesirable type and □ = desirable type of plants.

All susceptible plants are discarded, and resistant plants are again backcrossed to the recurrent parent. Each time 30–50 heterozygous plants (number will slightly vary for different crops) that resemble recurrent parent are selected for backcrossing. This process is repeated five to six times. After each backcross generation, selection pressure is imposed to identify and discard homozygous recessive plants (rr) and to identify heterozygous resistant (Rr) individual plants. At this time, phenotype of recurrent parent are recovered, and the selected plants will be homozygous for mostly all other alleles, except for the resistance gene, which will be in heterozygous (Rr) genotype. After the sixth backcross, in BC_6F_1 and BC_6F_2 generations, all the susceptible plants are discarded. In BC_6F_2 large number (300–500) of desirable individual plants, that is, resistant plants are selected. From these some 100 desirable homozygous dominant (RR) progeny rows are selected in BC_6F_3 and are bulked to form a new line, which along with parental cultivar is subjected for limited yield tests, if desired.

This is the simplest explanation of the character transfer. The gene is not influenced by modifier genes and not linked to other undesirable traits in the donor parent. If the gene or genes of interest (which will be moved comes from an unadapted or related species to recurrent parent), breeder has to be aware of genetic linkage and epistatic interactions to avoid inadvertently bringing in undesirable genes linked to the desirable target gene (i.e., linkage drag). Larger population sizes are used in this situation to identify recombinants. Recent innovations (i.e., marker-assisted backcrossing, marker-assisted recurrent selection, see Chapter 27: Molecular Tools in Crop Improvement and Cultivar Development) reduce the need for large population sizes.

Recessive gene transfer

The backcross method is also used to introduce recessive disease resistance gene(s). The progeny from backcross i.e., cross of F_1 (Rr) with a recurrent parent (RR) would segregate into two genotypes; RR and Rr and all the plants will be susceptible to the disease (Fig. 13.2). Therefore, phenotypic identification of resistance is not possible because the heterozygote cannot be identified without a progeny test. It is necessary to self the F_1 after each backcross to the recurrent parent to determine the genotypes of the backcross progeny and to select plants with homozygous recessive resistance gene. As shown in Fig. 13.2, the BC_1F_2 would segregate as 1 (RR): 2 (Rr): 1(rr). The plants with rr genotypes are selected. Selection pressure for disease reaction is imposed after each backcross generation.

Backcrossing procedures in different scenarios

Four different scenarios applicable in plant breeding programs looking to utilize backcrossing are described. These examples are provided to focus on phenotypic characterization, in the absence of molecular markers. Marker assisted backcrossing and marker assisted recurrent selection are described in Chapter 27 - Molecular Tools in Crop Improvement and Cultivar Development.

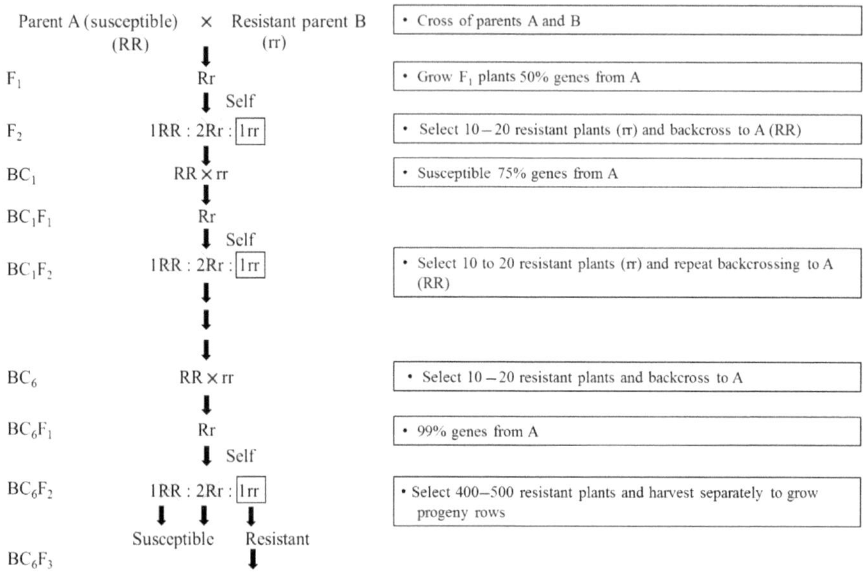

FIGURE 13.2

General out-line of backcross breeding for introducing a recessive gene (rr) into susceptible variety, where × = undesirable type and □ = desirable type of plants.

Scenario 1: single dominant allele where trait expression is observable pre-flowering (Fig. 13.3).

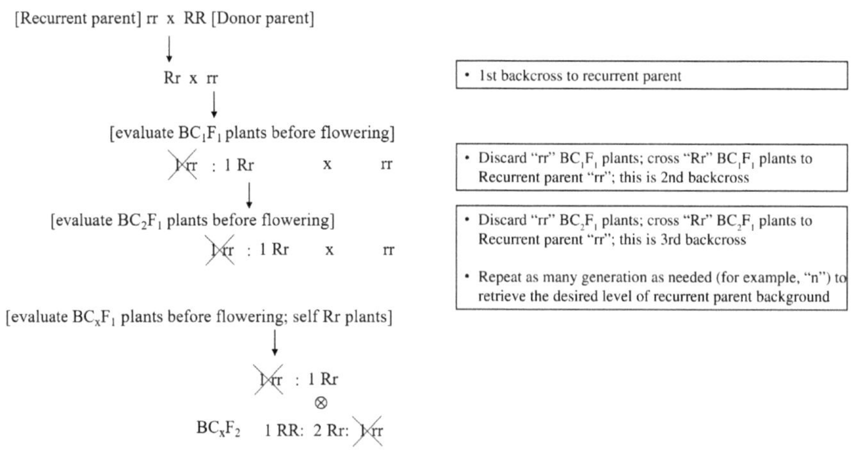

Evaluate BC_nF_2 plants before flowering. Discard all "rr" types, RR and Rr plants (that is all remaining resistant plants) are selfed, and each of these plants are harvest individually to form progeny rows. Using progeny testing in the following season, homozygous and segregating rows are identified. Segregating rows are discarded, while homozygous RR rows are kept and tested. These homozygous rows can be considered breeder row and testing and seed production can proceed.

FIGURE 13.3

Backcrossing single dominant allele where trait expression is observable pre-flowering.

Scenario 2: single dominant allele where trait expression is observable post-flowering (Fig. 13.4).

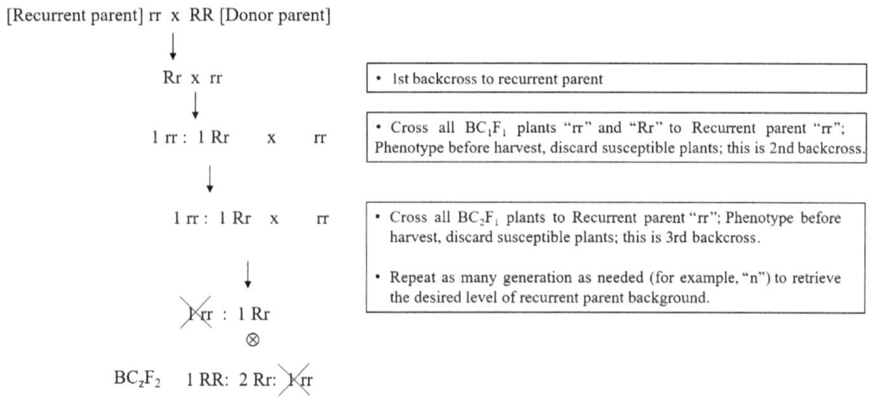

Evaluate BC_nF_2 plants post flowering. Discard all "rr" types. RR and Rr plants (that is, all remaining resistant plants) are selfed, and each of these plants are harvested individually to form progeny rows. Using progeny testing in the following season, homozygous and segregating rows are identified. Segregating rows are discarded, while homozygous RR rows are kept and tested. These homozygous rows can be considered breeder row and testing and seed production can proceed.

FIGURE 13.4

Backcrossing single dominant allele where trait expression is observable post-flowering.

Scenario 3: single recessive allele where progeny test is possible in same season (Fig. 13.5).

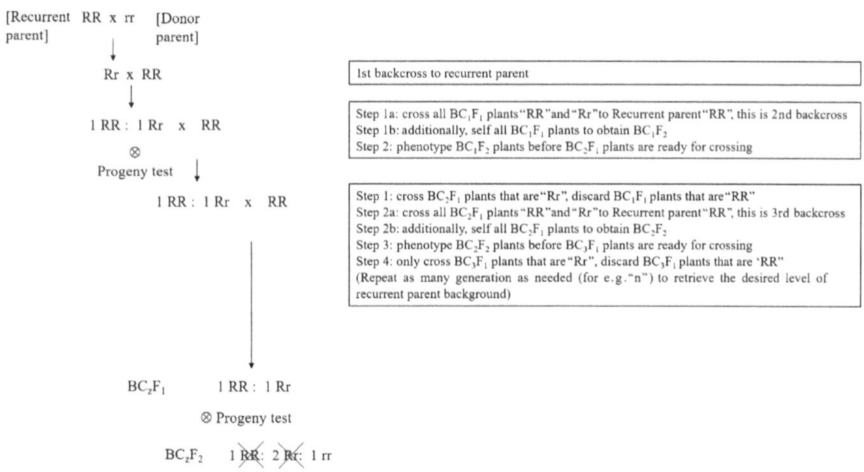

BC_nF_1 plants are progeny tested. Undesirable phenotype of $BC_{z-1}F_1$ plants' family are discarded. Progeny tests completed and selfed progenies of "rr" type are kept, while all other types are discarded. That is, only selfed families from "rr" BC_zF_1 plants are kept. These homozygous "rr" rows can be considered breeder rows, and further testing and seed production can proceed.

FIGURE 13.5

Backcrossing single recessive allele where progeny test is possible in same season.

Scenario 4: single recessive allele where progeny test is not possible in same season

In this scenario, cross recurrent parent (susceptible, "RR") to donor parent (resistant, "rr"), obtain F_1 (all "Rr" type), which is crossed to recurrent parent (this is 1st backcross). BC_1F_1 ("RR" and "Rr" types) plants are selfed, generating BC_1F_2 plants from each BC_1F_1 ("RR," "Rr," and "rr" types). BC_1F_2 are grown in progeny rows, and these are selfed. Within the segregating rows, "rr" types plants (resistant) are identified, and seed from each resistant BC_1F_2 plants is harvested. The BC_1F_3 plants are crossed to the recurrent parent to obtain BC_2F_1 seeds (1 RR: 1 Rr). This process is repeated as many generation needed to achieve desired homozygosity. Once the desired level of homozygosity is achieved, BC_nF_2 (where n is the number of generation) plants are selfed to obtain BC_nF_2 (1 RR: 2 Rr: 1 rr). These are self-pollinated. Resistant progeny rows are identified and harvested individually. Characterization and comparison to recurrent parent (and resistance reaction to donor parent) is confirmed, and rows that meet the criterion are bulked.

Modifications of the backcross method

The backcross method of breeding has been used either as a continuous (without progeny test) or as a discontinuous (with progeny test) series, and the latter versions allows for a more careful check of the transfer. In introducing foreign germplasm for the improvement of more complicated and not immediately measurable features such as yield, quality, winter hardiness, and straw strength, single backcross to the elite parent has proved helpful (the incomplete backcross). By this scheme, 75% instead of only 50% of the germplasm of the cross will be of the highly adapted type, and well balanced recombinations and/or transgressions will be more frequent in a resulting population, but this scheme requires a larger population size and a donor parent that has some acceptable levels of other characteristics.

Instead of strict backcross A/B//A, the cross A/B//C can be created with B being the foreign variety or introduction, and A and C are two well adapted varieties of somewhat different types (e.g. parental background) with complementary features. In such a three parent cross, A and B will constitute 25% each of the germplasm of the population, and C will constitute 50% of the germplasm. Hence, the choice of C will be more important than that of A. This scheme has proved to be efficient way to widen the gene pool in breeding programs. This is not backcrossing *per se*, and the example has been provided for clarification purposes only. This three parent cross has been used for both germplasm line and cultivar development.

Genetic basis of the backcross method

A heterozygous F_1 population between recurrent and donor parents contains 50% genes of each parent. The F_1 backcrossed to a recurrent parent will become homozygous for the genotype of the recurrent parent. The repeated backcrossing to the recurrent parent would shift the variation gradually towards one parent, instead of breaking up into 2^n homozygous genotypes as with selfing. The proportion of genes from the donor parent is reduced by one-half following each backcross generation and is given by the relationship $(1/2)^{m+1}$, where m is the number of crosses and backcrosses to the recurrent parent. For instance in BC_2, the germplasm of donor parent would be $(1/2)^3 = 12.5\%$ and in BC_4, it would be $(1/2)^5 = 3.125\%$. In other words following each backcrossing generation, contribution of genes (average recovery of genes) from recurrent parent would increase. The relationship for the recovery of the recurrent parent is given by Allard (1960) as $[1-(1/2)^{m+1}]$; therefore, in BC_2 the germplasm of recurrent parent would be 87.5%, and in BC_4 it would be 96.875% (Fig. 13.6).

When several genes are considered together, the proportion of homozygotes for all the genes of recurrent parent in any generation would be $[(2^m-1)/2^m]^n$, where n is the number of genes for which parents carry contrasting alleles and m is the number of generations of selfing or backcrossing. In a typical backcrossing program where a backcross progeny contains heterozygous resistant plants and homozygous susceptible in the ratio of 1:1, sufficient seed must be produced to obtain at least one resistant plant (Table 13.1). Linkage of the desirable gene with an undesirable one in donor parent adversely affect the success of a backcrossing program as the undesirable gene(s) may also be transferred along with desirable one (Table 13.2).

Since the recurrent parent is involved in backcrossing repeatedly, there will be number of opportunities for crossover to occur. If no selection is practiced except for desirable gene being transferred the probability of eliminating the undesirable gene is given as $1-(1-r)^{m+1}$, where r is the recombination fraction and m is the number of backcrosses. If undesirable gene is located 50 or more crossover units from desirable

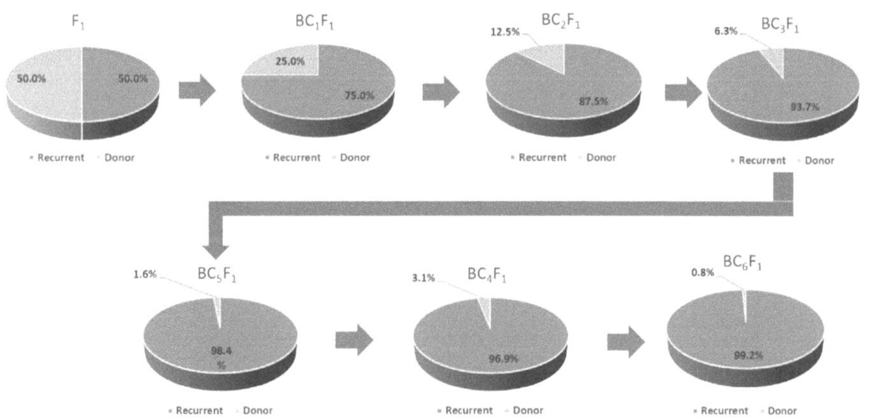

FIGURE 13.6

Proportion of homozygosity with each successive backcrossing cycle.

Table 13.1 The number of plants that need to be grown in a progeny in order to obtain at least one of the required genotype at three commonly used levels of probability.

Expected segregation ratio	Level of probability		
	0.95 (1 in 20)	0.99 (1 in 100)	0.999 (1 in 1000)
1:1	5	7	10
3:1	11	16	24
15:1	47	72	107

Table 13.2 Effect of linkage on the elimination of undesirable gene allele.

Recombination fraction	Probability of elimination	
	With five backcrosses	With selfing
0.50	0.98	0.50
0.20	0.74	0.20
0.10	0.47	0.10
0.02	0.11	0.02
0.01	0.06	0.01
0.001	0.006	0.001

From Allard, R.W., 1960. Principles of Plant Breeding. Toppan Co. Ltd., Tokyo, Japan, p. 485.

gene or is located on another chromosome, the probability of eliminating undesirable gene after five backcrosses would be 0.984; whereas in selfing series with selection only for desirable gene, the corresponding probability will be 0.50. These probabilities progressively become smaller with tighter linkages (Table 13.2).

The probability of getting the desirable recombination is clearly greater with backcross than with selfing with the assumption that no selection is practiced against the undesirable allele. However, if selection is applied against undesirable alleles, which are linked, a selfing series is more efficient. It is due to crossing over that is possible in both parents; whereas in backcrossing, effective crossing over cannot occur in the homogeneous and homozygous recurrent parent. In case of genes governing the characters with low heritability, backcrossing is better as selection is not effective. Recurrent backcrossing even in the absence of selection is a useful procedure for achieving homozygosity.

Number of backcrosses

For successful backcross breeding, it is essential that the genotype of recurrent parent must be reduplicated in its original form. The recovery of genotype of the recurrent parent is primarily a function of number of backcrosses. The selection in early backcross generations is effective in shifting the population toward the characteristics of that parent. Unless using a doubled haploid line, it is unlikely that the recurrent parent is composed of a single pure breeding line; therefore, sufficient plants of the recurrent parent are required to be used in making the backcross to recover the variability of recurrent parent. If the recurrent parent is still not fully homozygous (earlier filial generation), it may be advantageous to use the recent most version of the advancing recurrent parent in each round of backcrossing.

The proportion of homozygosity $= [(2^m - 1)/(2^m)]^n$, where m is the number of generations of selfing and n is the number of segregating genes. If the heterozygote is continually crossed to one of the homozygous parents, homozygosity is attained at the same rate of self-fertilization. Therefore, at five or six backcrossing, more than 95% of the population will be homozygous for the recurrent parent. With six rounds of backcrossing (at $n = 10$), 85% of the population will be homozygous for the recurrent parent. This contrasts with self-fertilization approach; because after six round of self-fertilization, 1024 different homozygous genotypes are present.

Six backcrosses compiled with rigid selection towards the recurrent parent in early generations (up to fourth backcross) were found satisfactory, with minimal probability to introduce undesirable characteristics of the donor parent. Briggs and Allard (1953) cautioned breeders about using small population sizes in backcross method. They reported that to transfer a monogenic dominant allele, theoretically, a minimum of 53 plants from backcrossed seeds, 96 F_2 plants, and 68 F_3 rows are required through 6 backcrosses. With these numbers, with 99.9% probability, breeder will have at least one Aa plant after each backcross, and at least one homozygous AA progeny in F_3. In case of incomplete dominance or recessive genes, slightly smaller total populations are required because homozygotes can be recognized in the F_2 generation removing the requirement to test F_3 progenies. If artificial hybridization is difficult, smaller population sizes can be utilized by growing F_2 and F_3 generations after each backcross and crossing on homozygote types. If the finished variety (from backcross method) is created from nonhomozygous recurrent parent, higher number of plants of the recurrent parent need to be used in the final backcross, as well as a larger number of F_3 lines seed should be bulked for seed increase of new improved variety. Sedcole (1977) suggested that a reliable approximate solution for number of plants to raise to recover one or more plants exhibiting that trait can be computed using this formula (assuming a normal distribution):

$$n \geq \{[(2) \times (r - 0.5) + (z^2) \times (1 - q)] + z \times [(z^2) \times (1 - q)^2 + (4) \times (1 - q) \times (r - 0.5)]^{1/2}\}/2q$$

where n is the number of plants necessary, q is the probability of the occurrence of the trait, and p is the probability of recovering r plants with the trait. z is the normal deviate that can be obtained from a standard statistical reference.

It is advisable to adjust the number of plants to raise on expected germination percentage. For example, if 15% germination is expected in the crop and seed samples, adjustment can be made as [$n/0.85$], where 0.85 comes from 100%−15% (=85%) from germination percentage.

Seasons needed for backcrossing

In Northern latitude based breeding programs, three to four generations of backcrossing can be accomplished in a year (field—if using tropical location, or controlled environment conditions). The advent of molecular markers has also dramatically changed breeding steps because "seed chipping" or "tissue sampling" can remove the need for progeny test. In the examples listed above, the examples of disease reaction are provided; however, any qualitatively inherited trait can be backcrossed to the recurrent parent. Molecular markers (flanking markers) have also enabled backcrossing quantitative trait loci (QTL) instead of single gene. If molecular markers are unavailable, replicated progeny row testing will be needed (instead of single plant progeny test as described in scenarios listed above) to successfully move the QTL in a backcrossing program.

Application of the backcross method

The backcross method is best suited for improving defect(s) of well-established cultivar(s) for one or two simply inherited specific traits. It is useful to transfer genes for biotic and abiotic resistance, and other useful gene(s) in both interspecific and intervarietal crosses; and to transfer entire sets of chromosome in the foreign cytoplasm to create cytoplasm male sterile lines to facilitate hybrid production in crops including corn, onion, and rice. The donor (of the chromosomes) is used as male, and crossing is continued until all donor chromosomes are recovered in the cytoplasm of the recurrent parent. Isogenic lines with different resistance non-allelic genes (race-specific genes) to a disease, insect-pest, and nematode can be produced. The donor parent with resistance is crossed, and five to six backcrosses are attempted with recurrent parent and desirable plants are selected under disease environment. The BC_6F_1 and BC_6F_2 are space planted, and homozygous resistant individual plants are selected. In BC_6F_3 plant to progeny rows are bulked to form an isogenic line. Several isogenic lines (8−10) with different resistance genes are prepared with the common recurrent parent mixed in equal proportion for evaluation of multiline and release. The breeder seed of each isogenic line is maintained separately.

Backcrossing is the preferred choice for converting maize inbred lines (non-genetically modified) to genetically modified inbred lines prior to use in hybrid seed production and sale. These genetically modified events, such as herbicide tolerance, insect resistance, disease resistance, can be transferred individually or in a gene cassette that contains multiple gene inherited together as a linked block.

Backcross breeding has been used to transfer simply inherited traits such as low erucic acid and glucosinolate content into adapted lines and breeding populations of *Brassica napus* and *Brassica campestris* (Downey and Rakow, 1987). The erucic acid content of the seed is not controlled by genotype of the mother plant but by the genotype of the embryo. The seeds borne on F_1 plants (F_2 seeds) have different erucic acid levels. The half seed method makes it possible to select individual seeds on the basis of their erucic acid content (Downey and Harvey, 1963). The selected seeds were grown into plants and backcrosses were performed. Two loci each with two alleles determine erucic acid level in *B. napus* and *Brassica juncea* (Kirk and Hurlstone, 1983). The heterozygous F_1 ($E_1e_1E_2e_2$) was backcrossed to the recurrent parent which have high erucic acid content, heterozygous BC_1F_2 seeds ($E_1e_1E_2e_2$) were selected, and plants obtained from the seeds were backcrossed. This permitted a backcross to be made every generation and after the last backcross, a selfing generation was grown to identify the zero erucic acid genotype ($e_1e_1e_2e_2$). This procedure was effective in transferring the zero erucic acid characteristic from spring types of *B. napus* into adapted and high yielding winter forms of this species.

A similar approach was used in *B. campestris* to transfer the high erucic acid characteristic from yellow sarson into "Tobin" a high yielding and white rust resistant cultivar. Two alleles at a single locus control erucic acid synthesis (Dorrell and Downey, 1964). After backcrossing with "Tobin," the zero erucic acid cultivar selection for heterozygous (E_1e_1) genotypes was carried out using half seed method. The selected plants were further backcrossed, and after the last backcross, a selfing generation was grown to select high erucic acid (E_1E_1) genotypes.

If backcross method is strictly applied, it offers no opportunity for general improvement; however, if applied in combination with pedigree or bulk method, it can also improve other characters. Backcross method of breeding complements forward breeding approaches for a breeding program focused on cultivar development and genetic gain improvement. No cultivar development program can run with backcross breeding alone, and will require a robust forward breeding component to convert lines for select gene(s) or QTL.

Merits of the backcross method

1. It provides a relatively inexpensive and rapid method of variety improvement. Two to four generation can be completed in a year in the greenhouse and off-season nurseries.
2. New variety can be identical to the original variety (recurrent parent), except for the gene(s) transferred from the donor parent. Through this procedure, the specific defect of a well-established cultivar is removed without the risk of disturbing the existing combinations of desirable genes.
3. Extensive yield tests of improved version (new variety) may be omitted as the performance of recurrent parent is already known. This may save up to 2–4 years and also considerable expenses.

4. Smaller populations of segregating generations are needed for selecting the desirable genotypes as compared to pedigree method of breeding.
5. The method is most suitable for interspecific gene transfer.
6. The backcross method could be combined with other methods like pedigree breeding or could be suitably modified for selecting transgressive segregants.
7. Molecular markers are easily integrated, and background selection can shorten the time to recover recurrent parent background.
8. Molecular markers can be used to detect rare recombination (and successfully select) to overcome linkage drag, which is difficult without molecular markers.

Limitations of the backcross method

1. Hybridization is attempted in each backcross generation, which is cumbersome.
2. New variety is superior only for the character transferred.
3. The backcross may not be used for improving a variety with respect to a number of characters.
4. By the time recurrent parent is improved, it may be outdated as new and superior varieties could be available.
5. Making crosses is a laborious and tedious in many crop species. As such method is not suitable in dealing with quantitative characters.

CHAPTER

Mutation breeding

14

Abstract

Mutations, the heritable change to an individual's genetic makeup, are one of the driving forces of evolution complementing selection (natural and artificial) and hybridization. Mutations can create evolutionary advantages or disadvantages leading to a preferential selection or deletion of genotypes from the gene pool. Lethal mutations do not carry their germline forward; however, non-lethal mutations over generations accumulate within the gene pool leading to an increase in genetic variation. Natural selection can reduce the abundance of some genetic changes within the gene pool, while adaptive changes can happen due to the accumulation of favorable mutations. This chapter describes main classes of mutagenesis, types of mutations, mutagens, their doses, and applications in breeding. Mutation breeding methodology is included.

Mutations can be somatic or germinal. Somatic mutation happens in somatic cell and therefore is not passed on to offspring; however, it has a role in vegetatively propagated crops. Germinal mutation happens in reproductive or germ cells and therefore is passed to the next generation; and is important in seed propagated crops. A germline mutation gives rise to a constitutional mutation (a mutation that is present in every cell) in the offspring and is passed on to descendants through reproductive cells. Somatic mutations on the other hand, involve non-reproductive system cells; and therefore, are not transmitted to descendants. Recessive mutations inactivate the affected gene. This causes a loss of function, due to the disruption of gene sequence and/or its expression leading to the altering of encoded protein structure. On the other hand, dominant mutations generally lead to a gain of function, i.e., upregulate the gene expression and increase the activity of the gene product.

Plant breeders have exploited mutations to create genetic variability and develop new cultivars. The term "mutation breeding" ("Mutationszüchtung") was given by Freisleben and Lein (1944) where they described it as the induction and development of mutant lines for crop improvement. In 2019, there were 3301 varieties and 236 species (separated as per their Latin name) in the International Atomic Energy Agency (IAEA) mutation variety database (https://mvd.iaea.org/) demonstrating the usefulness of mutation breeding.

When the other methods of creating variability are insufficient, mutation breeding becomes very important. For example, recombination and selection do not generally create a new gene or allele; however, mutation (spontaneous or induced) can

Plant Breeding and Cultivar Development. DOI: https://doi.org/10.1016/B978-0-12-817563-7.00025-8

create this change. This means that if the desired alleles are not present in breeding material, mutation breeding approach is a good alternative. While modern technologies of gene editing allow targeted base pair changes, this technology is currently limited to situations where gene sequence is known and different alleles need to be generated or gene needs to be turned off or on. Having said this, breeders are advised to consider gene editing approaches if resources and breeding objectives are achievable because regular mutagen-based approaches are non-targeted.

Main classes of mutagenesis

- *Induced mutagenesis:* induced by the use of radiation or chemical mutagens. It differs from site-directed mutagenesis as induced mutagenesis is a random process. This is a simpler process and can be accomplished without genomic sequence information. Outcome of induced mutagenesis are induced mutants. In 1936, the first induced mutant tobacco variety "Chlorina," using X-rays as mutagen was released in Indonesia. In this chapter, the primary focus will be on induced mutation. These mutations can be chromosomal mutations (deletion, duplication, inversion, and translocation) or point mutations (single nucleotide base pair change leading to a change in allelic form).
- *Insertion mutagenesis*: DNA insertions from genetic transformation, T-DNA insertion, activation of transposon elements or transposons, are considered insertion mutagenesis. Outcome of induced mutagenesis is insertion mutants or transposon mutants.
- *Site-directed mutagenesis*: creates specific and targeted changes in DNA. This is achievable through genetic transformation followed by homologous recombination between the T-DNA fragment and indigenous DNA molecules, zinc-finger nucleases, transcription activator-like effector nucleases, and clustered regulatory interspaced short palindromic repeat (CRISPR)/Cas-based RNA-guided DNA endonucleases. These lead to genetic modifications by inducing double-strand DNA breaks that stimulate error-prone nonhomologous end-joining or homology-directed repair at specific genomic locations leading to site-targeted mutation. The development of CRISPR–Cas9 technology allows for more efficient introduction of point mutations into the genome in a wide variety of crop plants. This method of genome editing has become more popular because it does not require a transposon insertion site, leaves no marker, and is efficient and simple with lower barrier to entry. More details are included in Chapter 27: Molecular Tools in Crop Improvement and Cultivar Development.

In nature, we observe spontaneous mutations that occur naturally with no intentional exposures to a mutagen. Spontaneous mutations include:

- Tautomerism: a base is changed by the repositioning of a hydrogen atom.
- Depurination: loss of a purine base (A or G) to form an apurinic site (AP site).

- Deamination: changes a normal base to an atypical base, for example, C → U.
- Transition: a purine changes to another purine or a pyrimidine to a pyrimidine, for example, A ↔ G or C ↔ T.
- Transversion: a purine becomes a pyrimidine or vice versa, for example, A or G ↔ C or T.

Types of mutations

While the above examples involve point or few nucleotide base pair changes, chromosomal changes are also relevant in mutation breeding. Novel genes are produced by duplication and mutation of an ancestral gene, or by recombining parts of different genes to generate new functions. Deletion leads to loss of the gene. Duplication can lead to gene addition and can result in additive effects. In inversion, linkage block changes occur, and other genes in close proximity cosegregate. In insertion and translocation, a gene moves to a new chromosome and can have a similar effect as duplication.

Traditionally, chemical or physical agents were used to induce mutations in crop genomes. These included radiation (X-rays, gamma rays, fast neutrons, etc.), chemicals [ethyl methane sulfonate (EMS)], and others. Physical mutagens are of two types: non-ionizing radiation (e.g., UV rays) and ionizing radiation (which are of two types: particulate and non-particulate, e.g., X-rays and gamma rays). The chemical mutagens are of three main types: base analogs (such as 5-chlorouracil), acridine dye, and alkylating agents (which includes EMS and MMS). These mutagens can disrupt chromosomes, causing deletions, insertions, breakage, etc., and will create genetic variation. The major disadvantage of this approach is the non-targeted mutation events. After receiving the M_1 seeds following mutagenesis (one has to send several thousand seeds of the same cultivar to be mutagenized), the plant breeder has to advance the generation to achieve homozygosity (mutant allele will segregate initially) and constantly phenotype for the traits of interest. This can be very resource-intensive, depending on the cost to phenotype the trait of interest (field for morphological trait or laboratory for quality trait or chemical component). At low doses, chromosomal changes are not as dramatic, and the mutation frequency is low, warranting large population sizes to be screened; however, it is not desirable to use high doses because major chromosomal aberrations and lethality can occur. Sometimes, it is very difficult to identify target mutant events, adding to the high expenses for phenotyping.

Spontaneous mutation and cultivar development

- Mexican teosinte evolved to become *Zea mays* L. due to spontaneous mutations.

- Semidwarf mutants of wheat and rice heralded the "Green Revolution." The height reducing gene in wheat (from variety "Norin-10") and rice (from variety "Dee-geo-woo-gen") were spontaneous mutants. These semidwarf mutants resulted in short statured cultivars that were more responsive to inputs including fertilizer and irrigation resulting in higher yield.
- Brown mid-rib mutant of maize (reported in 1924) that produces a low lignin content to improve digestibility for livestock.
- Bud sports are found in many flowers, ornamentals, and fruit trees including apples and grapefruit. "Sport" is a mutation in a part of a plant wherein unusual or singular deviation from the normal or parent plant is seen. This arises due to sudden variations in gene expression of somatic cells, leading to the phenotypically different shoots. Bud sports are mutations on a single branch. These bud strains can become an improved strain of the parent cultivar (red delicious apple from parent delicious variety), and if they are sufficiently different from the parental strain (tree) they can develop as new cultivars. Examples of bud sports include reduced internode types in ornamental tree species or red strain on apple varieties.

Induced mutation and cultivar development

- Barley: the short, stiffed-strawed barley variety Golden Promise (from United Kingdom) is an *ari-e.GP* mutant characterized as an elongation (*elo*) type of the semidwarf mutant. Golden Promise was developed by irradiating the seeds of cultivar Maythorpe. Golden Promise was a predominant variety in 1970s and 1980s due to its high yield, high malting quality combined with very desirable agronomic traits of short height, straw strength, and early maturity. It was preferred by distillers for Scotch whiskey production. Later research found Golden Promise to have good salt tolerance (better performance compared with other short height barley in soils containing high levels of sodium).
- Rice: Reimei was the first semidwarf rice cultivar developed through induced mutation. It was released in Japan in 1966. Reimei carries a semidwarf gene, which is allelic to the *sd1* gene present in Dee-geo-woo-gen (described earlier in spontaneous mutation); however, it has modifiers that increase the culm length. Reimei was developed from the cultivar Fujiminori and was extensively used as a parent in hybridization to develop 21 cultivars within 25 years of its release (Kawai and Amano, 1991). Calrose76 was the first semidwarf rice mutant variety developed in the United States in the late 1970s. It was widely used as a parent in breeding programs in the United States and internationally.
- Quality protein maize (QPM): the high protein content and better amino acid profile are achieved by the *opaque-2* single gene mutation. Maize endosperm protein is deficient in two essential amino acids: lysine and tryptophan. QPM varieties were developed by the combination of *opaque-2* mutant gene along with endosperm and amino acid modifier genes. Compared with regular maize, QPM varieties have about twice as much lysine and tryptophan, which makes it

suitable and useful for human and animal nutrition. These efforts started in early 1960s, a mutant maize with similar total protein content, but double the amount of lysine and tryptophan, was developed. Subsequent conventional breeding efforts generated numerous cultivars with improved agronomic characteristics, and these were referred to as QPM. Evangelina Villegas and Surinder Vasal were awarded the World Food Prize in 2000 for their work on the development and advancement of QPM cultivars in the world.

Important factors to consider in mutation breeding

Mutagen(s) and their doses

- The physical or chemical agent increase mutational events and thus mutation rates above the spontaneous background level. The important factors include the choice of mutagen (radiation, chemical, etc.), duration of exposure, and the plant material used (e.g., seed, plant vegetative part, pollen grains, and cells/tissues). In seed propagating crops, seed and pollen are commonly used. Pollen, if feasible, is useful for mutation breeding because they do not suffer from the problem of chimera (more common in seed) and both dominant and recessive mutation can be identified. Vegetative parts (cells and tissues) are common in vegetatively propagated crops and meristematic regions (e.g., apical meristem) are used. Mutation is maintained through vegetative part propagation.
- LD_{50} dose: Lethal dose 50 (LD_{50} dose) means that the rate at which 50% of the mutagenized seed produce viable plants (i.e., treated plants produce seed) in comparison with a control that consists of same seed (genotype and number) that undergoes the same procedure as described above except it is not mutagenized. This provides a reasonable comparison of lethal dose rate.

Chemicals

Generally, chemical mutagens leads to point mutations, although chromosomal changes may also occur. Examples of chemical mutagens are:

- nitrosoguanidine,
- hydroxylamine (NH_2OH),
- base analogs (e.g., BrdU),
- simple chemicals (e.g., acids),
- alkylating agents [e.g., *N*-ethyl-*N*-nitrosourea (NEU)],
- methylating agents (e.g., EMS),
- polycyclic hydrocarbons (e.g., benzopyrenes found in internal combustion engine exhaust),
- DNA intercalating agents (e.g., ethidium bromide),

- DNA cross linkers (e.g., platinum), and
- oxidative damage caused by oxygen (O) radicals.

As per FAO/IAEA (2018), the major chemical mutagen used to produce varieties are EMS (ethyl methanesulphonate), NEU (nitrosoethyl urea), MNU (N-methyl N-nitrosourea), and colchicine. The number of varieties (in brackets) developed by the use of mutagens include: EMS (106), NEU (57), Colchicine (46), MNU (53), EI (36), DMS (19), dES (14), NaN_3 (12) among other mutagens (FAO/IAEA, 2018). Mutagens are carcinogenic; therefore, it is important that mutagenesis is done in an approved facility suitable to handle the mutagen without detrimental health effects to a researcher. Compared to physical mutagens, chemical mutagens are simple to apply requiring less technical resources but are carcinogenic.

Radiation

Generally radiation leads to deletions and chromosomal changes.

- Ultraviolet radiation (non-ionizing radiation) excites electrons to a higher energy level.
- DNA molecules are good absorbers of UV light, especially with wavelengths in the 260–280 nm range.
- Two nucleotide bases in DNA cytosine and thymine are most vulnerable to excitation that can change base pairing properties.
- UV light can induce adjacent thymine bases in a DNA strand to pair with each other, as a bulky dimer.
- Ionizing radiations: Ionizing radiation induces large chromosomal alterations, for example, chromosomal loss, translocation and deletion. Ionizing radiation increases mutation rate by 1,000 to 1 million fold, and therefore has been used to induce heritable genetic changes. Examples include X-rays, γ-rays, etc.

Properties of radiation and its application in mutation breeding are described in Table 14.1.

Several factors play a major role in the success of radiation and chemical treatments for mutation induction. These include biological, environmental, and chemical factors, as these change the effectiveness and efficiency of mutagens in plant parts. While a detail listing is beyond the scope of this chapter, oxygen and water content are considered the two most important modifying factors for seed irradiation; while for non-seed tissue, stage of plant development and dose rate are ranked high.

For a plant breeding program, one of the first steps in the implementation of mutation breeding approach is determining the appropriate mutagen and the desired mutagenic efficiency. Plant breeder strives to obtain the highest mutagenic efficiency. This implies achieving maximum number of desired mutations (in the population), ensuring plant viability and keeping the background mutation rate to be low. Therefore, plant breeder must assess and decide the most appropriate mutagen dose. This is needed because too high or too low a

Table 14.1 Properties of radiation and its application in mutation breeding.

Types of radiation	X-rays	Gamma rays	Neutrons (fast, slow, and thermal)	Beta particles ($\beta--$ negatron; $\beta+-$ positron and EC)	Alpha particles (α) emitted by nucleus with atomic number >81	Particles from accelerators
Source	X-ray tube with cylindrical anode	Radioisotopes and nuclear reactions (^{60}Co and ^{137}Cs)	Nuclear reactors/atomic piles or accelerator [californium (^{252}Cf) and curium (^{248}Cm)]	Radioactive isotypes or accelerators (^{7}Be)	Radioisotopes from radionuclide (^{32}P)	Nuclear reactors or particle accelerators
Description	Electromagnetic radiation potent mutagen agent	Electromagnetic radiation like X-rays	Uncharged particles slightly heavier than a proton, not observable except through interaction with nuclei in the material it traverses	Emitted by the nucleus of radionuclide with neutron/proton imbalance efficient only when incorporated directly into cells	Two protons and two neutrons emitted as nucleus decay products, not efficient in mutation due to very low tissue penetration	Beam of fast-moving, electrically charged atomic or subatomic particles (i.e., quarks) Application ion implantation and/or mutation
Energy	1–500 keV. For mutation breeding: 50–300 keV	Up to several MeV	From less than 1 eV to several million	Up to several MeV	2–9 MeV	600 MeV to 2.75 GeV
Hazard	Dangerous, penetrating	Dangerous, highly penetrating	Very hazardous	May be dangerous	Very dangerous internally	Very hazardous

From FAO/IAEA, 2018.

Table 14.2 Steps in determining appropriate dose for use in mutation breeding applications.

Step 1: radiation source characteristics	Step 2: characteristics of the biological target	Step 3: prediction of dose effects
• High or low LET radiation • Energy distribution • Degree of contamination with other radiations • Dose gradients, requirements of dose homogeneity	• Seeds • Pollen grains • Gametophytes and zygotes • Criteria of radiosensitivity • Biological factors, environmental factors, etc. • Whole plants • Vegetative organs • Cells and/or tissue in culture	Early assessable criteria of primary damage in, e.g., seedling height of the first leaf, epicotyl length, etc. and their correlation with mutation frequency in M2, e.g., usually to chlorophyll indicator mutations

mutagen dose can have undesirable outcomes, for e.g., increased resource requirement to phenotype the population, human resource to handle the mutagenized population, and increase assessment time. Undesirable background mutations necessitate additional work with more crossing and backcrossing to weed them out. For plant breeders considering mutation breeding, preliminary experiments or literature search is needed to measure and decide optimum mutagen quantity applied and received by the target plant material. Table 14.2 is a useful starting point.

Choice of variety

This will depend on the objective of the breeding program. However, the options generally include a successful variety (recently released) or an upcoming promising variety in the final stages of evaluation. Previous research has shown that not all genotypes respond similarly to mutagenesis; therefore at times, breeders may be forced to choose a variety that has previously shown to be amenable to mutagenesis and is more stable. Other consideration may include availability of genome sequencing. As the cost of sequencing has gone down, more varieties are available in major crops with completed genome sequencing. Picking a variety that has its genome sequenced helps with genetic studies and characterization with easier implementation in breeding program. If minimal information is available to a breeder, it may be beneficial to include smaller seed sample of many varieties compared with a large seed sample of one or two varieties. This will circumvent the problems described above.

When to use mutation breeding

It is important to consider several factors when mutation breeding should be deployed. Primarily mutation breeding is used when recombination is insufficient to create genetic variation. For example, if there is a situation that a particular allele does not exist in natural population in the crop species, mutation breeding is a potential solution. As previously described, several important considerations need to be explored before mutation breeding approach can be implemented successfully. Breeder needs to consider the economic costs and the potential benefit (or reward) if the desirable allele can be created through mutation and selected for the benefit of stakeholders. This implies that breeding objectives need to carefully crafted, so that they are feasible, of sufficient economic value, and other ways to achieve these objective(s) are not available. Broadly, the goal of mutation breeding is to alter only a single trait and with major effect. While working to achieve their objective(s), plant breeder needs to be aware that other regions of the genome (i.e., other genes) may have also been mutated, and that one change may alter other aspects of the plant. Hence, extensive agronomic testing of that single mutant is required prior to commercialization or extensive use as a parent in the breeding program. Once a mutated gene (or its allele) are identified, genetic mapping is completed. This enables marker assisted backcrossing of that gene (or allele) in elite parents (See Chapter 13: Backcross method, and Chapter 27: Molecular Tools in Crop Improvement and Cultivar Development). This avoids the problems associated with the mutated line with multiple mutations (favorable and unfavorable).

Mutation breeding methodology

Here is one potential example of mutation breeding pipeline.

M_0: Expose seed to mutagens (through mutagenesis approached described above). Seed (M_0) of a pure line or inbred (homozygous and homogenous) genotype is treated (chemical, physical, etc.). The treated seed are grown, and plants grown from these seeds (after self-fertilization) form the M_1 population (where M stands for mutant, and "1" refers to the first generation in development, similar to the concept of filial generation explained in Chapter 11, Pedigree Method).

M_1 generation: close spacing to reduce tillering and branching, study of radiosensitivity, and selfing of each M_1 plant. Each M_1 plant may have a different mutation, or several mutations, in its genome. Each plant will also be heterozygous (or hemizygous) at numerous loci. An M_1 is similar to an F_1 developed from crossing non-inbred parents. The next step, self-fertilization of the M_1 plants to develop the M_2 generation allows the recessive mutants to be homozygous and produce observable phenotypes. Identifying mutants generally requires large populations of M_1 plants. Most mutations are recessive; therefore, are undetected in M_1 and need segregation analysis in M_2.

M_1 plants suffers from physiological disorders due to mutagenesis; therefore, phenotypic selection for mutant identification should not be done in the M_1 generation. Furthermore, mutant allele is generally recessive and is not expressed phenotypically, preventing mutant screening in M_1 (unless marker based screening is done). M_1 plants are weak; therefore, this generation is grown in stress-free conditions to ensure maximum growth and seed production for M_2 generation.

M_2 generation: should be space planted (each M_2 seed coming from a unique M_1 plant) and mother variety should be used as check. Epiphytotic conditions can be created for disease screening, if appropriate. Individual plants (mutants) are selected. If genomic DNA approaches are used for the detection of mutation, leaf or other plant part tissue can be collected from each plant for laboratory screening using molecular markers. Alternatively, instead of space planted M_2 seed from a unique M_1 plant, breeder can grow in head to row and select individual mutant plant separately.

M_3 generation: epiphytotic conditions can be used, if appropriate. Plant to progenies of selected M_2 plants are grown. Individual plants (mutants) are selected. The mutagenized population can undergo generation advancement through single seed descent or other breeding methods to develop near-homozygous lines.

M_4 generation: preliminary evaluation of the selected progenies is carried. Selection is carried out between progenies for desired traits.

M_5 generation: replicated yield trials at 1-2 locations are conducted. Selections are made on desired traits. Harvested seed is used as seed source for next year's multi-locational trial. If applicable, superior progeny(ies) can be used as parent in the crossing program.

M_6–M_8 generations: replicated multi-locational yield trials for 2-3 years are conducted. As per breeding objectives, quality trait test(s) or assessment of other traits are performed. Superior progeny(ies) can be used as parent in the crossing program. Seed increase is initiated, if commercialization potential is identified.

After several M generations, the mutant phenotype is confirmed and established as stable (non-segregating). Mutant lines can then be used in a forward-crossing or backcrossing program to transfer the favorable mutation to elite cultivars using multiple but separate crosses to ensure recovery of the elite phenotype together with the mutant trait. Simultaneously, inheritance of induced mutation is conducted along with mapping of the mutant gene. During the steps described above, other considerations are size of M_1 and M_2 families to recover the mutant type, ploidy level, genetically effective cell present in the germline (those cells of the germline that contribute toward formation of gametes to offspring), and frequency of chimeras.

Identification of mutations

Scenario I: desired allele is recessive "a"

M_0: All "AA."

M_1: predominantly "AA," rarely "Aa." Phenotypically both "AA" and "Aa" are indistinguishable.

M_2: "AA" plants will give all AA progenies, while the rare "Aa" plant will give AA, Aa, and aa genotypes (note: this is similar to F_2 generation).

At this stage (M_3 and onward), pedigree method, bulk method, or single seed descent method can be used to identify the mutation each having its advantage or disadvantage as described in Chapters 10–12. Information from Sedcole (1977) (see Table 8.4) can be used to determine the number of plants needed to identify the mutation. For example, at 95% probability, a breeder needs to test 11 seeds from each M_1 plant to identify the recessive "aa" genotype using no germination issues. Eighteen seeds from each M_1 plant are needed to find two M_2 progeny that are "aa" from a heterozygous "Aa" M_1 plant. If the capacity of a breeding program is to phenotype 10,000 M_2 plants coming from 18 seeds from each M_1 plant, only 555 M_1 plants can be progeny tested. While these numbers are for illustration purposes only, it brings forward the challenges associated with handling mutation breeding program through progeny testing. While pedigree approach is more cumbersome; however, progeny rows can be effectively used to identify the mutant allele early. Outcomes for other scenarios can be obtained using information from Sedcole (1977).

Scenario II: desired allele is dominant "A"

M_0: All "aa."

M_1: predominantly "aa," rarely "Aa." Phenotypically "aa" and "Aa" are distinguishable as "Aa" will have a different phenotype than "aa."

M_2: "Aa" plants will give "AA," "Aa," and "aa" genotype (note: this is similar to F_2 generation). "AA" and "Aa" are indistinguishable but selfing followed by progeny testing (M_3) can separate true breeding "AA."

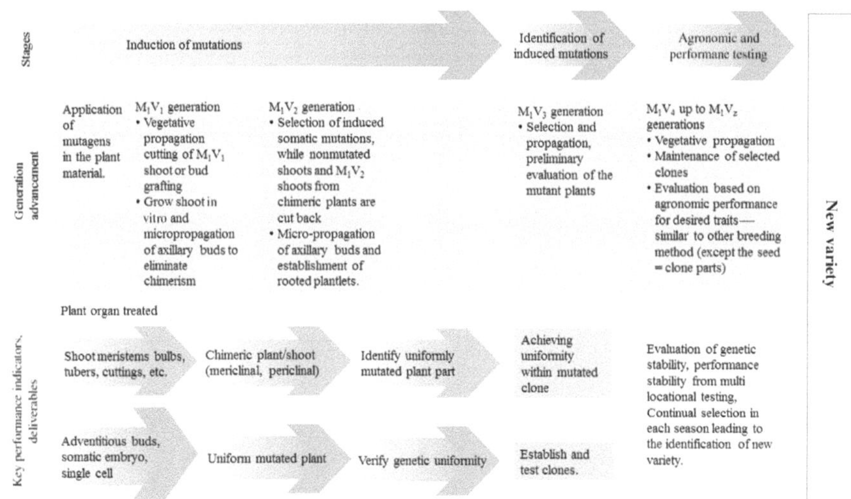

FIGURE 14.1

Mutation breeding in vegetatively propagated crops.

Adapted from: Sarsu, F., Penna, S., Kunter, B., Ibrahim, R., 2018. Mutation breeding for vegetatively propagated crops. In: Spencer-Lopes, M.M., Forster, B.P., Jankuloski, L. (Eds.), Manual on Mutation Breeding, third ed. Food and Agriculture Organization of the United Nations, Rome, Italy. 301 pp.

The above examples are for application for cereals, legumes, and oilseed. For vegetatively propagated species, the procedure can be as follows (Fig. 14.1). Variations in breeding method will exist, so this example is just to outline the approach.

Major differences between seed and vegetatively propagated crops

In seed propagated crops, germline mutation happens when mutant cells enter the germline and are transmitted to eggs and pollen. For vegetatively propagated species, mutation must be transmitted to the vegetative propagule, such as buds. Chimeras (definition: a chimeric plant is composed of cells harboring more than one genotype) are produced following a mutagen treatment and can be problematic in seed propagated crops. However, in ornamental plants with attractive mutations traits (e.g., new leaf or branching pattern and flower color), the mutant tissue can be isolated and using tissue culture, regenerate plants that carry the favorable mutation. These can be subsequently commercialized.

CHAPTER

Inbreeding depression and heterosis

15

Abstract

Inbreeding is observed as a result of mating of closely related individuals in cross-pollinated crops that are otherwise highly heterozygous. Inbreeding results in a decrease or loss of heterozygosity. Heterosis is commonly referred as "hybrid vigor" in evolution biology, and is defined as heterozygotes having higher fitness than the homozygotes in a population. In plant breeding terms, heterosis refers to increased size, shape, stature, fertility, biomass, and seed yield (or superiority) of hybrids over one or both parents or mid-parent value. Although inbreeding depression and heterosis were first described more than a century ago, no universally accepted theory exists. In this chapter, inbreeding depression and heterosis are described along with their calculation and proposed hypotheses. Types of heterosis, mechanisms of fixation of heterosis, and practical considerations in plant breeding are also described.

Since the earliest recorded history, inbreeding was considered to be associated with unfavorable biological effects, and that crossing of unrelated stock results in the restoration of vigor. As early as 18th and 19th centuries, scientific experiments were conducted to study the effects of out-crossing in plants. The findings of Koelreuter (1763) that hybrids were often possessed of most striking and unusual vigor, and Sprengel's (1793) work pertaining to relation between flowers and insects (demonstrating the importance of insects in fertilization of flowers), were very significant. It was also realized that there were adverse effects of continuous inbreeding including reduced fecundity and loss of vigor; therefore, outcrossing was necessary to maintain lines. Darwin (1868) opined that "although free crossing is a danger on one side which everyone can see, too close inbreeding is a hidden danger on the other." Later, Darwin (1876) reviewed the literature and also presented results of his own experiments with various species in "Cross- and Self-fertilization in the Vegetable Kingdom" and inferred that the offspring arising from self-fertilization were less vigorous than those obtained from cross-fertilization. Darwin summarized his findings by stating "The first and most important of the conclusions which may be drawn from the observations given in this volume, is that cross-fertilization is generally beneficial, and self-fertilization injurious. This was shown by the difference in height, weight, vigor, and fertility

Plant Breeding and Cultivar Development. DOI: https://doi.org/10.1016/B978-0-12-817563-7.00012-X

of the offspring from crossed and self-fertilized flowers, and in the number of seeds produced by the parent plants. With respect to the second of these two propositions, namely, that self-fertilization is generally injurious, we have abundant evidence." It is interesting to note that in late 1800s, experiments on maize hybridizations were done by Beal (1880), who reported on variety hybrids of maize that were more productive than the parent varieties; however, double cross maize hybrids were developed in 1917 by Donald Jones. Beal hybridized two strains (open-pollinated varieties) of maize that had been geographically separated for unknown number of years, by using one as a female (detasseled) and other as a male (non-detasseled). He reported that the hybrid seed harvested from the female population was more than 50% higher yielding than the parents. However, this discovery was not fully utilized until later when inbred lines were used as parents to develop hybrid maize cultivars.

This showed that before the era of Mendelian genetics, beneficial effects of out-crossing and adverse effects of inbreeding in cross-pollinated species were known to the plant breeders. However, the effects of continuous self-fertilization were not well understood until the first decade of 20th century when independent work of East (1908) and Shull (1909) on inbreeding corn were published. The most important effects of continued self-fertilization in cross-pollinated crops reported by these scientists are described in the next section.

Effects of inbreeding

1. Expression of deleterious genes, that is, the appearance of lethal and sublethal types (e.g., chlorophyll deficiency, etc.) are observed in early segregating generations of selfing.
2. Rapid segregation of population into distinct lines, that is, increase in homozygosity as a result of which uniformity for height, ear length, and maturity is observed.
3. Decrease in vigor and reduction in productivity (decrease in height, ear length, yield, etc.) under most favorable cultural conditions.
4. General decline in size and vigor of lines that survive.

The cross-pollinated species and species reproducing asexually are highly heterozygous. As a result of selfing or inbreeding, these species show severe reduction in vigor and fertility of the progenies. This phenomenon is known as inbreeding depression. The effect of inbreeding is more severe in early generation of inbreeding (five to eight generations) after which the performance stabilizes. Inbreeding is mating between individuals related by descent or having common ancestry (brother sister mating or sib-mating) that will lead to homozygosity. The highest degree of inbreeding is attained by selfing. Normally, cross-pollinated crops are not self-fertile; therefore, inbreeding is practiced by controlled self-pollination.

Inbreeding depression

Inbreeding is measured by the coefficient of inbreeding also known as inbreeding coefficient (F). The coefficient of inbreeding (F) is used to determine the strength of inbreeding, where F is the probability that two alleles at a locus in an individual are identical by descent (IBD). The estimate of F increases with an increase in homozygosity, or fixation of alleles, resulting from inbreeding. IBD refers to alleles that are descended from a common ancestor in a base population, while identity by state (IBS) refers to alleles that are the same, and do not need to be inherited from a recent ancestor. Two alleles that are IBD must be IBS, but not *vice versa* because a homozygous condition may be produced without inbreeding in recent base ancestor population.

The coefficient of inbreeding depends on the mating system and on the degree of inbreeding in the previous generation. Selfing is the most severe form of inbreeding as it attains homozygosity more quickly than any other form of inbreeding (sib mating; half- and full-sib matings, etc.). Selfing results in reduction of 50% heterozygosity (increase in 50% homozygosity) in each generation of selfing. In Chapter 4, Primer on Population and Quantitative Genetics, random mating population was explained. However, in case of selfing, the most severe type of inbreeding, the outcome is different. If we assume a population with P genotypes of A1A1, H genotypes of A1A2, and Q genotypes of A2A2 with complete selfing in the population, after one generation of selfing $P' = P + 0.25(H)$, $H' = 0.5H$, and $Q' = Q + 0.25H$.

If population was in HW ($P = p^2$; $H = 2pq$; $Q = q^2$) prior to selfing,

then $P' = p^2 + 0.5pq$, $H' = 2pq - pq$, and $Q' = q^2 + 0.5pq$

after one more generation of selfing, $P'' = p^2 + 0.75pq$, $H'' = 2pq - 1.5pq$, and $Q'' = q^2 + 0.75pq$

after one more generation of selfing, $P''' = p^2 + 0.875pq$, $H''' = 2pq - 1.75pq$, and $Q''' = q^2 + 0.875pq$.

Therefore, after "n" generations of selfing,

$$P^n = p^2 + pq(1 - 0.5^n)$$

$$H^n = 2pq(0.5^n)$$

$$Q^n = q^2 + pq(1 - 0.5^n)$$

The function $(1 - 0.5^n)$ is known as inbreeding coefficient. Inbreeding comes from mating individuals that are related by descent *i.e.*, IBD. As a consequence of inbreeding, homozygosity increases leading to a depressive negative expression of traits referred to as inbreeding depression. Inbreeding depression can be expressed as:

$$Xo - Xi = 2\,pqFd$$

where Xo and Xi are the mean of the population without and with inbreeding, respectively; p and q are the allele frequencies in the populations; F is the inbreeding coefficient (see above); and *d* the dominance deviation. The estimation and expression of inbreeding depression can be calculated as: $Xo-Xi$ in absolute units, where Xo is the mean of the trait without inbreeding and Xi is the mean of the trait with a given amount of inbreeding F $(0<F<1)$. Inbreeding depression is commonly reported as a percentage: $[(Xo-Xi)/Xo]\times 100$. In the case of mixed selfing and random mating, probability of random mating and probability of selfing needs to be factored in calculations.

The coefficient of inbreeding (F) for selfing is computed as:

$$F = (1/2)\times(1+F')$$

where F′ is the coefficient of inbreeding in the previous generation.

The F-value increases with each generation of self-fertilization. In randomly mated population the F-value is zero, which reaches to one as the heterozygosity goes toward zero. The less severe forms of inbreeding like sib-matings require many generations before inbreeding coefficient approaches one or complete homozygosity. In maize (cross-pollinated crop), three generations of full sibbing are equivalent to one generation of selfing (Stringfield, 1974).

In some of the cross-pollinated crops, the adverse effects of continued selfing (inbreeding) can be so pronounced that many of the lines may be lost during the process of inbreeding. In maize (corn), inbreeding depression is quite high. The very first generation coming from selfing is inferior to the corssbred variety or open-pollinated variety both in size of plants and yield. Jones (1939) reported the effects of continued inbreeding in maize for 30 generations. The experiment was started in 1904. Variations of generative nature were observed and caused extinction of one of the original lines. Selfing resulted in 25% reduction in height that ceased after five generations. Yield was reduced 75% between 16 and 20 generations of selfing (Table 15.1). After 20 generations of self-fertilization inbred lines appeared uniform for all the visible characters and homozygous for all loci that have any effect on hybrid vigor. The intercrossing of inbred lines restored vigor in the next generation.

Table 15.1 Performance (yield in bushel/acre) of three inbred lines isolated from open-pollinated variety and maintained by selfing for 30 generations.

Generation of selfing	Line 1–6	Line 1–7	Line 1–9
0	81	81	81
1–5	64	51	41
6–10	45	36	34
11–15	38	34	26
16–20	22	24	14
21–25	20	21	13
26–30	24	18	9

The parental variety yielded about 81 bushels per acre.
After Jones, D.F., 1939. Continued inbreeding in maize. Genetics 24, 462–473.

Genetic hypotheses for inbreeding depression

The main genetic hypotheses for inbreeding depression include the dominance hypothesis (recessive deleterious mutations), pseudo-overdominance (recessive deleterious mutations in linkage disequilibrium), or true overdominance (single loci with heterozygous advantage). The dominance hypothesis indicates that detrimental mutant alleles are generally recessive, and therefore can exist at low frequencies in breeding populations; however, after inbreeding, these will cause inbreeding depression (lower survival, lower fertility, and lower growth) because they are expressed in homozygous recessive state. The individual effect of the mildly deleterious mutations contribute insufficient effect on the phenotype; however, when aggregated in homozygous genotypes, the effect is noticeable. The true overdominance theory is also based on a single loci consideration, wherein heterozygotes have higher fitness due to nonsimilar alleles. The argument for this hypothesis is that overdominant alleles are maintained by balancing selection, keeping genetic frequencies in the population at intermediate level even when the homozygotes have low fitness. In meiotically suppressed regions, these mutations may accumulate with lower fitness (conditioning overdominance for the genomic region). For the dominance and pseudo-overdominance (mutational) hypotheses, the higher homozygote frequencies for recessive deleterious mutant alleles among inbred causes lower fitness compared with noninbreds or hybrids. The explanation of overdominance hypothesis suggests that inbred individuals have lower fitness because they are less heterozygous for the two alleles (A1/A2) than noninbreds or hybrids. Under these hypothesis, heterosis occurs due to the complementation between unlinked deleterious alleles in a hybrid genotype. As per the overdominance hypothesis, if two genes are linked in repulsion, in each of the two homozygote parent the performance will be reduced; however, in a hybrid between these same parents overdominance will be observed. Briefly these main hypotheses are presented in Table 15.2.

Plant species show differences in the degree of inbreeding depression. Among cross-pollinated crops maize, sorghum and pearl millet show moderate inbreeding depression. The decline in alfalfa is more severe, and lines could not be maintained beyond the third selfed generation. Carrot also deteriorates drastically upon inbreeding. However, onion, sunflower, and rye are fairly tolerant to inbreeding with minimal consequences of inbreeding depression. The cucurbits are monoecious and cross-pollinated, but decline in vigor on inbreeding is small. In general, self-pollinated crops are highly tolerant to inbreeding depression and most cases show no inbreeding depression.

Heterosis

The increased size, the excessive kinetic energy, the increased productiveness, resistance to disease, or to unfavorable conditions of the environments was observed in cross bred organisms when compared with corresponding inbred or pure inbred

Table 15.2 Genetic hypotheses proposed for inbreeding depression and heterosis.

	Recessive deleterious mutations (dominance hypothesis)		Recessive deleterious mutations at closely linked loci (pseudo-overdominance hypothesis)	Single loci with heterozygous advantage (true overdominance)
	Single locus	**Multiple loci**		
Parent genotypes	A/A × a/a ↓ A/a	A/A b/b × a/a B/B ↓ A/a and B/b	AAbb × aaBB (Ab,Ab aB,aB) ↓ AaBb (Ab and aB)	A_1/A_1 × A_2/A_2 ↓ A_1/A_2
F_1 genotypes and their fitness relative to the parents	Intermediate fitness (however, it is above the parental average; homozygotes shows inbreeding depression)	High fitness (heterosis in F_1; inbreeding depression in homozygotes)	Higher fitness than the parents	High fitness (heterosis in F_1; inbreeding depression in homozygotes)

Adapted from Charlesworth and Willis, 2009. The genetics of inbreeding depression. Nat Rev Genet 10 (11), 783–796.

organisms. The stimulating effects of hybridity were also called "hybrid vigor" by English writer and "Luxurieren" by German writers (Shull, 1948). For the first time, Shull (1914) proposed "heterosis" for this increase in vigor following union of dissimilar gametes and was used to replace word "heterozygosis." Shull (1948) quoted D.F. Jones in 1923 to have stated that "the increase in size and vigor resulting from crossing is called heterosis." Shull (1948) cited Merrian-Webster New International Dictionary (2nd Edition) which defined "heterosis" as the greater vigor or capacity for growth frequently displayed by cross bred animals or plants when compared with those resulting from inbreeding.

Heterosis refers to the developmental stimulation resulting by whatever mechanism from the union of different gametes, whereas hybrid vigor denotes the manifest effect of heterosis (Shull, 1948). Now these two terms: "heterosis" and "hybrid vigor" are being incorrectly used interchangeably or as synonym. Shull also emphasized that negative heterosis has no apparent or obvious relationship with the phenomena for which the word heterosis was proposed and has been adopted. Heterosis is the increase in size, yield, vigor, etc. of F_1 hybrid over average performance of the two parents. If there is no increase, there is no heterosis. Heterosis declines and disappears in F_2 and subsequent generations of a cross as a result of segregation and recombination that leads to reduction in heterozygosity at each locus. Not all F_1 hybrids will exhibit heterosis. In major crops, for example maize, heterosis has been exploited to develop commercial cultivar and generally heterosis is explained as the difference between the hybrid and the mean of its two parents (Falconer and MacKay, 1996), that is, mid-parent heterosis. Lamkey and Edwards (1999) defined inbred-mid-parent heterosis as the "difference between the mean of the F_1 and the mean of the parent populations when inbred to homozygosity," that is, mid-parent heterosis is the difference between the hybrid and the mean of the two parents used in developing the hybrid.

Mid-parent heterosis = $[(\mu_{F1}-\mu_{MP})/\mu_{MP}] \times 100$.

The trait mean of the hybrid is μ_{F1}, and μ_{MP} is the trait average of the two parents. High-parent heterosis is the superiority of a hybrid over the better parent. It must be noted that in cross-pollinated crops, such as maize, selfing leads to inbreeding depression and the creation of hybrid (F_1) restores the vigor, which is defined as "baseline heterosis."

Heterosis and inbreeding are opposite to each other and largely complementary. Theoretically, it is expected that heterosis would be equal to inbreeding depression if large number of crosses are considered among inbred lines derived from a single base population. However, in general, plant breeder(s) may be interested in heterosis expressed by crosses between selected parents derived from two or more differential populations (in case of maize, separate heterotic patterns), which may not be related and rather have no known common ancestry.

Heterosis and inbreeding depression are widely reported in plant kingdom but are not manifested equally in all the species and for all the characteristics. These phenomena are observed more frequently in cross-pollinated species than in self-pollinated species. The differential expression of heterosis and inbreeding depression among crops is attributed to the genic balance that a crop has acquired during the process of its evolution. The mating system of cross-pollinated species leads to maintenance of high degree of heterozygosity in different generations, whereas the mating system of self-pollinated species leads to maintenance of high degree of homozygosity in its population. Therefore, in cross-pollinated populations, survival under natural conditions is in heterozygous state which is advantageous for optimum fitness associated with strong heterozygous balance; and natural self-pollinated populations are composed of completely homozygous individuals and optimum fitness is associated with homozygous balance (Mather, 1943). However, it should also be noted that heterosis is not a reflection of the extent of heterozygosity between parents. An increase may be noted as the distance between parental maize inbred increases, this relationship does not continue unabated linearly. Additionally, among the same parental line crosses, heterosis for different traits vary providing evidence that same set of genes (or their alleles) do not control the heterotic response.

Heterosis is important and most relevant in crops where heterosis is present and seed can be produced economically to be sold to farmers. Heterosis is dependent on the presence of dominance and is the summation of allele frequency differences across loci. In maize and other cross-pollinated crops, heterotic groups have been created in such a way that they maximize the difference in allele frequencies in genes affecting target trait(s) and thereby, maximize heterosis. Examples of hybrid cultivars include commercial single-cross maize hybrids, commercial three-way cross maize hybrids, and sunflower hybrids. Hybrid cultivars are usually utilized for allogamous species, but some hybrids are produced for some autogamous species (e.g., sorghum and tomato). Single-cross hybrid cultivars are homogeneous and heterozygous. Three-way hybrids are both heterogeneous and heterozygous.

In general, all the self-pollinated species (except highly cleistogamous) have from slight to notable cross-pollination and it is also believed that during the course of evolution the inbreeders were evolved from the outbreeders (Stebbins, 1950). Therefore, in addition to strong homozygous balance in self-pollinated species, slight to negligible heterozygous balance could be associated with their fitness. The cross pollinating crops with strong heterozygous balance leads to high inbreeding on selfing (inbreeding) and crossing of inbred lines show high heterosis. The self-pollinating crops with strong homozygous balance and some heterozygous balance show as a result inbreeding leads to no reduction in vigor. At the same time, F_1 generation between diverse lines shows vigor and good amount of heterosis. Some of the cross-pollinated crops like cucurbits do not show much of the inbreeding depression but show similar population characteristic of self-pollinated species. This is because they have been maintained comparatively with small population sizes.

Heterosis is also manifested in higher intensity for certain characters which have fitness value. Hybridization of unrelated plants followed by selection, practiced by plant breeders has always made use of heterosis to some extent. However, hybrid cultivar development in crop species especially in cross-pollinated and asexually propagated crops exploited the phenomenon of heterosis. The adoption of hybrid varieties has increased yields of several crops, which had earlier not used/developed hybrid cultivars. The genetic basis of heterosis is explained as to how fitness lost during the process of inbreeding tends to be restored upon crossing in the F_1 hybrid. In the past, two most common theories have been proposed which are dominance and overdominance hypotheses. These are described as follows.

Dominance hypothesis

This hypothesis was first proposed by Davenport (1908) and was explicitly stated by Bruce (1910) and Keeble and Pellew (1910) with the assumption that cross-fertilizing species consists of large number of genetically different individuals which carry deleterious recessive genes in heterozygous condition. According to this hypothesis there is a positive correlation between recessiveness and detrimental effect, and dominance and beneficial effect. As a result of intercrossing of inbred lines the detrimental effect of recessive genes contributed into the heterozygote (hybrid zygote) by one parent are made ineffective (hidden) by their dominant genes (alleles) from the other parent. The result is an increase in vigor of the hybrid when compared with inbred line/pure line. For example, if inbred A has the genotype AABBccddEE and inbred B has the genotype, aabbCCDDee. The F_1 hybrid has the genotype, AaBbCcDdEe with dominant genes at all the loci and therefore, would show more vigor than either of the parent.

According to Crow (1948), the following assumptions were made in the dominance hypothesis:

- All the genes concerned with vigor are completely dominant and that in each case the dominant allele is advantageous while the recessive is deleterious from the stand point of survival.
- Each gene acts independently of the others and they are either additive or multiplicative in their effect.
- Crossing over would occur freely so that there is no tendency for balanced heterozygotes to accumulate in the population due to reduced recombination.
- An individual in which all the gene loci contain at least one dominant factor would show maximum vigor, that is, each parent would show maximum vigor. This would occur if each parent could supply all the dominant alleles lacking in the other and therefore, the hybrid would receive at least one dominant gene at each locus. The difference in vigor between any individual and its expected theoretical maximum would be determined by the number of homozygous recessive loci.
- That hybrid vigor is measurable in terms of selective value. The frequency of detrimental recessive factor in a population is determined by its selective disadvantage, mutation rates, migration and the size and breeding structure of the population. The selection could be natural or artificial.

Objections to this hypothesis were raised. Collins (1921) pointed out that even in the absence of linkage the chances of recovering homozygous types would be remote as the number of genes was large. Another objection was if heterosis were due to solely dominance of independent Mendelian genes, the F_2 distribution curve should be skewed rather than symmetrical distributions observed for the heterotic characters. However, Jones justified that symmetrical distribution could be due to groups of favorable dominant and unfavorable recessive genes. Collins (1921) opined that skewness would be difficult to detect as large number of genes were involved. If this hypothesis is correct then it should be possible to obtain individuals (lines) homozygous and true breeding for all the dominant genes which should be similar to the F_1 in vigor. However, high yielding homozygous lines have not been found. Jones (1917) rejected that the discrepancy with the hypothesis could be explained on the basis of linkage hypothesis rather than a dominance hypothesis that many genes could affect plant growth and each chromosome may contain many of these genes (dominant and recessive). Therefore, a series of well-placed crossovers would be required to obtain all the dominant alleles in one gamete. For instance, if one chromosome of a pair carries alleles AbCdEf and other carries the alleles aBcDeF, where the small letters stand for deleterious alleles and capital letters stands for favorable alleles. To get a chromosome with all the dominant alleles ABCDEF would require the occurrence of crossing over in just the right places in the chromosome, which is very improbable.

Overdominance hypothesis

A second hypothesis was proposed independently by Shull (1908, 1911) and East (1908) who assumed that there exists a physiological stimulus to development on crossing due to the genetic difference in the union of dissimilar gametes. This hypothesis was also advocated by Rasmusson (1934) and East (1936). According to this hypothesis the heterozygosis itself produces an increase in vigor. The heterozygote is superior to either homozygote (heterozygote lay outside the range of homozygotes) and the vigor increases with the proportion of heterozygosity. The term overdominance was coined by Hull (1945) for this phenomenon of the superiority of the heterozygotes. Jones (1945) provided evidence for occurrence of such loci by the finding of single gene mutations in corn which produce heterotic effects. This type of heterotic expression has been called single gene heterosis or super dominance hypothesis, however, the term most commonly applied is generally known as the overdominance hypothesis of heterosis.

East (1936) explained the idea further assuming heterozygous combination of alleles is superior to either of homozygous alleles. This infers that different alleles perform different functions individually and in combination the alleles give superior product. In an example he assumed that there are a series of alleles; a_1, a_2, a_3, $a_4, \ldots$, of a gene with gradually increasing divergence function (can do different things) as a result the heterozygote(s) is increasingly more efficient as their component alleles diverged more and more ($a_1a_4 > a_1a_3 > a_1a_2$, and so on).

Hull (1946) postulated the existence of overdominance to explain the results of his regression analysis of corn yields. His argument for overdominance was that usually the hybrid between inbred lines of maize yields more than the sum of the two inbreds. This would be possible if the dominant alleles were acting in a completely additive fashion and if inbreds with no favorable dominant alleles had a negative yield.

Dominance versus overdominance hypothesis

In most situations the two hypotheses: dominance and overdominance, are similar as inbreeding leads to loss in vigor and reduction in productivity, whereas outcrossing lead to recovery in vigor and fertility. The degree of heterosis would depend upon the genetic diversity of the parents. According to dominance hypothesis, heterosis results due to masking effects of dominant desirable alleles over harmful recessive alleles. A genotype with more dominant alleles would be more vigorous than one with a few dominant alleles. As per overdominance hypothesis, heterozygote is superior to either of the homozygotes and heterosis results due to complementation between divergent alleles of a gene. A genotype with more heterozygous loci would be more vigorous than one with less heterozygous loci for a particular quantitative trait. Even with advent of molecular tools, there is not yet one unified theory for heterosis. In a review, Schnable and Springer (2013) summarized that heterosis arises in crosses between genetically distinct individuals due to a number of mechanisms, and single simple explanation is implausible. They proposed that heterosis is the outcome of the action of multiple loci, with non-overlapping loci affecting heterosis for different traits; pointing to a need for multigene models to inform and explain heterosis mechanisms, and roles of genetic and epigenetic components.

Non-allelic gene interaction in heterosis

Sprague and Tatum (1942) demonstrated the importance of specific combining ability (SCA) as a source of variance among single crosses, which suggested that gene action could be driving heterosis. Hull (1945) proposed recurrent selection for SCA that would be more effective with respect to improvement driven by genes exhibiting overdominance; however, cautioned that it would be inefficient relative to genes that are dominant. Heterosis in excess of 5% of the mean of the random breeding population cannot be explained on the basis of simple dominance alone (Crow, 1947); however, it could be due to interactions of nonallelic genes or overdominance.

Allard (1960) opined that there is no reason to believe that both hypotheses cannot operate simultaneously in producing heterotic effects. In addition to dominance and/or overdominance gene actions, some genic interactions, that is, nonallelic interaction or epistasis (dominance × dominance) may contribute to heterosis. From current evidence, it may be postulated that the heterosis is largely due to dominance gene action but epistasis and overdominance are also involved, and likely epigenetic mechanism influence it too. Overdominance sometimes may be due to pseudo-overdominance arising from dominant alleles in repulsion phase linkage or epistasis of linked alleles.

Types of heterosis

The mid-parent heterosis is not as relevant to breeders, as the hybrid is only better than the average of the parents. For all practical purposes, heterosis may be calculated as the superiority of F_1 over the better parent or higher parent in a cross. Such heterosis is called better parent heterosis and is computed as:

High-parent heterosis $= [(\mu_{F1} - \mu_P)/\mu_P] \times 100$. The trait mean of the hybrid is μ_{F1}, and μ_P is the trait average of the better of the two parents.

If a released variety is used to compute the heterosis, it may be called as commercial heterosis or economic heterosis.

Fixation of heterosis

In some of the plant species, the peculiar genetic constitution of the genome and due to their nature of propagation the hybrid vigor is fixed. The superiority of F_1 hybrid is maintained indefinitely unless mutation occurs. This is advantageous to several species. These mechanisms are discussed as follows.

Asexual reproduction

Some plant species reproduce through aberrant methods of reproduction. They alternate at more or less regular intervals with usual sexual one. Through such alternation, the species exploits both sexual and asexual reproduction to their advantage. Sexual reproduction produces new combinations of genes and genotypes, which prove advantageous to the organism. These advantageous combinations or F_1 hybrids are then perpetuated by asexual reproductions in a state that protects them from producing new recombination. There are several cultivated plants in this category, for example, sugarcane, potato, and sweet potato. The heterosis is maintained through vegetative propagation indefinitely. They carry the same genes in heterozygous condition.

Apomixis

Apomictic species have the ability to circumvent sexual reproduction and its fundamental aspects, i.e., meiosis - recombination, and fertilization. Meiosis is not normal due to which diploid cells develop into seed without fertilization. Generally seeds develop from maternal diploid cells; therefore progeny from such seed is identical to the mother plant. The fixation of heterosis is through apogamy or apomixis, referring to the formation of sporophytes by parthenogenesis of gametophyte cells. In oranges, lemons, and other fruits pollination of the flowers stimulates the development of seeds mostly by apogamy. See Chapter 20: Breeding Methods Used in Asexual Crops.

Balanced polymorphism

In balanced polymorphism, the adaptive value (fitness) of the heterozygote (A_1A_2) is higher than that of either homozygote (A_1A_1 or A_2A_2). Heterozygote also show heterosis, that is, at least slightly superior to the best homozygote. The population will always be composed of heterozygous and homozygous individuals: A_1A_2, A_1A_1, and A_2A_2. In other words, the highly fit heterozygote and less fit homozygotes exist in the population. However, less fit are at a disadvantage and do not survive, or if survive then very poorly. The heterosis leads to a most interesting situation known as balanced polymorphism. In certain varieties of evening primrose (*Oenothera* spp.), genetic segregation is almost completely suppressed as the homozygotes are lethal and only heterozygous survives. This is called balanced lethal system. During evolution, this balanced lethal system developed due to complex translocations and this system also leads to fixation of heterosis.

Polyploidy

During the course of evolution in crop species, doubling of chromosomes have been observed. The doubling is observed in the genome of some species called autopolyploidy, and between different genomes after the hybridization of two species/genera (interspecific and intergeneric) called allopolyploidy. The F_1 hybrid is sterile, however, often becomes fully fertile after the doubling of chromosomes. Many of these polyploid species are propagated through vegetative propagation. The heterozygosity of the amphipolyploids is maintained generation after generation. In banana, maximum vigor is associated with the triploid state. The diploid bananas have hard seeds and make their fruit unacceptable; however, triploid bananas show maximum vigor and their sterility is advantageous as seedlessness is commercially acceptable. The triploidy of banana is fixed through vegetative propagation. Potato and sugarcane are polyploid species and are propagated vegetatively. In both species, after hybridization, superior clones are selected in F_1 generation and the superior clones having desirable combination of genes are propagated asexually and the heterozygous state is maintained indefinitely. See Chapter 7: Haploidy and Polyploidy in Crop Improvement.

CHAPTER

Population improvement 16

Abstract

The emphasis in breeding methods of self-fertilized crops is primarily on individual plants. On the other hand, the methods of breeding of cross-fertilized crops emphasize on the improvement of a population of plants. The populations of cross-pollinated crops (allogamous crops) are highly heterozygous and very heterogeneous. Such populations show a range of variation for most morphological and agronomic traits. For the genetic improvement of cross-pollinated crops, breeders seek to change the gene frequency such that new and desirable genotypes are produced. The breeding methods of cross-pollinated crops aim at preventing inbreeding or at least keep it to a minimum level to avoid undesirable effects. The procedures for population improvement in cross-pollinated crops may be described either according to the unit of selection, that is, individual plants or family of plants; or as according to the population undergoing selection, that is, interpopulation improvement (selection is practiced based on the performance of a cross between the two populations) and intrapopulation improvement (selection is practiced within a population). In this chapter various methods deployed for inter- and intra-population improvements are described.

Interpopulation improvement

As the term "interpopulation improvement" implies, the objective of these schemes is to improve two populations simultaneously. This scheme can also lead to the eventual development of hybrid if the two populations belong to different heterotic groups, unless the heterotic groups are not clearly established. These schemes are directed toward improvement in the performance of selected populations through the exploitation of interpopulation heterosis. Suboptimal (i.e., minimally selected) agronomically performing population are not suitable to include in this scheme. Prior intrapopulation improvement ensures that deleterious or undesirable alleles are eliminated before being subjected in an interpopulation scheme. During this interpopulation improvement process, some improvement in the *per se* performance of the two populations is also brought about along with an increase in the manifestation of heterosis. The two populations can belong to complementary genepools (heterotic patterns). In this chapter, for all breeding

Plant Breeding and Cultivar Development. DOI: https://doi.org/10.1016/B978-0-12-817563-7.00019-2

schemes, numbers of selected plants are given as an example only, and the actual number will depend on performance relative to the base population and checks, or breeder imposed selection percentages.

Reciprocal recurrent selection

The reciprocal recurrent selection (RRS) scheme was proposed by Comstock et al. (1949). "It involves two heterozygous populations each serving as tester for other and effectively exploits both general and specific combining ability. The performance of the F_1 hybrid between two broad-based varieties/populations could be improved over repeated cycles of selection." They were the first to suggest the use of RRS to simultaneously improve the two genetically different populations for both general combining ability (GCA) and specific combining ability (SCA). Two varieties, two synthetics or the F_2 from two single crosses could serve as source material. The main objective is to develop and improve two populations simultaneously leading to the development of superior inbred lines from each of the two populations that combine well with each other (across populations). The two major methods used for interpopulation improvement are: (1) RRS–half sib and (2) RRS–full sib. Through the use of this breeding procedure, all types of gene actions (additive, dominance, over dominance, and epistasis) are utilized in the crossed populations. The RRS differs from other schemes of recurrent selection because the second population serves as the tester parent to the first population and *vice versa*. The procedure is suitable when the objective is to improve complementary heterotic pool populations to develop hybrid varieties.

In brief, plants from one population are mated to plants of another population, and selection of individuals for the next cycle is based on the performance of the progeny in hybrid combination. Therefore, each population serves as a source material to advance and improve, and as a tester for the other population. For this breeding method, each cycle requires one generation for selection of individuals and a second generation for intermating of selected individuals to produce materials for the following generation.

As previously mentioned, RRS is a procedure to improve both the GCA and SCA of two populations simultaneously. Steps involved in allogamous species are listed below:

- *Step 1:* plants are selected in each of the two populations.
- *Step 2:* plants of population #1 are selfed and outcrossed as the tester to the selected plants in population #2 to generate testcross progeny. Plants of population #2 are selfed and outcrossed as the tester to the selected plants in population #1 (i.e., reciprocal).
- *Step 3:* the resulting testcross progenies are evaluated each season. Superior plants are identified based on their testcross performance. Selfed seed from these selected plants is used to intercross within each population to generate materials for the next generation.
- *Step 4:* cycle is repeated (i.e., recurrent and selection from step 3).

In addition to allogamous crop species, such as maize, this method can be used in autogamous crop species if male sterility exists. For example, in sorghum, two populations, each consisting of male-fertile and male-sterile, can be improved in a RRS scheme. Briefly, this can be accomplished as following: (1) population A - selected male-fertile plants (keep identity record) individually are crossed to several random male-sterile plants of population B; while population B - selected male-fertile plants (keep identify record) individually are crossed to several random male-sterile plants of population A. The resulting crosses are evaluated in a yield trial to make selections. Seed from selected male-fertile plants from each population are bulked and grown in isolation. At this stage, breeder needs to use male-sterile plants in A and B populations and allow random pollination of these male-sterile plants by selected male-fertile individuals (note: if male-sterile plants cannot be inserted, controlled pollinations can be made instead). These are harvested at maturity in each population, and bulked to constitute A′ and B′ that forms the new populations from which the cycle repeats, starting with: population A′ - selected male-fertile plants (keep identify record) individually are crossed to several random male-sterile plants of population B′; while population B′ - selected male-fertile plants (keep identify record) individually are crossed to several random male-sterile plants of population A′).

General outline of reciprocal recurrent selection—half sib

The steps of RRS half-sib are described as follows and presented in Fig. 16.1.

Step 1: the two genetically divergent populations, A and B are used as source materials. Some 200 plants from source A are self-pollinated and pollen from A is used to pollinate plants in source B. One plant could pollinate four or five randomly selected plants of source B and the progenies from these crosses would be used to evaluate the performance of selections in source A. Similarly, some 200 plants from source B are self-pollinated and pollen from B is used to pollinate plants in source A. For this, one plant from B could pollinate four to five plants randomly selected plants from source A and progenies of these crosses would be used to evaluate the performance of the selections in source B.

Step 2: the test cross progenies from step 1 are evaluated in the replicated trials; one consisting with progenies of source A plants as pollen parent, and the other consisting with the progenies of source B plants as pollen parent. All seed from each of the four or five crosses involving a single pollen parent are bulked to produce a single progeny from that parent. This way in each of the trial there would be at the most some 200 progenies. In practice, the number of progenies may be less as some of the progenies may have insufficient seeds and others could be discarded based on other traits like lodging and disease susceptibility.

Step 3: based on field trial data in step 2, plants from source A and B whose progenies were superior are selected. The remnant selfed seed of four or five pollen parents in the first year (step 1) of both sources is planted in rows in crossing

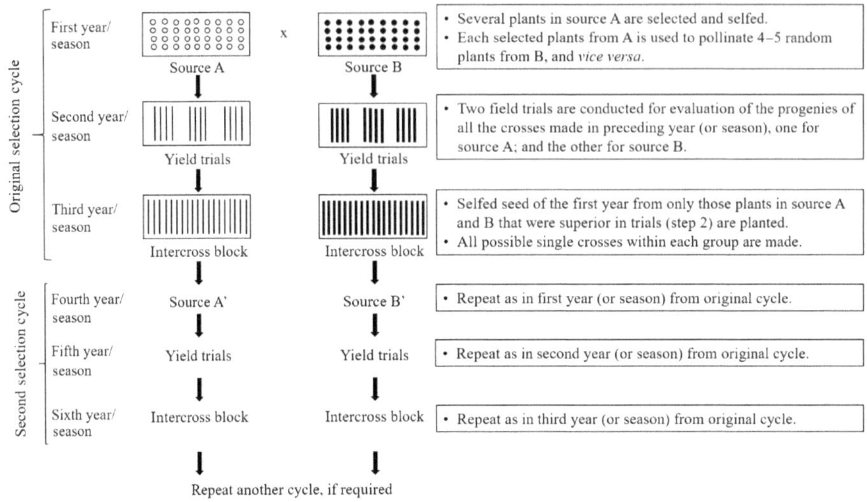

FIGURE 16.1

General procedure of reciprocal recurrent selection scheme (half-sib).

nursery. Within each source group, all or as many as possible single crosses are made between the plants from which seed was planted.

The seed of these crosses from two sources is composited to form two new populations, A′ and B′ to initiate next cycle of selection. Comstock et al. (1949) also suggested that sufficient plants should be selected in each generation to keep inbreeding at low level within the source population. The controlled pollination and record keeping of pedigree information during interbreeding would also help to avoid inbreeding depression within two populations. The evaluation of selfed progenies allows elimination of undesirable progenies.

Sprague and Eberhart (1977) reported 3%–5% of genetic gain per cycle for maize yield improvement of all recurrent selection schemes. However, for RRS scheme the genetic gain per cycle was 2.9%. This value included both populations A and B (1.8% genetic gain) and population crosses (5.1% genetic gain).

Genetic basis of reciprocal recurrent selection—half sib

In this scheme, two heterozygous populations are involved and each population serves as a tester for the other population. Through this scheme, both GCA and SCA variances are exploited. Comstock et al. (1949) compared RRS scheme and the other two schemes: selection based on GCA, and selection based on SCA. The comparisons were made on the assumption that (1) non-allelic gene interactions were non-existent, (2) only two alleles were possible at any locus, and (3) relative frequencies of genotypes at linked loci were at equilibrium. The RRS was considered as effective as the other two methods. In the case of loci at which there is overdominance, RRS was superior to recurrent selection for GCA. For loci with partial dominance, RRS was superior to recurrent selection for SCA. Therefore, in case, there is partial dominance at some loci and overdominance at others, RRS is more effective than the other two methods. If dominance is complete; the three methods are all essentially equal in efficiency. In addition, assuming no overdominance, RRS is suggested to be superior to selection based on GCA due to interactions of non-allelic genes or repulsion phase linkages between certain loci or both.

Russell and Eberhart (1975) suggested that if inbred lines taken from the recurrent selection populations were used as testers rather than recurrent selection populations themselves, genetic gains in RRS would be greater. This means that for population A, inbred line(s) extracted from previous generation of Population B should be used as tester rather than Population B itself; and for population B individuals, inbred lines(s) from previous generation of population A should be used as tester rather than population A.

Merits of reciprocal recurrent selection

1. Through the use of this scheme both GCA and SCA variances are exploited.
2. In general, this scheme is superior to other schemes of recurrent selections.

Full-sib reciprocal recurrent selection

This interpopulation improvement scheme was proposed by Hallauer (1967a,b). The scheme can easily be included in a breeding program to develop improved populations and useful single cross hybrids (Hallauer, 1967a). Both non-additive genetic effects (overdominance and epistasis) and additive effects can be selected using this method. When both populations under improvement produce two or more ears per plant, RRS proposed by Comstock et al. (1949) could be modified to evaluate full-sib rather than half-sib progenies for the improvement of two populations (Hallauer and Eberhart, 1970). The modification of using full-sib progeny evaluation for population improvement was termed *reciprocal full-sib selection* by authors. The scheme used to produce hybrid (full-sib progenies) and selfed seed on the same plants was described by Hallauer (1967a,b) and Lonnquist and Williams (1967) in maize. This breeding scheme serves dual purpose: (1) yield tests of the S_0/S_0 plant crosses for population improvement and (2) early testing of single cross combinations.

General outline of full-sib reciprocal recurrent selection

Different steps of full-sib RRS scheme are presented in Fig. 16.2.

Step 1: the two prolific populations A and B are grown in alternate rows (or in two blocks) with appropriate management practices to allow the expression of prolificacy. In the selected prolific plants of the two parental populations A and B, both ears are covered with silk bags. Full-sib crosses (progenies) are reciprocally produced by pairing plants from two populations. For this, top ear of a plant from population A is crossed with the plant from population B. The same operation is carried for the reciprocal plants from population B. The lower ear is selfed (using hand-pollination) to be saved as remnant seed. For convenience, selfing could be done one day and crossing on the other day. The tassel bags are properly labeled to identify male parent of the crosses and their respective selfs. About 200 full-sibs are made and seed of each full-sib cross is harvested separately.

Step 2: full-sib progenies (crosses) are evaluated in replicated yield trials at three to four locations in the main season. Some 20–30 desirable full-sib progenies (families) are selected based on data of different trials for yield and other traits. Pairs of S_1 lines (inbred lines) related to different full-sib crosses are evaluated separately for their *per se* performance.

Step 3: the remnant seed of the lower ears (selfed by hand-pollination), that is, S_1 plant from the best selected (using data from step 2) full-sibs is planted in ear-to-row (of the two corresponding populations) in two separate recombination blocks to constitute A1 and B1, the two improved populations, or selfed seed of S_1 plants of best full-sibs is bulked within each population (A × B and B × A crosses) and is allowed to random mate to provide new improved population.

Hallauer (1967a,b) recommended that selfed seed of the parents of the superior S_0/S_0 crosses should be planted in ear-to-row for continued selfing and crossing for rapid development of new single crosses. Selfed seed of the parents of the

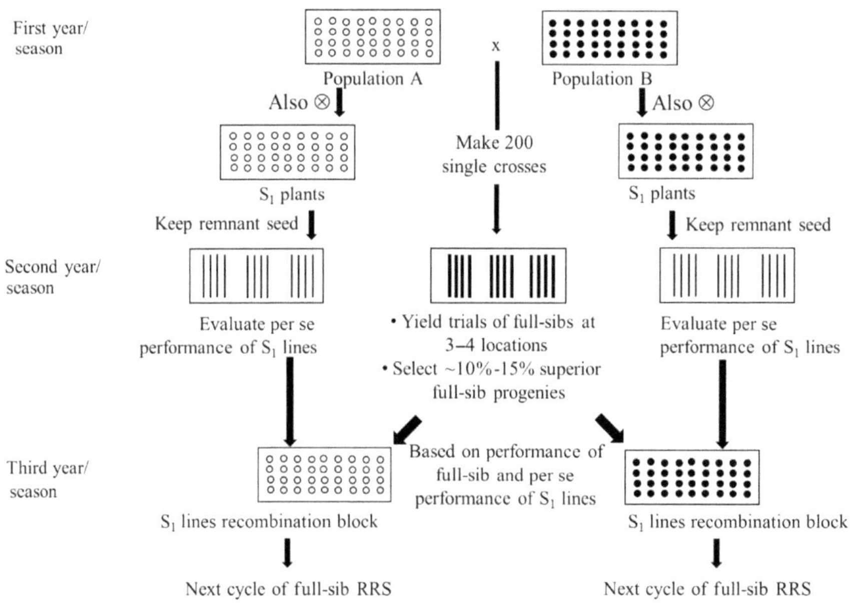

FIGURE 16.2

General procedure of reciprocal recurrent selection scheme (full-sib). In the third season, remnant seed of S_1 plants are used to plant the recombination block.

superior yielding S_0/S_0 crosses can be used for recombining to form two improved populations, which can be continually selected and improved as suggested in RRS by Comstock et al. (1949).

Genetic basis of full-sib reciprocal recurrent selection

The additive genetic variance of full-sib families is twice that of half-sib families. Relative efficiency of RRS and full-sib RRS was compared by Moll and Stuber (1971). Following six cycles of RRS, average gains per cycle for "Jarvis," "Indian Chief," and "Jarvis × Indian Chief" through RRS were 2.3%, 1.2%, and 3.5% per cycle and through full-sib RRS, 3.5%, 2.8%, and 2.5%, respectively.

Merit of full-sib reciprocal recurrent selection

1. The evaluation of each full-sib reflects the worth of two parental plants, that is, one from each population, therefore, only half of the progenies are evaluated when compared with the half-sib RRS scheme.
2. The superior S_0/S_0, S_1/S_1,..., $S_n \times S_n$ crosses could be used to simultaneously develop superior inbred lines and produce hybrids.

According to Hallauer and Eberhart (1970), there is considerable flexibility in the commercial product with the comprehensive breeding method. The development of single cross hybrids could be omitted, and a variety cross could be used as the commercial product in area where uniformity of the commercial crop is not important. An alternate would be to use single cross produced from S_2 or S_3 lines. However, it must be noted that now in row crops where hybrids are possible, single cross hybrids are the major type of cultivars.

Intrapopulation improvement

For intrapopulation improvement (i.e., improvement of one population rather than two populations simultaneously), selection is practiced within a population for intended objectives. In intrapopulation improvement, the fixable variance (i.e., additive and additive × additive) is exploitable. From studies conducted in a number of open-pollinated varieties (OPVs) and advance generation varietal hybrids of maize, it has been revealed that additive genetic variance is present for grain yield and other agronomic traits that could be exploited to improve the performance of the variety (referring to OPV) (Gardner, 1963). The population reconstituted after few cycles of selection would not only improve *per se* performance for the desired characters but also serve as the base population which could be used: (1) to develop a population for its cultivation as such (end product or synthetic variety), (2) to develop superior inbred lines for hybrid production, and (3) for isolating inbred lines with superior GCA and SCA. For the successful implementation of an intrapopulation improvement it is important to maintain adequate genetic variability for the trait under selection before initiating selection program and thereafter in each subsequent cycle of selection.

The proportion of additive variance is likely to be depleted when the population is subjected to some degree of inbreeding. This could be due to the use of small plant population size during regeneration or due to higher selection intensity. Therefore, a minimum population size may be maintained, which would depend on the number of genes and the nature of inheritance of the character(s) under consideration. Allard (1960) observed that with an increase in number of genes (characters) involved in selection there is an exponential increase in the minimum population size to achieve desired gains. Large population size is also desirable to overcome the problem of undesirable genetic linkages.

Mass selection

It is one of the oldest and operationally simplest intrapopulation improvement schemes. It is commonly used in cross-pollinated crops for the improvement and for maintenance of varieties. It is also used in self-pollinated crops for the maintenance of varieties, and improving landraces. Sprague (1955) observed that mass selection for maize improvement began with its domestication. Mass selection procedure implies phenotypic selection for some specific traits on an individual plant basis for the improvement of a population (Hallauer and Sears, 1969). Traits of higher heritability such as flowering time or morphological traits are suitable for this method. Mass selection is similar to phenotypic recurrent selection, except in the later scheme selected female plants are randomly mated only with selected male plants. This modification doubles the genetic gain when compared with phenotypic selection on individual plant basis, due to control of both male and female parents.

Differences between mass selection and phenotypic recurrent selection

Mass selection: female plants are selected, after pollination with unselected and selected pollen sources.

Phenotypic recurrent selection: males and females are both controlled. Only selected plants are intercrossed to obtain seed for the next cycle of selection. The expected genetic gain from the selection of the female parent only is one-half compared with the expected genetic gain when both parents are selected.

However, the terms 'mass selection' and 'phenotypic recurrent selection' are sometimes used interchangeably; although clear distinction exists. The difference between genotypic and phenotypic recurrent selection is that genotypic recurrent selection is based on progeny performance (combining ability). Phenotypic recurrent selection is based on the phenotype of the individual, and does not include progeny testing.

General outline of mass selection

Step 1: a large number of plants are selected based on their phenotype, and open-pollinated seed is harvested in bulk for growing the next generation.
Step 2: selection cycle may be repeated to increase the frequency of favorable alleles.

Detailed steps include:

- Population is space-planted to facilitate note taking on individual plants.
- Evaluate for the trait of interest and identify the best individuals.
- Harvest seed of the best deemed individuals ("best" implies, plants that meet all or majority of desired traits and phenotypes) and reconstitute seed to form the next cycle of recurrent selection.

Pollen control can be exerted if the trait can be evaluated prior to flowering. Undesirables can be removed before they contribute pollen to the rest of population, and this ability to control parental pollen helps improve the response to selection. [Genotypic selection (i.e., progeny selection) is usually more effective than phenotypic selection if the trait of interest is quantitatively inherited with low heritability. Genotypic selection of an individual plant is based on the performance of its progeny.]

Mass selection method is based on the phenotypic performance (appearance of the plant) of the female plant for its particular character; and without progeny testing and lack of control on pollen parent, the weaknesses of this method are obvious. Breeder can determine if the extra generation required for genotypic selection is possible, as gain per cycle is higher for genotypic selection (two generations per cycle) compared with a single generation cycle with phenotypic selection. The effectiveness of mass selection would depend on the accuracy with which the phenotype reflects the genotypes. Therefore, it is essential to grow the material under uniform environmental conditions, and traits with high heritability are targeted.

Genetic basis of mass selection

In this scheme, phenotype is a criterion of selection. Therefore, its efficiency would depend on the heritability of the character and additive genetic variance of the population. The method is effective for the characters that have high heritability on single plant basis; and the character is expressed before pollination such as duration of flowering, plant height, ear placement, and prolificacy, which are based on the female parent. As there is no control on male parent, selection practiced after pollination is not very effective for traits like seed yield, which has low heritability, controlled by more number of genes, and cannot be judged accurately based on the appearance of single plants.

Merits of mass-selection

1. It is simple and easy to carry out in field. The selection cycle is very short as it consists of only one generation.
2. It is quick to develop new variety (OPV) through this method.
3. The method is found effective for characters that have high heritability and are easily identifiable. In maize, it is found to be effective for the characters that express before pollination like plant height and duration of flowering, etc.

4. As the improved strain is likely to be similar or better to the original population in the range of adaptation, extensive yield testing is not required before release as cultivar.

Limitations of mass-selection

1. Selection is limited to the range of genetic variability already present in the population.
2. It is based on phenotypic selection of plants in a single location planting, and from unreplicated selection units.
3. The method is based on the female parent and there is no control on the pollen parent as the selected plants are allowed to open pollinate.
4. Strict selection leading to reduced population size leads in turn to inbreeding, reduced genetic variation and genetic drift.

Application of mass-selection

The taking of superior individuals from existing populations and using the increase from these without testing was the prevailing method of cotton breeding through the 19th and early 20th centuries (Brown and Ware, 1958). Cotton is an often cross-pollinated crop (5%–25% cross-pollination), and cross-pollination could be up to 50% if pollination promoting insects are abundant. In maize, mass selection has been a useful tool for changing gene frequencies of traits of relatively high heritability. In Iowa, after four cycles of mass selection for early silking in the "Eto" composite from Colombia, the interval from planting to silking decreased from 116 to 100 days and average ear height had decreased from 212 to 116 cm (Hallauer and Sears, 1972). A 6.3% average gain in yield per cycle over five cycles of mass selection was observed in "Hayes Golden," while 3.2% increase in grain moisture was also reported in some populations (Lonnquist, 1967). However, no significant improvement in either "Iowa Ideal" or "Krug" maize populations were reported after five and six cycles of mass selection, respectively (Hallauer and Sears, 1969).

Mass selection was successfully used to incorporate desirable traits from 25 Mexican maize races into a single population with days to maturity and plant architecture useable for temperate zone maize breeders (Genter, 1976). Except root lodging, 10 generations of mass selection produced significant changes in all measured traits. The rate of yield increase was 19.1% per cycle based on the yield of C_1 (first cycle composite) or 24.3% based on the yield of C_0 population; although linear, percentage increase per cycle decreased as the population became better adapted and more productive.

Modified mass-selection

The method was outlined by Gardner (1961) to overcome the problems of variation in soil fertility. The field was divided into small compartments or grids, each consisting of 40 plants with minimal environmental variation and selecting plants within each grid (selecting highest yielding 10% plants). Equal number of superior plants is selected from each of the plot, that is, selection is practiced within plots and not among plots. The seed from all selected plants is harvested and composited to grow next generation. It is also called as grid method of mass selection or stratified mass-selection. The method was used for the improvement of the open-pollinated corn variety "Hays Golden," a relatively early cultivar in eastern Nebraska (Gardner, 1961). They reported an average progress of 3.9% per generation. However, later publications found that this gain is short-term, and standardization per grid is required to remove a bias.

General outline of modified mass-selection

The operational details of modified mass-selection scheme in maize are as follows:

Step 1: some 7500–10,000 plants are grown on uniform plot at a population density of 30,000–40,000 plants/ha. To check contamination from foreign pollen and to reduce border effect, an isolation distance of 300–400 m and four to six border rows (on all four side of the field) may be provided.

Step 2: at maturity, the plot is subdivided into several grids or subplots of 20–40 plants (of two rows each). A selection intensity of 5%–10% is practiced. One to two plants per row from each subplot or grid could be selected.

Step 3: the seed from selected plants are harvested and bulked in equal number from each selected ear and used to raise the next generation (cycle).

Application of modified mass-selection

Hallauer and Miranda (1981) have reviewed the results of selection studies using mass selection in maize. Selection practiced for more than 10 cycles in three varieties of corn showed gains in yield per cycle were 3.0%, 0.9%, and 19.1%. Improvement in corn yield has also been realized through indirect selection for prolificacy (Gardner, 1978). Effectiveness of mass selection for yield can be further improved by detasseling weak and undesirable plants to provide for better parental control, by following simultaneous selection for prolificacy and grain yield, and by having additional cycles of selection for other agronomic traits. Mass selection has also been used successfully for changing leaf angle, plant height, and ear height, reducing duration of flowering, for increasing ears/plant, ear length, and seed size, and for improving resistance to diseases and insect–pests.

Both mass selection and phenotypic recurrent selection rely on selection of individual plants, which have several inherent issues that introduces bias, for example: (1) competition effect due to uneven plants and (2) microenvironment variability does not permit assessing breeding values. These can be overcome by the usage of gridding design (modified mass selection), and avoiding selecting plants that have missing neighbors.

Family selection methods (genotypic selection)

Mass selection suffers from several difficulties, but primarily challenges in measuring trait values on individual plant basis. Family based selection methods provide greater accuracy in predicting an individual's breeding value; and therefore, gives a large response to selection. Family selection methods also utilizes replication and broader (more environment) testing to choose higher performing genotypes compared with mass selection. In these methods, a family structure is created followed by evaluation and selection of best ones by progeny testing. In mass selection, individuals are phenotyped (for the trait or traits of interest) and those with the best phenotypic values are used as parents to form the next generation; therefore, the individual or group of individuals are common on which selection is based and recombination is done. In family selection methods the individuals used for selection decisions (e.g., half-sib or full sib, also called as "selection unit") can be entirely distinct from those used to form the next generation (i.e., individuals used as parents for the next generation also called as "recombination unit"). While the selection and recombination units are not the same, response to selection is observed due to the genetic correlation between them. In the text below, few examples of family selection methods are described.

A. Ear-to-row selection

Ear-to-row selection is the scheme of half-sib family selection or half-sib selection applicable to cross-pollinated crops. In this method, half-sib families are produced (from any broad-based population such as a synthetic, composite, etc.) for evaluation and recombination in one generation. The half-sib families (progenies) are created by random pollination of selected female plants (parents) as in ear-to-row method. Three basic steps are (1) production of families or progenies, (2) evaluation and selection of best families, and (3) recombination of families or plants within selected families to create a new population for the next cycle of selection. In modified ear-to-row method, seed from generation one families are evaluated in replicated trials in different environments for selection. The half-sib families could be selfed (S_1) for one generation and selection can be exercised on the basis of evaluation of their half-sib (HS) or selfed progenies (S_1) or a combination of both (HS and S_1). The two commonly known half-sib family schemes are: (1) ear-to-row method and (2) modified ear-to-row methods.

1. Ear-to-row selection method (half-sib)

This method was introduced in 1896 by C.G. Hopkins as a method for intrapopulation improvement in the chemical composition (oil and protein content) of the corn kernel at the Illinois Experiment Station (Hopkins, 1899). It is a half-sib progeny selection, which was used extensively by maize breeders in the early days of maize breeding in North America. In 1896, C.G. Hopkins started a long-term artificial selection experiment looking at oil (Fig. 16.3a) and protein

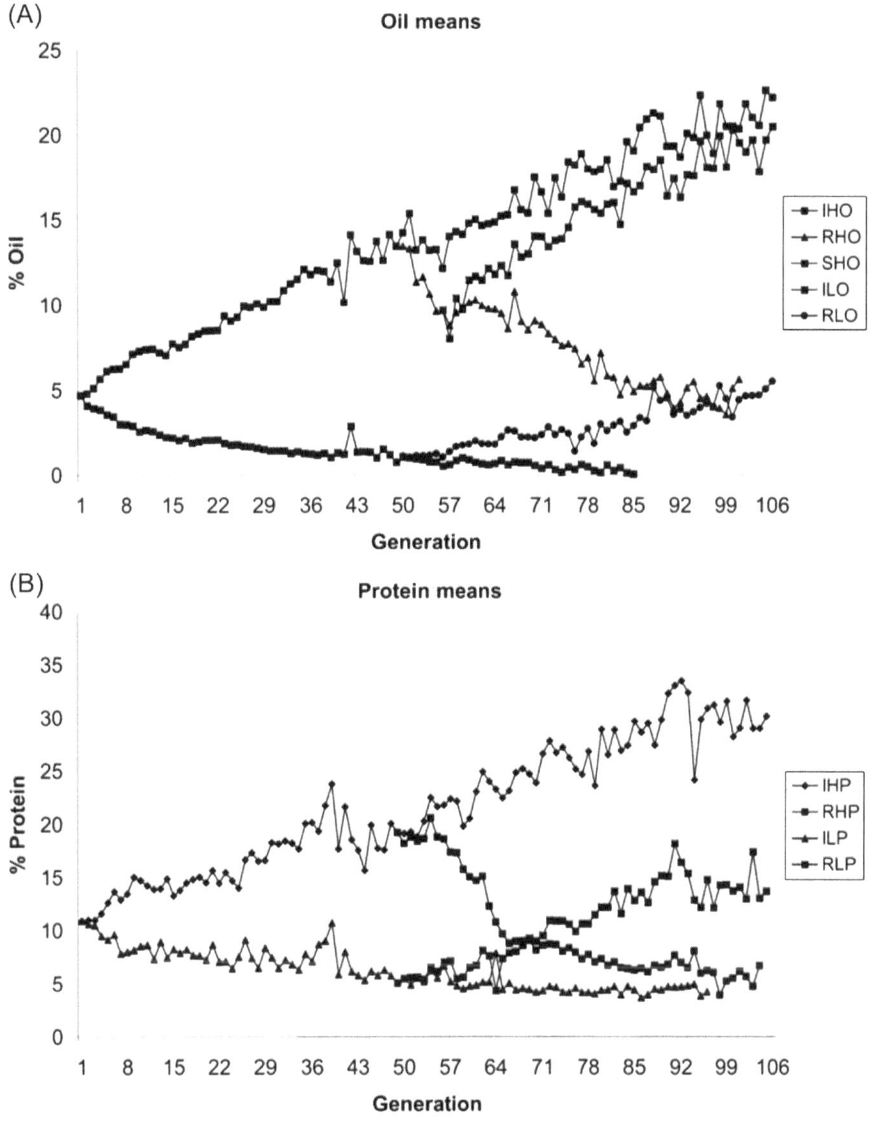

FIGURE 16.3

(A) Plot of mean oil concentration over generations for Illinois High Oil (IHO), Reverse High Oil (RHO), Switchback High Oil (SHO), Illinois Low Oil (ILO), and Reverse Low Oil (RLO); (B) plot of mean protein concentration over generations for Illinois High Protein (IHP), Reverse High Protein (RHP), Illinois Low Protein (ILP), and Reverse Low Protein (RLP). From Dudley JW (2007).

Table 16.1 Change in the chemical composition after 50 generations of selection in "Burr White," an open-pollinated variety of maize.

		Progress due to selection in	
Chemical composition	**Original lot**	**Positive direction**	**Negative direction**
Protein content (%)	10.9	19.5	4.9
Oil content (%)	4.7	15.4	1.0

(Fig. 16.3b) content in maize. The open-pollinated corn cultivar, Burr's White, was used as the founder population. Four strains were established: Illinois High Oil (IHO), Illinois Low Oil (ILO), Illinois High Protein (IHP), and Illinois Low Protein (ILP) with high and low referring to the direction of the selection. After 48 generations, reverse selection was started in each strain to establish the Reverse High Oil (RHO), Reverse Low Oil (RLO), Reverse High Protein (RHP), and Reverse Low Protein (RLP) strains. After seven generations of selection in RHO, selection was again reversed to create the Switchback High Oil (SHO) strain to study the effect of selection.

The effects of selection on oil content ceased at generation 85 in the ILO strain when the oil content reached a level that was no longer measurable with the analytical tools used in this experiment. Protein content reached a lower limit after approximately 65 generations, likely due to biological (i.e., physiological) limit in this crop species. An upper limit was not reached for oil content in IHO and SHO indicating that significant genetic variance still existed in these strains even after 100 generations of selection. The results of 50 generations of selection (28 generations of ear-to-row selection and 22 generations of mass-selection) are presented in Table 16.1.

General outline of ear-to-row selection. The general procedure of ear-to-row scheme is described as follows (Fig. 16.4):

- *Step 1:* superior ears are picked up from desirable plants of an open-pollinated population (source population)—can be a named variety or a breeding population. Seeds are harvested separately from each selected ear (plant).
- *Step 2:* in the next year, a portion of seed from each ear is planted in isolation as a progeny rows/ear rows (half-sibs) for evaluation of desirable character(s) for progeny test and the remnant seed is kept in reserve.
- *Step 3:* the superior progeny rows (5–10) are identified and remnant seed of previous year is then composited to form a new population (for its use in next cycle) or the bulk seed is used as a new open-pollinated cultivar for commercial cultivation.

Genetic basis of ear-to-row method. As in mass-selection, in ear-to-row method selection is based on female parent (maternal parent) without the control of male (pollen) parent. Ear-to-row selection was unsuccessful in improving grain yield in maize, which has low heritability. On the other hand, this method was effective in

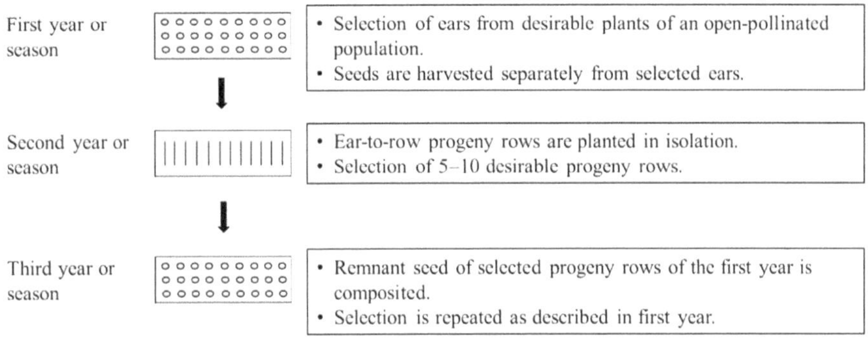

FIGURE 16.4

General procedure of ear-to-row method of selection (one selection cycle).

modifying the chemical composition (protein and oil content) of the maize kernel and other traits with moderate to high heritability.

Merits of ear-to-row method. It offers a means of increasing the accuracy of intrapopulation selection as it is based on progeny evaluation.

Limitations of ear-to-row method

1. There is no control in pollination of selected plants.
2. This method was of little value for the improvement of characters with low heritability.

2. Modified ear-to-row selection method

The two methods of intrapopulation improvement (mass-selection, where selection is based on individual plants, and the ear-to-row selection, where selection is based on progeny testing) were unsuccessful or were of little success to improve traits with low heritability like grain yield. The failure of these methods of selection could also be attributed to the utilization of poor field plot techniques among other reasons. Lonnquist (1964) suggested modification in ear-to-row method for better management of variation due to environment and genotype–environment interactions that would improve gains for yield. The method consisted of developing families (progenies), evaluation and selection of superior progenies and recombination conducted in one generation. The ear-to-row families (half-sib families) are evaluated in replicated trials in many environments. These modifications were found useful in breeding by Webel and Lonnquist (1967). Four cycles of modified ear-to-row selection were applied to "Hays Golden" (OPV of maize) at the Nebraska Experiment Station. The method involved among and within half-sib families selections. Individuals were selected based on phenotypic performance within selected families grown in a natural crossing block in isolation. Relative to the parental variety, an average yield increase of 9.44% per cycle was observed. A comparison with the earlier published results of using mass-selection in the same variety indicated that the modified ear-to-row scheme is more effective than mass-selection alone for yield improvement in "Hays Golden." The modified ear-to-row method would be more effective to use additive genetic effects in the base population before capitalizing on the dominance and epistatic effects in hybrid combinations.

General outline of the modified ear-to-row selection

- *Step 1:* the ears from a base population (OPV, composite or synthetic variety) of desirable plants are selected and are harvested separately (Lonnquist, 1964 selected 190 ears from an OPV.)
- *Step 2:* the seeds are held separately and are planted in ear-to-progeny row (or ear-to-rows) in replicated trials along with parental variety as checks. Each replicate is planted in a separate location and one replication is planted in main station with a crossing block in an isolation (4 females: 2 males ratio) superimposed on the lattice arrangement (Lonnquist used 14 × 14 triple lattice design and six checks were included. The source of seed of male rows was a composite of each of 190 ears). Detasselling is followed in this replication in female ear rows and check entries. This provides high level of recombination and the procedure is similar to polycross used in forage breeding. From each of the progeny rows, five superior plants based on phenotypic appearance were marked before harvest and five selected ear were placed in a bag and weighed together with rest of the ears-to-row progenies.
- *Step 3:* based on data of average performance for yield and other attributes from replicated trial 20% of the superior half-sib families or rows (ear-to-row families) were selected. The seed of the five ears from the selected families (progenies) from crossing block are used for the next cycle of selection and intercrossing.

As all comparisons of the selected populations, after each cycle of selection, are to be made with the parental population it should be maintained appropriately.

Genetic basis of modified ear-to-row selection. Modifications suggested by Lonnquist (1964) in ear-to-row scheme are performed at two levels; among ear–rows families (it is over environments) based on yield trial data further selection is practiced within ear–rows/families (it is within an environment) in crossing block in isolation, where plants are pollinated with bulk pollen from all the families. Therefore, both between and within family selection is based on one sex only. The predicted gain for such a procedure which completes selection cycle is given by Webel and Lonnquist (1967).

$$Gs \frac{k_1\,\hat{\sigma}_A^2/8}{\underbrace{\sqrt{\sigma_m^2+\frac{\sigma_{f(m)}^2}{f}+\frac{\sigma_{me}^2}{e}+\frac{\sigma_{f(m)e}^2}{fe}+\frac{(\sigma_e^2+\sigma_w^2)}{fer}}}_{\text{Among half-sib families}}}+\frac{k_2\,\hat{\sigma}_A^2\,3/8}{\underbrace{\sqrt{\sigma_{f(m)}^2+\sigma_{f(m)e}^2+\sigma_e^2+\sigma_w^2}}_{\text{within half-sib families}}}$$

or expressed as: $[((k_1 \times \sigma^2{}_A)/8)/(\sigma_{\text{PER}})] + [((k_2 \times \sigma^2{}_A) \times (3/8)/(\sigma_{\text{PW}})]$ where *Gs* is the expected response to 1 cycle of ear–row selection, k_1 is the selection differential in standard units among ear–row families, k_2 is the selection differential within ear–row families, σ^2_A is the additive genetic variance, σ_{PER} is the phenotypic standard error among ear–row families, and σ_{PW} is the phenotypic standard error within ear–row families.

For better genetic gain, Compton and Comstock (1976) suggested the completion of selection cycle in two growing seasons. For this, a part of the seed is kept before testing in multi-environments. One year is spent for evaluation and selection of progenies. Another year is used for recombination, where only selected families are grown and detasseled at flowering and pollen is provided by the bulk seed of selected progenies, that is both male and female gametes are from selected families (selection is in both sexes). This would double the genetic gain. The prediction equation is presented as:

$$Gs = [((k_1 \times \sigma^2{}_A)/4)/(\sigma_{PER})] + [((k_2 \times \sigma^2{}_A) \times (3/8)/(\sigma_{pw})]$$

Applications of modified ear-to-row selection. After effective utilization of additive genetic effects has been realized, the usual procedure of interpopulation improvement could then be applied with selection based largely on differences due to dominance and epistatic effects (Lonnquist, 1964). The extraction of lines for F_1 hybrid combinations can be used to further enhance productivity. This method can have numerous variations within and among crops based on what is used as a tester, parental control, and intercrossing. Tester can be a population *per se* or an inbred line. However, what identifies the half-sib or ear-to-row method is the evaluation of individuals through their half-sib progeny. Where possible, it is desirable to control both parents. This can be achieved by evaluating in one season and recombining in the next generation (in winter nursery or second season permitting agronomic performance assessment). This necessitates an extra season, but genetic gain will be higher. While half-sibs are being evaluated, the remnant seed of each individual should be kept as a reserve so that this seed can be used if the individual is selected to intercross and create material for the next selection cycle.

In crop species such as maize, obtaining selfed and half-sib seed from the same plant can be accomplished by self-pollinating the single ear on the individual to be tested, and use pollen from that individual to pollinate several individuals of the tester (population or inbred line). The ears on the tester are bulked together (these plants of tester received pollen from the individual to be tested) and represent the half-sib family to be tested for that individual. Recombining selfed progeny will require three seasons: (1) selfing and crossing to the tester, (2) evaluation, and (3) intercrossing selfed progeny. Here are three examples of the use of half-sib method applications:

Scenario 1: female parent selected; population used as tester.

- Season 1: start with a random mating population. Harvest ears of selected plants (~200).

- Season 2: grow 200 half-sib progeny plots (with checks) at multiple locations (can be replicated or unreplicated). The trait of interest can be any, including yield. At one location, grow plants in isolation for seed source for the next cycle. Select plants at this location within the half-sib row. Use other locations for testing.

At the isolation location, grow the male rows (bulk seed of all half-sib families) adjacent to female half-sib rows. Detassel the female rows. At the isolation location, harvest ears from each selected plant by hand. Make selections to pick the best half-sib families. These ears will form the seed for the next cycle.

Allow random mating of selected plants.

Repeat steps, if following recurrent selection. One can use an inbred line as a male tester instead of the bulked seed of a population.

Scenario 2: female and male parent selected; population used as the tester.

Season 1: intermate the population.

- Harvest ears from each plant. Selection may be performed.
- Divide the seed of half-sib plants into two: part 1 for next season field testing and part 2 for remnant seed to reconstitute the selected half-sibs.

Season 2: each half-sib (using part 1 seed) is a separate entry in replicated or unreplicated trials with two or more locations and including checks. Select superior half-sib families based on performance.

Season 3: remnant seed (part 2 of seed bag) of selected individuals are used for intercrossing to form the next cycle.

Repeat steps above if following a recurrent selection scheme.

One can use an inbred line as the tester instead of using bulked seed of the population.

Scenario 3: test cross progeny

Start with an intermated population.

- Season 1: plants of an intermated population are selfed. Pollen is used for selfing and pollinating a tester.
- Season 2: testcross progeny are evaluated in replicated tests. Selections are made to identify superior performing progenies.
- Season 3: selfed seed of selected families is used to form the next intermating cycle.

Cycle is repeated as above.

Merits of modified ear-to-row selection

1. It is a combination of both mass-selection and half-sib progeny testing.
2. Both between and within progeny selection are practiced.
3. Selection for more than one character can be practiced
4. It provides rapid increase of more favorable genes or gene complexes of additive sort without excessively high rates of inbreeding.
5. It would also retain much of favorable genetic material present in the original gene pool.

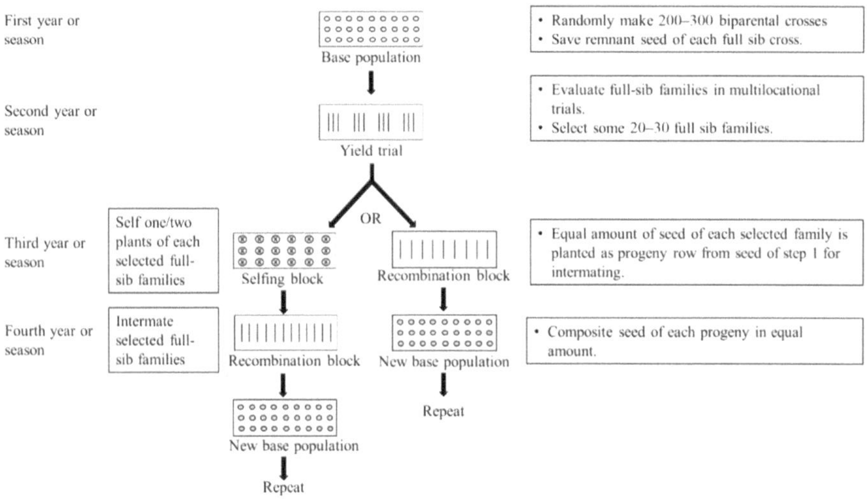

FIGURE 16.5

General procedure of full-sib family selection.

Limitation of modified ear-to-row selection. The parental populations are to be maintained throughout this improvement process as all comparisons would be made with this population.

B. Full-sib family selection

Full-sibs are individuals produced by mating between plants in pairs (reciprocal crosses such as A × B, B × A) (biparental crosses) using parents from the base population. In this method of selection, identities of both male and female parents are known. In other words, it is the controlled mating between two known parents. Therefore, their progenies (families) have common ancestry. In full-sib family scheme, families are evaluated in multi-location replicated trials and superior families are selected which are recombined to initiate next cycle of selection, therefore, requiring three seasons (Fig. 16.5).

General outline of full-sib selection

Step 1: from a base population, random pairs of plants are selected which are intermated (pollinating one with the other) and 200–300 biparental (i.e., paired) crosses are made. Half of the seed is used for planting full-sib progenies and remnant seed is saved.

Step 2: some 200–300 full-sib families are planted in single- or multi-location replicated yield trials. These families are evaluated for yield and other attributes. Of these, some 20–30 promising full-sib families are selected.

Step 3: the selected superior full-sib families of step 2 are recombined from remnant seed of step 1 for one or two cycles before the formation of families for next cycle of selection, that is, intermating of the selected families using remnant seed from step 1. Since progenies were already evaluated, the formation of full sibs could be taken up in an off-season nursery.

Each of the three steps is completed one per season. The three steps are repeated cyclically forming C_2 and later.

Example of the full-sib selection

- Season 1: paired crosses are made between pairs of selected plants in a population. Seed is divided into two parts: part 1 is for field testing, and part 2 is to reconstitute the next cycle.
- Season 2: Part 1 seed is used to plant field tests. Full-sibs are evaluated in field tests (single or multiple locations, replicated or unreplicated, with checks). Superior families are selected based on performance.
- Season 3: part 2 seed is used to intercross selected families. Intermated seed is used to form the next cycle.
- Cycle 2: seasons 4, 5, and 6.

Genetic basis of full-sib selection

The full-sib family selection scheme is more effective than half-sib family selection as it can mobilize twice the expected additive genetic variance and provides full control over both the parents providing selection for both sexes.

Merits of full-sib selection

1. Through this scheme improved variety for across the locations and/or a number of location-specific varieties could be developed.
2. Selection is based on progeny testing and not on phenotype of individual plants.

Limitations of full-sib selection

1. Each plant is tested by only one other plant, establishing interdependency of breeding values between mates.
2. The lack of a population-based or elite tester.

C. Selfed (S_1 and S_2) family selection

Self-pollination of plants from a base population for one generation gives rise to the S_1 progenies (families) and self-pollination of S_1 progenies gives rise to S_2 progenies. The key features of above schemes are the production of S_1/S_2 families followed by their evaluation in multi-location trials and recombination of the superior progenies from remnant seed to form next selection cycle. In the past, this scheme of selection has been used for the improvement of stalk rot and European corn borer resistance and for yield improvement in maize.

General outline of S_1 family selection

Step 1: from a base population some 400–500 S_0 plants are selected and self-pollinated. The selfed seed is harvested and half of the seed is planted as S_1 progenies for evaluation (see step 2) and other half is saved as remnant for intermating of superior progenies (see step 3; this step has some similarity to simple recurrent selection described in Fig. 16.6).

Step 2: evaluation of 400–500 S_1 progeny rows at one location in unreplicated trial for major diseases and in multi-location replicated yield trials (at three to four locations) for yield and other attributes. Based on across location yield data and all other trait data collected, some 40–50 S_1 progenies are selected.

Step 3: intermating of superior S_1 progenies is allowed to synthesize improved population, that is, C_1 cycle (Fig. 16.7).

The reconstituted population C_1 is used for the formation of S_1 progenies for the next cycle of selection. One cycle of S_1 selection is completed in three seasons (Fig. 16.7).

In the S_2 family selection, families are obtained by two generations of selfing and are used for evaluation. The superior S_2 families are intermated. The S_2 selection scheme would require one additional year to complete a cycle with one more generation of selfing. For this purpose, the S_1 progenies are evaluated in main season for biotic/abiotic stresses in unreplicated trials, and S_2 progenies are developed by selfing selected S_1 progenies. The S_2 progenies are evaluated in main season in replicated trials and same 20 to 30 promising progenies are recombined for regenerating C_1 cycle population. The recombination of selected progenies could be performed in off-season to produce population for second cycle selection. This can save 1 year per cycle.

Genetic basis of S_1/S_2 family selection

The additive genetic variance among S_2 progenies is twice that of S_1 progenies. The S_1 and S_2 progenies theoretically provide the highest expected gain per cycle for intra-population improvement.

Merits of S_1/S_2 family selection

1. Selection of individual plants is based on progeny test and not based on phenotype.
2. The deleterious recessive alleles are exposed and are eliminated in the selfed progenies.

Limitation of S_1/S_2 family selection

1. There is no control on pollen parent as selection is based on maternal parent.
2. Compared with mass selection the selection cycle takes additional years.

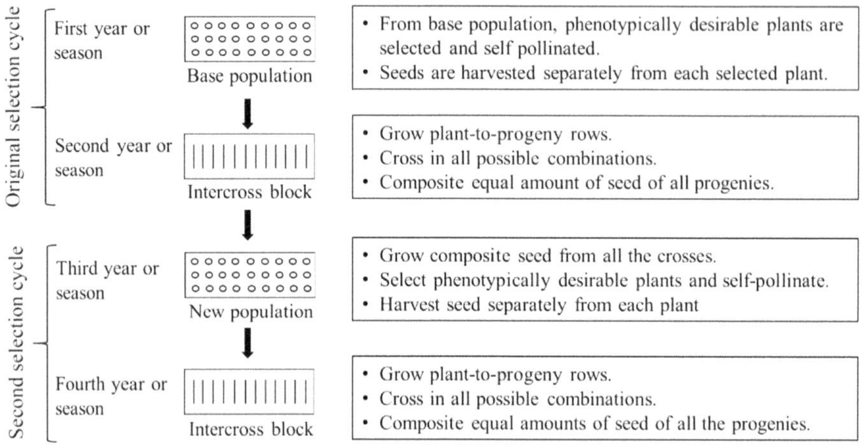

FIGURE 16.6

General procedure of simple recurrent selection scheme.

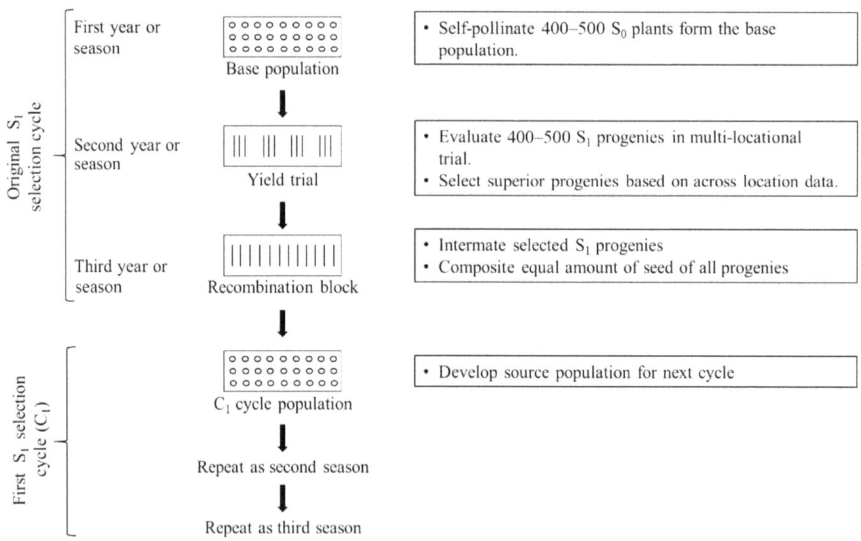

FIGURE 16.7

General procedure of S_1 family selection scheme as above would require one more season (or year) of selfing. For this, S_1 progenies could be evaluated in main season for biotic stresses in screening nursery and one to two plants in each selected S_1 progeny are selfed to produce S_2 seed/progenies.

Recurrent selection schemes

The above listed methods can be done in a single generation (as a breeding method) or cyclically (as recurrent selection method). In the recurrent selection breeding method, based on one or more selection criteria, and using within-family and among-family selection strategies, individuals are selected and intermated to produce the next generation. Selection can continue cyclically and repeatedly, hence the term, "recurrent." The recurrent selection methods are employed to achieve the following:

- The goal of recurrent selection is to improve the mean performance of a population of plants and to maintain the genetic variability present in the population.
- The underlying principle of recurrent selection is to increase the frequency of desirable genes that the breeder is attempting to improve.

Recurrent selection is used to improve populations in cross-pollinated species. OPVs are one type of cultivar developed using recurrent selection. One of the important objectives of recurrent selection is to concentrate the genes for a particular quantitative character(s) in a population of cross-pollinated crops without a marked loss of genetic variability. Recurrent selection provides a means of cyclical selection through which favorable genes are increased in improved population (new population) and it is superior to the original population (base or source population) in mean values for the trait(s) and genetic variability is maintained. The source population is heterozygous and it could be OPV, a synthetic variety, hybrids, etc., and the improved population could be used to produce varieties and/or inbred lines for producing hybrid varieties. Recurrent selection is a modified form of progeny selection. Hull (1945) named recurrent selection and for the first time used it. Allard (1960) quoted him, "Recurrent selection was meant to include reselection generation after generation, with interbreeding of selects to provide genetic recombination. Thus selection among isolates, inbred lines or clones is not recurrent until selects are interbred and a new cycle of selection is initiated." In recurrent selection schemes, intermating would provide greater opportunity for recombination and rate of inbreeding could be kept at a minimum though selection of large number of plants, hence genetic variability is maintained.

Many versions of recurrent selection have been developed as described earlier: S_0, S_1, S_2, half-sib, full-sib, ear-to-row, reciprocal recurrent, and reciprocal full-sib recurrent selection. Four different broad types of recurrent selection schemes have been reported. These are (1) simple recurrent selection, (2) recurrent selection for GCA, (3) recurrent selection for SCA, and (4) RRS. The first three schemes are used for intrapopulation improvement. The RRS scheme and its modifications, which are used for the interpopulation improvement, were described separately earlier. While the other methods of recurrent selection in intrapopulation improvement have been described earlier, recurrent selection for combining ability is presented below.

Recurrent selection for combining ability

These selection procedures are also called as test cross selection schemes. Through these schemes, genetic modifications of the population is achieved for better use in the exploitation of heterosis. The concept of recurrent selection for GCA was proposed by Jenkins (1940). The concept of recurrent selection for SCA was proposed by Hull (1945). In both of these schemes, testers are used to develop crosses which are evaluated in replicated trials. The tester parent is common parent mated to a number of lines in both cases. The choice of tester varies for source population (OPV, synthetic, hybrid, and inbred lines). Depending on the breeding program and its objectives, tester can be strong or weak, where a weak tester implies tester that carries many recessive, negative alleles at loci responsible for the traits of interest. The choice of weak tester in this case is supposed to reveal the positive alleles of the tested population or genetic material as reflects through the good performance of a progeny. However, no consensus exists on the optimal choice of strong or weak tester.

Recurrent selection for GCA: this is a half-sib progeny selection procedure in which a wide genetic base cultivar is used as a tester. The tester could be open-pollinated cultivars, synthetic varieties, and double cross hybrid. This scheme is more effective than other schemes when additive genetic effects are predominant, i.e., breeder is interested in general combining ability of lines, and therefore a broad genetic-base heterogeneous population is recommended as tester. The test cross progenies are evaluated in replicated trials in multi-location trials to identify desirable families which are the unit of selection. The variation in the performance in a group of test crosses is due to differences in GCA. This procedure has been effective in improving combining ability for grain yield in maize.

Recurrent selection for SCA: in this scheme, a tester with narrow genetic base (an inbred line or single cross hybrid) is used as a tester. This scheme is more effective than other schemes when overdominance gene effects are more predominant. A stable inbred line would be more effective tester than a single cross in building up a higher complementary selection between the tester and crossed lot under selection.

General outline of recurrent selection for combining ability

Different steps involved in recurrent selection for combining ability are described as follows and presented in Fig. 16.8.

Step 1: Some 300–500 desirable plants (S_0) from base population are selected. The selected plants are selfed and also test crossed to a tester with broad genetic base for testing GCA and tester with narrow genetic base for testing SCA in the first season/year. Selection for ear and grain characteristic after harvesting could be practiced.

Step 2: the test cross progenies (top crosses) are evaluated in multi-location yield trials at two to three locations in the second season/year. Based on these trials some 10% best test cross progenies (S_1 progenies) are identified, which are unit of selection.

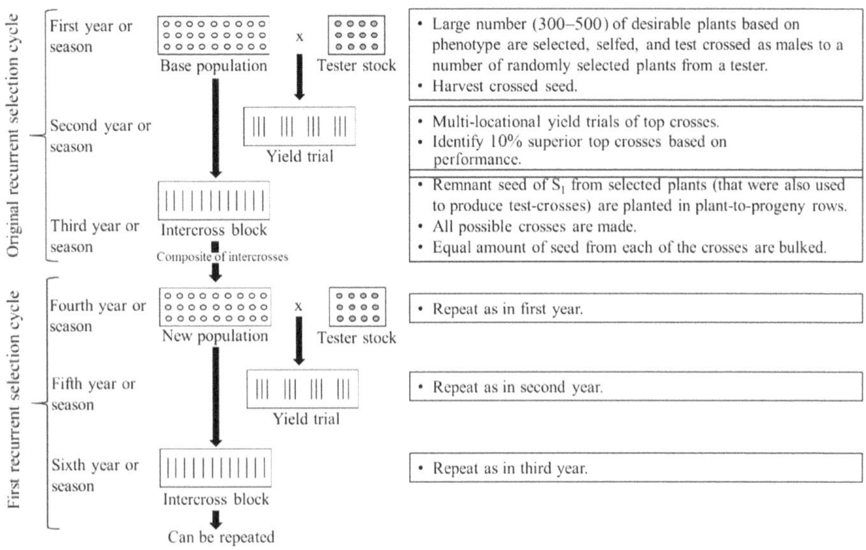

FIGURE 16.8

General procedure of recurrent selection for combining ability (for GCA using a broad genetic base tester and for SCA using a narrow genetic base tester).

Step 3: the progeny rows of selected families (from step 2) are grown by using the remnant seed of plants from step 1. These are intercrossed in all possible combinations and equal amount of seed from intercrosses are bulked to produce an improved population in the third season/year. This bulked seed could be used if required to grow the first recurrent selection cycle.

Alternatively, S_1 progenies and tester parent could be grown in the ratio of 4:2 in isolation block to produce top crosses. The S_1 progenies (female rows) are detasseled. To get adequate seed, the tester parent could be planted on two different dates. Production of S_1 progenies and crossing could be attempted simultaneously. The pollen collected from the S_0 selfed plant could be used for pollinating four to six plants of the tester parent, which is used as female parent. Additional cycle of recombination may be used depending on the specific needs of the program.

Genetic basis of recurrent selection for combining ability

The base population is heterozygous and that individual plants in a population differ for combining ability. Selection of large number of plants and the use of selfed seed for intermating helps in genetic improvement of population and also preserve its genetic worth. The use of limited selfing generation (S_0 or S_1 plants) for early testing of combining ability would help in identifying larger number of superior lines as the variance for combining ability among different S_0 or S_1

plants/families is greater than the within family variance. Therefore, selection among early generation of selfing (inbreeding) would be more effective than selection within progenies (families).

Merits of recurrent selection for combining ability

1. Recurrent selection for GCA is effective to concentrate genes for superior GCA. It also improves the yielding ability of the population at the end of a selection cycle and such a population is identical to the synthetic variety.
2. Recurrent selection for SCA would be useful in isolating superior inbred lines and their specific cross combinations suitable for the formation of hybrid variety.

Limitation of recurrent selection for combining ability

Some degree of inbreeding may be observed in recurrent selection.

CHAPTER

Recurrent selection in self-pollinated crops

17

Abstract

Recurrent selection is a cyclical improvement method aimed at concentrating desirable alleles in a population through a process of detecting better members of a population. Population breeding approach similar to the recurrent selection approach used in allogamous species is also available in autogamous species. In this chapter several methods are described, including phenotypic (mass) recurrent selection with or without recombination between cycles of selection, progeny evaluation with or without the use of male sterility, diallel selective mating system for broadening the germplasm of breeding programs, S_1 and half-sibs progeny recurrent selection with or without the use of male sterility, and single-seed-descent with cyclical selection procedures. This chapter also includes methods that integrate recurrent selection with genomic selection. Finally, advantages and limitations of recurrent selection are also presented.

The breeding methods of autogamous crops described earlier in this book either use existing variability for the selection of pure line or use artificial variability created through hybridization followed by selection of pure line from segregating generations. In hybridization followed by selection (pedigree/modified pedigree/bulk method), the F_1 hybrids from a cross of two (or more) desirable parents are allowed to self-pollinate to give F_2 generation that shows maximum variability; however, half of the genes are in homozygous state in a two-parent cross. After four generations of selfing about 94% of the genes would be homozygous, as a result the segregating population separates quickly from a hybrid into a large number of pure lines. The selection in such a segregating population would largely consist of gene combinations produced in early generation (F_1 in multiple parent cross; while in two parent cross, F_2, F_3 and possibly F_4, F_5 until single plant is selected for a progeny row creating the pure line). The rapid increase of homozygosity reduces or even eliminates the chances of desirable gene recombination when genes are linked in repulsion phase. There are few limitations in the routine breeding methods of self-pollinated crops based on hybridization followed by selection in selfing generations.

1. Relatively small gene pool is used for creating variability in the initial cross
2. Occurrence of linkage blocks/tightly linked genes restrict recombination.
3. Due to lack of intermating there is limited possibility for further changing the genotype of the segregants.

Plant Breeding and Cultivar Development. DOI: https://doi.org/10.1016/B978-0-12-817563-7.00007-6

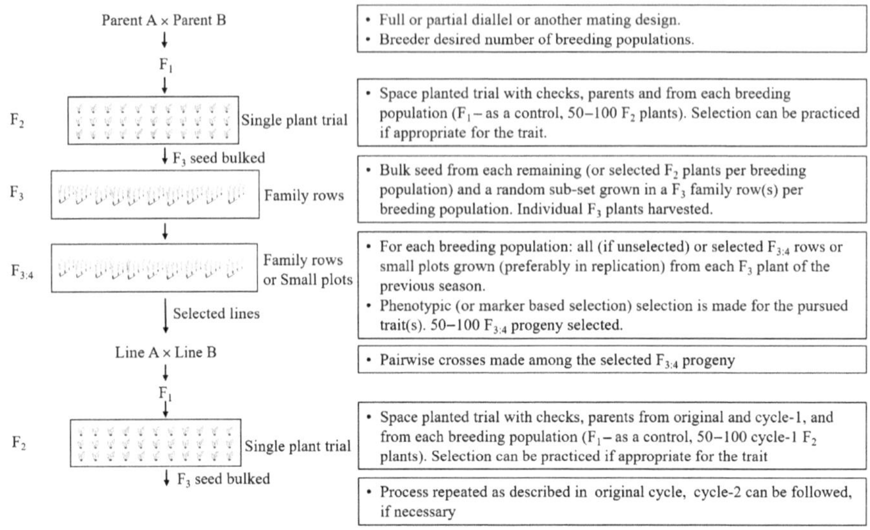

FIGURE 17.1

Flow diagram of recurrent selection scheme for breeding self-pollinating crops.

Adapted from Beeck, C.P., Wroth, J.M., Falk, D.E., Khan, T., Cowling, W.A., 2008. Two cycles of recurrent selection lead to simultaneous improvement in black spot resistance and stem strength in field pea. Crop Sci 48, 2235–2244.

To overcome these problems, recurrent selection in self-pollinating crop species were developed. These methods have been used extensively and effectively to improve important traits in cross-pollinated crop (maize) as explained in Chapter 16, Population Improvement. The three basic steps involved in a recurrent selection scheme are:

1. selection of parents,
2. intercrossing of selected parents, and
3. progeny testing, and repeating the cycle.

One example of the use of recurrent selection in autogamous species (field pea) for breeding disease resistance was provided by Beeck et al. (2008) (Fig. 17.1).

However, there is limitation of producing enough seed through hand crossing in each cycle of intercrossing in autogamous species. Despite these limitations, the use of recurrent selection procedures in these crops has been utilized by few breeders and seen a resurgence more recently (Gaynor et al., 2017). It is due to theoretical advantages as described earlier for this procedure. Several studies have been conducted using hand emasculation and pollination; and in some studies, male sterility system is also used. However, enough seed can be produced if male sterility system could be incorporated in a breeding program; and with the use of male sterility, natural crossing by wind/insects would substitute hand emasculation and pollination. Hallauer (1981) suggested the use of the following five cyclical selection methods in autogamous species:

1. Phenotypic (mass) recurrent selection with or without recombination between cycles of selection.
2. Progeny evaluation with or without the use of male sterility.
3. The diallel selective mating system for broadening the germplasm.
4. S_1 and half-sib progeny recurrent selection with or without the use of male sterility.
5. Use of single seed descent (SSD) with cyclical selection procedures.

These methods were proposed not for cultivar development *per se*, but with an objective to broaden the genetic base of population, which can lead to the development of improved cultivars (Hallauer, 1981).

Phenotypic (mass) recurrent selection with or without recombination between cycles of selection

Superior plants are selected from a segregating (F_2/F_3) population (base population from multiple breeding families) and seed of selected plants are bulked to grow next generation (C_1). From the segregating population desirable plants are reselected, and seed is bulked to grow next generation (C_2). Selection could be stopped any time if desired improvement is achieved. This scheme is effective for improving simply inherited traits, which are easily observable like disease resistance. As no intercrossing or selection between cycles of selection was performed for these studies, genotypes that resulted from mass selection were already present in the original population or segregated from those genotypes. Without recombination, this method should not be considered recurrent selection; however, it has a cyclical approach and is similar to the bulk method if the composition of segregating material is only from one breeding population.

One classical example of this method is in Oat. Frey (1967) conducted mass selection for seed width in oat. He measured indirect effect of mass selection on 100 seed width in a composite population originated by mixing 5 g seed samples of F_2 generation from 250 crosses. The mass selection for seed width increased mean 100 seed weight 9% over five cycles. This could be explained due to high genotypic correlation ($r = 0.69$) between seed width and seed weight, and average heritability value of 35% for seed width. Romero and Frey (1966) effectively used mass selection to reduce plant height in the same composite population, but noticed that mass selected populations were earlier and had higher grain yield.

Recurrent selection was applied by Miller and Rawlings (1967) for improving lint yield in upland cotton. F_1 plants of a cross between two upland cotton lines (one selected for high fiber yield and low fiber tenacity and the other for low yield and high tenacity) yielded 13% more fiber than the better parent. The F_1

was advanced to the F_2 and 81 F_2 plants were taken at random and increased by bulk self-pollination to the F_3 and F_4 generations. From these lines, six highest yielding lines were detected and 15 possible single crosses were made. The 15 cross populations together formed C_1 of recurrent phenotypic selection. Three cycles of phenotypic recurrent selection gave 29.7% increase in yield to average of parental lines, and 13% to the higher yielding parent. Selection was effective for fiber yield and not for fiber quality.

Matzinger et al. (1977) utilized broad genetic base population of tobacco (*Nicotiana tabacum*) in three independent studies with selection: (1) decreased plant height, (2) increased number of leaves, and (3) an index for increased leaves on shorter plants. Selections were made before flowering and intermating were made among selected individuals to provide seed for the next cycle as done for phenotypic recurrent selection, allowing them to study the effect of this selection method. Responses for selection was significant in all three studies and linear response to direct selection was observed over five cycles of selection (4.9% and 7.0% per cycle to selection for decreased plant height and for increased number of leaves, respectively) while maintaining genetic variability. These results suggest that this method can be useful for autogamous crop species; however, these methods are not routinely used anymore due to the extra resources needed for hybridization to intermate self-pollinating crops. Also, genetic gain in elite material is likely more in autogamous crops when breeding methods described for autogamous methods are practiced.

Progeny evaluation with or without the use of male sterility

In this approach, a F_1 hybrid is selfed to produce F_2 population and from this population outstanding F_2 plants are intermated (among themselves in pairs) or mated in some other mating design. This intermating restores heterozygosity in the progeny and creates more recombination. As a result, desirable genes from different F_2 plants are brought together (i.e., the accumulation of desirable/favorable genes) in the intermated population. This procedure would enhance the possibility of selecting transgressive segregants. This is repeated one or more times, that is, as per need. Breeder could also use intermating of plants in selected F_3/F_4 generation progenies. The intermating based on the desirable F_3/F_4 progenies is expected to be more effective for the improvement of quantitative traits like yield than intermating of the F_2 plants due to an increase in additive genetic variance with each generation. Different steps involved in this procedure are presented in Fig. 17.2.

One of the serious limitations of recurrent selection in self-pollinated crops is difficulty in getting required quantity of crossed seeds using hand emasculation and pollination. Therefore, use of male sterility is advocated in cereal, legume and other autogamous crops. When using genetic male sterility facilitated

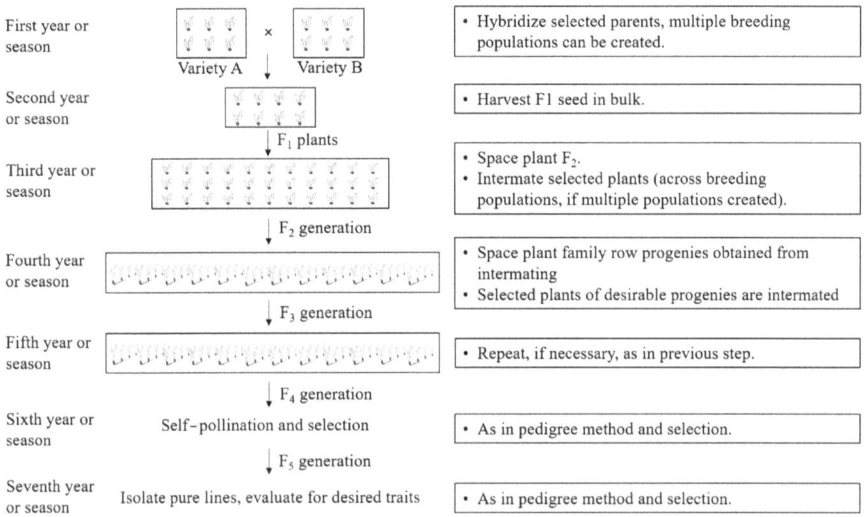

FIGURE 17.2

Scheme of recurrent selection without the use of male sterility in self-pollinated crops.

recurrent selection, crosses are made between selected plants (fertile) and male sterile plants of a population. F_1 is grown without selection and bulk harvested, and F_2 generation is grown in a space-planted nursery from which many male sterile (*ms*/*ms*) and male fertile plants (*Ms*/*ms*) are selected and crossed (intermating). Hybrid seeds are planted and no selection is practiced. The F_2 generation is again grown in a space-planted condition and crossing cycle is repeated (Fig. 17.3). The procedure is to plant a population segregating for male sterility that contains genes, which breeder wants to combine into one genotype. The method allows for recombination among desirable genotypes and prevents rapid inbreeding of the population. Individual plants can be selected at the end of any cycle and progeny rows are subjected to testing and evaluation. For this procedure, biparental or even composite crosses could be used.

DSM system for broadening the germplasm of breeding programs

The diallel selective mating (DSM) was designed by Jensen (1970) to broaden the genetic base of breeding populations and to supplement conventional breeding methods of autogamous crop species that are difficult to cross and produce few seeds per cross. DSM is a dynamic breeding procedure that simultaneously

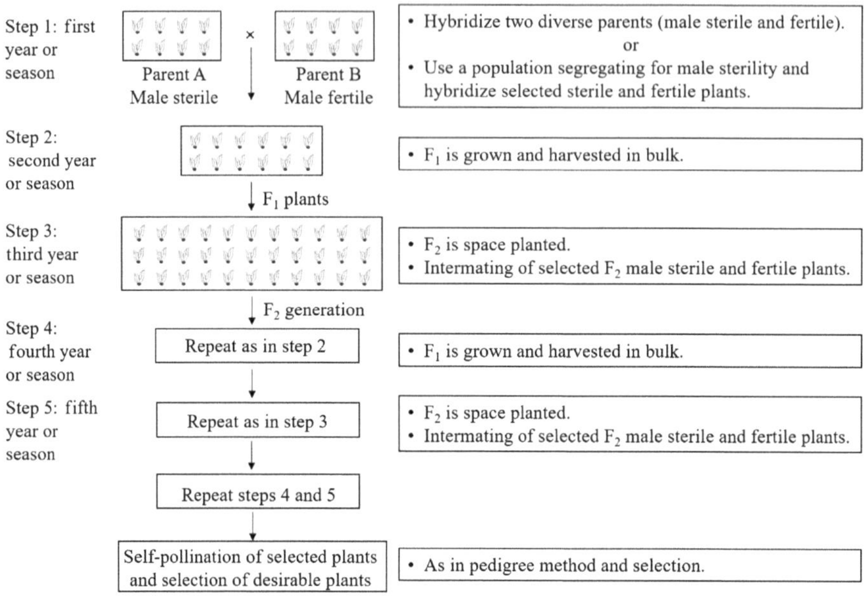

FIGURE 17.3

Scheme of genetic male sterility facilitated recurrent selection in self-pollinated crops.

permits the use of conventional breeding methods and also provides additional opportunities for intermating and helps breeder in three different ways (Fig. 17.4).

1. The development of F_2, F_3, etc. (selfing series, i.e., conventional breeding) at every stage of the breeding program, which helps in the isolation of pure lines for use as commercial varieties.
2. In the breeding program, intermating among selected lines/plants in each step (additional opportunities for intermating) and progenies from these intermatings form the next stage of the selfing series.
3. New germplasm can be introduced at any stage by intermating it with some of the select plants (in any of the series). The seed of F_1 plants in a selfing series is subjected to mass selection for seed characteristics. The selected seeds from all crosses of a series are bulked to establish the F_2 population. Further handling could be according to pedigree method or its modification(s).

The DSM scheme serves both short- and long-term breeding objectives. The short-term objective is fulfilled through selfing series leading to the isolation of pure lines (commercial varieties), and long-term objective is fulfilled through intermating of selected lines/plants, which produces recurrent selection series. Jensen suggested the creation of several breeding populations each with a rather narrow objective. He suggested the use of male sterility to overcome the difficulty of producing enough crossed seeds. The DSM system was effective for

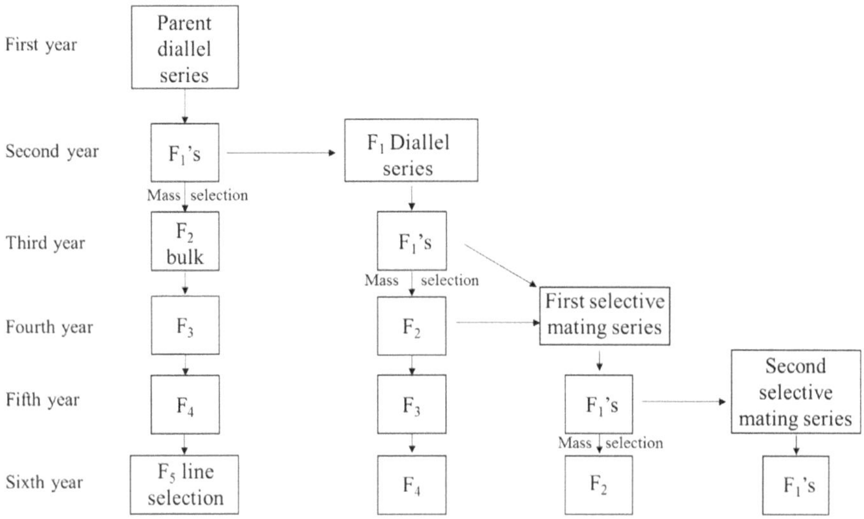

FIGURE 17.4

Scheme of diallel selective mating system for autogamous species proposed by Jensen (1970). The end product of the parent and F_1 diallel series are line variety or parent line, while the end product of first and second selective mating series are line variety, parent line or composite variety. While not shown, F_1 diallel, first, and second selective mating series are also advanced to higher filial generations.

Adapted from Jensen, N.F., 1970. A diallel selective mating system for cereal breeding. Crop Sci. 10, 629–635.

developing new cultivars in small grains. The general outline of DSM for cereal breeding is as follows (adapted from Jensen, 1970):

1. Planning phase
   - **a.** statement of project objective,
   - **b.** analysis of genes needed to reach objective, and
   - **c.** choosing of parent.
2. Implementation phase
   - **A.** Basic parent series (stated capacity 21 hybrid combinations):
     - **i.** Crossing unrestricted within single parent group:
       1. If $p \leq 7$: use complete diallel crossing (CDC) with formula $p \times (p-1)/2$, for example, if $p = 6$, total combinations = 15.
       2. if $p \geq 8$:
          - **a.** numbers small to intermediate (8–22): use partial diallel crossing (PDC) with same formula and requirement that crosses include at least one complete set.
          - **b.** numbers large (23–42): use modified PDC by making series of two parent crosses with requirement that each parent must be used at least once.

- **ii.** Crossing restricted between two groups:
  - **1.** if total possible between group combinations are:
    - **a.** ≤ 21: make all desired crosses between groups.
    - **b.** >21: make series of selective two-parent crosses (limit: 21) between groups with restriction that each parent must be used at least once.
- **B.** F_1 diallel series (to produce composite diallel population P_2):
  - **i.** ideal situation: use CDC (capacity 21 F_1 parents).
  - **ii.** other situations: use partial or modified PDC.
- **C.** Selective mating series:
  - **i.** Intra-populational:
    - **1.** mass selection plus selective mating in F_1 and F_2 of P_2: produces P_3.
    - **2.** mass selection plus selective mating in F_1 and/or F_2 of P_3: produces P_4, etc.
  - **ii.** Inter-populational:
    - **1.** mass selection plus selective mating (e.g., $P_1F_3 \times P_2F_2$): produces supplementary hybrid seeds for P_3.
    - **2.** backcross to project reserve adapted genotypes (e.g., P_3F_1 or $P_3F_2 \times$ adapted F_1s): produces P_5.
- **D.** End product series, from populations 1–5:
  - **i.** pure line variety after conventional selfing, selection, and testing,
  - **ii.** composite variety from selective mating series, and
  - **iii.** parent line for further crossing.

S_1 and half-sibs progeny recurrent selection with or without the use of male sterility

Burton and Brim (1978) used S_1 and half-sib recurrent selection methods for improving seed protein (due to higher heritability and trait importance) in soybean. Population I was derived from a cross of two highly adapted breeding lines. Population II was generated from backcrosses of nine unadapted plant introductions with high seed protein percentage to a highly adapted breeding line. In addition to study the effect of recurrent selection in these two populations, the effects of effective population size and number of lines tested per cycle were also evaluated (populations IA, IB, IIA, and IIB). The cycle of selection consisted of: (1) intermating among selected lines, (2) testing S_1 lines from the matings in step "1," and (3) selection of lines with highest seed protein levels as parents for the next cycle. If composite of parents chosen from the testing generation in each cycle is used as entries in a replicated field trial (multi-location), selection progress can be assessed. This allows comparisons in each generation. Their results showed that seed protein (%) significantly increased in a linear manner in all populations: in population IA from 46.3% to 48.4% with six cycles of selection; in population IB from 46.4% to 47.6% with four cycles of selection, in population IIA from 42.8% to 46.1% with five cycles of selection, and in population IIB from 43.2% to 45.9% with four cycles of selection. Due to known seed protein - seed oil inverse relationship, significant linear decrease in seed oil % was observed. Seed yield did not show a consistent trend. While population size and number of lines tested per cycle had little effect on genetic progress in this scheme, recurrent selection was shown to be useful method to increase population mean and maintaining genetic diversity.

Holbrook et al. (1989) used a recurrent selection program that began with the formation of a population (RS4) with genetic potential for high yield and high seed protein. The C0 (cycle 0) test population originated from multiple matings of 10 high protein lines. Seven or eight matings for each parental combination resulted in 234 F_1 hybrids. The high protein parents were F_3 lines from C7 of population 1A in a recurrent selection program for increasing seed protein defined in Brim and Burton (1979). A restricted index (suggested by Holbrook et al., 1989) that allows maximum genetic gain in one or more traits while restricting genetic gain to zero in another set of one or more traits was followed.

$$I = \text{Yield} - \left(\text{Var}G_{\text{Y-P}}/\text{Var}G_{\text{P}}\right) \times \text{Protein}$$

where I is the index, $\text{Var}G_{\text{Y-P}}$ is the estimated genetic covariance between yield and protein, and $\text{Var}G_{\text{P}}$ is the estimated genetic variance of protein.

The 29 S_2 families were randomly intermated through hand emasculating and pollination, and the seeds were used to generate S_1 families for the next cycle of selection. Modified pedigree selection was applied to S_1 families chosen in the first restricted index selection cycle. Two additional cycles of intermating were performed and lines were selected. Two most desirable lines were bulked for further testing and release as cv. "Prolina" (Burton et al., 1999). Both seed yield (cycle 2 parents had 171 kg/ha more yield than cycle 0) and total protein (cycle 2 parents had 93 kg/ha more protein than cycle 0) were significantly increased after two cycles of index selection, while seed protein concentration was unchanged. Authors postulated that direct selection for yield is expected to result in significantly lower protein concentration in later selection cycles; therefore, it was proposed that restricted index selection as an attractive method in conjunction with recurrent selection as a long-term approach to increase yield while maintaining high protein concentration.

In the previous studies listed above, intermating of selections was done using hand pollination which is time consuming and tedious as well. Genetic male sterility has been used to facilitate random mating among selected progenies for genetic recombination and reselection. The spontaneous male sterile allele (ms_1) was reported in soybean by Brim and Young (1971). All seeds production on male sterile plants results from pollen contributed by a male fertile plant (Ms_1Ms_1 or Ms_1ms_1) through insect vectors (various bees). The ms_1ms_1 plants are male sterile as a result seed set is low and pods contain only one seed which is larger than seeds of normal plant of the same genetic background. Male sterile plants show incomplete senescence at maturity; therefore, they are easily distinguished from male fertile plants. Brim and Stuber (1973) described its use in recurrent selection program. They listed variations of a basic scheme, depending on selection objectives, heritability of the character or characters under consideration, selection unit, and availability of greenhouse or winter nursery facilities. They proposed an outline of a basic scheme:

Generation 1. Allow natural pollination to intermate population segregating for male sterility. Seed is harvested from only male sterile plants in this generation. Thus, male sterile types are maintained automatically in the population without the need for reintroduction of the male sterility gene, even after repeated cycles. If deleterious factors associated with male sterility arise, remedial action needs to be taken.

Generation 2. Advance (by natural selfing) progenies of male sterile plants from generation 1 to provide adequate seeds for field testing.

Progenies of male sterile plants from the intermating block will consist of both fertile and sterile plants. Proportions of each type will depend on the frequency of the male sterile allele in the intermating population. The primary purpose of this generation is to produce sufficient selfed seed on male fertile plants for replicated trials, in generation 3. Seeds produced on the generation 2 fertile progeny of a plant harvested in generation 1 will form an entry that is tested in generation 3. The length of selection cycle can be reduced by the usage of winter nurseries or greenhouse facilities. The selection units can be S1 progeny (if seeds are harvested from a single fertile plant chosen randomly from the progeny of each generation 1 plant), or selfed half-sib family (if seeds are harvested from all or several fertile plants from the progeny of each generation 1 plant) that provides large seed amount for testing and is operationally simpler.

Generation 3. Test progenies of male fertile plants produced in generation 2. Remnant seeds from parents of selected progenies are composited for planting the crossing block (generation 1) of the next cycle of intermating. The total number of plants needed in the intermating block will depend on the source of seeds composited, because if inadequate seed amount are obtained in generation 2, seed produced on male fertile plants in generation 3 test progenies can be substituted changing the proportion of male sterile plants in next cycle necessitating changing population size.

Burton and Brim (1978) used three cycles of recurrent mass and half-sib family selection to increase soybean seed oil percentage in a population segregating for male sterility, facilitating intercrossing of selected individuals and production of half-sib families. Seed oil increased linearly from 18.4% to 19.5% at an average rate of 0.3%/cycle leading them to conclude that both recurrent mass selection and within family selection contributed to the observed gain.

Use of single seed descent with cyclical selection procedures

Brim (1966) presented a line extraction system, now popularly called the SSD method. The SSD is a method of sampling, frequently used in autogmous species to assure that the range of genotypes in original population also will be present in the future generations. Essentially, this method helps to: (i) maintain the broadest possible representation of genotypes in the base population (genetic variation of the generation of maximum variability) until selection is practiced in some inbred generations (when sufficient homozygosity is reached), and (ii) increase the genetic variation (maximum among line additive genetic variation) among selfed progenies (such as in, F_4, F_5, or F_6) as a result of selfing or inbreeding.

The SSD can be used as a supplement to cyclical selection in autogamous species. Compton (1968) discussed how the obstacle of making large number of crosses for interpollination/intermatings (recombination phase of selection) can be reduced in recurrent selection programs by the use of SSD in self-pollinated crops. He presented a procedure that allows breeders of self-pollinated crops to conduct selection recurrently in population without large numbers of pollinations. Recurrent selection can be conducted with only 200–400 pollinations. The procedure is based on the use of SSD method, but each derived line being a descendent of a different F_1 rather than from different F_2 plants.

Payne et al. (1986) reported on gain of grain yield after the first three cycles of recurrent selection in oats. They selected 12 parents for high grain yield with diverse pedigree and diverse maturities for cycle C_0 which were intercrossed to provide 66 crosses. SSD was used to advance F_2 through F_4 followed by a year for yield increase. The highest yielding line from each of 21 best yielding crosses was selected to establish 21 parents for cycle C_1. Similar procedures were used to develop cycle C_2 and C_3 followed by replicated trials were conducted for 2 years along with 12 parents of cycle C_0 and 21 parents of cycle C_3. Gain in yield of 3.8% per cycle for three cycles was obtained. Almost 1% gain per year was found over cycle C_0 parents. This was a rapid gain since the parents of cycle C_0 were higher yielding genotypes.

Crops with doubled haploid production can utilize the same scheme as described for SSD.

Integration of recurrent selection with genomic selection

More recently a two-part breeding strategy was suggested (Gaynor et al., 2017): part 1—product development component that is aimed at development and screening of homozygous lines, and part 2—population development component aimed to increase the frequency of favorable alleles through rapid recurrent genomic selection. The two-part strategy uses genomic selection (GS) to perform recurrent selection on F_1 plants in the population improvement component and to perform selection in either the preliminary yield trial (PYT) or head-row stage of the product development component. Using simulation, authors showed that increased genetic gain can be expected with the use of recurrent genomic selection compared with conventional genomic selection and no-genomic selection scenarios. These were achieved due to rapid cycling and use of headrow data (from genomic prediction) as parents compared with waiting longer until a detailed characterization is completed. This follows the principles of genetic gain, where more rapid and steeper genetic gain is achieved with earlier cycling of best performing lines as parents. However, Gaynor et al. (2017) made 100 biparental population from 50 parental lines in the crossing (of the 1225 possible parental hybridization) using optimal contribution selection (Woolliams et al., 2015) approach for parental crossing combination decisions. While approach presented by Gaynor et al. (2017) is not "recurrent" selection *per se*, as it does not follow the three steps of intermating, evaluation and selection; however, it can be considered as recurrent selection method because the objectives match: to systematically increase the frequency of desirable genes in a population to create and enhance opportunities to extract superior genotypes. Empirical evidence is now needed to determine whether this approach works broadly, and further refinement may be needed to extend the application in recurrent selection. While a traditional pure line development breeding program uses longer cycles but are essentially following a recurrent selection type scheme - (a) make crosses among superior parents, (b) advance generations using a breeding method, (c) testing and evaluation to select superior lines for release as cultivar or develop elite lines, and (d) use lines from step c to generate the next cycle of crossing, advancement and selection. Intermating is not followed in the same manner as described in recurrent selection schemes; therefore, it should not be confused with recurrent selection methods.

The GS method can be integrated in a recurrent selection method outlined by Beeck et al. (2008), where they proposed a trait improvement in pea for higher heritability traits. However, with GS, lower heritability traits can be selected for intermating using the same method for self-pollinating crops. The method proposed by Beeck et al. (2008) combines a yearly cycle of crossing with selection on F_1, followed by selfing and further selection in F_4 lines derived through SSD. If GS is integrated in this system, one scenario can be as follows: the cycle starts with 1000 heterozygous individuals, and after making selection for high priority traits, the top 10%–20% (depending on desired selection intensity and/or comparison to checks) are crossed in a pairwise manner generating 50–100 breeding families. In each breeding family generate 20–40 plants, selection can be made on single plants using GS and phenotypic selection at 10%–20% selection intensity leaving 200–400 F_1. These will be selfed and advanced to F_2 (maintaining ~50–100 F_2 per breeding family), where another round of GS (and or phenotypic selection) can be practiced, and through SSD or single pod descent generate F_4 families (progeny rows). Phenotypic and GS will be practiced and the remaining lines can be utilized to start the next cycle by first crossing the 50–100 elite lines in pairwise crossing to create heterozygous F_1. However, the selected F_4 lines are also moved forward for regular testing to develop cultivars. This approach will achieve: (1) cultivar development and (2) rapid recurrent selection for multiple traits with varying heritabilities. The numbers and selection intensity provided are just an example, and the actual numbers will depend on the crop and breeding program's infrastructure. The GS models need to be robust and accurate, and continually improved in the breeding program for a successful outcome.

Advantages of recurrent selection

1. It provides greater opportunities for recombination and accumulation of desirable genes. It is made possible by restoring heterozygosity through intermating of selected plants in the segregating generation.
2. Due to repeated intermating, recurrent selection also provides additional opportunity to break linkage blocks. As a result, the frequency of genetic recombination in a population increases, thereby maximizing the exploitation of genetic variability. This approach is more relevant when unadapted or plant introduction is used to bring certain useful traits into a breeding program.
3. The method can be useful in field crops including cereals and grain legumes for improving quantitative traits.

Limitations of recurrent selection

1. In some of the autogamous species such as grasses and grain legumes, flowers are small and fragile, hence hand emasculation and pollination are tedious and time consuming. Therefore, few seeds are obtained per cross. As a result, for intermating, sufficient seeds are not available. These limitations could be overcome by the use of male sterility facilitated crossing but it still increases complexity, and it is not conclusively shown that recurrent selection approach will be equal or better and more practical in deployment than traditional methods of generation advancement and selection.
2. Identification of superior plants in the F_2 generation is difficult for quantitative traits, for example, yield.
3. Repeated intermatings prolongs duration of breeding program. Therefore, time taken to develop a variety through recurrent selection would always be greater than even pedigree selection.
4. There exists a possibility of breaking linkages between genes for desirable traits. Therefore, recurrent selection (specifically, the intermating part) is not an ideal strategy for cultivar development where most crosses are elite x elite types, but rather it is useful for germplasm development.

CHAPTER

Synthetic and composite varieties

18

Abstract

Synthetics are early or advanced generations of cross-fertilized seed mixture of desirable parents selected on the basis of general combining ability, whereas the composites are produced by mixing together of the breeding materials on the basis of some desirable agronomic traits. Synthetic and composite varieties utilize appreciable amount of heterosis due to additive genetic variance. These cultivars are favored in fringe/marginal areas where soil fertility is low and management practices are also less specialized. Synthetic and composite varieties are propagated for a limited period and synthetics can be reconstituted from the parental stocks. Synthetics are particularly useful in forage crops, where hybrids are either infeasible or uneconomical to create. This chapter includes: general outline of producing synthetic varieties, genetic basis of synthetic varieties, synthetic varieties in forage crops, tests to measure combining ability, broad- and narrow-based synthetics, applications of synthetic varieties, merits, and limitations of synthetic varieties. Development of composite varieties, their merits, and limitations, are explained.

In cross-pollinated crops, the focus of breeders is on population improvement, whereas in self-pollinated crops, the focus is on single genotype for genetic improvement. However, in numerous cross-pollinated crops (along with several self-pollinating crops), target cultivars are hybrids, where inbred lines are first developed followed by crossing them across heterotic pools or male/female reproductive groups to create hybrids. The methods of population improvement are covered in Chapter 16 (Population improvement), and hybrid cultivars are covered in Chapter 19 (Hybrid varieties). In cross-pollinating crops, another important types of cultivars are synthetic and composite varieties.

Synthetic cultivars consist of a mixture of heterogeneous and heterozygous individuals; parental lines are generally clones or inbred lines. These parental lines are maintained so that synthetic cultivars can be reconstituted when needed. These parental lines (clones or inbreds) are assessed for their general combining ability (GCA), and lines exhibiting superior combining ability are crossed in a polycross configuration to produce S_0. The S_0 plants are allowed to intermate to produce S_1, which in the case of asexually propagated crops such as alfalfa, can

Plant Breeding and Cultivar Development. DOI: https://doi.org/10.1016/B978-0-12-817563-7.00005-2

be sold as a synthetic cultivar. Sometimes, S_2 seed may be sold as a commercial cultivar, and mostly this choice is driven by the amount of seed that can be produced and expression of heterosis. Because maximum heterosis is observed in the S_1, it is a more favorable generation for cultivar development; however, this decision is made due to crop reproductive biology and ability to produce sufficient seed for commercial sales. In annual crops such as maize, asexual propagation is not feasible; therefore, the progression from Breeder Seed to Foundation Seed to Certified Seed production is from the S_2 to S_3 to S_4 generations, respectively.

A synthetic cultivar differs from an open-pollinated variety (OPV), which is developed by mass selection. A cultivar developed by mass selection is composed of genotypes bulked together without having undergone preliminary testing to determine their combining ability. This makes an open-pollinated cultivar the same as a landrace cultivar. Hybrids are preferred over synthetic cultivars in crops where hybrids can be created economically and commercialized. However, in crops that show heterosis, but in which hybrid production is difficult (infeasible and/or uneconomical), synthetics are important and preferred. Synthetic varieties are known for their hybrid vigor and for their ability to produce usable seed (in seed propagated crops) for succeeding seasons. Because of all these advantages, synthetic varieties have become increasingly favored in the cultivation of many species, for example, forage crops. The procedure of synthetics and composites, their genetic basis, merits, and limitations are described in this chapter.

Synthetic varieties

For the first time the commercial utilization of synthetic varieties (cultivars) was given by Hays and Garber (1919) in maize. They suggested the production of maize varieties through the recombination of several selfed strains. The advantage of synthetic varieties would be that the farmers can save their own seed for planting the next crop. Such varieties would be useful in areas of crop cultivation in marginal soil and marginal areas, which are poor in nutrient status. The early synthetic varieties in maize yielded about the same as that of OPVs; however, this is a reflection of relatively early stage of hybrid breeding and not due to genetic potential. Hays et al. (1944) reported the performance of synthetic varieties made up of parental lines selected on the basis of GCA. They evaluated $n(n-1)/2$ single crosses of 20 inbred lines of maize and synthesized a variety from eight inbred lines which showed the best performance in single crosses. Equal number of seeds of each of the 28 single crosses from eight inbred lines were bulked together and planted in isolation plot and seed were harvested. No selection was practiced for ear or plant characteristics. The variety was maintained in isolation in subsequent years. The synthetic variety yielded equal to Min-hybrid 403 and was superior to OPV. This showed the importance of GCA of the selected material in producing synthetic varieties (Allard, 1960). Synthetic varieties could also be produced from short-term inbred lines after their evaluation through top crosses (Jenkins,1940). This procedure is known as "recurrent selection of GCA." This procedure utilizes a broad genetic based tester. The important steps are (1) isolation of one generation selfed (S_1) lines, (2) evaluation of these lines for yield and other traits using top crosses, (3) intermating the superior lines to produce a synthetic variety, and (4) above steps could be repeated after one or two generation of open-pollination.

A synthetic cultivar (variety) may be defined as an advanced generation of open-pollinated (random mating in all combinations) seed mixture of a number of genotypes (parents) selected based on their GCA. The genotypes (parents) could be inbred lines, clones, hybrids, OPVs, etc. that have been tested for their GCA. The genotypes that combine well with each other in all combinations are included in the synthetic variety. A parent that has high GCA produces high performing progeny when mated with a series of other parents. Consequently, a commercial synthetic variety is an advanced generation of a population initiated by crosses among selected parents (based on GCA) and multiplied by a number of random outcrossing or intermating in isolation. The seed from these intermating is used as synthetic variety (see, Wright, 1981).

In other words, synthetic variety is derived after a set number of random mating generations and are progeny of intermating a limited number of select constituents (chosen due to good GCA). In the first generation, it is considered a random mated population of chosen founders (i.e., parents). As previously explained, parents can be inbreds, clones, or OPVs. These synthetic varieties can be seed or vegetative part propagated, but breeders need to continually maintain them. The pollination is controlled in seed production and maintenance of both synthetics and composites, so that frequency of genes of the selected materials is maintained. The prior testing of hybrid performance distinguishes synthetic with that of composite. Due to the nature of how composites are created, they have greater variability than synthetics.

Synthetic cultivars are preferred in cross-pollinating species where heterosis is expressed, and is foundational for the success of synthetic variety. However, due to the open pollination mating, selfing also happens (except in self-incompatible plant species) therefore, heterosis is not fully exploited. Farmers can use a synthetic cultivar for several generations (as an open-pollinated population), but once inbreeding depression causes yield reduction, farmers need to obtain seed from the breeding institution or company. Therefore, synthetic cultivars are reconstituted regularly by the breeder.

The panel of parental lines can be assembled from a previous synthetic cultivar or from other experimental populations. Parental lines can consist of different clones (forages) or inbred lines (in seed propagated crops, eg., maize). Clones will be highly heterogeneous and heterozygous, and each clone will be unique. These clonal lines are used to establish a *source nursery* with several thousand individual plants generally grown in a space-planted grid system to reduce environmental variance and enable meaningful comparisons between experimental clones amongst each other or to a check within smaller grids. A breeder selects for highly heritable traits, such as disease reaction, and morphological traits. Once the superior clones are identified, they are grown in a *polycross nursery* to either facilitate random pollination among clones or, with careful arrangement, to facilitate an equal chance for each clone to contribute pollen to other clones. Clones may be replicated to ensure uniform pollination. Further evaluation may be done, and an equal amount of seed is harvested per clone per replication. In a perennial species, clones may be grown for more than 1 year, and seed may be harvested each year. This polycross nursery is used to produce seed for progeny testing in performance tests. In corn, replicated testing can be done at several locations in 1 year; in forages and perennials, testing can be done ≥ 1 years in several locations and can be replicated. Based on the progeny testing, superior clones from polycross nursery are identified and crossed to each other to produce the synthetic. Syn1 or Syn2 may be released as a synthetic cultivar in a forage species (with clonal propagation). In maize, two or three more rounds of pollination may be needed to obtain sufficient seed for the commercial launch of a synthetic cultivar. *Half-sib selection* is widely used for breeding perennial forage grasses and legumes. A polycross mating system is used to generate the half-sib families from selected clones maintained vegetatively. The families are evaluated in replicated rows for 2–3 years. Selection of traits with high heritability, for example, oil and protein content in maize, is effective.

The difference between a synthetic and an open-pollinated cultivar is the ability to reconstitute the seed in a synthetic because the parents (inbred, clones, and hybrids) are used in a predetermined manner and configuration. In an OPV, the original population cannot be created because there is no control on the parent configuration. After a limited number of generations, seed needs to be reconstituted for a synthetic using the breeder-selected parental stock; while in OPV, random mating happens in each generation and the population can be propogated indefinitely although performance will deteriorate with each generation.

General outline of producing synthetic varieties

The general outline of production of synthetic cultivar in maize is as follows.

Step 1: the large number of germplasm resources of broad genetic base are screened in nursery for biotic and abiotic stresses and evaluated for desirable agronomic traits in replicated trials. Several of the superior lines from above trials are selfed simultaneously and some 10–20 lines could be selected according to the breeding objectives.

Step 2: these selected lines (10–20) are evaluated for their GCA (as per the mating design chosen by the breeder); and based on the GCA and their *per se* performance, 6–10 best parental lines are selected for intermating.

Step 3: equal amount of seeds from each of the selected lines (parents) that combine well with each other are mixed (Syn 0) together and seed is planted in isolation. The open pollination would produce the crosses in all possible combinations. Alternatively, all possible single crosses could be manually

made in the crops like maize, if feasible. The seed is harvested in bulk. This step is the synthesis of a synthetic variety.
Step 4: the composite seed of step 3 is planted in isolation plot. The generation is known as Syn 1, which is equal to the F_1. If enough bulk seed of Syn 1 generation is available it could be used for distribution.
Step 5: the seed of Syn 1 generation could be multiplied for one or two generations in isolation (Syn 2 or Syn 3). This seed may be distributed for commercial cultivation.

The seed of the synthetic varieties is maintained through open pollination. It would be desirable that synthetic varieties are reconstituted (resynthesized) from parental lines at regular interval to ensure hybrid vigor is maintained and not diluted by selfing that will happen in each cycle of crossing.

Genetic basis of synthetic varieties

Allard (1960) equated mathematical relationship on the performance of synthetic varieties given by Wright (1922) as:

$$\hat{F}_2 = ((\overline{F_1} - \overline{P})/n)$$

where $\hat{F}_2$ is the estimated performance of F_2 generation, $\overline{F_1}$ is the average performance of all possible single crosses among the parental lines involved, $\overline{P}$ is the average performance of the parental lines, and n is the number of parental lines involved.

This implies that performance of synthetic varieties would depend on three factors, the number of parental lines (n) involved in the synthesis of the variety, the mean performance of the parental lines ($\overline{P}$), and the mean performance of all possible F_1 hybrids ($\overline{F_1}$) among the parental lines. When F_1 hybrids are intermated among themselves, a single cross in the F_2 generation will lose 50% of excess vigor of the F_1 over the parents. If 3, 4, 5,..., n lines are involved as the parents of the synthetic variety; 1/3, 1/4, 1/5,..., $1/n^{th}$ of the vigor will be lost, respectively. If all assumptions of Hardy-Weinberg law hold true, there would be no further decline in vigor in succeeding generations because zygotic equilibrium for any gene is reached in one generation of panmixis.

A close agreement was noted between predictions made from Wright's formula and actual observations based on the studies conducted on maize in the United States. Of the three factor listed above, the yield of a synthetic variety could best be increased from combination of relatively high yielding parents; and five to six high combining parental lines would be optimum for constituting a synthetic variety. Increasing number of lines would improve yield of the synthetic variety sacrificing high combining ability. Low yield of inbred lines is a limiting factor to their use in the formation of synthetic variety. However, use of non-inbred lines or lines developed from limited inbreeding may overcome these limitations. One generation selfed (S_1) lines could also be used in developing a synthetic variety (Jenkins, 1940). These lines could be tested for combining ability in top-crosses for yield and other traits followed by intercrossing the better lines to produce a synthetic variety.

The above formula given by Sewall Wright is for describing the relative performance of a synthetic variety relative to the performance of the parents and number of parents used to produce a synthetic variety. However, this relationship is applicable when the crop species is a diploid and the parental genotypes are inbred lines. For non-inbred materials, such as in forage crops, "P" is replaced by "S_1," the selfed progeny of the parents; and with assumptions of allogamy, no selection, absence of epistasis and diploid genome. More details are provided in the next section.

Synthetic varieties in forage crops

Forage crops (legumes and grasses) have small but perfect flowers and some of these have self-incompatibility, hence production of hybrid seed is difficult. Many forage species are increased through asexual means. Selected plants from open-pollinated population can produce progeny as high yielding as F_1 hybrid from inbreds. In view of the above facts, it is necessary to rely on natural crossing to test combining ability of the plants. Expected yield of a synthetic variety in forage crops:

$Y = C - [(2k - 1)/k] \times [(C - S)/n]$, where $k = 1$ for diploid, 2 for tetraploid, 3 for hexaploid, and so on; where Y is the expected yield of a synthetic variety in equilibrium, C is the mean of all possible crosses between "n" parents, and S is the mean of all intraparent progenies (i.e., S_1 per se). $[(C - S)/n]$ is the inbreeding depression arising from intermating or related plants from Syn-1 to Syn-2.

Performance of a synthetic variety from a selected number of possible clones (non-inbred) can be predicted using the following equation:

$$Y_{\text{select}} = Y + \left[(2(n-1)/n) \times \sum \text{GCA}\right]$$

The number of clones used as parents should be larger to counteract inbreeding depression; however, larger number of parents will result in lower estimated GCA. This necessitates that breeders need to determine the optimum number of parents to use for developing synthetic varieties. Six to 12 parents are generally optimal but this may change for crops depending on levels of inbreeding depression and availability of high GCA clones for parental material. If all clones are used in creation of a synthetic, it will have similar performance as an OPV, which are always lower yielding than optimally created synthetic varieties.

Tests to measure combining ability

- *Open-pollinated progeny test*: this test relies on the progenies derived from open-pollinated seed produced on selected plants (or clones) by out crossing with other plants (or clones) present in the nursery.

- *Top-cross test:* the seed is obtained from selected plants (or clones) that are planted in alternate rows with a single tester (OPV or a broad-based synthetic). The test cross seed may consist of both selfs and intercrosses with other selected clones. Planting of more number of plants of tester variety compared to selected plants in nursery increases the proportion of top-crossed seed.
- *Polycross test*: it refers to the progeny from the seed of a line, which was subjected to outcrossing (random pollination) with selected lines growing in the same nursery (Tysdal et al., 1942). The efficiency of this test would depend upon the proper layout of polycross nursery to ensure random pollination. The selected clones are grown in isolation, and can be randomized in a small plot. Large number of replications (10 or more) is essential. An equal amount of seed of a clone from each replication is collected and bulked. This seed represents the seed of all possible single crosses among the clones included in the nursery. If pollen from all parents is used as a topcross tester, polycross and topcross tests will be equivalent and GCA estimates from the two tests will be highly correlated. Polycross test is most commonly used for evaluating GCA in forage crops when compared with other tests.
- *Single cross test:* every selected clone is crossed with a number of other clones by growing together (each pair of clones) in isolation. This could be obtained by enclosing the inflorescences of two different clones in same bag. If all possible, single crosses among n selected clones are attempted $[n(n-1)/2]$, i.e., diallel cross. The average combining ability of any clone can be calculated from single cross data as the mean performance of that clone in its crosses with all other clones with which it was crossed.

General procedure of synthetic variety production for forage crops particularly those propagated clonally

Step 1: some 100–200 superior individual plants (clones) from broad-based source nursery are identified based on biotic and abiotic resistance and other agronomic traits as desired by the breeding objectives.

Step 2: from step 1, seed of each of the 25–50 most desirable plants (or clones) are harvested separately and used for planting in polycross nursery to generate seed for progeny testing. The seed from each clone is harvested separately.

Step 3: the progeny test (GCA) is conducted to evaluate yield and other traits according to the breeding objectives. Some 5–10 best performing clones (from step 1) are identified and will be used as parents in a synthetic in a polycross nursery.

Step 4: the selected clones are randomly planted in vegetatively propagated nursery in a small isolated plot for cross-fertilization among themselves. The seed is harvested in bulk and is called Syn-0 seed. This can be done in cages also. This is called Syn-0 generation because it reflects the parent generation of a synthetic variety.

Step 5: equal amount of Syn-0 seed from each parent is mixed and planted to ensure random mating in field. The bulk seed is harvested from Syn-0 to produce Syn-1 seed for commercial multiplication.

The cross-fertilized seed from the Syn-1 synthetic variety is known as Syn-2 seed. Between Syn-1 and Syn-2 generations in forage crops (both diploid and polyploid cross-pollinated), generally the maximum yield reduction occurs. This is similar to maize synthetics based on inbred lines. The yield of Syn-1 would be 10%–12% higher than in the Syn-2 or Syn-3 generations. Since the total number of synthetics that can be produced with few clones is quite high; breeders are dependent on GCA estimates to pick parental clones. Total number of possible synthetics can be deduced from the formula ($2^n - n - 1$) (Wricke and Weber, 1986). The two general types of synthetic cultivars are; broad based and narrow based.

Broad-based and narrow-based synthetics

Broad-based synthetics are based on a larger number of selected clones, while narrow-based synthetics are based on a relatively few selected clones. There is no specific number of clones that define the two types; however, broad-based synthetics generally would have between 20 and 100 parental clones and narrow-based synthetics generally have less than 20 clones. The parental clones of broad-based synthetics generally are selected only by phenotypic evaluation, while the parental clones chosen for narrow-based synthetics generally are based on phenotypic and genotypic evaluation. It is important to balance combining ability against inbreeding depression in narrow-based synthetics. When fewer clones are used to initiate the variety, inbreeding depression in advanced generations probably will be higher.

The merit of broad-based synthetics over narrow-based is an increased stability over environments. However, the breeding of board-based synthetics requires less testing for combining ability than does the breeding of narrow-based ones. Therefore, to identify these superior parents, precise evaluation for combining ability must be made. An evaluation sequence is recommended, beginning with phenotypic selection. It is not essential to select a large number of plants, provided that several adapted sources of germplasm are sampled. Phenotypic selection should be followed by clonal evaluation at more than one location so that the Genotype × Environment interaction can be evaluated.

Applications of synthetic varieties

In the United States and Europe, the number of synthetic varieties of forage crops (timothy, orchard grass, red fescue, ryegrass, grasses, and legumes) have been developed and released for commercial cultivation. Alfalfa is a self-incompatible and is naturally cross-pollinated by bees and other insects. Large number of synthetic varieties have been developed and released. Most of these released cultivars are broad based and developed from a large number of parents (>40).

Genetically, these consist of heterogeneous populations of heterozygous individuals. More recently, narrow-based synthetics are becoming more prevalent, likely because better phenotypic and genotypic selection methods are available as well as germplasm is now better characterized. While broad-based synthetics can mask poor performing clones, narrow-based synthetics need to use well-characterized clones as parental stock.

Merits of synthetic varieties

1. Seed of synthetic varieties can easily be produced from open pollination and maintained by the farmers. Therefore, seed of synthetic varieties is less costly compared with the hybrids.
2. Synthetic varieties could be recommended for areas (and in crops) where seed industries are not well developed.
3. When compared with F_1 hybrid the synthetic varieties have wider genetic base; therefore, can adapt early to the different growing conditions and are stable over the years and seasons.
4. It can also serve as a reservoir of germplasm.

Limitations of synthetic varieties

1. Synthetic varieties in general are lower yielder than hybrids, but sometimes equal yielders to the OPVs; and therefore, at times have limited scope and usefulness.
2. Synthetic varieties primarily exploit additive genetic variance.
3. Average yield of synthetic varieties may not be good if parental lines of cross combinations do not have high GCA.

Composite varieties

The composite varieties are produced by mixing seeds of phenotypically superior lines followed by their open pollination to produce crosses in all possible combinations among the mixed lines. Like the synthetic cultivars, the concept of composite varieties is also based on the exploitation of additive genetic variance in a population. In a composite variety, mixture of genotypes from several sources is maintained in bulk from one generation to the next using open pollination. A composite may consist of all types of hybrids, populations, and other less similar genotypes. The components of composite varieties should have common characters such as maturity duration, plant height, and tolerance to diseases that are important for the uniformity of variety. Ten to 20 phenotypically superior entries may be included in a composite and new entries could be added any time. For the development of composite varieties, the F_1 and F_2 generations of intervarietal crosses are generally tested at two to three locations in a particular region or area. Based on the F_2 performance, the crosses that do not show significantly decline in yield are selected and their seed is composited in equal amount which are further tested in advance generations for release. The component lines/varieties are generally not tested for GCA. Open pollination is followed to produce crosses in all combination among the mixed lines. Unlike synthetic varieties, composites varieties can not be reconstituted because the parental sources and the their contributions are not fixed.

General outline of composite variety development

The general outline of composite variety development in the crop like maize is described as follows.

Step 1: parents of composite crosses could include varieties, advance generations of single crosses, double crosses and other forms of hybrids, which are evaluated at two to three locations for their *per se* performance. The materials which are unadapted, low performers, highly susceptible to major biotic and abiotic stresses, and do not meet the locally acceptable grain characters are rejected.

Step 2: the desirable parents selected from step 1 are randomly pollinated or intermated, or crossed in all possible combinations (excluding reciprocals). Alternatively, instead of open pollination to create and maintain the variety, chain crosses could be attempted. The number of chain crosses would be equal to the number of parents. If eight parents are involved A to H, the F_1 crosses would be A × B, B × C, C × D, D × E, E × F, F × G, G × H, and H × A.

Step 3: the F_1 and F_2 generations obtained from step 2 are evaluated along with parents and standard check (commercial variety) in replicated yield trials at two or more locations.

Step 4: candidate composite variety(ies) that are higher yielding in F_1 than the best parents with little or no reduction in F_2 and possess other desirable plant and grain characters are selected.

When open pollinated, seed are harvested to create the composite variety. If manual crossing is used (modified composite), at harvest equal number of F_1 seeds are bulked from each cross to create the composite variety. Therefore, the advance generation seed from such crosses is the composite seed. Subsequently, the composite varieties are maintained by open pollination in isolation and variety may be grown in half hectare (or in an area as desired for seed production quantity of the composite variety) in isolation for three to four cycles to avoid inbreeding depression and maintain the desired level of genetic variability in the composite variety.

Merits of composite varieties

1. Development of composite varieties and their seed production is simple and less time consuming than hybrids.

2. Farmers can save their own seed for growing the next crop in isolation coupled with mild negative selection that is, removal of undesirable plants.
3. Composite varieties due to their wider genetic base are less vulnerable to biotic stress(es). Such varieties can also serve as useful base population for the development of superior inbred lines for the production of hybrids.

Limitations of composite varieties

1. As the parental lines are not tested for their combining ability, the yield performance cannot be predicted.
2. Unlike synthetic varieties, composite varieties developed through open pollination cannot be recreated as the parental source is not always known and their proportion is not equal.

Composites are less technically demanding compared to hybrids and synthetics, and are a reasonable compromise if synthetics cannot be generated. They will be better than OPVs in yield, if proper methods are used to create a composite variety.

CHAPTER

Hybrid varieties

19

Abstract

Hybrid cultivars are one of the most important types of cultivars, particularly for crops that have majority breeding efforts happening in the private seed sector. A hybrid cultivar is the F_1 offspring of a planned cross between inbred lines, cultivars, clones, or populations. Hybrid may be the product of a single cross, a three-way cross, or a double cross. Absolute requirements for a hybrid cultivar are superior performance over the parents (heterosis) and an ability to economically generate seed for commercial seed sale. This chapter contains: steps in the development of hybrid varieties, hybrid seed production, production of hybrids, including through the use of cytoplasmic male sterility. We also include information on hybrid varieties in horticultural crops and in self-pollinated crops, including mechanisms that promote the development of hybrids, such as male sterility and self-incompatibility. Approaches to overcome self-incompatibility are described, along with its implications in plant breeding.

The basic investigations on inbreeding and outbreeding were carried out in maize by East (1908) and Shull (1908, 1909). Shull (1908) conducted experiments to study genetic variability within an open-pollinated maize cultivar, and concluded that the cultivar consisted of a number of heterogeneous genotypes, which upon inbreeding can produce a number of pure lines. These pure lines (commonly and more appropriately are called inbred lines in cross-pollinated crop species) in single crosses, restored vigor. Shull (1910) suggested that, from several crosses made using pure lines of the heterogeneous population, the most productive cross could be determined. However, these findings were not accepted widely at the time due to poor yielding ability of pure lines, and the cost of hybrid seed was not acceptable to growers. In maize, the suggestion of double cross hybrid rather than single crosses was made by Jones (1918), and this removed constraints of producing hybrid seed from pure lines, which were low yielding. The double crosses were tested in late 1920s and early 1930s and were compared with open-pollinated cultivars. In general, compared to open-pollinated varieties (OPVs), double cross hybrid had better standability, higher productivity, and were more tolerant to stress conditions. These findings laid the foundations of maize hybrid cultivars.

Plant Breeding and Cultivar Development. DOI: https://doi.org/10.1016/B978-0-12-817563-7.00015-5

A hybrid cultivar is the F_1 offspring of a planned cross between inbred lines, cultivars, clones, or populations. The hybrids may be the product of a single cross, a three-way cross, or a double cross. Absolute requirements for a hybrid cultivar are superior performance over the parents (heterosis), and an ability to economically generate the seed for commercial seed sale. For realizing its high yield potential, F_1 hybrid seed is produced and used afresh each year, enabling each seed sold is in a heterozygous state and expressing full heterosis. In the case of maize, farmers moved from OPVs to hybrids due to several advantages offered by hybrids, including higher yield and standability.

The development of hybrids in maize is easier due to the ability to follow a configuration of males and females for seed production; female rows are detassled, and seed is only collected from female rows. Therefore, the role of male rows is to serve as the pollen source with the requirements that the male must combine well with the female [higher specific combining ability (SCA)], must possess good pollen shed, and must nick well with the female, that is, must flower close to the flowering time of the female. Similarly, female inbred line parent should combine well with that male, and have suitable agronomic characteristics including high yield potential for good hybrid seed production.

Steps in the development of hybrid varieties

Similar to other types of cultivar development program, breeder of a hybrid variety needs to follow the four steps - step 1: define objectives; step 2: form genetic base by creating segregating population (to exploit heterosis, breeders keep separate male and female heterotic pools); step 3: perform selection by testing performance of F_1 [inbred line crossed with tester(s)]; and step 4: create new crosses between suitable female and males with good combining ability, conduct yield trials (testing of experimental hybrids) and seed multiplication (inbred and hybrid).

Production of inbred lines

In cross-pollinated crops it is necessary to develop true breeding lines by self-pollination methods or doubled haploid (DH). Initially, inbred lines were developed from the heterozygous source material (population) such as an OPV, synthetic, composite variety, or advanced generation population of hybrid. For the development of high yielding hybrids, inbreeding should be taken up simultaneously in two or more genetically diverse base populations that are known to show high heterosis in cross combination in the F1 generation. In modern maize hybrid era, inbred (and more recently, DH) lines are developed from breeding populations created by crossing lines from within heterotic pools and then pairing (crossing) them with inbred or DH lines from a dissimilar heterotic pool to obtain maximum expression of seed yield. This implies that complementary heterotic groups are improved simultaneously, with new breeding populations made within each heterotic group. Inbred or DH lines developed from these crosses are crossed with tester inbred lines from complementary heterotic group(s). Selections are made on inbred lines for their suitability for male or female parent traits, and on hybrid performance from crosses with tester inbred lines.

For inbred line development, usually five to seven cycles of inbreeding (selfing) are considered desirable to achieve true breeding lines. With every additional cycle of inbreeding there is a loss of vigor that is more rapid in the first few cycles and stabilizes after five to six generations of inbreeding. However, the level of heterosis, that is, increased performance of the hybrid over the parents is largely independent of the degree of inbreeding. As a result of inbreeding, the inbred lines become more uniform and homozygous and the variation between the inbred lines increases. Doubled haploids do not suffer from further inbreeding depression from generation to generation.

An efficient method for rapidly developing inbred lines from the doubling of monoploids was suggested by Chase (1949) in maize. In this procedure, the desired maize population is crossed to a marker line. The kernels that do not exhibit the dominant expression of xenia were selected and grown in glasshouse. The chromosome number of the seedlings was studied. The doubling of the chromosome number through colchicine was attempted in selected plants. The resulting lines may possess the homozygous genetic component of the monoploid. The frequency of such homozygous diploids may be low, that is, 1/10,000 plants. In more recent times, DH system (through in vivo haploid induction and colchicine treatment for genome doubling) has become the predominant choice of inbred line production in commercial maize programs. This has primarily been adopted because the time required to produce inbred line can be reduced by 2–3 years. DH lines can be used more quickly in test cross to assess their potential as a parent in hybrid development. The DH process has been significantly improved since its early discovery days, but research is still on-going on the development of inducer lines and increasing induction rate to make the DH system continually more economically feasible. Haploid inducer lines with induction rates greater than 10% have been developed; however, there is still a genotype effect (i.e., induction rate vary depending on the genotypic background).

DH line development method also provide an advantage of better response to marker-assisted selection (and also genomic prediction and selection) because markers can be used to identify and select for desirable genes, at the haploid stage, which can be fixed upon chromosomal doubling. This provides an efficient and rapid tool to eliminate unfavorable genes and to enrich favorable genes to improve the genetic pool. These DH lines, once created, do not show inbreeding depression in the following generations. In the absence of spontaneous gene mutations or transpositions, that may cause certain deleterious influences and segregation, DH lines provide a powerful tool in maize breeding. Because only one round of recombination happens in the DH process, the breakage of desirable linkages is minimized; consequently, chances to break undesirable linkages are reduced. Therefore, breeder needs to recognize the need for optimum population sizes and may find larger population sizes (per breeding family) to be useful.

Testing of inbred lines

It is desirable to simultaneously evaluate inbred lines for their female and male suitability traits. For example, for inbred lines that will be used as a female, it is important to assess their yield *per se* along with agronomic traits and should have reasonable male characteristic (tassel and pollen); while inbred lines to be used as males should have good tassel architecture, pollen viability and longer time dispersal along with suitable agronomic traits, although it should also have reasonable seed yield for seed production steps. Good combining ability is a must for inbred lines. Inbred lines during the early generations of inbreeding (including S_1) can be evaluated, and the procedure is equally effective. It helps to eliminate the poor combiners in S_1, and larger populations of promising lines can be studied in later generation.

1. *Evaluation of per se performance of inbred lines:* it is necessary to know the *per se* seed yield performance or potential of inbred lines(s) for commercial hybrid seed production particularly for female lines. Inbred lines are grown in breeding blocks for further inbreeding or seed increase, and are visually evaluated for seed yield (potential) and other agronomic traits. Some of these traits may include ear placement, ear size, kernel fill, etc. The superior selected lines should also have adequate pollen shedding ability for seed production purposes, and resistance to the major diseases and insect–pests of the area. It is possible that in few examples, no reciprocal differences are present if a line is used as male or a female parent. However, in other cases, male inbred lines are evaluated for their utility as a pollen source parent. This is common in maize hybrid production, where different heterotic pools are used for male and female inbred line development. Regardless, a line used as female should produce high grain yield across environments (particularly where seed production plots will be grown), while the male parent should have good tassel and pollen shedding in those environments. For the proper nicking of the two lines and better seed setting (and for earlier detasseling of female rows) it is desirable that the female parent is 2–3 days earlier than the male parent. To summarize, it must be noted that both female and male inbred lines need to have adequate yield *per se* as well as tassel-pollen traits, for seed production feasibility; however, female heterotic group need a more rigorous selection for yield *per se*, while male heterotic group need a more rigorous selection for tassel and pollen traits.
2. *Evaluation of combining ability of inbred lines:* it is desirable that lines are initially evaluated for their general combining ability (GCA). The selected

Table 19.1 Calculation of general and specific combining ability from a North Carolina Design II in maize.

Female\male	W	X	Y	Z	Half-sib means
A	8.5	11.3	12.8	10.7	10.83
B	9.2	9.8	10.7	11.3	10.25
C	7.9	11.9	10.1	11.7	10.4
D	9.9	9.7	11.4	12.3	10.83
Half-sib mean	8.88	10.68	11.25	11.5	10.58

Yield unit is metric tons per hectare.
Example of calculation of GCA, Heterosis, SCA.
General combining ability (A) = half-sib mean of A − overall half-sib mean = 10.83 − 10.58 = 0.25.
General combining ability (W) = half-sib mean of W − overall half-sib mean = 8.88 − 10.58 = − 1.7
Heterosis (AW) = (yield of AW hybrid) − ((half-sib mean of A + half-sib mean of W)/2) = (8.5) − ((10.83 + 8.88)/2) = −1.355.
Specific combining ability (AW) = yield of AW hybrid − overall half-sib mean − GCA of A − GCA of W = 8.5 − 10.58 − 0.25 − (− 1.7) = −0.63.

inbred lines are subsequently tested in single cross combinations for their SCA (for NC design example, see Table 19.1).

A. *GCA*: in maize, generally breeding programs will estimate both GCA and SCA. This can be achieved by using a line × tester mating design by crossing experimental DH or inbred lines with testers. In the first stage of testing (first test crosses after inbred line development), testers may be used to generate GCA (by the use of two or more testers across the experimental inbred or DH lines). In the female heterotic groups line development, experimental lines are used as female and male tester will belong to the male heterotic group; while in the male heterotic group line development, experimental lines are used as female and the male tester will come from the female heterotic group, that is, reverse isolation. The male heterotic group scheme is developed using reverse isolation, due to the feasibility to use inbred lines as female with testers being pollen donors. Generally for testing the GCA of the inbred lines, top crosses (inbred line × tester crosses) are produced by growing the inbred lines in the ratio of four to six columns of female rows: two rows of male tester rows under isolation. Each isolation block uses just one tester (pollen source) to avoid pollen mixing. Female rows (experimental inbred lines) are detasseled and also visually evaluated for their agronomic performance. Natural pollination helps create hybrid seed on female rows. Based on their yield and agronomic traits, poor lines are rejected. The seed of top crosses harvested from female parents is used for evaluation in yield trials along with tester parent and commercial hybrids (eg., 5 × 5 or 7 × 7 or other simple lattice design, alpha-lattice) at multiple locations. Data are recorded on yield, yield components, resistance to major diseases (s) and insect–pest(s), and based on across-location data, the lines with superior performances in top crosses are retained for further evaluation. Estimation of GCA assists breeders to pick next set of testers and additionally allow for an enhanced understanding of more specific inbred combinations for potential commercial hybrids.

B. *SCA*: the line × tester mating design will allow breeders to estimate SCA. Additionally in several instances it is desirable that inbred lines selected on the basis of their superior GCA are divided into groups (representing separate heterotic groups). For example, each group could consist of 10 inbred lines and all possible 45 crosses are assessed for their SCA. From each group, one or more of the well established inbred lines already existing in the on-going breeding program are included. It is necessary that inbred lines in each group have similar maturity duration and adequate level of resistance to major diseases. These inbred lines are picked after extensive testing to fewer testers in earlier stages of hybrid yield testing. Crosses are made by hand pollination and full record of the parentage may be shown on the tassel bag. The seed of inbred or DH lines is maintained by sib- or self-pollination within the row. The single cross hybrids between a set of inbred lines, along with commercial hybrids are evaluated at four to six or more locations for 2–3 years. Number of locations may be substantially higher with well resources breeding programs. Data obtained from the single cross hybrids is used to determine the GCA of the parental lines and the SCA of the hybrids. If applicable, single cross data could be used for predicting the performance of double cross and three-way cross hybrids. The selected inbred lines can be used in the synthesis of various types of hybrids as described in the following pages and also summarized briefly in Table 19.2.

a. *Single cross hybrids:* the high yielding single crosses can be developed more efficiently because only two best inbred lines are involved. Also, high yielding single cross hybrids are more uniform and their seed production is easier than double cross and three-way cross hybrids. Therefore, this is the predominant type of hybrid in commercial programs.

b. *Double cross hybrids:* a double cross hybrid involves four inbred lines and their two single crosses. The seed production of a double cross hybrid is more economical, as a single cross hybrid is used as a seed parent. Because of this factor, the double cross hybrids were popular in the early days of maize hybrids. The performance of the double cross hybrid could be predicated from the mean of the single crosses, as explained later in this chapter.

c. *Three-way cross hybrids*: a three-way hybrid involves three inbred lines in its pedigree, that is, a single cross (A × B) and an inbred line (C). Due to its higher yield potential single cross hybrid is used as female parent and inbred line is used as male parent. The inbred line "C" should have good pollen producing ability. The performance of the three-way hybrid cross

Table 19.2 Examples of various types of hybrids.

Hybrid	Seed parent (female parent)	Pollen parent (male parent)
Single cross hybrid (A × B)	Inbred (A)	Inbred line (B)
Double cross hybrid [(A × B) × (C × D)]	Single cross hybrid (A × B)	Single cross hybrid (C × D)
Three-way cross hybrid [(A × B) × C]	Single cross hybrid (A × B)	Inbred line (C)
Multiple hybrid or six-way cross [(A × B) × (C × D)] × [(E × F)]	Double cross hybrid [(A × B) × (C × D)]	Single cross hybrid (E × F)
Multiple hybrid or eight-way cross [(A × B) × (C × D)] × [(E × F) × (G × H)]	Double cross hybrid [(A × B) × (C × D)]	Double cross hybrid [(E × F) × (G × H)]
Top cross hybrid (A × V)	Inbred line (A)	Composite variety (V)
Double top cross hybrid [(A × B) × V]	Single cross hybrid (A × B)	Composite variety (V)
Modified single cross hybrid or sister line cross hybrid [(A × A′) × B]	Cross among sister line (A × A′)	Inbred line (B)
Modified double cross hybrid or sister line cross hybrid [(A × A′) × (B × B′)]	Single sister line cross hybrid (A × A′)	Single sister line cross hybrid (B × B′)

could be predicted from the mean of single crosses as explained later in this chapter.

d. *Multiple cross hybrid*: the multiple crosses may involve six or eight inbred lines for producing six-way or eight-way crosses. Eight-way cross is produced by crossing a double cross with a double cross. The multiple crosses are useful in producing synthetic varieties rather than using them as F_1 hybrids commercially.

e. *Top-cross hybrids*: top crosses or inbred variety (A × V) is used for testing the GCA of the inbred lines. In addition, top crosses could be used as commercial hybrids. The inbred line (A) may be used as female parent and the variety (V) may be used as male parent as the variety provides viable pollen over a prolonged period of time, however, top cross could also be produced using variety (V) as female parent and inbred (A) as male parent, if the inbred has capacity to shed good amount of pollen.

f. *Double top-cross hybrids*: a double top-cross hybrid involves a single cross hybrid (A × B) and a variety (V) for its synthesis. The variety is used as pollen parent and single cross hybrid is used as seed parent.

Prediction of hybrid performance

All possible single crosses are made between selected inbred lines. The basis of selection of the inbred lines is *per se* performance and GCA. The performance of the double cross and three-way cross hybrids is predicted using yield data of the single crosses involved in these. The number of possible double and three-way cross hybrids

Table 19.3 Number of possible hybrids among a set of inbred lines.

Number of inbred lines used	Number of possible hybrids			
	Single crosses	Top crosses	Three-way crosses	Double crosses
4	6	4	12	3
6	15	6	60	45
10	45	10	360	630
15	105	15	1365	4095
20	190	20	3420	14,535
n*	$n(n-1)/2$	n	$n(n-1)(n-2)/2$	$n(n-1)(n-2)(n-3)/8$

increases rapidly with an increase in the number of inbred lines (Table 19.3). Therefore, the prediction of the hybrid performance is necessary to reduce the number of hybrids that may be selected for the synthesis and multi-location testing. A close relationship between actual and predicted performance has been observed.

1. *Prediction of double cross hybrid performance:* the performance of a double cross hybrid [(A × B) × (C × D)] may be predicted from the single cross yield data by one of the four methods proposed by Jenkins (1934) which are described briefly here.

a. *Method I:* the performance of a double cross hybrid is predicted based on the average performance of all possible six single cross hybrids among the four inbred lines as: 1/6 [(A × B) + (A × C) + (A × D) + (B × C) + (B × D) + (C × D)]. As per this method, the predicted performance for the three possible double cross hybrids: [(A × B) × (C × D)], [(A × C) × (B × D)], [(A × D) × (B × C)] among four inbreds would be same.

b. *Method II:* the performance of a double cross hybrid is predicted based on the average performance of the four non-parental single cross hybrids. In this method, the predicted performance of each of the three possible double cross hybrids would be different. Performance of [(A × B) × (C × D)] would be predicted as: 1/4 [(A × C) + (A × D) + (B × C) + (B × D)].

c. *Method III:* the performance of a double cross hybrid is predicted based on the average performance of the four parental inbred lines over a series of single crosses. For instance, if four inbred lines A, B, C, and D are crossed to other four inbred lines E, F, G, and H; the performance of the double cross hybrid [(A × B) × (C × D)] would be predicted as: 1/16 [(A × E) + (A × F) + (A × G) + (A × H)... + (D × H)].

d. *Method IV:* the prediction is based on the average performance of lines in top crosses, as: 1/4 [(A × V) + (B × V) + (C × V) × (D × V)]. The variety (V) is a broad-based variety used as tester parent.

Of the above four ways for the prediction of performance of the double cross hybrid, the method II is well accepted. In addition to above methods, the performance of a double cross hybrids could also be predicted from the three-way cross data. The predicted value of the double cross hybrid [(A × B) × (C × D)] is equal to the mean performance of the two three-way cross hybrids: (A × B) × C and (A × B) × D.

2. *Prediction of three-way hybrid performance:* The performance of the three-way hybrids may also be predicted from the yield data of the single cross hybrids. The performance of three-way hybrid (A × B) × C would be equal to the mean of the two non-parental single cross hybrids (A × C) and (B × C).

Improvement of existing parental lines or their replacement

For any positive increment in the yield of the hybrid for seed yield and stability of performance, requires a corresponding improvement of existing parental lines or the replacement of weaker lines by the superior ones.

1. *Improvement of available inbred lines*: if inbred lines of a commercial hybrid lack in one or more characters such as resistance to a specific disease or in protein quality, which may be improved through the use of one of the following methods.
   a. *Back cross method:* This method has been used in the past for improvement of cross-pollinated crops as well. The method is useful for the improvement of a simply deficient character. As the name suggests, F_1 hybrid among two inbred(s) is crossed back to one of the parents. General scheme(s) of operation for the backcross program for the transfer of a dominant and a recessive character is described in Chapter 13: Backcross method.

      This method has distinct advantages:
      - The frequency of the desired homozygotes is twice as high in the backcross method compared with the selfing method. Therefore, population size of the progenies required in the backcross program is much smaller than needed in the forward breeding based method.
      - The chances of breaking the close linkage between a desirable and undesirable gene through back crossing are higher because of frequent crosses to recurrent elite parent.
      - The level of attainment of homozygosity under backcrossing is similar to selfing.
   b. *Convergent improvement:* the method was suggested by Richey (1927) for increasing the *per se* productivity of inbred lines. It involves reciprocal improvement of two inbred lines of a single cross. The F_1 is crossed back to each of the parental inbred lines separately. Plants with desirable characters are selected in both series of backcrosses and used in developing next cycle of backcrosses. This process is repeated in each cycle of backcrossing. After three to four backcrosses, the lines are selfed to recover the recurrent parent.
   c. *Multiple convergent improvement:* this method was also proposed by Richey for improving the vigor and *per se* productivity of the parental

inbred lines of a hybrid, utilizing various diverse lines. It involves making separate repeated backcrosses with a series of new elite materials using the inbred line as the recurrent parent. In each cycle of backcross, selection is practiced for character for which the recurrent parent is deficient. After five to six backcrosses, recovered lines are merged together to extend the genetic base for further improvement of the line.

2. *Selection of more productive lines:* methods have been developed to identify inbred lines to replace existing ones in the pedigree of a hybrid. Some of these methods are described as follows.
   a. *Gamete selection:* it consist of pollination of inbred lines with a variety or a population with a goal of selecting gametes, that is, carriers of genes with which the combining ability of the inbred lines can be improved. Stadler (1944) proposed this method to develop a superior line from a heterozygous source (V) to replace an inferior line of a hybrid. The following steps are involved in replacing for instance line A of the double cross hybrid [(A × B) × (C × D)].
      Step I. F_1 crosses between selected plants of a desirable heterozygous source (V) and line A, using the latter as female parent are attempted (A × V). Any variation among the F_1 plants may be due to different gametes contributed by the variety (V).
      Step II. F_1 plants are self-pollinated and also use its pollen to cross to the tester cross, (C × D) as a female parent. Also cross tester (C × D) with line A to develop (C × D) × A.
      Step III. Yield trials are conducted with test crosses (C × D) × (A × V) and those crosses are selected which perform better than (C × D) × A.
      Step IV. Selfed progeny similar to the selected top cross is further selfed to derive the inbreds for replacing line A in the hybrid.

The frequency of desirable gamete and zygote would occur in the ratio $\sqrt{x}$ and x, respectively; hence, there are greater chances of identifying better gametes available in elite line x variety (A × V) cross rather than an inbred line.

   b. *Recurrent selection for SCA*: compared with gamete selection, the method of recurrent selection is more used in practical breeding. This scheme was proposed by Hull (1945). The method is used for isolation of better combining inbred lines and consists of the following steps:
      i. *Step I.* Some 200–300 plants are self-pollinated in a broad-based population and crossed to an elite inbred line as tester parent. Inbred line B of the single cross A × B is used as tester parent in case the inbred line A is to be replaced in the hybrid.
      ii. *Step II.* The top crosses are evaluated in yield trials including single cross (A × B) as one of the check entries. Inbred lines are also evaluated for their *per se* performance simultaneously in a separate nursery. Based on their top crosses and *per se* performance, inbred

lines are selected and remnant seed is used to reconstitute the population.

The above steps are repeated for two to three cycles of selection and suitable lines can be selected to replace the weak line in the hybrid program.

c. *Step ladder breeding for the hybrids*: this is an approach to improve the performance of a double cross hybrid by replacing a pair of inbred lines of one of the parental single cross hybrids. In this scheme, in two cycles, both single crosses are replaced to further improve the performance of double cross hybrid. The better of the two single cross hybrids (SC_1) is used as a tester parent to develop a series of three-way cross hybrids, which are tested against the standard double cross hybrids ($SC_1 \times SC_2$). The performance of three-way cross hybrids is used to predict the likely performance of a series of outstanding new double cross hybrids. Through this scheme, pairs of new superior inbred lines are used to make a superior single cross, SC_2 A which would replace single cross, SC_2. Furthermore, single cross, SC_2 A would be used to identify superior three-way cross hybrids and its performance could be used to find a replacement for SC_1.

The above schemes were developed during the times when double cross maize hybrids were still the norm; however, the concept explained above can be useful even in modern era for maize and other cross-pollinated crop species where single cross hybrids are common.

Seed production of hybrids

In case of hybrid varieties, the farmers require F_1 hybrid seed regularly. Therefore, a regular supply of certified seed to the farmers is ensured. The production of a double cross hybrid seed involves four distinct stages: maintenance of inbred lines, seed production of inbred lines, seed production of single cross hybrids, and seed production of double cross hybrids; however, for single crosses only the first three stages listed are applicable.

Maintenance of inbred lines: the basic or nucleus seed of the parental inbred lines is grown as ear-to-row at breeder station and is maintained by the breeder (or breeding organization unit involved in seed production) through hand pollination to rigidly control its purity. Inbred lines can be developed by inbreeding selected heterozygous plants until sufficient homozygosity is reached without severe inbreeding depression. Sib-mating may be used to maintain inbred lines. With the advent of DH technology, this problem is minimized.

Seed production of inbred lines: this is the first stage of seed multiplication from basic or nucleus seed and is called breeder seed of inbred lines. This is produced by hand pollination or in isolation under the direct supervision of breeder. The foundation seed of the inbred lines is produced from the breeder seed under isolation by the breeder or seed producing agency. In maize, a minimum of 400 m of isolation from other fields is necessary.

Seed production of single cross hybrid: the single cross seed is produced in isolation from foundation seed of inbred lines. The female and male parents are grown in rows in the ratio of 4:2 to 6:2. An isolation distance of 300 m or more is required from other maize fields. To avoid self-pollination, all plants of female rows are detasselled before they shed pollen. Seed plots are regularly inspected to remove possible off-types. The single cross seed are harvested from the female rows. If applicable, this single cross seed are used for the production of double cross hybrid. If the single cross seed are to be used by the farmers for growing commercial crop, then it should be registered with certification agency. After processing the seed is bagged and tagged.

Seed production of double cross hybrid: the seed of double cross hybrid [(A × B) × (C × D)] is produced in isolation from single crosses and an isolation distance of 300 m or more is required from other maize fields. The female and male single crosses are sown in rows in the ratio of 6:2. The plants/rows of female single crosses are detasselled and seeds from these rows are harvested. This is certified seed of the double cross hybrid. The seed production plots are inspected and certified by the seed certification agencies. After processing the seed is bagged and tagged.

Use of cytoplasmic male sterility in the seed production of hybrid

Both cytoplasmic male sterility (CMS) and mechanical detasselling are used in commercial seed production. The detasselling, that is, removal of entire tassel, the male inflorescence (or mechanical emasculation) of the female rows in the hybrid seed production of maize is a labor consuming operation. To facilitate hybrid seed production, incorporating CMS in the female inbred line(s) eliminates the process of emasculation. A maize plant with CMS in the absence of specific restorer gene(s) when pollinated by normal fertile plant that also lacks restorer gene(s), would produce only male sterile progenies. Male sterility gene could be transferred to an inbred line by repeated (six to seven) backcrossing. 'A' line has male sterile cytoplasm (S) with non-restorer gene (*rf rf*), and 'B' line has male fertile cytoplasm (N) with non-restorer gene (*rf rf*) and is used as recurrent parent in the conversion program (Fig. 19.1). The new male sterile inbred is maintained by pollination from its fertile counterpart. B line (maintainer line) is maintained by selfing or sibbing or open-pollination in isolation. The A and B lines are isogenic lines. Another parent, fertility restorer parent (R line) is required in the CMS system.

When the male sterile plant is pollinated by pollen from a plant containing restorer genes (*Rf Rf*). The R lines are also initially developed using backcrossing and maintained like the B line; however, once the restorer pool is well established

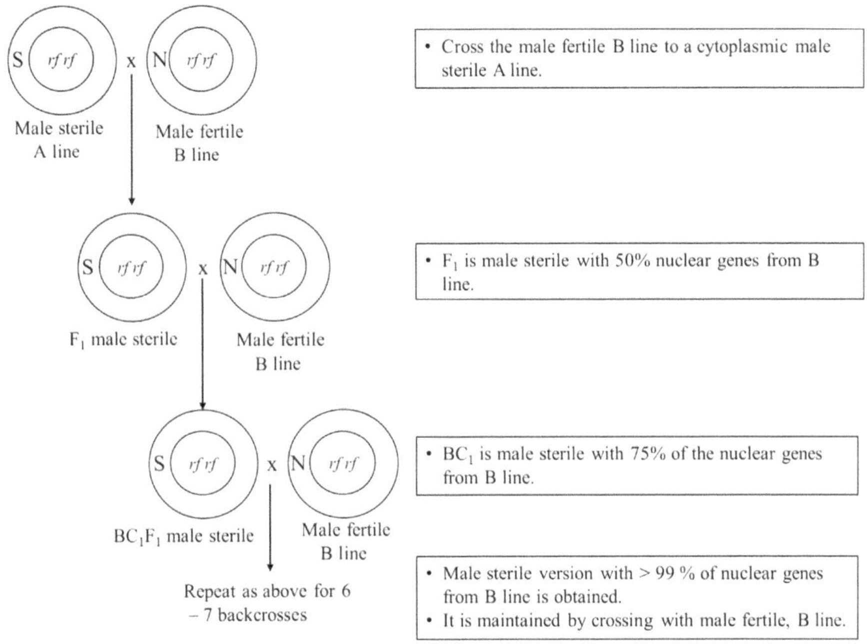

FIGURE 19.1

Transfer of cytoplasmic male sterility (CMS) through backcrossing, where A line is male sterile (used as non-recurrent parent) and B line is male fertile used as recurrent parent. N is normal cytoplasm and S is sterile cytoplasm. Breeding programs keep/maintain an A line stock for donor source of CMS.

there may be minimal need for backcrossing as the restorer pool consists of parents that have the restorer gene already. Single cross hybrids have been used for the exploitation of heterosis. The male sterile A line is used as female parent and R line possessing the restorer gene is used as the male parent. The seed set on the male sterile line (the female parent) is the hybrid seed and the seed produced on the male parent is selfed seed. If male inbred produces sufficient pollen, then single cross hybrid seed of maize is produced by planting female and male inbreds in the ratio of 4:2.

The CMS system is commercially used in other cross-pollinated crops such as pearl millet (*Pennisetun typhoides*), sorghum (*Sorghum bicolor*), onion (*Allium cepa*), sugarbeet (*Beta vulgaris*), and sunflower (*Helianthus annuus*). The CMS can be utilized in the production of double cross hybrid maize seed. The double cross seed (ABCD), will be male sterile as none of the inbreds contain pollen restoring genes. As such a hybrid could be used in case of forage crops or in crops where grain or seed is not the commercial product. However, in seed crop like maize, an identical fertile double cross seed with pollen restoring genes is blended with male

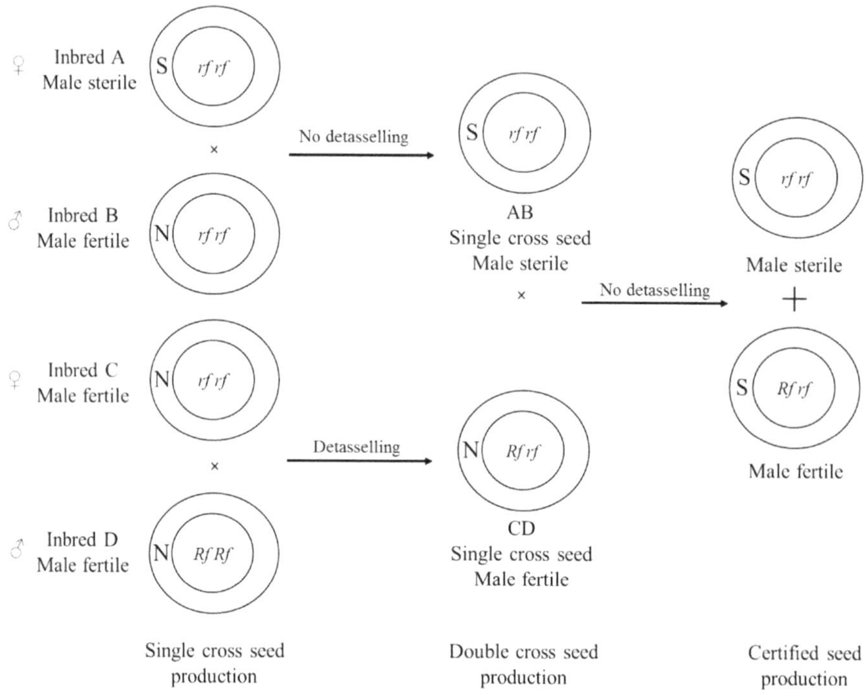

FIGURE 19.2

Certified seed production of a double cross using one inbred male sterile and one or two inbred with dominant restorer genes.

sterile double cross seed (one part fertile to two to three parts sterile) to ensure adequate pollination in the farmers' fields. The seed production of a double cross maize hybrid using one inbred line as male sterile and either one or two inbred with dominant nuclear genes as restorers (Fig. 19.2). If one of the inbred (D) of the second single cross (C × D) is having dominant pollen restoring nuclear genes then 50% of plants in the certified seed would be male fertile; however, if both C and D inbred have pollen restoring genes then all plants in certified seed plot would be male fertile. The other procedure of seed production in double cross maize hybrid could be through the use of two inbreds male sterile and one inbred with dominant restorer genes. In certified seed, only 50% of plants would be male fertile. In commercial maize breeding, only single crosses are marketed. However, information on other types of hybrids is provided for educational purposes.

In the three-line CMS system breeding scheme (Chapter 2, Mode of Reproduction in Crop Plants, Figure 2.7), breeders concurrently improve "B" and "R" lines pools. For example, any of the breeding methods described in pure line method can be followed to develop improved "B" and "R" lines. "B" and "R" lines may be a part of different heterotic pool or female–male groups. Once an improved "B" line is developed, a CMS line (with CMS source seed stock) is crossed to the "B" line and through successive backcrossing, an "A" line version of "B" line is developed. Breeders can save time to develop hybrid varieties by starting to assess GCA and SCA (crossing BC_2 or BC_3 lines to testers) while backcross conversion is on-going. At the end of backcrossing-based line conversion, new "A" line is identical to "B" line except having the CMS gene (i.e., it is male sterile). Since selfing of "A" line will not give any seed, "A" line is crossed (used as a female) to "B" line (used as a male) to produce seed of "A" line for hybrid seed development (by crossing "A" line with "R" line). Examples of the use of this scheme are rice (three parent system), sorghum, and canola.

Hybrid varieties in horticultural plants

Manual emasculation and pollination have been used successfully in vegetable crops such as tomato, sweet pepper, eggplant, and ornamentals such as begonia. In tomato, a single pollination may produce some 200 or more seeds. Self-incompatibility has also been used in hybrid seed production of some vegetable crops. The self-incompatible lines, which are otherwise cross-compatible, could be planted in alternate rows. The seed produced on both lines would be hybrid seed. This system has been used for hybrid seed production in broccoli and cabbage.

The genetic male sterility (GMS) system is used for production of F_1 hybrid seed in flowering plants in geranium, carnation, and dianthus. The CMS system has been used in vegetables such as onion, carrot, and sugar beet, and in flowering plants such as petunia. The monoecy (male and female flowers are on the same plant but on different parts of the plant) has been used for hybrid seed production in muskmelon and cucumber.

Hybrids have been produced and exploited in species that have capacity to produce seed and also are propagated through vegetative means, for example, sugarcane and coconut (*Cocos nucifera*). In these species, once superior hybrids are developed these can be multiplied without progeny testing and are maintained indefinitely asexually. This is covered in Chapter 20: Breeding methods used in asexual crops.

Double hybrid system in asparagus was created for situation where tissue culture methods were not available. Double hybrid creation involves crossing a single male (say, variety A) and a single female plant (say, variety B), which is followed by selecting only male hybrid plants as parents for the double hybrid. At the same time, in a different cross, a single male (say, variety C) and a single female plant (say, variety D) are crossed followed by selecting only female hybrids as parent for the double cross. This scheme helps to achieve a high yield potential, good uniformity compared with clonal hybrids, and avoids inbreeding depression related issues experienced in development of inbred. The second type of varieties in asparagus is all-male varieties. The first step is to produce a super male parent (YY) by selfing andromonoecious plants (XY) that produces segregation ratio of 1 XX: 2 XY: 1 YY. The YY and XY types are identified through test crosses with XX females and observing the progenies. The YY super males are crossed to XX females, producing all-male hybrid XY progeny. In Asparagus, due to higher yield observed in male plants, this is a preferred scheme compared with the use of female plants.

Hybrid varieties in self-pollinated crops

Hybrid varieties have also been developed and successfully exploited in self-pollinated field crops such as cotton (*Gossypium* spp.), rice (*Oryza sativa*), and pigeonpea (*Cajanus cajan*). The GMS system has been used in these crops. The male sterile line with a single recessive gene (*ms ms*) for male sterility is allowed to be pollinated with the male fertile line, which possesses one dominant gene (*Ms Ms*). The seed produced on male sterile line is the hybrid seed and F_1 is male fertile (*Ms ms*). The male sterility is maintained by crossing male sterile plants with heterozygous fertile plants. CMS system is used in multiple crops including rice, sorghum, and pigeonpea.

For the development of hybrid varieties, it is *a priori* that F_1 hybrid exhibits a significant heterosis, that is, the F_1 performance exceeds the better parent and pure line cultivars for grain yield and other selected attributes. The expression of heterosis or hybrid vigor is maximum in the F_1 generation and is expected to be halved in each subsequent generation. In view of this, new seed stocks have to be produced each year; therefore, the F_1 hybrid seed would be more expensive when compared with the pure line or synthetic varieties. The work of Shull and East between 1905 and 1912 laid the foundation for the exploitation of heterosis in cross-pollinated crop, maize and subsequently in several other cross-pollinated crops such as pearl millet, sweet corn, cole crops, onions, sugarbeet, curcurbits, spinach, carrot, castor, sunflower, and often cross-pollinated crops such as cotton, sorghum, and pigeonpea. The possibilities of exploiting hybrid vigor were made in 1970s onwards in autogamous crops such as tomato, wheat, and rice.

There is no difference in the hybrid breeding of self- or cross-pollinated crops, however, the success of hybrids in self-pollinated crop would depend on the following three main considerations:

1. The main concern is the magnitude of heterosis in a hybrid variety.
2. The effective and efficient mechanism of cross pollination in a crop. The availability of male sterility coupled with free movement of pollen grains would make exploitation of hybrid vigor easier.
3. In some of the self-pollinated crops such as wheat and barley the seed requirement/unit area is very high, which should be compensated by the yield advantage of a hybrid variety. The production of hybrid seed could be facilitated with the use of non-genetic tools or with the help of genetic factors.

Table 19.4 Examples of male gametocides found effective in self-pollinated species.

Male gametocide	Crop(s)
Ethrel	Wheat and rice
FW 450	Cotton,[a] tomato, and groundnut
Gibberellins	Rice and lettuce
Sodium methyl arsenate	Rice
Zinc methyl arsenate	Rice

[a]*Often cross-pollinated.*

The F_1 hybrid seed can be produced either through the use of chemical hybridizing agents (CHAs) or male sterile lines. Male sterility can be temporarily induced by a range of chemicals known as male gametocides or CHAs. Some of the male gametocides which have been used in self-pollinated species are given in Table 19.4.

An ideal gametocide should have the following characteristics:

1. Selective induction of complete male sterility in both early and late flowers in a treated line and not be influenced by environmental conditions.
2. No adverse effect on plant growth and development, on female fertility and should be devoid of biological toxicity.
3. Its application should be easy and cost effective. It should be safe to humans.

The genetic factors which have been used in the production of hybrid seed in self-pollinated crops are described as follows.

Male sterility

In case of male sterility, plants fail to produce fertile pollen. In seed crops, two types of male sterility are important. These are the GMS and cytoplasmic–GMS. The application of male sterility in commercial plant hybridization is discussed in the following text.

1. *GMS system:* when male sterility is conditioned by nuclear gene(s) only, it is referred as GMS. It is also called nuclear male sterility. It is governed by a single recessive gene, *ms* and originates through spontaneous mutation. The GMS is reported in rice, barley, sorghum, brassica, soybean, pigeonpea, tomato, etc. Male sterile (*ms ms*) line is maintained by crossing with heterozygous fertile (*Ms ms*) line. In the F_1 generation half of the heterozygous fertile plants are to be removed.

   GMS (*ms ms*) and fertile (*Ms Ms*) plants are grown in alternate rows in an ideal ratio and F_1 seeds are produced on male sterile line either through natural pollination or through hand pollination. Fertile plants occurring in the male sterile parent are rouged out before flowering. In a hybrid seed production block, an unnoticed fertile plant may fertilize the male sterile plants. As a result, the fertile plant would produce 25% sterile and 75% fertile plants. This may cause contamination, and ultimately would result in reduced heterosis and loss of uniformity.

The GMS can substitute manual emasculation and pollination in crops such as tomato, eggplant, chillies, and cotton where several seeds per cross are produced. GMS could also be used in crops where suitable CMS system is not yet developed. The efficiency of GMS could be increased if the fertile and/or sterile plants can be identified with the help of markers in early seedling stage. Male sterility has been reported to be closely linked with a recessive gene for no-anthocyanin pigmentation in vegetative parts of the plants.

GMS has been used to make two-line hybrid rice (Virmani et al. 2003). Male sterile (A line) line has nuclear gene(s) and male sterility is expressed under certain environmental conditions, such as photoperiod and/or temperature. These male sterile lines can be environmental-conditioned GMS, photoperiod sensitive GMS, thermosensitive GMS, or photoperiod and thermosensitive GMS lines. On the other hand, restorer lines have restorer gene that can restore fertility in F_1 when crossed to the male sterile A line. Seed is produced on female inbred lines that are homozygous for environmentally sensitive (photoperiod or temperature or both) recessive male sterility genes. For seed increase of female lines, environment (such as short day and/or lower temperatures) are chosen so that there is repressed expression of male sterility genes. This allows the female lines to reproduce through self-pollination. For hybrid seed production, plants (male and female) are planted in an environment (such as long day and/or high temperature) that promotes and ensures the expression of male sterility gene in the female. Therefore, successful hybrid seed production can be accomplished.

2. *The XYZ system of GMS:* this system was outlined by Driscoll (1972) for producing hybrid wheat. It involves three lines: X, Y, and Z which are homozygous for male sterility (*ms ms*) and possess two, one, and zero doses of an alien chromosome 5R, respectively which is derived from a related species, rye. The alien chromosome does not pair with any of the wheat chromosome and bears the corresponding gene for male fertility (Ms) and a marker gene (hairy peduncle). The X and Y lines are useful only for the production of female parent. The X line contains 21 pair of wheat chromosomes and one pair of rye chromosome and produces only one type of gamete hence is pure breeding. The Y line has one dose of alien chromosome and produces two types of egg and pollen on the male side. However, only pollen excluding 5R chromosome is functional due to certation. The selfed progeny of Y plants produces sterile Z plants (75%) and fertile Y plants (about 25% of which exhibit hairy peducle). The Z plants resulting of selfing of Y plants are male sterile, hence produce only one type of egg which are devoid of alien chromosome. Limited amount of rouging is required in one selfed generation of Y line. The final product is devoid of the alien chromosome.

3. *Cytoplasmic–CMS system:* it results from interaction of cytoplasm with nuclear genes and is commonly referred to as CMS. The CMS sources could be derived from intraspecific crosses as in sorghum and flax, or from interspecific crosses as in rice, wheat, barley, and pigeonpea, or from intergeneric crosses as in wheat and brassica. Backcross breeding is used for the development of CMS lines (see Chapter 13: Backcross method). An effective restoration system is required for the use of CMS system in the development of hybrids in grain crops. Plants possess gene(s) in the nuclear genome that interact with the sterile cytoplasm to produce fertile F_1 hybrids. In general, the restorer genes are sought in the cytoplasmic donor species itself and/or in cultivated varieties. The CMS for commercial hybrid seed production involves the production of seed of three type of lines: A, B, and R. As explained earlier in this chapter, the A line is a female parent with male sterile cytoplasm, hence would not set seed unless cross fertilized with a maintainer parent B. The B parent is genetically similar to the A parent but is lacking male sterility and is maintained through selfing. The restorer line R contains genes that restore fertility. The A and R lines are created through backcross breeding, although R lines can be created from an R line gene pool that involves making cross between R lines and developing improved inbred lines progenies. The production of commercial hybrid seed involves following steps:
    a. multiplication of CMS lines (A × B),
    b. multiplication of maintainer and restorer lines, and
    c. production of hybrid seed (A × R).

The CMS system has been used successfully in commercial hybrid seed production in many cross-pollinated crops (previously described in this chapter), as well as in self-pollinated crops for example, rice. For seed production of A × B, the female (A) and male (B) are planted in a 4:2 row ratio and for hybrid seed production, A and R are planted in 6–8: 1–2 row ratio. The isolation distance of 100 m in A × B cross and 40–100 m in A × R cross is maintained (these distances vary as per crop species. These isolation distance requirements can vary, and information can be easily obtained from a certification agency (state or national level). The synchronization of flowering is critical to obtain adequate cross pollination, which can be obtained by staggered planting and with the use of growth hormones. In case of rice, the planting of the parent rows across the prevailing wind direction and clipping of flag leaves of A, B, and R lines at booting stage has given good results. One or two applications of GA (gibberellic acid at 20 ppm) after clipping have been found to improve panicle exertion of A, B, and R lines. Rope pulling and agitating the pollen parents with a pole are used to improve out-crossing in hybrid rice production program in China.

Self-incompatibility

It is the failure of the pollen grains of a plant to fertilize the ovules of the same plant. This happens even though both pollen and ovule development are normal and viable.

However, the pollen can effectively fertilize ovules of most other plants of same species as a result all plants produce seed in self-incompatible species. Self-incompatibility is genetically controlled (by a locus *S* with multiple alleles) physiological hindrance to self-fertilization. It is one of the outbreeding mechanisms and occurs due to interaction of products from specific alleles at one or more incompatibility loci in the style and those from identical alleles in the pollen or sporophyte. It is widespread in nature and occurring in families of poaceae, cruciferae, compositae, and rosaceae. More details can be seen in Chapter 2: Mode of reproduction in crop plants. Depending on the species incompatibility could express in the following ways:

1. The germination of the pollen may be decreased, for example, in broccoli. Removal of stigma allows normal pollen germination.
2. The pollen germination is normal but pollen tube growth is inhibited in the style, for example, in tobacco.
3. The incompatibility reaction occurs after fertilization, for example, in Gesteria. This mechanism is of rare occurrence.

The self-incompatibility could be gametophytic or sporophytic. The gametophytic self-incompatibility reaction is expressed when both pollen and style have identical alleles. Through this system only plants heterozygous for S-alleles are produced. All the pollen of a plant behaves in the same way in case of sporophytic self-incompatibility. This is because the reaction is between the pollen grain coating (sporophyte) and the papillae of the stigma and the expression is the germination of the pollen tube.

How to overcome self-incompatibility?

Doubling of the chromosomes and use of physical and chemical mutagens have been used to overcome self-infertility. Both of these approaches are useful in species with gametophytic system than in species with sporophytic system. The incompatibility could also be overcome temporarily using techniques such as removal or mutilation of the stigmatic surface, bud pollination (hand-opened and self-pollinated 3–4 days before anthesis, that is, before inhibitory protein form) in gametophytic and sporophytic systems or lowering the temperature (to slow down the production of the inhibitory substance or enzyme). Irradiation of style with physical mutagens could also breakdown single locus gametophytic self-incompatibility.

Implications of self-incompatibility in plant breeding

Self-incompatibility hinders self-fertilization and promotes cross fertilization, hence produces heterozygosity. Selfing of self-incompatible plants can create large amount of variability, and consequently superior recombinants could be selected. Self-incompatibility could be used to produce homozygous lines and for the production of F_1 hybrids (single, double, or triple cross hybrid varieties). The system is useful in species where male sterility is unavailable. Sporophytic incompatibility has been used in cabbage and other cole crops for hybrid seed production. The procedure consists of self-pollination of individual plants through bud pollination from source population. Selection is applied for desirable characters and strong level of self-incompatibility for few generations. Several self-incompatible but cross-compatible inbred lines having different S-alleles are developed to produce hybrid seed.

Single cross hybrid: two self-incompatible and cross-compatible lines S_1S_1 and S_2S_2 planted in alternate rows in isolation. The seed set on each line will be mostly hybrid seed, where cross fertilization is brought about by pollinating insects, mostly bees. Alternatively, a self-incompatible line may be interplanted with a self-compatible line, and the seed from self-incompatible line would be hybrid seed. Single cross hybrids are more uniform, and show maximum heterosis.

Double cross hybrid: the inbred lines used in the production of single cross hybrid are expected to be less vigorous which inflates the cost of F_1 hybrid seed, in crops with fewer resources or less extensive breeding. Therefore, double cross hybrid seed is produced that involves two single crosses developed using four inbred lines, for example, $(S_1S_1 \times S_2S_2) \times (S_3S_3 \times S_4S_4)$.

Top cross hybrid: this is a cross between a single self-incompatibility inbred line as female and a good open-pollinated cultivar as pollen parent.

The gametophytic self-incompatibility has been used in vegetatively propagated crops for the productions of synthetic varieties and double cross hybrid (e.g., red clover). The clones are planted in adjacent rows for hybridization purposes.

CHAPTER

Breeding methods used in asexual crops

20

Abstract

Some of the cultivated species that produce seeds very poorly or produce seeds only under special conditions, are propagated through asexual reproduction or are vegetatively propagated and not by seed. These include some of the major food crops, for example, several root and tuber crops, fruit crops, ornamental plants, and grasses that are not multiplied through seed but are vegetatively propagated by tubers, cuttings, grafting of scions on root stocks, etc. In this chapter, breeding approaches used in asexual crops are described, including clonal selection and hybridization. Considerations in micropropagation and its usefulness in asexual crops are explained. Mutation breeding approaches and apomixis are also described.

In sexual reproduction, the male and female gametes fuse after meiosis. It is called *amphimixis* which is opposite to *apomixis*, that is, without the fusion (or mixis) of the two gametes. In the absence of sexual reproduction, the genetic composition of plant material being multiplied remains essentially the same as its source plant because at cell level mitosis leads to multiplication of somatic tissues. Much before the discovery of laws of inheritance (using hybridization) were known, Thomas Knight developed varieties of apples, pears, and grapes during 1811–38. He selected superior plants from the heterogeneous population of F_1 cross in the asexually propagated species and developed superior varieties. Several fruit species produce flower regularly and produce satisfactory seed, but they are preferred to be propagated asexually. It is done to avoid segregation and recombination, allowing uniformity in the commercial produce.

All the plants of a vegetatively propagated variety are genetically identical (may be homozygous or heterozygous), that is, they form clone, which is one individual divided into many parts with the same genetic constitution. Due to this, such crops are also referred to as clonally propagated species or clonal crops, and the cultivars are called clonal cultivars (or varieties). A clonal variety maintains genetic purity indefinitely, unless somatic mutations occur which are called sports.

A clone is defined as "the group of plants produced from a single plant through asexual reproduction." Clones are reproduced through mitosis. As meiosis

Plant Breeding and Cultivar Development. DOI: https://doi.org/10.1016/B978-0-12-817563-7.00006-4

Table 20.1 Some important differences among the three terms: clone, pure line, and inbred line.

Clone	Pure line	Inbred line
Clone is a progeny of a single vegetatively produced plant.	Pure line is a progeny of a single self-fertilized plant.	Inbred line is produced through selfing or other form of inbreeding in cross-pollinated crops.
Clones are highly heterozygous and phenotypically homogeneous.	Pure lines are homozygous and homogeneous.	Inbred lines are homozygous or almost homozygous and homogeneous, except doubled haploid lines which are homozygous and homogeneous.
Clones are maintained by vegetative or clonal propagation.	Pure lines are maintained by natural self-pollination.	Inbred lines are maintained through artificial selfing or close inbreeding.
Superior clones are used as variety for cultivation.	Superior pure lines in self-pollinated crops are used as variety for cultivation.	Inbred lines are used as parents in the production of hybrid and synthetic cultivars.
Genetic variation in a clone could arise through somatic mutation in the buds. Occasional sexual reproduction and mechanical mixture may also give rise to mixture in a variety.	Genetic variation in a pure line could arise through mutation, natural crossing and mechanical mixture from other material.	Inbred lines are contaminated through cross-pollination and spontaneous mutation.

is not involved in the propagation, there is no recombination and segregation within a clone as it is propagated. Therefore, all individuals from a clone have the same genotype as that of the parent plant. In some plant species, fruit develops without fertilization, that is, *parthenocarpy*. The advantage of parthenocarpic fruit over seeded fruit is longer shelf life and higher consumer preference.

The three terms: clones, pure lines, and inbred lines have been used to describe materials developed by plant breeders to denote sameness of their genetic constitution. However, there are some differences among these terms, which are described in Table 20.1.

In clonal crops, pollination failure and/or non-functional sex cells (egg and sperm) are the most frequent reasons for the lack of seed development. Self-incompatibility is one example of pollination failure. For example, citrus seedless fruit cultivars are self-incompatible and do not set seed when planted in commercial fruit orchards. Due to parthenocarpy, these clonal cultivar trees still produce fruits. Clonal trees are propagated by grafting where root stocks are generally cultivars with healthy root traits, while the stem stock is elite fruit producing clone.

Features of vegetatively propagated crops

1. Majority of perennial horticultural species are asexually propagated. The perennial species are in the ground for a long time, at least for years and often decades, for example, mango, coffee, citrus, banana, rubber, tea, oil palm, coconut, cacao, etc. However, some of the asexually propagated species, especially major food crops complete their breeding cycle in a season/year, for example, sugarcane, potato, sweet potato, yams, etc.
2. Many of them show reduced flowering and seed set. Many varieties do not flower at all. Only fruit crops have shown regular flowering and seed set.
3. They are largely cross-pollinated, for example, mango, citrus, sugarcane, rubber, tea, oil palm, coconut, cacao, etc.; however, some are self-pollinated for eg. coffee, and some are both self- and cross-pollinated for eg. banana.

   In banana, chromosomal imbalance prevents successful fertilization. Commercial banana clones are triploid—possessing two sets of chromosomes from one parent and one set of chromosomes from the other parent. Sexual reproduction (due to chromosomal imbalance) is rarely successful and fruits are produced through *parthenogenesis*. Each pseudo stem produces only one banana cluster before it dies off. New stalks are continuously produced from the rhizome to replace them in each fruiting cycle. These side shoots or suckers at the base of the main stalk can be removed and replanted to continue the cultivar. Tissue culture is another way to propagate bananas.
4. They are highly heterozygous and highly heterotic. They also show high inbreeding depression on selfing. Heterosis is fixed and is maintained as long as the cultivar is propagated asexually.
5. Majority of them have been derived from interspecific hybridization and polyploidy, for example, banana, sugarcane, and *Rubus*.
6. All progenies from an individual plant propagated asexually are genetically identical (clone) and phenotypically uniform. Any phenotypic variation among the progenies is environmental in origin. Only source of natural genetic variation in a clone is somatic mutation.

Unlike sexually propagated crops, in asexually propagated crops, flowering and seed set are restricted to certain environmental conditions at a few locations. For instance, the optimum conditions for floral induction in sugarcane include a photoperiod of just 12 hours at high light level and a day–night temperature regime of 27°C–31°C and 22°C–23°C, respectively, when the plant has two to four elongated internodes exposed. Therefore, crosses cannot be made successfully at all research stations, hence introduction is more relevant in such a crop. In India, in the case of potato and sugarcane, established clones as well as crossed seed may be introduced from other centers in the country as well as from abroad. The crosses (to create breeding families) are made for potato in Shimla (Himachal Pradesh), and for sugarcane in Coimbatore (Tamil Nadu). Crossed seed is introduced to other breeding stations across India. From the crossed seeds,

F_1 seedlings are grown and clonal selection is followed. This scheme can be implemented in other similar situations globally, where initial crossing (to create breeding families) is cumbersome or infeasible necessitating a centralized crossing nursery. Alternatively, in sugarcane, where pollen and seed production does not occur naturally in latitudes above 15°N and below 15°S, crossing is conducted in controlled environment chambers and greenhouses where daylength and temperature can be manipulated.

In a crop with relatively few years of breeding efforts, the divergence between introductions and landraces compared with improved germplasm is not as wide as in crops with a more extensive breeding history. As a result, introductions and/or landrace accessions play a more relevant role in a clonal crop with few years of directed breeding efforts compared with other crops such as potato that have been extensively bred. In sugarcane, breeders have exploited genetic variation from multiple *Saccharum* species, primarily *Saccharum spontaneum* which has provided adaptation, disease, and pest resistance traits.

Breeding approaches

The two main breeding approaches in asexual crops are; (1) clonal selection, and (2) hybridization. For hybridization-based approach, after objectives are clearly defined and are considered biologically feasible, genetic improvement in clonal crops starts with the assembly and evaluation of a broad germplasm base (i.e., choice of parents). This is followed by the production of new recombinant genotypes derived from hybridization between selected elite clones, and careful evaluation in a set of representative environments. In principle, the breeding approaches used in asexually propagated species is similar to that used for sexually propagated species. Asexually propagated species are highly heterozygous. The F_1 developed from the crossing between two genotypes is a segregating generation, hence selection starts in the F_1 generation rather than in the F_2 generation in sexually propagated species. The steps involved for breeding uniformity and stability in seed propagated crops/species can be omitted, as each superior selection from a segregating generation is constant in vegetative multiplication.

Clonal selection

The clonal selection is similar to pure line selection in self-pollinated crops. The clonal selection is followed to utilize the genetic variability already present in the variety (population). Superior clones are selected from a mixed population (a genetically variable population), which could be an old unimproved variety or an improved variety that has become variable over time or allowed to produce seed. Selection for superior clones based on single plants may not be reliable; therefore, the progenies of single plants propagated vegetatively are evaluated for biotic and other stresses followed by their testing in replicated trials for 2–3 years. The best clone is multiplied and released as variety. In theory, such a variety would maintain its genetic ability indefinitely. However, clones may show decline in vigor, yield, etc., with time. This is called clonal degeneration. It may arise due to mutation, for example, in some clones of potato, bolter mutations may occur at a frequency of 10^{-3}. Some pathogens, for example, virus and bacterial diseases, are transmitted easily to progeny through vegetative propagation, and is the most important cause of degeneration.

The scheme for clonal selection is presented in Fig. 20.1. A brief description is as follows:

First year: the mixed population is spaced planted and exposed to disease(s) or abiotic stress(es) of interest. Selection of large number of superior plants is practiced for plant type, maturity duration, and resistance to diseases and abiotic stress(es).

Second year: all selected plants are grown separately as plant to progeny (this forms clones). The clones are evaluated visually for various traits. About 50–100 clones with superior traits are selected and rest are discarded. Selection is practiced between clones and not within clones.

Third year: selected clones are subjected to evaluation in preliminary yield trials along with checks. Selection is made for morphological traits and for resistance to biotic and other stresses. The clones which are superior compared with the released/check cultivars are selected.

Fourth to sixth years: the superior clones are tested in multi-location replicated trials along with check varieties. The clones are evaluated for yielding ability as well as for quality traits and to location specific pest and disease resistance. Since clones in year 1 onward are the same; therefore, resource-permitting, more locations can be used in year 3 and onward to identify clone(s) for commercialization. It can also lead to a reduction in the number of years of evaluation; however, breeder has to determine the optimal combination of number of locations and years to determine feasibility of commercial success. For example, in less varied target commercialization environment, fewer years of evaluation is justified.

Seventh year: the best clone is multiplied and released for cultivation.

Advantages of clonal selection

1. The method is applicable to the existing variability.
2. It avoids inbreeding depression and retains some of the original gene combinations.
3. The method is also applicable for maintaining the genetic purity of clones.

Limitation of clonal selection

1. Improvement is within the range of genetic variability of the parental population.

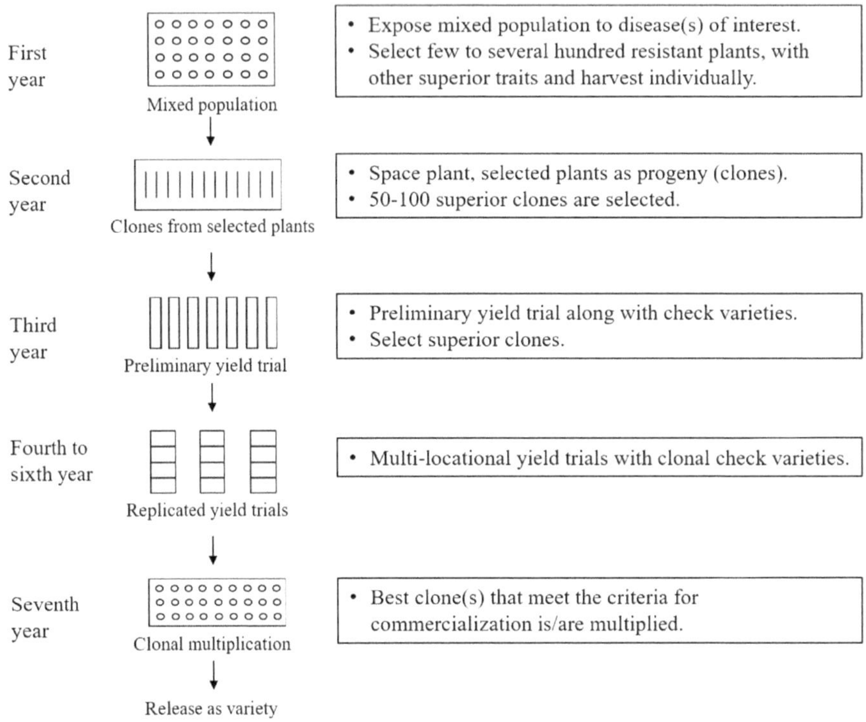

FIGURE 20.1

Scheme for clonal selection in asexually propagated species.

Hybridization

This procedure is applicable to species that are able to produce flowers and seeds in good quantity, for example, potato and sugarcane. The interspecific crosses between cultivated species and wild species have been made to transfer specific genes, for instance resistance to specific biotic and abiotic factor(s). Asexually propagated species are highly heterozygous. Heterosis can be fixed in clonal propagation. The selection of parents based on phenotype is not sufficient. The identification of superior clones to be used as parents (in single cross, multiple-parent cross) is based on their specific or general combining ability. Mating designs can be used for this purpose to conduct combining ability analysis and determine the best combiners to use in hybridization. Breeder should have information on potential parent strain (from previous experience or experiments) prior to deciding which parents to choose for crossing. The crosses are attempted among selected parents. Once F_1 is produced, the breeding procedure is essentially the same as clonal selection. Selfing of F_1 to produce F_2 is seldom practiced as selfing would lead to a reduction in vigor which is not desirable. Each F_1 plant is a potential source for a new clone.

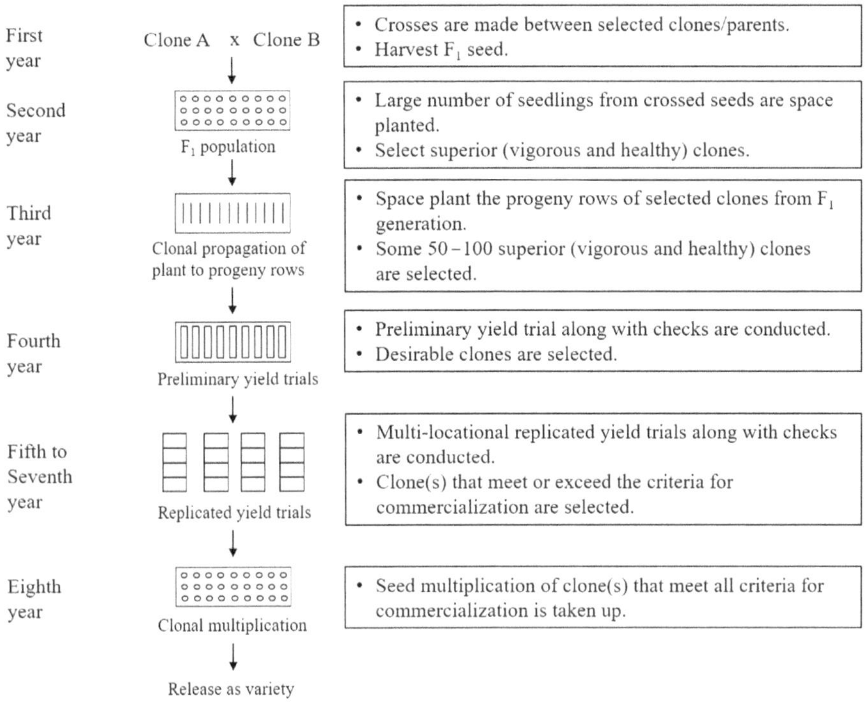

FIGURE 20.2

Scheme for hybridization followed by selection of annually asexually propagated species. This example is provided for one breeding family in the breeding program.

The segregation starts in the F_1; therefore, superior plants are selected in this generation. Clones are tested for resistance to biotic and/or abiotic stresses, yield, and quality traits, is applicable. The that meets all criteria for commercialization clone is multiplied and released for cultivation. It is important to recognize the difference between biological seed and vegetative (clonal) parts (used for propagation). The F_1 seeds obtained from crossing are biological seeds, whereas the plant parts used from that F_1 for further testing in the next generation, generates a clone (identical to the F_1 plant).

The scheme for hybridization and selection is presented in Fig. 20.2, and is described as follows:

First year: the selected parents/clones are crossed to produce crossed seed. The crossed seed is harvested. Hybridization of more than two parents could be attempted in polycross nursery as in case of clonally propagated grasses and legumes. The crossing block can be in field for over a year. To overcome flowering synchronization issues, sufficient length of time needs to be used for the crossing block. Additionally, some crops may produce very few seeds per hybridization so proper planning is essential to obtain a sufficient breeding family size.

Second year: a large number (several hundreds to thousands) of F_1 seedling from crossed seed of a breeding population are space planted. Superior F_1 plants, which are vigorous and healthy, are selected from each population. Selection pressure can be applied for stresses in the nursery. Several cuttings or plant parts taken for testing from a single F_1 plant will be identical to each other and to the F_1. Plant parts taken from different F_1 plants will be dissimilar to each other. Due to low correlations between the performance of individual plants and yield plots, selections should be done only on traits with high heritability. Traits with higher heritability include plant type, reaction to diseases, etc. At this stage, size of the nursery is quite large (say, >10,000 F_1 plants), breeder may use spray paint, tags, or another method to mark discards because some of the traits are visually assessed by breeder, and digital data collection may not be feasible in all cases. Up to 20%–30% of plants are retained.

Third year: plants to progenies of the selected plants are space planted (or grid planted) using vegetative propagation (clonal propagation). The clonal progenies are evaluated visually. In each trial, care is taken to ensure that the same number of plants is grown for each entry to avoid bias in performance, traits expression, and selection. Some 200–300 superior progenies (clones) are selected. Each plot is of the same size (length and width) to minimize nongenetic variation biasing selection. For example, uneven number of plants or plot size gives unnecessary advantage because the extra space gives it uneven advantage for water, sunlight, and nutrients. Either a GPS planter is used so that each plot is same size, or a plot-trimmer is used to cut the plots into equal size. If uneven plant stands or plot size is observed, this information should be recorded and used to make better selection decision.

Fourth year: the superior clones are tested in preliminary yield trial along with checks. The clones are evaluated for resistance to disease(s) of interest and other traits including quality aspects. Based on selection criteria (performance, comparison with checks) outstanding clones are selected for multi-location yield trials. Considerations described for plot size and plant stand is very important for this stage. Moreover, breeder can either try to group more similar entries into the same test or select for high heritability traits to minimize error in selection. This ensures that variability due to plant architecture traits, such as canopy coverage, plant height, stem branching is reduced.

Fifth to seventh years: multi-location yield trials along with checks are conducted at several locations for at least 2 years. The evaluation is made for yield, maturity duration and location specific host–pathogen and/or host-parasite interaction. Best clone(s) is/are identified for release as variety. As previously explained, since clones in year 1 onward are the same, resource-permitting, more locations can be used in year 3 and onward to identify the clone for commercialization. More detailed processors' and consumers' preference (i.e., end-use quality) traits are assessed at this stage. These traits are more expensive or time-consuming to generate data, only the most advanced entries are tested for these traits. As the most promising material is getting evaluated for processing and end-use quality, breeder needs to start steps to multiply the stock for commercial planting.

Eighth year: best clone is multiplied vegetatively and released as variety for cultivation. Multiplication can start in sixth or seventh year to prevent delay in commercial sales. The "best" clone implies that it meets all criteria of selection and merits commercialization.

To reduce the problem of interplot competition, the distance between rows can be increased and the plant–plant distance can be reduced, or an empty row can be left between plots to increase the competition among plants from the same genotype and reduces the competition between plants from different genotypes. Breeders may also consider dividing the field into smaller blocks and conducting selection within a block (with commercial or elite checks present regularly across the field within each block). Each smaller block has same or similar breeding families. Adequate data collection of relevant traits can help a breeder to assess family performance within all clones from a cross and among different crosses, and gives an indication of the performance and potential of selected clones for use as a parent in crossing.

If there is interest in transferring one or two traits, backcrossing can be used. However, backcrossing in clonal cultivars is problematic because of inbreeding depression. One way to overcome this is to use different clones instead of the recurrent parent, but this results in introduction of new genetic variability with each backcross.

In some root or tuber crops (e.g., Cassava), the botanical seed obtained after crossing either is planted directly in the field or is first germinated in greenhouse conditions and then transplanted to the field when growth is suitable to transplant. At this stage, selections should not be made for traits whose expression differs between biological seeds and vegetative cuttings due to low trait correlation. Since the commercial product is derived from vegetative cuttings, root data from biological seeds should not be used to make selection unless trait correlation is high for the two growing conditions. On the other hand, transplanted F_1 (biological) seed does not generally suffer from this problem as their root system is similar to what will be observed with vegetative cuttings. In clonal crops, the area of the plant used to obtain cuttings also influences the performance of the resultant plant. Therefore, plant breeders should be aware of similar issues that exist in their crop. In cassava, vegetative cuttings from the mid-section of the stems usually produce better performing plants than those taken from the top or the bottom of the stems. This variation in the performance of the plant, depending on the physiological status of the vegetative cutting, is responsible for larger experimental errors and undesirable variation in the evaluation process. Breeders look to keep consistency in vegetative cutting as it is important to remove this unwanted source of error. Some species may have limitations on the number of cuttings (or vegetative parts) that can be obtained per plant, and this influences or limits the size and number of multi-location trials.

Advantages of hybridization approach

1. New genetic variability is created.

2. The gene of interest can be transferred from wild species to the cultivated type.
3. The heterozygosity is maintained indefinitely and heterosis is also fixed.

Limitation of hybridization approach

1. Crosses cannot be attempted at all breeding stations.

Achievements of hybridization in asexual species

Several of the present day cultivars of sugarcane have been developed through interspecific hybridization from crosses of *Saccharum officinarum* with *S. spontaneum* and *S. barberi*. Disease resistance from *S. spontaneum* has been transferred to *S. officinarum*. Similarly, most of the modern cultivars of potato are derivatives of interspecific hybrids. The resistance to late blight, leaf roll virus has been transferred from *Solanum demissum* to *S. tuberosum*. In strawberry, resistance to aphids was transferred from *Fragaria chiloensis* to cultivated strawberry. Resistance to abiotic stresses has also been transferred through hybridization in asexually propagated species, for example, frost tolerance is transferred from *Solanum acaule* to *S. tuberosum*. Cultivated varieties of commercial bananas are the product of interspecific hybridization. Triploid ($2n = 33$ with ABB genomes) seedless bananas is a product of interspecific crosses between wild species *Musa acuminata* ($2n = 22$, AA) and *Musa balbisiana* ($2n = 22$, BB). *M. balbisiana* is hardier and more drought tolerant than *M. acuminata*. Vegetative parthenocarpy and female sterility appeared in *M. acuminata*, allowing the production of pulp without seeds. *M. balbisiana* also provided much needed genes for resistance to Panama disease, leaf spot, and nematodes. Crosses within *M. acuminata* or between *M. acuminata* and *M. balbisiana*, coupled with female restitution and haploid fertilization, led to cultivated varieties as we see today: (1) homogenomic hybrids which are essentially AAA dessert and highland bananas, and (2) heterogenomic hybrids comprising the AAB (plantains) and the ABB (cooking bananas). From the triploid subgroup AAA, "Cavendish" variety is most common worldwide. Cavendish, originating from a single clone, contributes to almost half of the world's banana production.

Micropropagation

One of the biggest challenges in clonal crops production is to keep parental lines and breeding stocks free of virus and other diseases that can be transmitted through vegetative propagation. *In vitro* methods (such as tissue culture), which can be used to keep stocks disease-free, are often used to rapidly increase clonal stocks. The tissue culture methods of plant propagation, known as micropropagation, utilize the culturing of apical shoots, axillary buds, and meristems on suitable nutrient medium to grow new clones. Micropropagation offers an ability to continually produce clones (all year round operating nurseries and laboratories), produce and propagate hybrids, and produce disease-free plants. It is a cost-effective technique and requires small space.

Steps in the micropropagation method:

1. Selection and maintenance of stock plants for culture initiation,
2. initiation and establishment of culture - from an explant, like a shoot tip, on a suitable nutrient medium,
3. multiple shoot formation from the cultured explant,
4. rooting of *in vitro* developed shoots, and
5. transplanting and hardening, that is, acclimatization, before transplanting in the field.

Micropropagation or tissue culturing can lead to somaclonal variation (SV). SV can be defined as genetically stable variation generated through plant tissue culture (Larkin and Scowcroft, 1981). It has been used by breeders as an approach to create and exploit greater genetic diversity and provides a mechanism to expand the germplasm pool for plant improvement and cultivar development. While the resultant success of SV has not been as great as initially promised, it has led to the identification of valuable genotypes, for example, aluminum tolerance in rice (Jan et al., 1997). SV can also cause negative effects; therefore, a mechanism of genotype purity is desirable in clonal crops.

Clean clonal material is the essential starting material for cell multiplication for propagation. It is very important to maintain disease-free clonal crops and/or to purify an infected cultivar. Infection can occur due to bacteria or viruses. Viruses are more damaging due their systemic nature. Plant material is visually inspected for the presence of pathogens; however, this is not the most effective method for viruses because there may be no obvious symptoms (asymptomatic). Therefore, virus indexing is performed, which includes testing of plant material to confirm the presence or absence of transmissible viruses or diseases with an aim of preventing spread or establishment of infected plant material. Virus indexing can identify previously characterized viruses.

Two main methods are used to detect the presence of specific pathogens. These are:

- serological, such as enzyme-linked immunosorbent assay, and
- nucleic acid based, such as real time-PCR.

These techniques can detect latent viruses and other pathogens. However, a negative test is not conclusive evidence of the absence of pathogens and could just be due to ineffective assay. If diseases or viruses are detected, it is important to eliminate them using following methods:

- Tissue culture: even when the pathogen is systemic, tissue from the terminal growing points can be used for further propagation as it is often pathogen-free.
- Heat treatment: works well for fungal, bacterial, and nematode infections. For viruses, a longer treatment is required relative to other pathogens.

- Chemical treatment: surface sterilization with chemicals can be used to eliminate pathogens.
- Use of apomictic seed.

Mutation breeding approaches in asexual crops

Mutations arise spontaneously or are induced deliberately. Whether natural or artificial, somatic mutations are characterized by tissue mosaicism called *chimerism.* A chimera consists of two or more genetically different types of cells. The chimeras are maintained by vegetative propagation and cannot be transferred to progenies by sexual means. The chimeras are of four types: secotorial, mericlinal, periclinal (each of these are of genetic origin), and graft chimera (nonheritable, consisting of mixture of tissues after grafting is made).

The physical mutagens have been used more frequently than chemical mutagens for inducing mutations in asexually propagated species. The vegetative part of the plant that contains bud from which a plant can develop, are useful materials for mutagen treatment (i.e., irradiation). These could be tubers, rhizomes, shoot apex, cuttings, bulb, etc. It is desirable that the exposure of mutagen is given in the early developmental stage of the bud. Mutations are unicellular event. Mutagenic treatment of multicellular somatic tissues leads to the induction of mutation in few cells to form chimera. The chimeras are usually unstable. A chimeric plant represents a mixture of mutated as well as normal tissues; and in such situations, the diplontic selection resulting from competition between original and mutated cells does not allow the mutations to be expressed. The clonally induced mutations are chimeral, and start as sectorial or mericlinal which are mostly unstable. A stable mericlinal chimera then forms a so-called periclinal chimera. Periclinal chimera is characterized by tissue of two thin layers of different genetic make-up. In this type of chimera, a single cell in a layer undergoes a mutation, and by anticlinal cell divisions an entire layer becomes genetically different from the adjoining (two) layers. The periclinal chimera gives rise to mutated shoot and remains stable when propagated vegetatively. Variegation pattern observed in dicot leaves is an example of periclinal chimera. Variegated plants can revert to all green or all non-pigmented leaves; therefore, chimeral variegation is preserved by periodic pruning and removal of non-chimeral stems. Since all-green stems have more chlorophyll, they can quickly grow over variegated stems necessitating pruning and removal of non-chimeric stems.

The apex of adventitious buds is initiated by one or few daughter cells of an original cell, often an epidermal cell. Plantlets obtained from such buds are used in mutation breeding to produce non-chimeric stable mutants in many vegetatively propagated plants such as *Begonia*, *Streptocarpus*, *Chrysanthemum*, and potato. However, breeder needs to grow two to three clonal generations for desirable mutations and vegetative stability.

Handling of mutation-induced segregating generations

In vegetatively propagated crops, the generation raised from the treated propagules is called the VM_1 generation. Chimeric plants are selected and propagated to produce VM_2 and subsequent generations. Since in the clonally induced mutation breeding, the VM_1 and succeeding generations are multiplied by asexual reproduction, only dominant mutations are selected. The general procedure of mutation breeding in asexually propagated species is briefly described below.

VM_1: a large number of cuttings are irradiated with mutagen and are space planted. The plants showing mericlinal and periclinal chimeras are selected.

VM_2: selected VM_1 plants are grown in separate rows using vegetative propagation. The solid mutants or homozygous mutants are identified and selected in this generation.

VM_3: plants showing stable mutations selected in VM_2 are vegetatively propagated and grown in separate rows to confirm their breeding behavior. Identical progeny is bulk harvested.

VM_4: selected mutants are evaluated in on-station trials along with check varieties.

VM_5 to VM_6: selected best mutant clone is evaluated in multi-location replicated trials.

VM_7: the best clone(s) is/are multiplied and released as cultivar.

Achievements of induced mutations in asexual species

Induced mutations have been very widely used to improve certain traits in asexually propagated species. These traits are simply inherited, easily observed, or selected in the mutated plants. Several superior cultivars have been released in case of horticultural and ornamental species. In general, ionizing radiations have been used as an acute or chronic dose, sometimes plants have been exposed to irradiation in gamma field. In case of apple, mutants with improved skin color and resistance to *Podosphaera leucotricha* and *Venturia inaequalis* were selected using gamma-rays. Fruit of bigger size with deeper red skin color and early maturity were selected in peach using gamma field. True dwarf (1/4 to 1/5 size of normal cherry) plants were picked using X-ray treatment. In case of ornamentals, eg., Rose, Carnation, *Chrysanthemum*, *Dahlia*, Azalea, *Streptocarpus*, and *Alstroemeria* mutant plants with change in density of color, color pattern on petals, large flower size, and rarely with smaller petals have been selected. In general, potted plants (in Rose), rooted cuttings (in Carnation), leaves (in *Streptocarpus*), and dormant tubers (in *Dahlia*) were used for X-ray or gamma-ray irradiation. A classical experiment was conducted in peppermint in the United States (Murray, 1969). All the peppermint (*Mentha piperita*) varieties were susceptible to verticillium wilt. He used neutron or X-ray for irradiations of peppermint cuttings. About 1,00,000 plants derived from irradiation treatment were grown in wilt infected soil. The mutants were presumed to be dominant. Murray reported the successful use of irradiation breeding to obtain *Verticillium* wilt resistant strains of peppermint.

Advantage of induced mutations

1. Somatic mutations would be useful in cases where an outstanding clonal variety has to be improved for a specific characteristic.

Limitations of induced mutations

1. Handling of clonal propagules is cumbersome.
2. Induced mutation breeding in asexual crops is time consuming.

Apomixis

In flowering plants meiosis and fertilization are involved in the development of embryo and allow genetic recombination, segregation, and production of different genotypes. However, plants of some species form their seed by an asexual process called apomixis. Apomixis is the asexual formation of a seed from the maternal tissues of the ovule, avoiding the processes of meiosis and fertilization, leading to embryo development. The word apomixis is derived from two Greek words: apo (away from) and mixis (fusion). The progeny of apomictic plants are exact replicas of the female parent. Apomixis is widely distributed among higher plants. More than 300 species belonging to 35 families are apomictic. It is most common in Graminae, Compositae, Rosaceae and Rutaceae. Apomictic relatives have been found in major food crops like wheat, maize and pearl millet.

In broad sense, apomixis signify all forms of asexual reproduction in higher plants which replace or act as substitutes for sexual reproduction (Stebbins, 1941). These include reproduction by means of bulbs, rhizomes, runner, etc. In restricted sense, it refers to asexual seed formation which is called agamospermy, that is, seed are formed without fertilization of egg cell. Agamospermy is divided into two main categories: adventitious embryony and gametophytic apomixis (Fig. 20.3).

1. *Adventitious (sporophytic) embryony:* an embryo is formed directly from a somatic cell in the ovule without any intervening formation of embryo sacs and egg cells. It was earlier called nucellar embryony. It is reported to occur in *Citrus* species and in Orchids.
2. *Gametophytic apomixis:* such apomicts usually show a complete sporophyte–gametophyte (alternation of generation) but unreduced embryo sacs are formed whose egg cells (rarely some other cells) develop parthenogenetically. It must involve a substitute of meiosis and for fertilization (i.e., apomeiosis), not found in flowering plants. Parthenogenesis

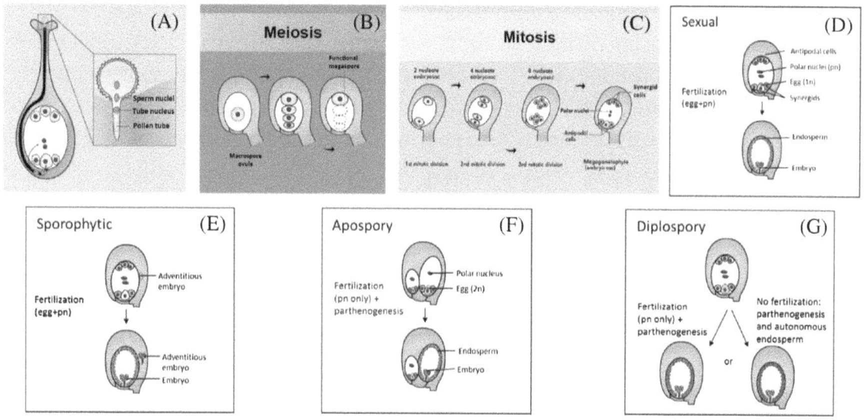

FIGURE 20.3

Sexual and apomictic seed development pathways. Embryo sac prior to fertilization and after fertilization is depicted. (A–D) Process of sexual reproduction that creates 2 maternal: 1 paternal endosperm and 2*n* sexually derived embryo. (E) Sporophytic apomixis that creates 2 maternal: 1 paternal endosperm, 2*n* sexually derived and 2*n* clonally derived adventitious embryo. One embryo is genetically identical to mother and other a product of sexual reproduction. (F) Apospory apomixis that creates 2 maternal: 1 paternal endosperm and 2*n* clonally derived embryo. (G) Diplospory apomixis that creates (left) 4 maternal: 1 paternal or 4 maternal: 2 paternal endosperm and 2*n* clonally derived embryo and or (right) 2 maternal: 0 paternal endosperm and 2*n* clonally derived embryo. In apospory and diplospory, embryo is derived from the egg cell of unreduced embryo sac, specifically nucellar cell in apospory and megaspore mother cell in diplospory.

Adapted from Conner, J.A., Ozias-Akins, P. 2017. Apomixis: engineering the ability to harness hybrid vigor in crop plants. In: Schmidt, A. (Ed.), Plant Germline Development: Methods and Protocols, Methods in Molecular Biology, vol. 1669. Humana Press, New York, NY, pp 17–34. DOI 10.1007/978-1-4939-7286-9_2.

cannot be identified cytologically but is detected by the presence of haploid offspring. It gives rise to unreduced gametophytes in apomicts. They fall into two categories:

a. *Apospory:* an embryo may develop from unreduced egg in an embryo sac derived from an aposporous or nucellar cell (somatic cell). Twin and sometimes triplet seedlings are common in the progeny of plants that produce by apospory. This occurs in some species of *Rubus* (raspberries, blackberries), *Allium* (onion, leek), and *Opuntia* (prickly pear).

b. *Diplospory:* an embryo may develop from mitotic division of the megaspore mother cell (archesporial cell). It is an important method of reproduction in genus *Eragrostis* and in *Tripsacum.*

For an anatomical viewpoint, the limit between diplospory and apospory is difficult to draw in certain genera. However, in certain grasses of the family

Panicoideae, the aposporous embryo sacs are monopolar and four nucleated, and sexual ones are eight nucleated. In most apomictic plants, fertilization of the polar nuclei by a sperm (pseudogamy) is necessary for endosperm development and seed viability. This is also a deviation from normal reproductive behavior and characterizes pseudogamous apomicts, an aberrant endosperm formation. Cytological examination of developing ovaries during megasporogenesis and embryosac development is required to distinguish among the four mechanisms of a gamospermy (adventitious embryony, apospory, diplospory, and parthenogenesis).

Obligate and facultative apomixis

Regardless of the mechanism involved, the gametophytic apomixis may be obligate or facultative. The apomictic plants that have no capacity for sexual reproduction are called *obligate apomicts*. The sexual process is replaced completely by asexual methods of reproduction. Both sexual and apomictic reproduction coexist in a plant of *facultative apomixis*. This term has often been used to describe an entire species. Species without some sexual reproduction are very rare, but there are ecotypes that seem to be obligate apomicts. A species may consist of some biotypes that are obligate apomicts and others are completely sexual. In bahiagrass (*Paspalum notatum*), the sexual strains are diploid and apomicts are polyploid.

The progeny test may be conducted to determine the method of reproduction in plants. The progeny of an obligate apomict will be completely uniform, and each plant will be identical to the maternal parent. The progeny of a facultative plant is capable of producing identical offspring that resemble the maternal parent; and as a result of some sexual reproduction, variable plants are usually present among their progeny. The proportion of apomictic progeny produced by facultative apomicts (partially sexual) has been shown to vary with several factors including the chromosome number of the pollen parent (Stebbins, 1941), and the environmental conditions under which seeds are produced (Knox, 1967). Two methods of reproduction could be observed occasionally in facultative apomicts (Asker, 1979). These are fertilization of unreduced egg cells and parthenogenetic development of reduced egg cells (Table 20.2). The hybrids formed by fertilization of unreduced egg cells and reduced egg cell (parthenogenetic) are called U-hybrids and R-hybrids, respectively.

Table 20.2 The four methods of reproduction possible in facultative apomixis.

	Offspring formed by	
Egg cells	**Parthenogenesis**	**Fertilization**
Unreduced	Uniform maternal offspring	Autotetraploids (U-hybrids)
Reduced	(Poly-) haploids	Variable sexual offspring (R-hybrids)

After Asker, S., 1979. Progress in apomixis research. Hereditas 91, 231–240.

Identification of apomixis

In general, the deviations from normal breeding behavior may suggest the possibility of apomixis in both cultivated and wild species. However, these deviations should be checked with more detailed crossing, progeny testing and/or cytological methods for confirmation of apomixis (see, Hanna and Bashaw, 1987). Some of the indications of apomixis are:

1. Limited or no genetic variation in the F_2 progeny of a cross between two distinct parents. It first needs to be established that F_1 was not a self.
2. When the F_1 progeny resembles the mother plant, confirmation of apomixis through cytological studies of maternal parent is required.
3. Recessive genotype from a cross of recessive apomict crossed to homozygous dominant pollinator.
4. Structural heterozygosity or aneuploid chromosome number that remains constant from parent to progeny.
5. Unusually high seed fertility in crosses that are expected to be sterile.
6. Multiple seedlings per seed, multiple stigmas, multiple ovules per floret and double or fused ovaries.
7. The presence of more than one embryo sac in an ovule is a confirmatory test of apomixis. The screening involves a careful and systematic tracing of steps for development of embryo sac and embryo through microtomy of ovule right from megaspores to embryonic development.
8. Identifying and mapping the genes regulating apomixis may be useful in transferring them to crops through genetic engineering. Apomixis is closely related to supergene (a group of genes that are always transmitted together as a package). Researchers reported that plants with mutation for omission of second division (*OSD1*) gene lacked the second round of cell division that occurs during meiosis, and *OSD1* mutant in combination with mutations in two other genes that affect meiosis, resulted in a genotype in which meiosis is totally replaced by mitosis without affecting subsequent sexual processes (d'Erfurth et al., 2009). Due to the lack of reduction division of meiosis, cells were diploid similar to the body cells of the plant, rather than haploid as expected in normal reproductive cells.

It may be noted that it is only recurrent apomixis, namely, diploid forms of *apospory* or *parthenogenesis* or *apogamy* or *adventitious embryony* and *vegetative propagation* are useful in plant breeding as they produce viable diploid embryo without fertilization and continue to perpetuate over generations. Non-recurrent apomixis is of academic use only.

Use of apomixis in plant breeding

In this section, apomixis term is used to denote asexual formation of seed, that is, agamospermy. Apomictic reproduction may vary from obligate to facultative (partially sexual) with varying degrees of apomixis both within and between species. Apomixis could offer an unique opportunity in developing superior cultivars in some species. However, the prerequisites are availability of cross-compatible sexual and apomictic biotypes, the recovery of apomictic progenies from the F_1 and/or in succeeding generations. The crosses of sexual × apomictic biotypes would produce large genetic variation. This is due to the heterozygous nature of apomictic parent.

Breeding procedure of apomictic varieties of buffelgrass, *Pennisetum ciliare* has been outlined by Taliaferro and Bashaw (1966) (Fig. 20.4). Two obligate apomictic biotypes of buffelgrass; blue and common and clones from the sexual variant of blue buffelgrass plant discovered in 1958 were used (Bashaw, 1962). The chromosome number of $2n = 36$ has been reported for each of the above mentioned biotypes. The blue, common, and sexual buffel grass types were referred as BB, CB, and SB, respectively. Immense variation was observed for mode of reproduction in the selfed (S_1) progenies and in the F_1 hybrids indicating that tremendous heterozygosity must exist in this species. Classification of S_1 and F_1 plants for mode of reproduction was achieved by embryo sac analysis or study of progeny variability or both. In both cases (S_1 and F_1), the female parent was a completely sexual plant that was heterozygous for genes controlling (method of) reproduction and the male parent in the crosses were obligate apomicts.

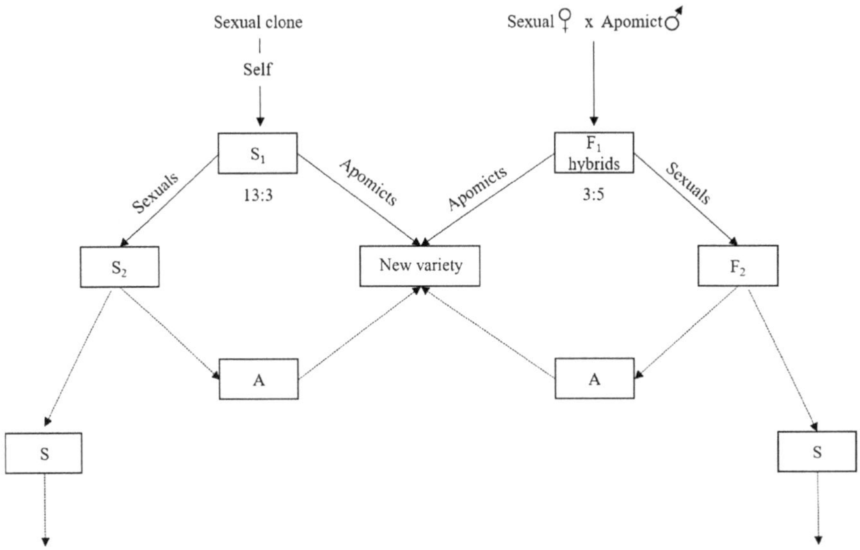

FIGURE 20.4

Scheme for developing new apomictic of buffelgrass, an obligate apomictic. A = Apomictic, S = Sexual.

After Taliaferro, C.M., Bashaw, E.C., 1966. Inheritance and control of obligate apomixis in breeding buffelgrass, Pennisetum ciliare. Crop Sci. 6, 473–476.

Therefore, the progeny of S_1 also provide a source of apomictic plants for selection. Two types of plants, completely sexual and obligate apomictic, were produced in the ratio of 13: 3.

The hybrids were obtained by using a heterozygous sexually produced plant (SB) as female and pollinated by an apomictic male BB/CB. The progenies of the F_1 hybrid(s) segregated into three apomicts and five sexual plants. The superior apomictic plants were selected. Such plants take maximum advantage of hybrid vigor. Selection of apomictic plants is also possible in the F_2 generation but with a considerable loss of vigor. Since obligate apomixis fixes the genotype and prevents all genetic variation except by mutation, any apomictic buffelgrass line superior to existing cultivars could be increased and used as variety. Selection of superior apomictic plants among S_1 or its progeny offers most rapid means of developing new varieties. One very promising strain of buffelgrass was selected. It was a rhizomatous, green leaf type with an excellent forage producer equal in seed production to CB. However, continued improvement may be possible by repeated crossing of superior sexual plants with superior apomictic male pollinators. Superior obligate apomictic plants selected at each generation would represent potential new cultivars. More recent examples can be perused in *Journal of Plant Registrations*.

Maintenance of apomixis

If an apomictic plant is detected, its inheritance should be studied by crossing part of the flowers with pollen of sexual plants or normal plants and seed from remaining flowers are collected. The segregation pattern of selfed and F_1 plants should be used for understanding the inheritance of apomixis. A true apomictic plant would produce progenies like parental mother apomict plant.

Exploitation of apomixis

In sexual crops, to exploit apomixis, it is required to detect/to identify spontaneously occurring apomictic plant and to transfer it artificially through crossing the apomicts and sexually producing plants (amphimicts). Obligate apomixis has been found in wild relatives of some cultivated plants for instance in *Elymus rectisetus* and *Tripsacum dactyloides*, wild relatives of wheat and maize, respectively, as well as in wild species of genus *Pennisetum*. Apomixis is also reported in many grass genera. The transfer of genes controlling apomixis from wild to cultivated species and to successfully produce hybrids involves; manipulating ploidy levels of either wild or cultivated species, handling of large populations and efficient screening methods. Sometimes, production of bridging hybrids may be required to obtain better crossability or to improve fertility. Molecular techniques may be applied for transferring genes between genera that are otherwise not crossable.

Superior apomictic varieties have been developed and released in case of Kentucky bluegrass (*Poa pratensis*), bahiagrass (*P. notatum*), and buffelgrass (*Pennisetum ciliare*) that demonstrates the potential use of apomixis in plant breeding. It is also an active research interest area for cereal crops where hybrid cultivar production is difficult.

Advantages of apomixis

1. Hybrid vigor or heterosis can be fixed (regardless of heterozygosity) through the use of apomixis.
2. The exploitation of maternal effect is possible generation after generation.
3. In crops where cytoplasmic genetic male sterility or restorer systems are not available, apomixis would make production of commercial F_1 hybrids possible.
4. Superior obligate apomict could be increased by open pollination for number of generations without loss of vigor. Outcross contamination is eliminated in obligate apomicts and isolation is not necessary to produce F_1 or to maintain and increase parental line.
5. New apomictic genotypes could be easily produced by hybridization of sexual and apomictic plants.

Limitations of apomixis

1. Ideal reproductive behavior may not be realized in all species.
2. Before the use of apomixis in cultivar development, it is necessary to understand its facultative behavior, inheritance, effect of environmental factors, ploidy, and seed sterility.
3. Apomixis could reduce biodiversity as it fixes genetic recombination, promotes asexual reproduction, and sometimes could create genetic combinations that are not even viable for sexual reproduction.
4. Sudden environmental changes and evolution of new and more virulent strain (s) of pests and diseases would be disastrous to apomictic crop cultivars because there would be few apomictic varieties and a smaller available gene pool.

CHAPTER

Breeding for resistance to abiotic stresses

21

Abstract

Abiotic stresses are non-living types of conditions, including environmental and edaphic factors etc., that impact crop production. These include water stress (flooding and drought), temperature (extreme cold and heat), nutrients deficiency and toxicity (major and minor elements), soil pH (acidic or salinity), excess light, and mechanical stress. In this chapter, major abiotic stresses in crop plants are described including causal agents, symptoms, types, and mechanisms. Several breeding approaches (direct and indirect) are included, with emphasis on the development of suitable selection criteria, parental selection, hybridization and selection, mutation breeding, doubled haploid, and tissue culture. Integration of molecular breeding methods in the breeding pipeline is also included.

Abiotic stresses include water, temperature, nutrients deficiency and toxicity, soil pH, etc. With the movement of agriculture outside of riverbanks and in an ever-increasing climatic variability, cultivars need to have an in-built in package of tolerance against abiotic stresses. The worst affected regions are in the semi-arid zones with most rapid population growth and most vulnerable food production due to over grazing, over cultivation, and the large seasonal variability in rainfall and indiscriminate use of ground water irrigation. With the increase in area under irrigation, salinization too increases. Salinity forms a complex of problems such as sodicity and high pH and creating problems of soil structure, water permeability, and nutrient imbalances. Due to erratic weather events (infrequent but large rainfall events), waterlogging is also an issue that plants have to bear. Weather patterns are characterized by intense rainfall followed by long drought periods in the crop growth cycle subjecting plants to cycles of flooding and drought in the same season. The spectrum of abiotic stresses constraints plant growth and development resulting in low crop yield and reduced profitability.

The food production must be increased regularly if the demands of an increasing world population are to be met. This could be achieved either by production from expanded land areas (utilizing marginal lands) or by increased production from higher yields of existing good arable lands. Genetic solutions (i.e., improved cultivars) can be a boon for marginal lands if they can withstand environmental challenges of that region. Extension of cropping to marginal land areas will

Plant Breeding and Cultivar Development. DOI: https://doi.org/10.1016/B978-0-12-817563-7.00017-9

present many non-living stress problems. Mineral iron deficiency or toxicity, drought, salinity, acidity, alkalinity, high and low temperatures, and air pollutants are factors that limit crop productivity in these environments. The limitations to increase yields on existing cropped areas can be partially overcome with greater production inputs and more usage of prescriptive cultivars that can produce higher yield in such regions and farmlands. Increasing production inputs are costly in labor and energy and are beyond the reach of small and marginal farmers, as well as reduce the net farm income in high income countries. The research on the genetic control mechanisms of plant responses and breeding tools will lead to the development of new crop cultivars specially bred for adaptation to stress situations. Before undertaking a breeding program for abiotic stress tolerance, the following points should be considered:

1. The specific problem(s) must be identified for the region for which the breeder is attempting to develop cultivars.
2. The problem must be severe and widespread to merit plant breeding attention, and investment of resources by the breeding organization.
3. The problem must not be amenable easily and economically through other means.
4. The problem must be solvable through breeding, which requires screening techniques to assay plant response to stresses, and useful genetic variation for the trait to exploit.

Therefore, from the very inception of a program to develop crop cultivars to fit specific stress environments, the expertise of several disciplines including soil science, agronomy, plant physiology, molecular genetics, and plant breeding are needed for the problem definition and goal development.

Mechanisms of resistance

Plant resistance to an environmental stress refers to a measure of the plants ability to survive and even to grow productively in the presence of stress. A stress is recognized by its effect on the plant. The degree to which an individual or genotype is able to withstand or recover from a period of stress is a measure of its resistance.

Response to abiotic stresses is a complex trait, which involves numerous metabolic pathways, cellular, and molecular components. Mechanisms for several abiotic stresses are interconnected that leads to cellular damage and secondary stresses including osmotic and oxidative stresses. Briefly, the initial stress signals are perceived by membrane receptors that transmit these signals downstream to trigger transcription, regulated by hormones, transcription factor (TF) binding proteins, miRNAs, and TFs to activate stress response mechanisms. The aim is to reestablish homeostasis, protect, and repair damaged proteins and membranes.

Differential response is observed among plants pointing to varying response, driven by main and epistatically networking genes, and interactions between genetics and environment. The two main mechanisms used by plants exhibiting stress *tolerance* or stress *avoidance* are *acclimation* and *adaptation*. These have evolved primarily through natural selection as several species show conserved response when subjected to similar environments. Upon stress recognition, regulatory responses work to reestablish cellular and organismal homeostasis or lessen episodic shock effects. These functional conservations are beneficial for a breeder, because these can be utilized to develop more tolerant cultivars for multiple stresses and across crops.

Types of abiotic stresses

Water deficit stress

Water stress is one of the most important abiotic stress factors affecting plant growth and development. Both water deficit (drought) and water excess (flooding) can be a source of plant stress. Drought is a meteorological term, and is commonly defined as a period without significant or insufficient rainfall in rainfed crop production. Under rainfed agriculture with insufficient rainfall and where means to irrigate crops do not exist, water deficit stress can be quite damaging. In literature, water stress and drought stress are used synonymously; however, water deficit stress is more appropriate to describe a situation where plants do not have required water to function normally. Term drought stress can also be used for water deficit conditions, but the word "drought" has complex meanings and definitions; therefore, drought stress should be clearly explained including the context it is used. Generally, drought stress is used to describe a situation of prolonged period of water deficit. Drought stress disrupts the soil–root–plant–air continuum and water potential gradient, causing ionic and osmotic constraints on cells causing yield losses and other trait deviation from normal expression.

The amount of damage done to the plant's growth and development is dependent on the duration and severity of water deficit period. The variability in plants response to drought conditions refer to the ability of a genotype to be more productive given the same soil moisture as another genotype when grown together, and the expression of trait variability in a breeding population. Nearly all plants are subjected to some degree of water stress during their life cycle. Plants may encounter mild or severe drought, and this water stress affects the plant metabolic processes. More seed yield reduction is observed in severe or prolonged stress. In the absence of artificial irrigation, possible solutions include; usage of cultural practices that increase the availability of stored soil moisture or development of cultivars (pure line, clones, hybrids, and synthetics) that better tolerate water deficit stress. Different mechanisms are used by crops to mitigate losses due to drought, for example, radiation shedding and stomatal closure would be more beneficial than a reduction in leaf area from leaf senescence or shedding as it will lead to higher yield loss. Many other mechanisms are deployed by plants, such as root system architecture with better sensing and response, leaf pubescence with trichomes assisting with foliar absorption of water in semiarid regions, stomatal closure to regulate leaf water potential, cuticular wax to reduce water loss, canopy temperature, morphological responses such as leaf rolling, curling, and drooping.

Based on seasonal soil water availability patterns, drought could be of three types: (1) terminal, (2) unpredictable, and (3) unpredictable and terminal. The mechanisms of drought tolerance/resistance can be divided into three categories: phenological, morphological, and physiological. Escape and avoidance are important for drought resistance. In addition to these, tolerance and recovery may also play an important role in drought resistance. These mechanisms are briefly described below.

Escape

In regions where growing season is short and terminal drought stress is present, drought escape is an important strategy. In drought escape, the phenological development matches the period of soil moisture availability, which decreases the impact of drought stress on crop yield and production as plants will complete their development, including reproductive cycle, before severe water deficit comes. Early maturity is therefore one of the most important factor of drought resistance through escape mechanism. As an example, local cultivars of cowpea [*Vigna unguiculata* L. (Walp.)] flower progressively earlier going from south to north through the Sudanian and Sahelian zones of Africa toward the Sahara Desert.

Avoidance

Avoidance is the ability of a plant to maintain high water status during drought periods. This mechanism is suggested on the premise that while plants can not alter external environmental stresses, it can prevent or decrease the penetration of the stress into its tissues. Stress avoidance is broadly divided into two groups: maximizing water uptake through improving the root structure and function that allows plant to acquire sufficient water, and optimize the use of absorbed water for the production of dry matter and seed yield.

Plants with avoidance mechanism are able to exclude stress, either partially or completely, by a physical barrier (insulating its living cells from the stress) or by a steady state or exclusion of the stress (a chemical or metabolic barrier). Examples of physical barrier are leaf cuticular waxes, as plants with thick layers of cuticle and stomata are likely to resist the effects of drought. The amount of cuticular waxes correlates with water permeability. Stomata act as plants protective mechanism by decreasing water loss through closure during periods of plant water deficits as well as influencing the rates of photosynthesis and respiration. Low stomatal frequency may be associated with drought tolerance. Stomatal resistance has been investigated as a tool to determine drought tolerance in multiple crops. Traits that provide drought avoidance can be useful for breeding targets. Breeders can conduct research to determine that there are no unintended negative outcomes of breeding for such traits.

Adaptive mechanisms, which enable a crop to maintain water uptake, would be combinations of a denser root habit (greater root–soil interface/cm^3 of soil) with deeper rooting, as well as more efficient root system architecture traits that allow easier access and efficiency of water uptake. The traditional upland rice (*Oryza sativa* L.) of Southeast Asia and West Africa has deep and thick roots and they compare well with those of wheat cultivars. Drought avoidance through deeper roots (greater root growth and coverage) and extraction of soil moisture is an important drought tolerance mechanism, but its usefulness is limited if soil conditions restrict root growth. As the depth, width and branching of root systems increases, plant water stress decreases. Deep roots can access moisture reserves. It is often being speculated that materials tolerant to poor soil fertility conditions should have greater root growth, and thus perform well under drought conditions in similar sites. In some newer cultivars, the effect of drought on growth of tap roots in drought resistant and sensitive cultivars has been shown to be smaller and not significant, that is, there have been reports of drought-resistant cultivars with smaller shoots and larger root systems than drought sensitive cultivars. In wheat, it has been shown that some semidwarf cultivars have better yield in water limiting conditions than other elite normal height cultivars; therefore, this old adage of "taller plants have deeper roots" or "above ground height equate below ground root length" are no longer a universal rule, and breeders have broken these associations.

Leaf rolling is perhaps the most universally obvious symptom of drought. The onset of leaf rolling (indicating a decrease in turgor pressure potential), the extent of rolling and the elasticity in unrolling among rice cultivars during a diurnal cycle of water stress have been used as an indirect score of drought tolerance at International Rice Research Institute (IRRI) breeding program and other programs worldwide. Wilting score has been used as a proxy for drought tolerance in soybean, and genetic variation has been exploited to develop drought tolerant cultivars. Measurements of leaf water potential and stomatal diffusive resistance indicates that the physiological reasons for the differential drought response among genotypes may be due to more effective protection of the level of tissue hydration and due to increased stomatal diffusive resistance in tolerant genotypes. These genotypes have a tendency to a more complete return to the level of control plants after the recovery period. Reduction in photosynthetic rate, growth, transpiration rate, and internal CO_2 concentration are all physiological responses to drought stress.

Different abiotic stresses can have conflicting or antagonistic responses. For example, under heat stress, plants open their stomata to cool their leaves by transpiration. However, in situations of heat and drought stress, plants would not be able to open their stomata resulting in higher leaf temperature. Nutrient stress can compound or aggravate the problem of plants facing temperature and water stress. Plants' ability to acclimate to abiotic stress conditions requires energy and resources.

Tolerance

Tolerance is the ability of a plant to withstand water stress, which is measured by the degree and duration of low plant water potential. Due to the nature of assessment and sometimes disagreement between internal water stress and external symptoms, it is a difficult trait to breed. Both dehydration avoidance and tolerance are responsible for water deficit tolerance. Mechanisms that protect cell structure, organelle membranes, and proteins are implied in tolerance.

Drought recovery

Drought vary in duration and when rainfall commences, rapid recovery and return to active growth and development are important features of drought recovery, which is defined as the ability of a plant to satisfactorily resume growth and yield after experiencing drought stress. For example, in rice drought recovery is correlated with vegetative growth vigor, high tillering ability, and ability to maintain high leaf water potential.

Chemical

Both abscisic acid (ABA) and proline accumulation have been detected in drought stressed plants. ABA appears involved in stomatal function under drought but since it accumulates in other tissues, additional functions may be involved. Increased expression of ABA biosynthetic genes and ABA responsive genes are observed. This phytohormone has been associated with stomatal closure to reduce transpiration as well as it has been shown to play a role in the induction of genes that enable drought acclimation in various plant species. Overexpression of ABA receptors has shown to improve drought tolerance but in non-stress conditions yield reduction has been reported. Accumulation of proline, nitrate and ammonium ions may contribute to osmotic adjustment and reflect inhibition of protein synthesis. The biochemical and molecular responses include reduced photochemical efficiency, decreased RuBisco activity, accumulation of metabolites (e.g., proline and monodehydroascorbate reductase), increase in other antioxidative enzymes, and reduced reactive oxygen species (ROS) accumulation.

Waterlogging stress

Atmospheric stress in the soil has been referred to oxygen deficiency, waterlogging, and anaerobiosis. All involve a build-up of carbon dioxide, ethylene, and other potentially toxic gases, and depletion of oxygen that leads to a reduction of aerobic respiration. The response of plants to anaerobic soil conditions has been described as flooding damage and excess moisture energy. Levitt (1980b) defined flooding as the presence of water in soil in excess of field capacity.

With the exception of rice, most agricultural crops experience only transient or short term water logging. The duration of water logging may range from few hours to several weeks. The stress due to flooding occurs in the below ground part of the plant. Only root is exposed to flooding stress, the injury to the shoot is not a primary effect of excess water. The flooding replaces gaseous air by liquid water, leading to the gas stress. The following are the effects of water logging:

Morphological and anatomical effects: most agricultural crops require a suitable growing environmental condition that is neither too dry nor too wet ensuring maximum growth and productivity. In waterlogged soils, air space is filled with water. The oxygen remaining in the soil, either dissolved in water or trapped in air cavities, is quickly depleted by respiration of plant roots and soil microorganisms. There is an imbalance between this depletion of oxygen and the replenishment of oxygen in soil because atmospheric O_2 slowly diffuses into waterlogged soil, plunging roots into an anaerobic condition. If such conditions persist for a long duration, the reducing processes in the rhizosphere further aggravate plant's condition. Waterlogging conditions are worsened if fields have poor drainage, which causes standing water or water saturated soils. Standing water visible in field is not the only waterlogged soil condition; and even if there is a lack of standing water but soil air spaces are filled with water, waterlogging exists.

Typical effects of waterlogging on the growth of shoots include reduced elongation, chlorosis, senescence, abscission of the lower leaves, wilting, hypertrophy, epinasty (downward growth of petioles), adventitious root formation on the lower portion of the stem, lenticel formation, aerenchyma formation, leaf curling, and decline in growth rate. Higher root porosities have been reported in tolerant varieties of wheat, maize, and sunflower. Plants that form increased intercellular air spaces (aerenchyma) in the cortex provides canals parallel to the axis of the root through which gases can diffuse longitudinally. Also, plants that form adventitious roots close to the soil surface where oxygen tension is usually higher, are more quickly restored after transient water logging, for instance in tomato.

Physiological and biochemical effects: the physiological and biochemical effects of waterlogging include; changes in respiratory metabolism, root permeability, water and mineral uptake, nitrogen fixation, and endogenous hormones. Oxygen deficiency in root systems disrupts root metabolism by forcing plant to switch from aerobic to anaerobic respiration resulting in a decrease of adenosine triphosphate production, accumulation of toxic end products of anaerobic respiration, and rapid depletion of organic compounds. The limited supply of available energy reduces the absorption and translocation of water and nutrients. The disruption of root metabolism caused by inadequate oxygen supply also adversely affects the hormonal balance of the shoot and suppresses the synthesis and translocation of hormones in the root.

Of the various plant hormones studied, one of the most striking hormonal changes found in waterlogged plants is the dramatic increase in ethylene concentration. Information on the regulation of ethylene biosynthesis provides further tests of the hypothesis that 1-aminocyclopropane-1-carboxylic acid (ACC) from the anaerobic roots is responsible for accelerating the shoot ethylene production. The pathway of ethylene biosynthesis from methionine has been demonstrated in both reproductive and vegetative tissues. 1-Aminocyclopropane-1-carboxylic acid synthase (ACS), the key enzyme that converts *S*-adenosyl-methionine to ACC, has been extracted from tomato fruit and characterized. Since this is a rate-limiting step in the sequence, factors such as indole acetic acid, wounding, anaerobiosis, or senescence which accelerate ethylene synthesis, have their primary effect on the activity of ACS. In contrast, inhibitors of ethylene synthesis can act at either of the final two steps of the pathway. Aminoethoxyvinylglycine (AVG) and aminooxyacetic acid both inhibit ACS; while cobaltous ions, uncouplers such as 2,4-dinitrophenol, and high temperatures, as well as anaerobiosis, block the synthesis of ethylene from ACC. When AVG is supplied to anaerobic roots, the accumulation and transport of ACC is inhibited, ethylene production is reduced, and epinasty is prevented. On the other hand, CO^{2+} exerts little effect on the level of ACC in the xylem sap, but effectively inhibits both ethylene production and epinasty. These observations are consistent with the proposed role of root-synthesized ACC in the etiology of the flooding syndrome. One of the best characterized ACC transport systems is in tomato plants when subjected to flooding (or root hypoxia), and the translocation of ACC from roots to shoots in response to stress. In rhizosphere, lack of oxygen induces the expression of ACS in roots resulting in an increased ACS activity. Due to a lack of oxygen and absence of ACC oxidase (ACO) in roots, excess ACC is not directly converted to ethylene in roots. Instead, ACC is loaded into the xylem and transported to shoots and subsequently converted into ethylene by ACO present in the leaves. Differential expression of both ACS and ACO during hypoxia has been reported in tomato, sunflower, rice, maize, and *Arabidopsis*. Long distance transport of ACC has also been shown in presence of other root stress conditions, for example, drought, nutrient stress, and salinity. ACC deaminase containing bacteria have shown to reduce stress susceptibility of plants during drought, flooding, and salinity further demonstrating its role.

A famous example of submergence tolerance in rice comes from Indian landrace "Dhalputtia" that is a source of *Submergence1* (*SUB1*) (Xu et al., 2000, 2006) locus that provides submergence tolerance by restricting underwater elongation growth. Dhalputtia can survive more than 14 days of submergence compared with 7 days for other varieties. Fine mapping of *SUB1* revealed three genes encoding ethylene-response factor (ERF) DNA-binding proteins; and out of these only *SUB1A-1* allele was uniquely found in submergence-tolerant landraces. Subsequent studies confirmed the role of this gene, and reported the functional role of *SUB1* genes that under flooding gaseous phytohormone ethylene trapped within cells promotes *SUB1A-1* expression. This enhances the accumulation of two TFs: *SLENDER RICE 1* (*SLR1*) and *SLR1-LIKE 1* (*SLRL1*) that impede the response to phytohormone gibberellin (GA) resulting in restriction of cell elongation and less rapid consumption of energy reserves. When flooding recedes (i.e., de-submergence), *SUB1A-1* carrying genotypes show less tissue damage from

ROS. Also, dehydration associated with reaeration is less than non-*Sub1A-1* genotypes. The presence of healthy tiller meristems enables *SUB1A-1* genotypes to resume development when flood waters recede.

Temperature stress

Temperature stress is one of the major factors affecting plant growth and development; and therefore, crop yield. The temperature stress either could be due to high temperature (heat stress) or due to low temperature (freezing or chilling stress).

High temperature stress

The term thermo-resistance or thermo-avoidance may also be used for heat resistance and heat avoidance, respectively. The term thermo-resistance is used as synonymous for heat resistance/tolerance, which is an ability of a plant to survive a heat stress within its tissues. Based on response of the plants to high temperature it may be possible to classify genotypes as tolerant and sensitive. As a general rule, the acceleration of plant development by heat constitutes a detrimental effect on yield. High temperature stress is associated with death of cells, tissues, organs, or whole plants. Sun scald or bark burn is common in trunks or twigs of trees exposed to intensive solar radiation and temperature. Similar burns may be a plaguing problem in some genotypes of fruits and vegetables such as grapes, tomatoes, etc., which have poorer transpirational cooling. Such symptoms may occur when the organ's surface temperature reaches 48°C–50°C. Seedlings and reproductive organs are more liable to be injured by heat. The effect of heat is known to cause death of seedlings and flower abscission, pollen sterility, and poor fruit set at reproductive stage, and cause yield reduction. Heat stress can affect the structural integrity of proteins in cytoplasm and cause membrane protein denaturation and aggregation. Even post fertilization, heat stress can affect crop quality as high night temperatures have shown to reduce seed maturation reducing seed quality. Heat shock protein (HSP) is an important part of the acclimation mechanisms. Tolerant plants have the ability to maintain key functions of respiration and photosynthesis during heat stress by rapid changes in gene expression, increased HSP levels, and membrane modification. Plant responses to high temperature stress are closely linked to the water status of the plant. Wilting, leaf burn, and leaf folding or abscission serve as indicators of high temperature-induced damage. Reduced soil moisture can potentiate these symptoms, especially when the high temperatures occur during periods of rapid leaf growth and expansion. During growth and development, plants may avoid high temperature injury by mechanisms such as dramatically increased transpiration rate, higher leaf reflectance, or angle or orientation to the sun.

The temperature which kills 50% of plants/seedlings could be the heat killing temperature. The heat avoidance should be measured at a temperature just below the heat killing temperature and may be measured in growth chambers as the ratio of air temperature to leaf temperature. This is helpful in establishing useful screening nurseries.

Low temperature stress

Low temperatures in the growing season may reduce germination, may retard vegetative growth by inducing metabolic imbalances, and can delay or prevent reproductive development. Chilling temperature can damage tissues of sensitive plants, while freezing temperature will damage most tissues during active growth. Low temperature during winter seasons may cause the death of over wintering plants during their dormant phase. The over wintering habit, for example, winter wheat in North America, results in higher yields due to the abundance of early spring moisture from snowmelt available to the crop. Death of plants may occur during winter by: cell membranes disorganization during severe freezing exposure, anaerobic stress due to flooding and ice encasement, heaving of plants from soil by the formation of ice lenses, or due the activity of low temperature active pathogens under snow cover. The mechanisms of low temperature tolerance have been dealt in two subheadings titled as chilling stress, and freezing stress. The two groups may overlap in terms of stress and the response to stress at 0°C, when water in plant may freeze. However, by and large, the two groups differ in that, one is normally subjected to above-freezing (chilling) temperatures and the other subjected to freezing temperatures.

Chilling stress

A chilling temperature can be defined as any temperature that is cool enough to produce injury but not cool enough to freeze the plant. The theoretical definition of a chilling stress would be the number of degrees that environment temperature is below optimum for the plant activity being measured (Levitt, 1980a). However, in practice, this definition would be unworkable, because the optimum temperature for any plant process is not a constant, but may vary with other conditions. For the majority of plants, chilling stress refers to any temperature below 10°C–15°C and down to 0°C. The injury increases with the degrees of chilling and stages of plant development. Symptoms of injury include tissues necrosis, pitting, and discoloration of crops. There have been a number of reports of variability to chilling injury within chilling sensitive species. The chilling injury generally occurs in species that originated in tropical and sub-tropical regions and are grown in temperate regions; but not in plants that originated in temperate zones.

Freezing stress

Freezing stress, in its widest meaning, involves complex stresses and strains that occur when plants are subjected to sub-zero temperatures. Tolerance to freezing stress embraces terms such as winter hardiness, winter survival, frost resistance, etc. Therefore, freezing stress involves plant reaction to a situation when the temperature of the environment and the plant or its organ drop below 0°C, and various disorders consequently develop. The severity of stress varies depending on rate of cooling, minimal temperature, duration at the minimal temperature, and rate of thawing. Interactions with other stresses are common in a field environment, and plant growth and development stages are important as well.

A freezing low temperature stress or freezing stress may be defined as the freezing potential of the low temperature stress (Levitt, 1980a). Freezing injury consists of two main types: (1) direct injury due to intracellular freezing, and (2) freeze dehydration injury due to extracellular freezing. The former is explained to a direct physical effect of the intracellular ice on protoplasm, and resistance is due to avoidance of intracellular freezing. The latter is ascribed to the mechanical stress and solution effects, both of which increase with the degree of cell dehydration and contraction. The simplest measure of freezing stress is the number of degrees the environmental temperature is below the freezing point of pure water at atmospheric pressure (i.e., $-T$°C, where T is the temperature in °C). The freezing (frost or cryo) injury may occur in all plants; and therefore, it is more prevalent than the chilling injury. It may appear as tissues, organs, or plant death, or it may lead to specific symptoms such as frost cankers. Plants that are exposed to chilling injury are usually killed by the first touch of frost. The plants native to cold climates are frozen solid at the lowest temperatures without injury. Between these two extremes all degrees of freezing resistance occur in flowering plants. For mitigation, osmolyte accumulation and hydrophilic proteins synthesis are useful strategies against ice formation.

Soil nutrient stress

Nitrogen, phosphorus, potassium, calcium, magnesium, and sulfur are major elements (macronutrients) and normally the subject of routine fertilization. Many secondary elements (i.e., micronutrients) may be limiting; and therefore, need fertilization for proper crop growth. Similar to other previous stress described, genetic variability has been observed in most crops for their response to both excess (toxicity) and deficiency, while other genotypes grow well in the same soil.

Briefly, several nutrient deficiency symptoms are described for maize and soybean.

1. Nitrogen (N) deficiency: maize plants symptoms are pale, yellowish-green with spindly stems. Symptoms begin on the older and lower leaves and progresses to the newer leaves, because N is plant mobile nutrient. In soybean, less commonly observed symptoms are that lower leaves show chlorosis (yellowing) and may fall off with progression of symptoms. Newer leaves show a lighter coloration, and poor shoot and root development.
2. Phosphorus (P) deficiency: maize plants show dark green color, with older leaves showing reddish purplish leaf tips and margins. P deficient plants show stunted growth. New emerging leave do not show the typical deficiency coloration. Newly emerging leaves do not show such coloration. Deficiency symptoms are more prominent in younger plants. In soybean, purple color is

observed at the base of stem. Older leaves show a bluish dark green color and are smaller in size. Pod formation is negatively impacted.

3. Potassium (K) deficiency: symptoms in maize plants include yellowing and necrosis of leaf margins in the lower leaves. Symptoms are typically not seen in younger plants. Since K is plant mobile, symptoms progress up the plant as K translocates from old to young leaves. In soybean, chlorosis (yellowing) is observed at the border of leaflets. As the chlorosis progresses toward the center, necrosis is seen at the border and that too eventually progresses to the center. Pod formation is negatively impacted.
4. Calcium (Ca) deficiency: in maize, leaf tips stick to the next lower leaf, giving a ladder-like appearance which is typical for Ca deficiency in maize. Plants show stunting, as Ca is plant immobile. In soybean, Ca deficiency causes a reduction in meristem growth. The terminal bud is negatively impacted and petiole collapses. Newer leaves are wrinkled and chlorosis is observed (moves from border of leaflets to the center).
5. Magnesium (Mg) deficiency: in maize, first symptoms are yellow to white interveinal striping of the leaves in lower canopy. Beaded streaking is observed due to dead parts of leaf tissue. Mg is plant mobile, and in older leaves, reddish-purple color and, in severe deficiency tips and edges show necrosis (browning). In soybean, yellowing of leaf borders with interveinal chlorosis is seen, leading to drying of the border leaf tissue.
6. Sulfur (S) deficiency: in maize, seen on younger plants as an overall general yellowing of the plants, with overlapping symptoms to N deficiency. Unlike N deficiency, yellowing of leaves in upper younger leaves is seen as S is relatively plant immobile. Plant stunting is observed. In soybean, symptoms are typically stunted plants, and pale green leaflet color. Symptoms are similar to nitrogen deficiency; however, chlorosis is generally more apparent on upper (newer) leaves.

Breeding approaches will require usage of appropriate genetic material with trait expression, genetic variation and appropriate selection sites where uniform stress can be applied and differential response observed. Improving nutrient use efficiency involves increasing nutrient uptake, nutrient utilization efficiency, and nutrient harvest index, each involving many crop physiological mechanisms and agronomic traits. For example, to breed for improved nitrogen use efficiency approaches have relied on targeting nitrogen uptake efficiency, as well as nitrogen utilization efficiency, which are subdivided into nitrogen reallocation, nitrogen harvest index, and stay green in cereal crops. In each crop, physiological understanding is required to determine optimal trait targets for specific nutrients.

Salinity and salt stress

Saline soils are described as those that have an electrical conductivity (EC) of the saturation soil extract (ECe) of more than 4 dS/m at 25°C (Richards 1954).

However, the most widely accepted definition of a saline soil is proposed by the Soil Science Society of America, is one in which the ECe is greater than 2 dS/m. The EC is directly proportional to the salt concentration in solution. It is measured in deciSiemens/meter (dS/m), which has replaced milli or micro-m hos cm.

Salt-tolerance means the ability of plants to grow satisfactorily on saline soils. Levitt (1980b) used the term "salt resistance" in a broader sense, which includes: (1) salt tolerance—when plants respond to salinity either by accumulating salts generally in their cells or glands and (2) salt avoidance—when plants avoid salt stress maintaining their cell salt concentration unchanged either by water absorption or salt exclusion, that is plants with salt avoidance can grow on salty soils, whether these accumulate or exclude the salts. Although the effects of a salt are due to its ions, a distinction is made between a salt stress and an ion stress. If the salt concentration is high enough to lower the water potential appreciably, the stress is called a salt stress. If the salt (or acid or base) a concentration is not high enough to lower the water potential appreciably, the stress is called an ion stress (Levitt, 1980b). Salinity in soil or water presents a stress condition to crop plants that is of increasing importance in the sustainability and stability of crop production in semiarid regions of the world. These soils contain soluble salts in amounts high enough to interfere with plant growth, but not enough exchangeable sodium to alter the soil characteristics appreciably. These soils have good physical structure with satisfactory permeability and can be reclaimed easily through the simple process of leaching with good quality water if the water table is also lowered simultaneously.

Chlorides and sulfates salts (soluble) are most common, and are of sodium, calcium, and magnesium types. Sodium and chloride are the predominant ions in highly saline soils. The most common symptoms of saline soils in field conditions are spotty crop growth and often white salt crusts on the surface. Plant growth and development are impaired due to salinity via water stress, excessive uptake of ions, for example, Na^+, Cl^-, causing nutritional imbalance and cytotoxicity, and oxidative stress due to generation of ROS. In mild salinity conditions, growing plants often have a blue-green tinge. In cereal species, barren spots and stunted plants are observed; and severity and frequency can be used as a proxy to determine the level of salts in soil. In moderate salinity, if uniformly present in the field, restricted growth is the only indication without soil testing. Plants in saline soils can be confused with drought stress symptoms; however, wilting is less common in salinity due to plants ability to adjust their internal salt content to maintain turgidity. This happens due to gradual changes in the osmotic potential of soil solution allowing plants to adapt to an extent, although yield reduction will happen. Research and breeding efforts have led to the identification and utilization of genes that provide improved performance in plants under saline stress.

A sodic soil is defined as a soil in which the sodium absorption ratio [SAR = $Na^+ / [(Ca^{2+} + Mg^{2+})/2]^{0.5}$] is greater than 15 (Bresler et al., 1982). The sodic condition is often associated with a high pH ($>$9.0 in 1:2 soil/water suspension) because of precipitation of Ca as $CaCO_3$ at a high pH. Thus, sodic and alkali soils are considered to be synonymous. The soils have a hard clay layer or cemented bed of amorphous or concretionary lime in the subsoil. These soils are often low in organic matter. The organic matter, which dissolves in this highly alkaline medium, is deposited on the soil surface and imparts a dark black color of the soil. The soils are known as black alkali soil. The reclamation requires application of amendments. In general, the plant suffers from low availability of nitrogen, calcium and zinc, and other trace elements cause problems either of toxicity or of deficiency. In several crops, the ability of plants to retain potassium ion in plant tissues has been reported to play an important role in salt tolerance, including in cell signaling during stress. Genetic mechanisms and gene sources may be different depending on the type of salt causing stress; therefore, breeders need to understand the types of salt impacting crop they are developing cultivars.

Acid mineral stress or acid soil stress

An acid soil normally forms when the basic ions supplied to the soil or from the parent material are less than their losses by leaching. These soils are frequently low in available P, low in base exchange capacity, and high in leaching capacity. Thus, nearly all nutrients have to be added to maintain fertility in these soils. Acidity increases the availability of Fe, Mn, and Al, and at high levels these elements are often toxic to plants. Liming is usually needed to raise the soil pH to overcome toxicity problems.

Plants adapted to acid mineral soils utilize a variety of mechanisms to cope with adverse soil factors. Both tolerance and avoidance are most probably required simultaneously, although to varying degrees. Of the tolerance mechanisms, better recirculation are presumably less important than high tissue tolerance to toxic mineral elements, Mn and Al in particular. In most acid soils, Al and Mn occur together because of similarity in the pH dependent solubility of Mn oxides and Al containing soil minerals. Mn toxicity, unlike Al ($pH < 5.0$), can occur at higher pH (> 5.0) in poorly drained or compacted soil. The principal effect of Al toxicity in plants is severe restriction on root growth because Al directly inhibits cell division in the root apical meristem. Plant tops affected by Al toxicity show symptoms similar to those of P deficiency.

Boron toxicity stress

Boron (B) is essential micronutrient for the growth of vascular plants, but may also result in toxicity when present in excessive amount. The range between a deficient and a toxic level of B is very narrow, hence an imbalance in B nutrition are widespread. B deficiency is most prevalent in cool, humid regions and has been described as the most widespread of all known trace element deficiencies. High concentrations of B may occur in the ground water and soil of arid and semi-arid environments (low rainfall areas), and high B concentrations are most frequently observed with soils formed in parent material of marine origin and deep seated fault systems. The sensitivity to B toxicity in barley and wheat is governed by the ability of cultivars to exclude B. A wide variation in tolerance to boron toxicity has been observed in different crop species with variation across cultivars in the same species. The tolerance mechanism is mediated by active efflux of boron from roots, and efflux transporters (genes) have been identified. Breeding success has been limited due to a suite of stresses that accompany boron toxicity.

Iron deficiency chlorosis stress

Iron (Fe) is involved in many important biological processes in plants including photosynthesis and respiration, and is indispensable for chlorophyll synthesis. In soil, Fe exists in ferrous (Fe^{2+}) and ferric (Fe^{3+}) forms. Ferrous iron is soluble in soils but is readily oxidized by atmospheric oxygen. In calcareous soils (at pH 7.4–8.5), ferric iron is more common and has low solubility in the soil. There are two strategies in plants to take up Fe^{3+}: (1) Gramineae species secrete phytosiderophores (PS) and transport Fe^{3+} into roots in Fe–PS complex, and (2) Non-gramineae species plants reduce Fe^{3+} through membrane-bound reductases before being transported by a Fe^{2+} transporter. When plants are unable to take up iron from the soil, Fe deficiency occurs and iron deficiency chlorosis (IDC) is one such stress. IDC is a complex physiological disorder that manifests in high pH calcareous soil due to low iron availability in the soil, inadequate uptake or non-optimal, or limited transport and metabolism of iron. The unavailability of Fe at less than optimum level impairs chlorophyll development, leading to interveinal yellowing (chlorosis) particularly in young plants, although in extreme cases plant death can occur. Genes that provide tolerance against IDC have been identified, and are used to develop IDC tolerant soybean varieties.

Breeding approaches

For developing superior cultivars of crops for tolerance/resistance to abiotic stresses, it is imperative that the nature of stress condition(s) is understood and defined, and the diverse range of germplasm needs to be tested for response to stress. Several studies in the past have been conducted to study and identify variation against major abiotic stresses in crop species of economic importance. Genetic studies against major abiotic stresses have been conducted, leading to identification of several important quantitative trait loci (QTL), gene, and TFs controlling and or influencing the traits. In response to abiotic stress, plant cells sensing the non-normal conditions activating variations in gene expression, metabolism and physiological responses. The process of this recognition involves sensing mechanism (in plants) that recognize the first instance of stress, thereby initiating signal cascades to different cell organelles and activate TFs inducing the expression of genes involved in abiotic stress response and protection. For example, these include genes: (1) encoding proteins that protect plant cells against tolerance to flooding [*SUB1A*, Group VII ethylene response factor (ERF-VII)], heat (HSP), drought (late embryogenic abundant proteins, LEA), cold (antifreeze protein), and free-radical scavengers; (2) involved in signal cascades and in transcriptional control such as in heat (heat shock transcription factors), drought responsive gene (e.g., ABA) that activate adaptive responses to osmotic stress, salinity [plasma membrane Na^+/H^+ antiporter salt overly sensitive 1 (SOS1)/ Na^+/H^+ exchanger 7 (*SOS1/NHX7*)], and transcriptions factors (important in every stress signal and gene activation), kinases (involved in plant development and signaling); and (3) ion homeostasis imbalance that is reported to be mediated via the SOS pathway involving calcium signaling.

Direct approach

It is the selection for absolute performance (growth rate or yield), under actual stress or selection for only a small reduction in performance under stress. A superior variety at the threshold stress level will also yield relatively well under sub potential levels. Tolerance to drought may be present in such a variety, which is expressed as an unidentified component of stability in performance over various locations expressing that stress. In the process of breeding, yield and stability are tackled as one complex. The accumulation of environmentally stable yield genes equates with better performance under stress situations. This approach has been successful in several crops, for example, sorghum, wheat and maize. Testing over a large number of sites with varying moisture availabilities, although expensive, should enable the elimination of those genotypes which may have yield suppression traits under moisture stress. The multi-location evaluation of rice lines/cultivars conducted by the IRRI under the International Rice Testing Program and CIMMYT coordinated international wheat nurseries and tests are two of the best examples of the direct approach for breeding stress tolerance. These provide an opportunity for germplasm evaluation by national or international breeding organizations in a wide spectrum of environmental stresses across regions. Other larger organizations, such as large seed companies, can manage similar types and infrastructure of testing.

Indirect approach

In this approach, experimental populations and lines must be selected, developed, and tested under relevant conditions, implying that screening and selection for morphological or physiological characteristics that contribute to stress resistance. Specific traits and genetic factors that enable the plant to better withstand stress are identified and introduced into superior genetic background using conventional and molecular tools. At CGIAR institutes (specifically ICRISAT), the crop improvement strategy for drought tolerance incuded the following three phases: (1) identification and characterization of genetic sources of drought tolerance, (2) improvement of genetic resources, by bringing together useful genetic variation from germplasm into a smaller number of lines, and (3) production of agronomically acceptable types by combining yield and adaptability with drought tolerance and other traits as per the breeding objective.

Reyniers et al. (1982) suggested following two approaches to improve drought tolerance in rice:

First approach: breed varieties for each factor of drought tolerance, in particular deep rooting, high transpiration tolerance, and good translocation of carbohydrates from source (stems) toward the sink (grain), and then make hybridization among identified useful genotypes showing such good characters to accumulate them in the same cultivar.

Second approach: select the most tolerant genotypes at each stage in a series of field trials, that is, each with drought stress at a different development stage, and then make hybridizations between these identified genotypes producing the best drought tolerance for each test and screen the lines accumulating tolerance at different development and growth stages.

These two approaches (breeding for each factor of drought tolerance, and breeding for expression on drought tolerance at each developmental stage) are still relevant in the modern plant breeding era; however, molecular tools (such as markers) can assist in the selection process. Moreover, due to the commonality of genetic control of factors, there may be some overlap; however, a combined approach may be quite useful if dissimilar genetic control exists.

In the ensuing pages several breeding approaches are described. The classical approach uses the genetic variation already available and also uses the sexual cycle to recombine DNA through independent assortment of chromosomes and through recombination. Induced mutations are also a part of classical breeding approaches. These breeding approaches have made and continue to make important contributions to crop improvement by providing plants that can tolerate environmental stresses.

Development of suitable selection criteria

The lack of a unified abiotic stress resistance mechanism and processes for abiotic stress complicates the screening methods and development of universal breeding approaches. It needs to be kept in mind that traits associated with avoidance and tolerance can be constitutive (differing between genotypes) or adaptive (vary with the stage of the life cycle); therefore, each breeding program needs to assess their target environmental conditions and appropriate genetic resources for their variety development efforts. Water and temperature stresses are quite variable between genotypes and locations, particularly for timing, duration and severity of trait expression; therefore, utmost care needs to be given to the establishment and maintenance of screening nursery, and measurement protocols. These managed stress environments can be very effective in observing genetic variation that needs to be used in conjunction with established and well-characterized check varieties to determine selection cutoffs. For example, salt affected regions are typically inconsistent in their salinity, and areas with no salinity as well as areas with high salinity are present in the same field exhibiting spatial and temporal differences. This unevenness may occur over short (sub-meter) or long distances, and also vary in its timing. This variation is caused due to varying water table, surface topography, and variation over time. In both scenarios and other similar situations, a long-term map of the field can be used to create zones prior to the establishment of a screening nursery. The stresses such as drought and waterlogging are sporadic and rare events; therefore, it is difficult to make a direct selection for stress tolerance in the field. However, if areas of consistent low rainfall are available, supplemental irrigation can be used to create stress conditions. Temperature stress nurseries can be established in field at locations where a more predictable weather pattern (for freezing or heat stress) is expected.

Screening nursery size has to match the program requirement; however, the nursery size may be the physical limit. Moreover, certain traits such as root system architecture that have been suggested as a selection criterion, are among the most difficult to select in field (and even in controlled indoor conditions). This is due to limited availability of convenient methods to extract and evaluate for root trait characteristics for breeding purposes. More recently, computer vision and machine learning based phenotpying systems have simplified root trait data collection (e.g., Falk et al. 2020a, 2020b); however, field based root extraction in a higher throughput is an open area of research and tool development. This precludes traits as well as limit population sizes, which need to be large. The possibility of evaluating for root characters at seedling stage may be explored in crops for the establishment of at least a reasonable rank correlation between controlled conditions (indoor) can be used as a proxy for field trait expression.

An alternative to selection in stressed environments is to adopt indices based on morphological, biochemical, and physiological traits associated with tolerance (Blum, 1987). These include diagnostic criteria by which salt tolerant genotypes can be identified easily in large segregating populations. For example, ion accumulation and ion exclusion selection criterion indicates tolerance in salt includers and salt excluders, respectively. Membranes can be studied wherein leakage of electrolytes indicates sensitivity, while vital staining indicates disturbed or undisturbed metabolism. In salt screening nurseries, germination percentage, emergence percentage, seedling survival, chlorophyll fluorescence, and other parameters can be used to phenotype differential response of tolerant to susceptible genotypes.

Selection

There could be natural selection or artificial selection. Early domestication of crop plants concentrated their attention on species adapted to natural selection to their locality. With the advance of trade and human migration, crop selection was tested in new environments, and those performing satisfactorily were retained and those not adapted were discarded. Natural selection would not only be useful for temperature stress but for other stresses as well, if population have been grown at locations where the probability of a stress near the critical level is high. The natural selection in heterogeneous populations of crop plants should increase the frequency of genes for temperature extremes such as winter hardiness and for other abiotic stresses. On the other hand, artificial selection involves sorting out and propagating individual plant(s)/genotype(s) or group of genotypes from mixed/segregating populations. For selection to be effective there must be genetic variation, which can be identified and distinguished from environmental variations.

Selection for abiotic stress tolerance can be done in the field or controlled environment conditions. Early generation segregating population in self-pollinating crops, clones, inbred lines, hybrids, or other types of cultivars can be selected and/or characterized for abiotic stress in field nurseries. Although the control over environmental factors increase from growth cabinets to greenhouse to controlled field conditions to uncontrolled field, the correlation with target agroecological zone of cultivation of variety move in the opposite direction (uncontrolled field to controlled field to greenhouse to growth cabinets). Therefore, breeders have faced a conundrum in deciding on what type of environment to utilize to make selections. Ideally, extensive experimental and validation needs to be done to ascertain that the target trait is well correlated with plants response in commercial field.

As previously explained, this includes an understanding of the time and severity of stress at various growth and development plant stages. The target trait under selection needs to have good repeatability and lower measurement error (improved accuracy and precision). Ideally, pre-commercialization testing should include a field stage screening for the abiotic stress. Moreover, reliance on screening for just one trait is risky, as rarely a single trait is responsible for the control of tolerance against the stress. Additionally, seldom does a field experience only a single abiotic and biotic stress as plants are exposed to multiple stress at different phenological stages; although at times the sequence of stress may be somewhat consistent helping in cultivar development strategies, and providing focus on important objectives rather than convoluting selection of target traits. Regularly, heat and drought stress are noticed more commonly, similarly drought stress with certain diseases exist together (concurrently or sequentially).

Hybridization

For inbreeding species, the following breeding methods are being frequently used for the improvement in stress resistance and yield.

Intraspecific hybridization

If the level of tolerance in the agronomically improved cultivars is not adequate, the collections of the crop species may be screened for sources of desired variability. For example, other programs (within and outside of organization) can be contacted (or use information from published papers) to identify and retrieve seed of parental strains (with abiotic stress tolerance) to utilize in crossing scheme. These types of screenings are also useful for genome wide association studies. After planned hybridization, the segregating populations can be handled using one of the following methods.

1. *Pedigree and bulk-pedigree selection*

   In this method, one parent is chosen for genetic properties that are useful (e.g., salt tolerance), while the other parent is chosen for complementing desirable agronomic characteristics. Sometimes, three-way and double crosses may be more effective than single crosses for combining traits from diverse parents especially if two parents cannot provide the combination of traits desired in the targeted cultivar, or if one parent (with abiotic stress tolerance) is unadapted or agronomically poor. For example, three parent cross may be useful if the one of the parent used for its abiotic stress tolerance is unadapted. This will allow to increase the frequency of elite genetic background by giving 25% elite background from the single cross, and an additional 50% elite background from the third parent. Selection for plants with desired combination of characters starts in the F_2 generation and the progenies of the selected F_2 plants are reselected in succeeding generations until genetic purity is reached. These methods are explained in more details in Chapter 10, Bulk method, and Chapter 11, Pedigree method. A disadvantage of the system is that early generations can be expected to be heterogeneous and quite unstable in their response to environmental interactions.

   In the bulk population method, selection is delayed until a later generation, commonly F_5 or later. At later generations, segregation and homozygosity will increase the correspondence between genotype and phenotype. The method is suitable for problem soils or marginal climates. It serves as a source of natural selection. A modification of either of these methods or a combination of these two and/or with single seed descent method could be used. For example, population bulks can be grown along with checks in screening nurseries and only plants that show a better response compared with checks are used to create next generation bulk for that breeding population. This process can be repeated to keep increasing the frequency of desirable plants in the bulk followed by plant pull and progeny row evaluation. Another strategy is to test early generation bulks in screening nurseries to decide which populations show merit to continue. Breeding populations that are worse in stress reaction than checks, are discarded. Marker-assisted selection (MAS) can also be used instead of phenotypic evaluation only, if robust molecular markers are available. Basic segregating population can be developed by crossing large number of diverse donor parents and advancing these through the pedigree populations to F_4 or F_5 without selection. The selection among such large segregating populations under appropriate environments would produce a full spectrum of genetic materials adapted to various edaphic stress conditions.

The plants/progenies are evaluated against check (continuous control) which are planted after few or alternatively after each row in the segregating generation. When the lines/progenies are partially fixed, these may be starting with single replication tests and then moving to replicated tests [See Chapter 26: Field Plot Design in Plant Breeding].

2. *Backcross method*

   The method is a form of recurrent hybridization by which a gene (or two to three genes) for a superior characteristic may be added to an otherwise desirable variety. One parent is an elite variety, which lacks a gene for superior characteristics that is present in a second genotype (donor). Beginning in the F_1, the hybrid is successively backcrossed to the adapted parent variety for several generations - ideally complemented with MAS. The purpose of this method is to incorporate gene(s) from donor parent into the elite line (recovering the genetic composition of the recurrent parent, except for the addition of a gene (s) for the superior characteristic from donor parent). The backcross method will be most successful if molecular marker linked to abiotic gene and QTL are known, as this allows for a more accurate phenotypic selection in backcrossing rather than phenotypic selection for a complex trait.

3. *Mass selection/recurrent selection*

   For out-crossed species the production of F_1 hybrids takes a prominent place. Genetic advances among hybrids are usually the result of making crosses between selected inbred lines chosen based on combining abilities. The mass and recurrent selections are important in accumulating genes of tolerance/resistance in the populations/genotypes against abiotic stresses. A mass selection involving relatively controlled sib crosses between selected genotypes within a genetically diverse population may be effective. It may be possible to identify elite populations from test crosses. The stress tests could be applied at a level that will eliminate 50% individuals. The survivors can be sib-crossed. Seed from the resulting F_1 hybrids can be bulked to form the F_2 bulk population for the next cycle of selection. The remnant seed from each cycle could be increased for use in breeding program and to assess genetic gain for stress tolerance.

Interspecific hybridization

If genes for stress tolerance are not available in cultivated types, these can be transferred from the wild relatives/species which are the rich source of stress tolerant genes. The commercial cultivars could be improved by the transfer of genes for abiotic stresses from their wild or weedy congeneric species and their cospecific subspecies, for instance the increased drought tolerance from *Zea mexicana* transferred to *Zea mays*. For more details, refer to Chapter 6, Wide Hybridization.

Mutation breeding

Mutagenesis may be employed in an effect to produce the desired variability in cases where such variation is not readily available. A large number of mutant varieties have been developed from crosses using mutants as one of the parents, and also several mutant cultivars have been developed from the parent variety through induced mutations. Induced mutations have been exploited both directly and indirectly for the development of stress-resistant cultivars in an array of species including cereals, oilseeds, grain legumes, fiber crops, fruit trees and vegetables. Gamma-rays, a physical mutagen, has been frequently used in the production of a stress-resistant mutants and numerous cultivars have been developed in wheat and rice crops. More details are presented in Chapter 14, Mutation breeding.

Production of doubled haploid

The doubled haploid (DH) technique of obtaining homozygous lines of self-fertilized crops is becoming popular, particularly in rice, wheat and barley, and in cross-fertilized crops such as maize (for inbred development) as it represents a saving of 2 or more years in cultivar development and commercialization. Haploid plants can be screened with molecular marker linked to the genes of interest, or a genome wide selection approach can be used to only advance haploids that merit selection to be developed into DH lines. Additionally, DHs can be grown in abiotic stress nursery, if available, to make directed selections before deciding which DHs to move into yield and performance testing.

The cell/tissue culture approach

The tissue culture is a cyclic procedure whereby types of hormones incorporated into a medium and their concentrations can be modified to force either differentiated or undifferentiated growth. Differentiated growth involves the production of either shoots or roots. Callus usually can be produced from any differentiated structure (e.g., leaf, stem, and root) by placing explants on media containing relatively high levels of auxin and low levels of cytokinin. Once produced, callus can be grown either as large, multicellular masses on solid media or as small aggregates in related liquid media. By using high levels of cytokinin and low levels of auxin in the media, it is sometimes possible to stimulate the undifferentiated callus into producing differentiated structures which eventually turn into shoots. Auxin, on the other hand, is a growth hormone that promotes the elongation and growth of root tissues.

The tissue culture can also involve the production of protoplasts, wall less plant cells. The presence of a cell wall can make many genetic and biochemical studies nearly impossible. Protoplasts are produced easily by incubating either differentiated or undifferentiated tissue in a mixture of cellulose + macerase. These enzymes digest the cell wall (cellulose) and surrounding pectins and hemicellulose. The protoplasts then are cleaned of cellular debris and enzymes by differential centrifugation. Cell walls usually reform on the individual cell within one to two days when protoplasts are planted on a callus forming medium. Cell division begins within 1–2 weeks. The callus which forms from the protoplasts then can be shifted to media stimulating redifferentiation.

The potential of tissue and cell culture for effecting genetic modification has been stressed in number of research papers. One approach to such genetic modification involves the isolation of stable variant cell lines from established cell cultures either directly or following mutagen treatment, regeneration of plants from such lines, and study of the phenotypic expressions of the genetic variants. The isolation of variant cell lines requires that the cells be subjected to appropriate selection pressures, that is, the exposure of high salt concentration, high temperature, and chilling temperatures and test for retention of tolerance/resistance following growth in the absence of the environmental stress. More than 45 year ago, it was suggested that selection from cell or tissue cultures may lead to a rapid increase in stress resistance of tobacco lines (Nabors et al., 1975).

Molecular approach and its integration in breeding pipeline

Although, morphological markers such as leaf rolling for water stress have been useful as predictors of genetic response to selection, they are not very useful for selection based on root traits which are crucial in drought and heat tolerance but difficult to phenotype in the field (without destructive sampling and large resource input). Also, phenotypic expression of complex traits, such as in drought tolerance - leaf water potential, relative water content, stomatal conductance, photosynthesis, metabolites, enzymes, etc. are not easily measurable. For the improvement of stress traits, DNA markers provide a larger benefit and use. Gene and QTL governing stress response can be efficiently selected if molecular markers on the gene or closely linked to the gene or QTL of interest are identified. These can be identified through linkage or association mapping approaches, and functionally validated prior to implementation. Additionally, interlab and interbreeding family validation provides confidence for a wider implementation. The selection based on molecular markers helps in overcoming difficulties associated with low heritability, recessiveness, and difficult screening assays during gene transfer and selection. Several genes or QTL have been targeted by breeders in MAS approaches.

Stay green QTL that enable a stay-green phenotype provides the maintenance of functional green leaf area during grain maturation through the regulation of cytokinin and ethylene metabolism and the amelioration of ROS has been used in cereal species. Other genes or QTL include: for drought tolerance (*NAC10*, *SNAC1*, and *AP37* in rice, *NF-YB2* and *ACS* in maize, and *TASG1* in wheat), temperature tolerance (*MYSB3* in rice), flooding tolerance (*SK1*, *SK2*, and *SUB1* in rice), *Saltol* QTL in rice for salt tolerance, *Pup1* QTL in rice for enhanced phosphorus update, and *DRO1* (deep rooting 1 QTL) in rice - where a single nucleotide deletion results in a reduced root angle or asymmetric cell elongation in root tip region. The *DRO1* QTL causes deeper roots providing benefits in non-stress conditions due to better ability to uptake nutrient from soil. From a breeding perspective, it is essential for a breeder to keep abreast of developments in molecular genetic tools for selection aid.

As climatic variability is becoming more pronounced globally, study of phenotypic plasticity is becoming increasingly more important. Plasticity is inherent to a given trait in response to environmental cues and stimuli, and both adaptive (fitness benefit) and non-adaptive (responses to physical processes or resources limitation) plasticity are important for cultivar development. Molecular, physiological and biochemical studies can unravel genetic loci and mechanisms that allow plants to use multiple avenues to express adaption to climatic changes for short- and long-term benefit.

Examples of breeding schemes are presented in Fig. 21.1 (non-DH) and Fig. 21.2 (DH). In a non-DH method, breeders have to make several decisions: (1) choice of germplasm, (2) choice of breeding method, and (3) choice of trait screening. The choice of germplasm and breeding method has been previously covered; however, one important aspect is the decision to use backcrossing if one to two genes or QTL are being introgressed (from donor) to elite recipient parent.

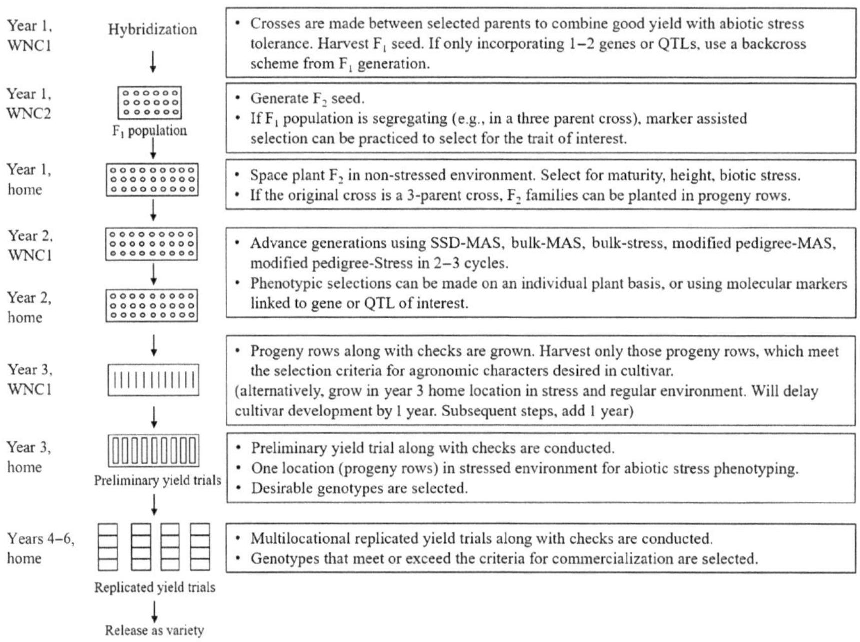

FIGURE 21.1

Potential scheme to breed for abiotic stress tolerance trait in self-pollinating crops. SSD-MAS, bulk-MAS, bulk-stress, modified pedigree-MAS, and modified pedigree-stress are different combination of generation advancement and selection. The combination of winter nurseries and home environment are presented as an example and will be standardized by the breeder to reduce the time to develop cultivar without compromising on sufficient and good data quality. *Home*, Home environment; *MAS*, marker-assisted selection; *SSD*, single seed descent; *WNC*, winter nursery cycle.

Otherwise, a forward breeding approach is used. The choice of breeding methods in abiotic stress trait focused approaches depends on resources (space, time, personnel, etc.); however, efforts should be taken to ensure selection (phenotypic or genotypic) is happening at each stage. This ensures that undesirable material is not carried into progeny row stage. If the QTL under selection does not fully control a trait but provides improvements (higher percentage of genetic variation explained), it is imperative that MAS is complemented by phenotypic screening. It cannot be over emphasized the importance of good experimental protocols, as good quality data in breeding program supersedes almost everything else about the breeding pipeline. In each of the breeding method, population advancement can be complemented with selection by growing in specialized nurseries. Replicated plots (of breeding population family) can be grown along with checks, to select breeding families that have merit to be advanced for more thorough resource investment for abiotic trait breeding. These approaches are applicable to both abiotic and biotic trait improvement. Progeny rows can be split into breeding test and grown in specialized nursery to make selection. The specialized nursery may not give sufficient seed for next season, and in this case it is very important to grow the same variety in breeding test (which will be harvested for seed source next season or year).

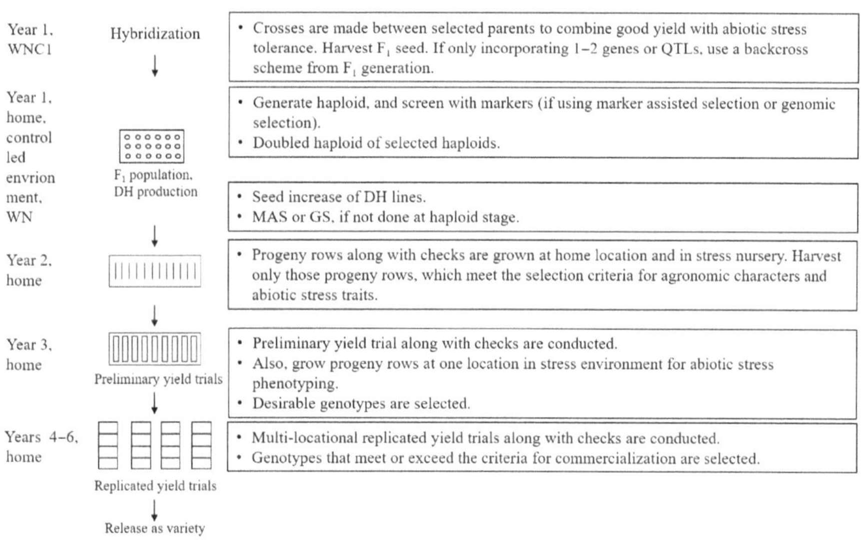

FIGURE 21.2

Potential scheme to breed for abiotic stress tolerance trait in self-pollinating crops where doubled haploid method is available. The above scheme is only provided as a framework example of breeding pipeline. The year-season combination will be dependent on the length of time taken to develop DH line from the initial F_1 seed. *WN*, Winter nursery.

The choice of trait screening is complex. For example, the choice includes; the usage of home and off-season nurseries, and screening nurseries. In both scenarios, indirect selection is practiced. Indirect selection means selecting in one environment (e.g., in non-stress environment and home environment) to gain a response in another (e.g., stress environment and off-season winter nursery). Indirect selection is more effective than direct selection if the product of the heritability in the stress (or off-season) environment and the genetic correlation between the two environments is higher than the heritability at the home (or non-stressed) environment. Therefore, prior experiments are needed to ascertain this relationship for the target trait and environments used at the home and off-season or stress environment. A lower genotype × environment interaction will ensure success. Additional consideration includes identification of traits that will facilitate indirect selection for abiotic stress tolerance leading to the improvement of one trait by selection of a correlated trait (see Chapter 4: Primer on Population and Quantitative Genetics). Indirect selection will work if genetic correlation between traits is high, and secondary trait has higher heritability than primary trait or there is an expectation of higher selection differential for secondary trait than primary trait. Breeders try to select for stable genotypes in both stress and non-stress environments and include selection for yield and abiotic stress in early filial generations.

The schemes presented in Figs. 21.1 and 21.2 can be followed without marker tools. However, the availability of MAS (see Chapter 27: Molecular Tools in Crop Improvement and Cultivar Development) allows breeder to fix the gene or QTL in early generation; therefore, later generation can focus on characterization for yield and abiotic stress traits. Although, phenotypic selection in generation advancement for abiotic stress will enable accumulation of minor effect alleles providing a better overall tolerance against the abiotic stress. If genomic selection (see Chapter 27: Molecular Tools in Crop Improvement and Cultivar Development) is available, its utilization in earlier stages of breeding pipeline will have the most impact as selection needs to happen as early as possible to ensure only genotypes and population with merit are advanced freeing up resources for the program. MAS (and genomic selection) is even more useful in a DH breeding method, as haploid plants can be screened and only those with merit are advanced to DH. This combination of DH and markers usage reduces time to develop cultivar and also frees up resources to reduce financial footprint or expand performance testing.

The breeding of abiotic stress traits is complex but successful is achievable as evidenced globally in multiple crops by high yields and other desired traits including quality obtained in challenging environments. Breeders need to extensively collaborate with scientists from multiple disciplines to achieve success in breeding for abiotic stress traits, and to meet the needs of farmers and other stakeholders.

CHAPTER

Breeding for resistance to biotic stresses

22

Abstract

Crop production globally is severely impacted by biotic stresses, primarily by diseases and insect-pests. Therefore, a breeding pipeline invariably is focused on the integration of genetic resistance to protect yield and quality losses. The main topics in this chapter include; types of genetic resistance to diseases and insect-pests, inheritance of resistance, inheritance of virulence, breeding for simple and complex resistance, and multiple disease and insect-pest resistance in seed and vegetatively propagated crops. Several methods that can prolong the life of resistance gene and reduce catastrophic losses are also discussed.

A plant is healthy or normal when it carries out its physiological functions to the best of its genetic potential. Whenever plants are disturbed by pathogens or insect-pests, one or more of the physiological functions is interfered beyond a certain deviation from the normal that leads to sub-optimal performance, invariably reducing biomass and yield. This may happen in either of four ways: (1) by killing of plants, leaving a gap in the crop stand beyond the capacity of neighbors to compensate (e.g., vascular wilts, various soil borne fungi, and some boring insects), (2) by general stunting caused by metabolic disruption, nutrient drain or root damage (e.g., many viruses and aphids), (3) by killing branches (e.g., many boring insects and some fungal diebacks), and (4) by destruction of leaf tissues (e.g., many rusts and fungal foliar diseases such as blights, mildew, and leaf spots). Other effects of diseases and insect-pests attacks are more general. These include damage to the crop economic product, often more apparent after harvest (e.g., cereal smuts, various rots, and borers of fruits and tubers), and effects on quality (e.g., insect or fungal blemishes of fruits and tubers). Human history has recorded events of significant destruction by insect-pests and diseases, for example:

- One of the most devastating epidemics occurred more than 175 years ago in Ireland, namely, the potato late blight epidemic caused by *Phytophthora infestans*. "The Irish Famine" of the 1840s caused by late blight epidemics over several years caused at least one million deaths.
- In the 20th century, in India, the great Bengal famine of 1942-44 caused a minimum 1.5 million deaths. This was due to a combination of factors such as cyclone, flooding and war related events, in addition to rice brown spot disease caused by *Helminthosporium oryzae* = *Cochliobolus miyabeanus*,

Plant Breeding and Cultivar Development. DOI: https://doi.org/10.1016/B978-0-12-817563-7.00010-6

which destroyed the major food crop (rice) of that region (Padmanabhan, 1973). This disease was a significant triggering factor for the famine.

- In the United States, there was a serious plant disease problem known as the Southern Corn Leaf Blight epidemic (Hooker, 1972). This disease is widespread in tropical and subtropical areas, but was relatively minor in the US Corn Belt prior to 1970. In that year, a newly emerged race (race T) of *Cochliobolus heterostrophus* (*Helminthosporium maydis* = *Bipolaris maydis*) combined with favorable weather, especially for the pathogen, in the southern United States caused major damage to the corn (i.e., maize) crop. Race T can attack maize plants with the T cytoplasm that, at the time, was in at least 85% of the hybrid maize seed planted.
- The scab of wheat and barley caused by *Fusarium graminearum* (teleomorph *Gibberella zeae*) was a major problem in midwestern United States early in the 20th century and as a result, wheat acreage shrunk considerably (McMullen et al., 1997). For a long time, this disease was not prominent until early in the 1990s, when it reemerged due to a combination of very favorable weather for the pathogen and quite susceptible varieties of the wheat and barley. Both seed yield and quality are affected, and in years of high disease levels many fields are not even harvested. Additionally, the fungus produces a deoxynivalenol (DON) toxin that is harmful to humans and animals.
- In recent times, *Helicoverpa armigera (Heliothis)*, is one of the most damaging insect of cotton in India. It is a polyphagous pest that has become a big nuisance causing serious crop production challenges especially in cotton and pulse crops (chickpea and pigeonpea). In several cases, this noxious pest has caused the failure of cotton crop and is suggested for farmer committing suicides due to heavy crop losses. Susceptibility of host cultivars and favorable weather conditions were responsible for the insect-pest epidemic.

Two major plant disease outbreaks on non-food crops that were most likely caused by pathogens introduced into Europe and North America are chestnut blight (*Cryphonectrica parasitica*) and Dutch elm disease (*Ceratocystis ulmi*). Although neither are directly involved in the food supply, both diseases have caused considerable financial losses over wide areas of two continents with serious damage to native flora. In both cases, genetic uniformity in the host plants via vegetative propagation aided in the spread and destruction caused by the disease (Stuthman, 2002). There are numerous diseases and insects-pest on a myriad of crops that severely limit yield and quality, causing economic and social problems.

Disease triangle

The optimum conditions for a disease to occur and develop are a combination of three factors (Fig. 22.1): susceptible host, infective pathogen and favorable environmental conditions. A change in any of the factors causes a corresponding

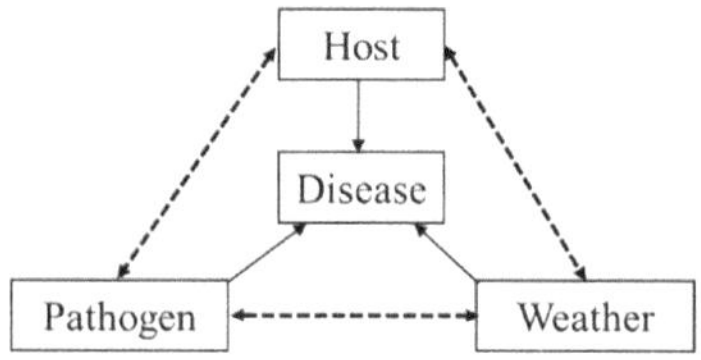

FIGURE 22.1

A combination of three factors required for the disease to appear.

change in the expression of disease. Such a triangular interaction also holds for plant/pest relationship in general.

Natural populations and wild plants rarely show epidemics, due to genetic heterogeneity and natural biological control (e.g., hyperparasitism). Hosts (i.e., crop) and pathogen (fungi, bacteria, viruses, insect-pests, and nematodes) live together in a complex equilibrium, wherein neither dominates in a natural ecosystem. However, modern agricultural technology has introduced important changes: (1) it has narrowed down the genetic base of cultivars (bred cultivars rather than landrace populations), which alters the dynamic balance between host and pathogen resulting in epidemics, and (2) it has generated more or less continuously disturbed populations and has changed the whole ecosystem, creating profoundly altered habitats for both host and pathogen. The extreme cases are observed in clonal cultivars (see Chapter 20: Breeding Methods Used in Asexual Crops). This does not mean that all pathogens create epidemics; they obviously do not, and the majority remain what they were before, i.e., unimportant. However, some of the minor pathogens may increase due to favorable environment, host susceptibility (due to more virulent race or biotype), and continuity in large populations. For instance, *Helminthosporium* (*Bipolaris*) leaf blight in maize in the United States, and hoppers in rice in India and the Philippines became major pests due to change in one or more of the above factors.

The prevention of epidemics and ultimately the reduction of losses in yield have been of great interest in crop production. Insect-pests and some of the pathogens may be controlled by the use of chemicals; but chemicals can create hazards to human health and can produce undesirable side effects on non-target insects, animals, and plants. These also add input costs; and if not applied timely, can be wasteful. The development of resistant varieties is the most effective method of controlling many diseases and insect-pests of row crops or plantations, which are grown on extensive acreages with a large number of plants per hectare.

The task of developing resistant varieties is difficult because many species of diseases/insect-pests have a large number of genetically diverse physiological races with differing degree of virulence for various kinds of crop plants or for varieties within a single crop species. New races are continually produced by mutation and recombination; and migration and gene drift cause population changes. Additionally, some of these biotic factors have tremendous power of multiplication and rapid dissemination. Due to the vast and shifting populations of pest and pathogen races, the value of resistant varieties varies with variations in the geographical distribution and prevalence of races to which they are either resistant or susceptible. Accordingly, they may be resistant at some times and in some areas but not in others. Therefore, in establishing and running/organizing a breeding program for the development of disease and/or insect-pest resistant varieties of any plant, some of the following should be considered:

- What is/are the most important disease(s) and/or insect-pest(s) of the crop species in the target area of cultivars? Prioritization of targeted traits is crucial, because it is near impossible to design and run a breeding program to simultaneously incorporate resistance to all known diseases, and insect-pests.
- Which techniques/methods and germplasm sources are best to identify gene(s) and the genomic regions for resistance to targeted disease(s) and/or insect-pest(s) of the crop species? How effective and durable are these genes and/or quantitative trait loci (QTL)?
- How to incorporate gene(s)/QTL for resistance to targeted disease(s), insect-pest(s) into germplasm to develop cultivars, without simultaneously introducing adverse (and unintended) effects related to agronomic ability and quality of the product, for example, through linkage drag?
- How to incorporate gene(s)/QTL for resistance to a targeted disease(s) and/or insect-pest(s) into commercial variety without simultaneously incorporating increased susceptibility to other disease, and insect-pests?
- How to maintain the incorporated resistance in useful form for a long period of time? This is a problem of special significance to clonal crop cultivars and forest geneticists, where long rotations covering many years are the rule. These are also important in situation if only one or few sources of genetic resistance are known; therefore, there is a fear that genes/QTL conditioning resistance may become ineffective.

Definition of resistance

To measure resistance or susceptibility levels, we measure the extent to which a plant prevents the entry or subsequent growth of the pathogen within its tissues or the extent to which a plant is damaged by a pathogen (Fig. 22.2) (Eenink, 1977).

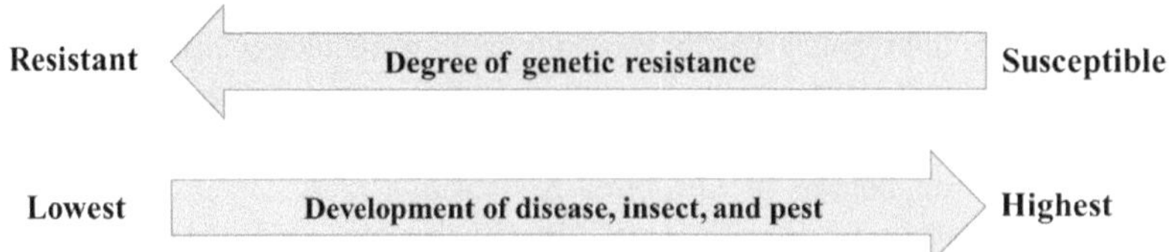

FIGURE 22.2

The degree of genetic resistance and the development of diseases, and insect-pests.

Degree of resistance may be measured by using susceptible cultivars of the same plant species as checks. Immunity is the extreme expression of resistance, and anything less than immunity is resistance. Resistance may be qualified by such words as high, intermediate, or low because there may be gradations between extreme resistance and extreme susceptibility. Resistance may be due to a number of external and internal factors and the infection/damage due to a number of favorable physical and chemical environments in the host from the contact until the completion of its life cycle.

Types of genetic resistance to diseases

Varying types of reactions are noticed on crop plants in response to pathogen attack. Breeders and pathologists have used multiple terms to classify and differentiate types of resistance, although similarities and overlap exists between these main classifications: for example, (1) *monogenic* (trait under one gene control), *oligogenic* (trait under few genes control) and *multigenic* (many genes control the trait), (2) *race specific* (*vertical*) and *race non-specific* (*horizontal*) resistance, and (3) *qualitative* and *quantitative* resistance. The monogenic/oligogenic/polygenic classification is based on the genetic control, the race specific/race non-specific is based on gene for gene interactions, and the qualitative/quantitative is based on genetic control and pathogen reactions. Plants also employ a more complex and evolutionarily more robust system of "non-host resistance" against a broad range of pathogenic species.

A race non-specific resistance implies that the organism is resistance to all isolates of the pathogen and inheritance is polygenic. In race non-specific resistance, infection takes place but its colonization, and spread of disease is reduced. Compared with qualitative resistance, quantitative resistance generally has lower narrow sense heritability. Quantitative resistance are more difficult to breed as genetic control is more complex and influenced by modifiers and epistatic interactions, as well as have more pronounced interaction with the environment. From a plant pathology viewpoint, differences exist between these terms, but from a breeding perspective we will refer to simply inherited, race specific, vertical, and qualitative resistance as similar; and complex inheritance, race non-specific, horizontal, and quantitative resistance as similar. However, the demarcation is not as clear cut in all cases.

Horizontal resistance (HR) provides an alternative to vertical resistance (VR), and is evenly spread against all races of pathogen (Van der Plank, 1963). Its stability is postulated to be due to its polygenic inheritance. It is also presumed that HR acts to reduce the effectiveness of one or more components of pathogen fitness in the pathogen, through accumulation of resistance genes that act quantitatively in the host (Van der Plank, 1968). Nelson (1973, 1978) did not agree with this interpretation; and therefore, redefined HR as resistance that reduces the apparent infection rate. *Field resistance* or *generalized resistance* or *adult plant resistance* (APR), etc., have been used in many contexts as an alternative to non-specific resistance. Broadly speaking, HR to diseases is polygenic (many genes control), studied using mean and variances, race (pathotype) non-specific, and durable. HR manifests itself as lower infection rate, latency period and sporulation potential. On the other hand, VR to diseases is mono- or oligo-genic (few gene control), studied using ratios, and race (pathotype) specific. VR exhibits itself as presence versus absence reactions and generally gives complete control.

Breeders and pathologists utilized oligogenic (race specific, vertical, and qualitative) resistance in the earlier years of breeding because of the ease to incorporate into varieties, due to its simple inheritance (often a single gene) and easily recognizable character. However, with few exceptions, it is generally acknowledged that qualitative resistance is unstable and not durable. The pathogen usually requires a single genetic change to overcome a single gene resistance. Because of this, many resistant varieties have had short-lived usefulness. In race specific resistance, single Mendelian (major effect) resistance genes are matched with the pathogen's "avirulence" genes implying that gene will have resistance to some pathogen races but not to others. *Hypersensitive* reaction falls in this category. The qualitative resistance is characterized by a hypersensitive reaction and provides a high level of disease resistance, including at times complete resistance. *Seedling resistance* genes (e.g., some of the *Lr*, *Sr*, and *Yr* genes in wheat for leaf rust, stem rust, and stripe rust, respectively) mostly give qualitative resistance. Although they provide a clear resistance, for most plant breeding applications APR genes are more useful because they are more durable, and protect during the growing season. However, if crop production is threatened due to disease infection at seedling stage, seedling resistance genes (multiple independent genes) are needed for crop protection. For a breeder, seedling resistance and qualitative resistance are simpler to breed because these are oligogenic, generally with easier phenotyping and availability of DNA markers for forward or backcross breeding.

The use of a single gene for the protection against insect-pest or disease is not an ideal strategy because changes in pathogen (race) or insect (biotype) population can overcome the resistance genes. Large farmer income loss happen when genes are "defeated" and for these reasons multiple independent genes are pyramided to give a stronger resistance package in a variety. Forward selection and backcross breeding or a combination of each can be utilized to pyramid multiple genes with qualitative reactions. However, if only a single gene is known to exist, breeding effectiveness options become limited. For example, orange wheat blossom midge (OWBM) [*Sitodiplosis mosellana* (Géhin)], lays its eggs on the emerging heads of wheat (*Triticum spp.*), and causes losses due to damaged kernels, poor quality and lower yield. The genetic resistance to OWBM is due to antixenosis (deterrence of oviposition) and antibiosis (inhibition of larval growth or death); however, only a single gene (*Sm1*) has been identified for antibiosis. As the insect-pest causes large losses in all temperate wheat growing regions of the world, *Sm1* usage is in a delicate position. Farmers and breeders need to use genetic resistance provided by *Sm1* for a better crop production, however, a partnership between them is the only way to increase the lifeline of the *Sm1* gene's

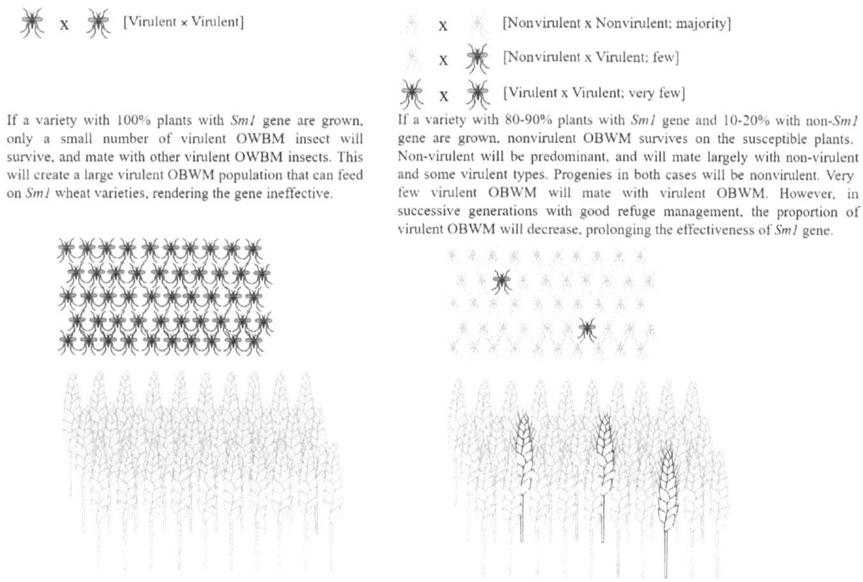

FIGURE 22.3

Diagrammatic representation of using variety refuge to prolong the effectiveness of a single gene controlling orange wheat blossom midge (OWBM) in wheat (*Triticum aestivum* L.) crop. The OWBM insect is not drawn with actual morphological features and shape. Gray-shaded wheat contains the *Sm1* gene that controls the OWBM, while lighter shaded wheat with black outline are plants of a susceptible wheat variety (non-*Sm1*).

effectiveness (Fig. 22.3). In Canada, a wheat Stewardship Agreement is used wherein farmers agree that they will use the technology (*Sm1* variety) responsibly by limiting the use of farm-saved seed to one generation past Certified seed. Certified seed of the OWBM variety with *Sm1* gene is sold as a varietal blend, with 90% made up of a midge tolerant variety and the remaining 10% midge susceptible seed of a non-*Sm1* wheat variety (Fig. 22.3). As a result, the longevity of *Sm1* gene's effectiveness is increased by several decades compared with a few years if a variety refuge is not used, and farmers can achieve $36/acre in yield and grade benefits.[1]

Breeding for quantitative resistance

The qualitative resistance alone may be ineffective or unable to provide necessary protection therefore, breeders develop varieties with qualitative and

[1] Source: https://midgetolerantwheat.ca/.

horizontal/quantitative/race non-specific/field resistance. In a seminar on HR to rice blast disease, Van der Plank (1975a) made the following suggestions -

- HR may be determined as field resistance. In the absence of VR, resistance is HR. He suggested that if one can exclude all VR genes, one can simply compare cultivars or lines in the field, and the comparison will measure HR alone.
- The method then is to expose lines (or cultivars) to infection in the field by virulent races to which the lines are (vertically) susceptible. The resistance that remains is horizontal. The difficulty arises here that the lines must be exposed to races virulent on all of them or to a mixture of several races, some of them virulent on some lines but avirulent on others (however, VR confounds the results).
- The lines/varieties that are more difficult to infect, for example, those that develop the fewest lesions per plant, per leaf or per centimeter square of leaf area may be selected.
- The lines/varieties in which period from inoculation to sporulation is longer, may be selected.
- The lines/varieties in which sporulation is less abundant, may be selected.
- HR may be enhanced by breeding.
- The HR and VR may be combined.

While numerous examples exist in the literature for HR, *Fusarium* head blight (FHB) in wheat is an interesting case. This wheat disease is one of the most damaging in North America and Europe. The Fusarium complex consists of almost 20 different species, *F. graminearum* is one of the most damaging as it leads to early senescence of the wheat spike causing shriveled grains (therefore, lower yield) and produce the mycotoxin DON, which is a virulence factor as well as toxin for food and feed. All resistance discovered for FHB are quantitative and controlled by numerous smaller effect genes. Several resistance mechanisms have been described: (1) Type I: invasion, (2) Type II: fungal spread, (3) Type III: toxin accumulation, (4) Type IV: kernel infection, and (5) Type V: yield reduction. Several genetic loci have been reported in wheat, with Sumai 3 wheat from China, and Frontana wheat from Brazil reported as most used sources of moderate resistance to FHB. In FHB and other diseases where different resistance mechanism exists, breeders have to identify the appropriate phenotyping assay to obtain correct measurements for trait phenotyping.

The *Fusarium* species cause major losses in multiple crops in several families including cereals, oilseed and legumes. In soybean, *F. virguliforme*, causal fungi of sudden death syndrome (SDS), is a major disease in the United States and researchers have been working extensively on studying the genetics of resistance to SDS. Many genes (*Rfs*) and QTL (115, although often overlapping) have been reported by soybean researchers (source: Soybase). However, these genes/QTL have a small effect on trait expression (i.e., each gene or QTL explains a small percentage of the phenotypic variation) demonstrating the complex inheritance. Furthermore, phenotyping is difficult because it is a root and stem disease with toxin symptoms visible on the leaves.

The race non-specific resistance in rust may involve; exclusion of fungus, limitation of pustule size or possible slow growth and development of the fungus. The joint action of these host characters may drastically slow down a disease epidemic to the point of insignificance. Characters such as these, involved in slow rusting, were observed with wheat leaf rust (*Puccinia recondita*), maize rust (*P. sorghi*) and barely leaf rust (*P. hordei*). If components of HR are independent and under separate genetic control, it may be possible to recombine them to increase the level of resistance. The smaller pustule size and latent period component of slow rusting in wheat for leaf rust are governed by more than one gene. It suggests the possibility of selecting for these components in segregating populations, and the host genotypes with more stable resistance by the assembly of these components may be developed. Additionally, breeding of HR is not looking to breed for an absolute resistant. The idea is to have a delayed onset, and/or reduced infection (by pathogen) or infestation (by insect-pest).

In breeding for protection against wheat stem sawfly in the North American northern plains, variable level of stem solidness is observed among cultivated wheat species. As incomplete expression of solid stem pith is evident in hexaploid compared with tetrapoid wheat. However, the reduced expression (or incomplete solidness) behaves as a quantitative trait but the benefits are still present; that is, reduced infestation (and feeding) by the larvae. Interestingly, solid-stem locus 1 (*SSt1*) controls stem solidness and the QTL interacts epistatically with other minor QTL to synergistically enhance the expression of stem solidness compared with *SSt1* alone (Nilsen et al., 2017).

Breeding for field resistance with qualitative resistance

The severe breakdown of true resistance for rice blast observed in varieties Kusabue, Yuukara, Pi5, and others is caused by lack of field resistance that results from the Vertifolia effect (defined as the erosion of HR in a variety to disease due to the presence of strong qualitative/race specific/vertical resistance, characterized by the presence of R genes). True resistance itself appears to have no or little effect on field resistance (Asaga and Yoshimura, 1969). Toriyama (1975) stated that field resistance was equivalent to HR. The breakdown of field resistance was conspicuous in Chugoku 31. The resistance of this variety was governed by a single major gene *Pi-f*; however, its expression was quantitative. The moderate level of field resistance is generally accepted as race non-specific, that is, horizontal. This type of resistance is governed by a polygene system and is hardly influenced by the shift of races or mutation of the fungi. If a useful level of quantitative/field resistance is incorporated with qualitative resistance into a variety, that variety will be more stable in its resistance than the varieties possessing qualitative or quantitative resistance separately.

Soybean Cyst Nematode (*Heterodera glycines*) (SCN) is one of the most devastating pest of soybean crop. Breeding for SCN resistant soybean variety is a major breeding target in the northern US soybean growing regions. While different sources of tolerance have been reported the two main sources used by breeders are PI 88788 and Peking. In PI 88788, only the *rhg1-b* allele (7−10 copies) needs to be functional for resistance that reduces the severity but does not include a hypersensitive reaction, while in Peking *rhg1-a* (1−3 copies) and *Rhg4* alleles together give a rapid and strong localized hypersensitive response. The traditional method utilized by plant breeders to select SCN tolerance lines include a greenhouse screening and field testing (with soil sampling). Both methods are labor intensive and require longer duration. Functional SNP markers (also for copy number) to select for *rhg1* resistance allele and *Rhg4* resistance allele, along with other major *Rhg* genes have been developed and routinely used by breeders. Since *rhg1-b* in PI 88788 requires multiple copies to provide resistance, molecular assays have been developed to determine the *Rhg1* copy number in breeding lines.

The phenotyping for SCN is also unique. Soybean varieties carrying the resistance genes reduce the amount of SCN reproduction when grown in the field. However, repeated sowing of the varieties carrying the PI 88788 resistance source (named PI 88788 resistance for the plant introduction that is the main source of resistance used by US breeders), the SCN population dynamics changes and SCN population capable of reproducing readily on the resistant varieties develops. This renders the resistance source ineffective, challenging breeding efforts. Earlier, the term "race" was used for SCN quantification; however, now "HG type" is used. It is based on a greenhouse assay using "HG type indicator" soybean check lines with SCN resistance genes. The HG type test provides information on the reproduction ability of the SCN population in the test variety. Each of the indicator check line serves as the benchmark of SCN population, enabling testing the effectiveness of SCN resistance in breeding lines relative to the indicator check lines (Tylka, 2006).

In wheat, marker-based selection for FHB resistance is a success story. The major QTL for FHB resistance have been mapped on chromosomes 3B (*Fhb1*), 5A (*FHb5*), and 6B (*Fhb2*) have been under most active consideration by breeders although many other QTL with minor effects have been identified. FHB QTL particularly *Fhb1* and *Fhb3*, have been studied in different genetic backgrounds and environmental conditions and their effects are deemed large enough to be a candidate for MAS of these QTL. In Canada and United States, wheat breeders have developed cultivars with improved FHB resistance using MAS in their programs. Interestingly, CIMMYT wheat varieties did not contain *Fhb1* possibly because this locus is in repulsion with *Sr2* (stem rust gene), and breeders at CIMMYT likely selected for stem rust resistance (including *Sr2*) as breeding for stem rust is of higher importance than FHB for their program. With the identification of these two genes in coupling phase, breeders have another avenue to select for both genes. Since only *Fhb1* has been cloned, gene-based markers (i.e., diagnostic/perfect markers) are only available for this QTL, while other FHB QTL are selected using flanking or linked marker. Due to the genetic variability, and need to select for and combine multiple QTL for FHB resistance, genomic selection is a very attractive strategy. With a robust prediction equation, which can be developed by using large population sizes for training, diversity of germplasm and phenotypic variation, higher marker density, and minimizing unequal sample size (between phenotypic classes, i.e., resistant to susceptible), genomic selection can make a large impact in a breeding program. Due to trait complexity, backcross breeding is not an ideal method for FHB, except when the objective is to transfer 1–2 FHB QTL into a susceptible variety.

Gene islands

Among the gene islands, *Lr34-Yr18-Sr57-Sb1-Pm38-Ltn1* is a well-known example from wheat. This provides resistance to wheat leaf rust (*Lr34*), stripe rust (*Yr18*), stem rust (*Sr57*), spot blotch (*Sb1*), and powdery mildew (*Pm38*) pathogens and has remained effective for more than a century providing partial resistance against the pathogens and no virulence yet reported. *Ltn* stands for leaf tip necrosis and has been used a phenotypic marker for selection. Other examples include *Lr67-Yr46-Sr55-Pm46* and *Lr46-Yr29-Sr58-Pm39-Ltn2* gene complex in wheat. These genes have very close linkage, but pleiotropy has also been suggested. What makes them unique is the expression of partial resistance (slow rusting and slow mildewing) associated with race non-specificity, which provides more durable resistance (i.e., APR) seen in adult plants. Slow rusting have several features, such as slow disease progress even with compatible infection type, lower infection frequency, longer latency (period between infection with microorganism and start of symptoms), and smaller spore size or less sporulation.

Inheritance of resistance to diseases

The rediscovery of Mendel's laws of inheritance in 1900 provided the foundation necessary for the analysis of the differential reaction of varieties to diseases. The first reported genetic study of disease resistance was published in 1905 by Biffen in England. He obtained a 3 (susceptible): 1 (resistant) ratio in the F_2 populations of crosses between resistant wheat variety (Rivet) to yellow rust (*Puccinia striiformis*) and susceptible varieties (Michigan, Bronze and Red King). The F_3 families appeared in the ratio of 1/4 true-breeding susceptible, 1/2 segregating, and 1/4 true-breeding resistant. He also suggested that resistance to ergot was governed by two factor pairs. Since then countless researchers have published on the mode of inheritance of resistance, and generally described it as monogenic/oligogenic and polygenic (see Singh and Singh, 2005). Inheritance and allelic relationship of resistant genes have been studied in major food crops against important pathogens and insect-pests (Singh and Singh, 2005). The inheritance of resistance could be studied by the use of controlled pathogen cultures, with appropriate crop controls/checks. The parents, F_1's, F_2's, backcross populations, F_3 progenies, recombinant inbred lines, and/or doubled haploid are inoculated with single-spore cell culture. If single spore culture is unavailable, an inoculum representing a mixture of spores can be used; however, it will not be able to provide information on race specific resistance. The scoring of individual plants (or progeny row) is recorded. The plants/progenies are classified (into appropriate phenotypic classes or categories, for example, resistance and susceptible) based on the reactions of parents and F_1's.

If monogenic or oligogenic resistance (one or two genes) is expected, inheritance studies are conducted by crossing the resistant line (variety or germplasm line of interest) with a susceptible line. The F_1 generation is studied to determine whether the gene(s) is expressed as a dominant (F_1 will be resistant) or recessive (F_1 will be susceptible). The genetic ratios of F_2 and later generations are studied to determine the number of genes controlling the traits. Chi-square goodness of fit values is calculated to determine whether the ratios deviate from the expected Mendelian segregation ratios (Fig. 22.4). In case of two or more genes controlling the traits, expected genetic ratios can be calculated and observed versus expected class numbers can be tested to determine the best fit genetic ratio. For example, a 9 A_B_: 3 aaB_: 3 A_bb: 1 aabb ratio is expected in the F_2 generation where two genes are segregating independently. Epistasis can also be determined. However, to determine whether the two lines contain the same or different functional alleles control the resistance, allelic studies are conducted (Fig. 22.4). In allelism study, different races are used to inoculate the plant material (developed by crossing the two resistant parents) and the expected ratios are tested using the chi-square goodness of fit test. Similarly, if two lines are segregating for different resistance genes; for two genes, the genotypic ratio will be 9 A_B_ (resistant): 3 aaB_ (resistant): 3 A_bb (resistant): 1 aabb (susceptible), if both dominant forms (A and B) of the two genes condition resistance. Different ratio are tested if the expectation of resistance genes are different (say, A and b resistant, or a and B resistant, or a and b resistant). In sufficiently homozygous state, say F_5 onwards, a 1 AABB (resistant): 1 aaBB (resistant): 1 AAbb (resistant): 1 aabb (susceptible), that is, 3 resistant:1 susceptible ratio will be observed. For additional information, see the "gene for gene concept" section.

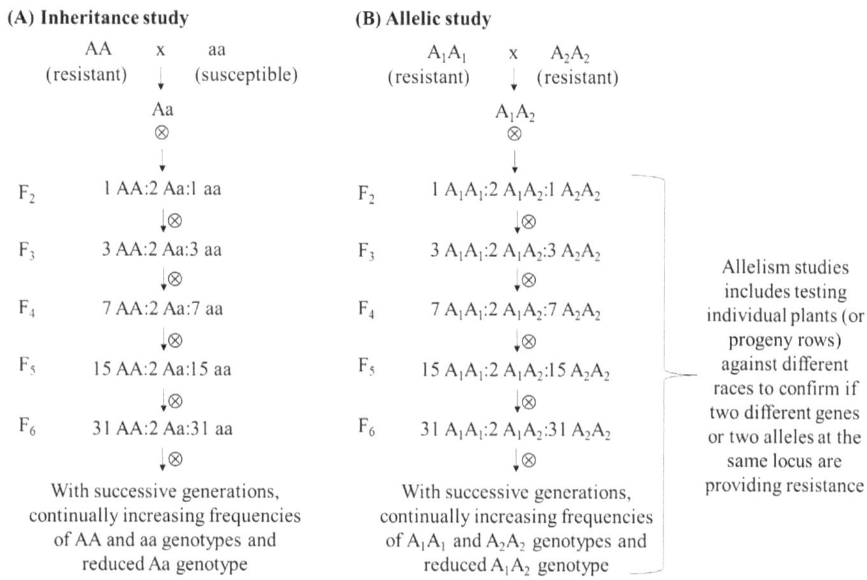

FIGURE 22.4

(A) Inheritance study where a single dominant gene controls the resistance. (B) Allelic study where two alleles at the same locus controls resistance with a differential race reaction (resistance is not controlled the same way by A_1 and A_2 against all races).

Differential sets

A differential set is used by plant pathologists to identify and distinguish different species of a pathogen and races/strains/biotypes/pathotypes within a species (i.e., race analysis). It consists of a set of plant varieties or germplasm lines with known susceptible and resistant reactions to races/strains/biotypes/pathotypes. Levin and Stakman (1918) used the term "differential host" in their work on stem rust (*Puccinia graminis*) and used three biologic forms of stem rust pathogen on six differential hosts (wheat varieties). This laid the foundations for the usage of differential sets to establish the identity of the rust forms and thereby distinguish between different biological forms. Earlier research by Stakman and colleagues used different plant hosts to separate the stem rust pathogen, for example, *P. graminis tritici* and *P. graminis secalis* to determine wheat and rye hosts differential reaction (Stakman and Piemeisel, 1917). However, the concept of varieties with different reaction types to distinguish biological strains or pathotypes was given even earlier by Barrus (1911). The inheritance and allelic studies helped in creating clearer differential sets for the host–pathogen systems. Each host–pathogen system has their unique differential set. The development and use of differential set was useful to develop varieties with disease resistance that was effective over different locations and years, and the cooperative work provided an established differential set for various host–pathogen systems.

The differential set consists of varieties, ideally each with different single resistance gene enabling races to be differentiated due to qualitative differences in reactions to different pathogen strains. A differential set evolves with new varieties added (or deleted), to reduce the problems associated with ratings, for example, to obtain clear differentiation between high and low infection types rather than more complex infection types (say three or more classes of infection types). Varieties in differential sets are included where each variety contains a single resistance gene rather than multiple genes. Also, as new races are identified new varieties may be added to the differential set. For most clear results, backcrossing should be used to create a series of differential varieties that have the same genomic background except for the resistance gene (isogenic lines). For example, if 10 resistance genes are known, backcrossing (molecular enabled or phenotypic selection based) can be used to create a series of varieties with the same recurrent parent crossed to donor parent (selected for a single gene in each variety). The creation of these near-isogenic lines and their inclusion in the differential sets minimizes background effect from modifier and epistatic interactions. Finally, a standard international differential set is preferred as it allows for global comparison of race analysis.

The knowledge on which pathotypes (i.e., races) are present in the area for which cultivars are being bred, is important. However, in addition to determining the races present in the area, their frequency is also important so breeding efforts can be changed or modified accordingly, and cultivar recommendations can be made to farmers. For example, in race analysis of wheat stem rust pathogen, bioassays under controlled conditions are needed, and require appropriate differential set. The brief outline of the race analysis for rust pathogen is as follows (source: FAO):

- Samples of infected stems are collected from field where the disease is found. Stems are processed and stored as per established protocols.
- Subculturing the spores to increase the rust sample; rust spores are transferred from infected tissue to healthy leaves of a seedling of an extremely susceptible variety for this purpose. Post-inoculation, appropriate disease conditions are created with manipulation of temperature, moisture, and light to achieve high disease levels.
- Spores from the subculturing steps (from infected leaves of susceptible variety) are collected. These spores are used to inoculate the differential set. Bulk inoculum or single pustule isolates can be used; however, the use of single pustule isolates increases the chance of working with a single race.
- If single pustule isolate is used, in a new experiment, spores are again subcultured and multiplied on the susceptible variety. Thereafter, the differential set is inoculated.
- Seedling infection types are rated on the differential set using standardized rating scales. Races are identified based on the combination of virulence/avirulence

patterns exhibited, where virulence is the degree of severity of disease produced by the pathogen in the host. With the use of a differential set, preferably each variety of the set with a different single resistance gene, races can be distinguished by their qualitative differences in reactions to different pathogen strains.

Varieties in the differential set are also useful for allelic studies. To determine the novelty of resistance in a breeding line or accession, crosses are made between line containing purported "new" sources of resistance and with varieties (or lines) in the differential set. The filial generations (F_1, F_2, onwards) are screened against the known races appropriate for the variety in the differential set. For example, if variety A in the differential set is effective against race 1, then filial generations can be screened against race 1 to determine whether the breeding line or accession has the same gene as in variety A. This allelic relationship for resistance genes allows plant breeders to effectively and efficiently utilize different resistance genes and accessions to incorporate "novel" resistance gene(s) or QTL into their program. These genes/QTL can be subsequently pyramided with other resistance sources in a single cultivar for a broad-based resistance.

Inheritance of virulence to plant pathogens

During the period following Biffen's work, the knowledge of the inheritance of resistance to diseases was expanded and parallel advances were made towards the understanding of variation in fungi. Early studies on the inheritance of virulence were carried out in the early 1930s with the fungus *Ustilago* (Nicolaisen, 1934). He showed that the capacity to incite a host, and the expression of either a susceptible or a resistant disease reaction was under a Mendelian control. Inheritance of virulence can be studied by crossing the isolates/races with differing pathogenicity. The parents, F_1, F_2 cultures are inoculated on the tester host genotypes.

Gene-for-gene concept

The gene-for-gene hypothesis was proposed by Flor (1942) as the simplest explanation of the results of studies on the inheritance of pathogenicity in the flax rust, *Melampsora lini*. On the varieties of flax (*Linum usitatissimum*) that had one gene for resistance to the parent race, F_2 cultures segregated in monofactorial ratios. On the varieties that had two, three or four genes for resistance to the parent race, the F_2 cultures segregated into bi-, tri-, or tetra-factorial ratios. This suggested that for each gene that conditions resistant reaction in the host there is a corresponding gene in the pathogen that conditions pathogenicity. Each gene in either member of a host–pathogen system may be identified only by its counterpart in the other member of the system. The incompatibility between a host and pathogen is due to the interaction between products of at least one of host resistance gene and at least one of its corresponding pathogen avirulence gene. Also, if more than one interacting gene pairs are involved, the incompatibility is similar or lower than the incompatibility gene pair acting alone.

The correspondence between the number of genes for resistance in the host and the number of recessive genes for pathogenicity in rust is presented in Table 22.1. The total number of genes concerned with infection in both the pathogen and the host is not necessarily the number indicated by these tests. For instance, a variety of flax may have two dominant genes for resistance, LLNN, and may be susceptible to one race of the rust and resistant to another. These races could have the genetic constitution $a_La_La_Na_N$ and $A_LA_La_Na_N$. The hybrid between these two races would segregate to give a ratio of 3 (nonpathogenic): 1 (pathogenic) isolates in the F_2 generation when tested on the tester with LLNN genotype. This shows that a hybrid between two races shows a monogenic segregation on a digenic variety. This complementary relationship of host and pathogen is called the gene-for-gene hypothesis for disease resistance.

A further example of the operation of the gene-for-gene hypothesis system in host and pathogen is given in Tables 22.2 and 22.3. The two varieties Ottawa 770 B and Bombay, each having one dominant gene for resistance, and their hybrid

Table 22.1 Reaction of flax varieties to F_2 segregates of a hybrid between race 22 and 24 of *Melampsora lini.*

Variety of flax	Number of resistance genes in the host	F_2 ratio of segregates of rust		No. of virulence genes in race 22 and 24
		Pathogenic	Nonpathogenic	
Ottawa 770B	1	1	3	1
Newland	1	1	3	1
Bombay	1	1	3	1
Akmolinsk	1	1	3	1
Italia Roma	2	1	15	2
Bolley Golden	2	1	15	2
Morye	3	1	63	3

Table 22.2 Digenic inheritance of resistance in flax varieties (Flor, 1956).

Rust race	Genotypes of parental flax varieties		F_2 genotypes			
	Ottawa 770 B (LLnn)	Bombay (llNN)	LN	Ln	lN	ln
22	S	R	R	S	R	S
24	R	S	R	R	S	S
	Nos. observed		110	32	43	9
	Expected on the basis of 9:3:3:1		108	36	36	12

Table 22.3 Digenic inheritance of pathogenicity in races of *Malampsora lini* (Flor, 1956).

Flax variety	Genotypes of parental races		Genotypes of rust segregates in F_2			
	Race 22 ($a_La_L A_N A_N$)	Race 24 ($A_L A_L a_N a_N$)	A_LA_N	a_LA_N	A_La_N	a_La_N
Ottawa 770B (LLnn)	S	R	R	S	R	S
Bombay (llNN)	R	S	R	R	S	S
	Nos. observed		78	27	23	5
	Expected on the basis of 9:3:3:1		75	25	25	8

segregated in the F_2 generation to give a digenic ratio when tested with two rust races 22 and 24. Similarly, the F_2 segregates of the hybrid between races 22 and 24 showed a digenic ratio when tested on the varieties Ottawa 770 B and Bombay. Only the rust races possessing recessive alleles for pathogenicity in respect of all the dominant resistance genes present in the host genotype are capable of producing infection.

Flor's gene-for-gene hypothesis does not tell us anything about the gene quality, that is, if the gene remains resistant or becomes ineffective. A second gene-for-gene hypothesis proposed by van der Plank assumes Flor's hypothesis tell us about the quality of resistance gene. The quality of resistance gene in the host determines the fitness of the matching virulence gene in the pathogen to survive, when this virulence gene is unnecessary. Unnecessary gene means that the matching resistance gene in the host is not present. Reciprocally, the fitness of the virulence gene in the pathogen to survive when it is unnecessary determines the quality of matching resistant gene in the host, as judged by the protection it can give to the host.

Types of genetic resistance to insect-pests

Painter (1951) wrote the first book on insect-pest resistance in crop plants and made agricultural scientists cognizant of the fact that the use of resistant crop plant was an ideal way to protect crops against insect-pests. He proposed genetic mechanisms of resistance, which were grouped into three main categories: non-preference (or *antixenosis*), *antibiosis* and tolerance.

Antibiosis refers to a situation where the biology of the insect is affected such that pest abundance and ensuing damage is reduced when insect infests the resistant variety of the host plant for food when compared with abundance and subsequent damage on a susceptible variety. Antibiosis type of resistance manifests as increased insects' mortality or reduced longevity and reproduction. Antibiosis (physiological response) is used in its usual sense as the tendency to prevent,

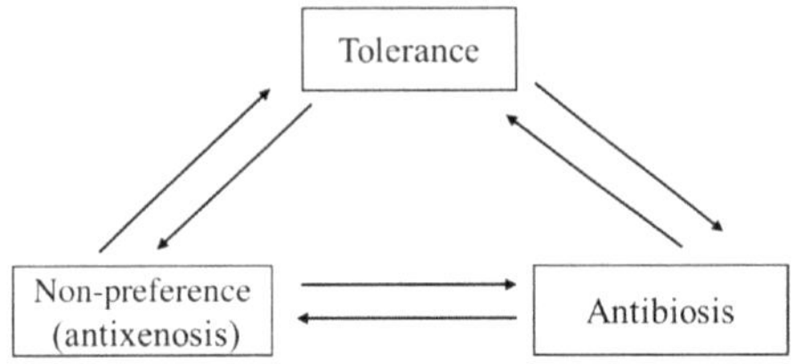

FIGURE 22.5

Three types of resistance mechanisms that are interrelated and one or more of these are frequently present in insect-pest tolerant/resistant varieties.

injure or destroy (insect) life, when it uses a resistant variety of the host plant or species for food. Later, another term was used: Antixenosis, where insect-pests have non-preference for a resistant plant compared with a susceptible plant in the same crop species. Non-preference is insect's response (behavioral response) to plants that lack the characteristic to serve as hosts, resulting from negative response or total avoidance during search for food, oviposition, or shelter or for the combinations of the three. Tolerance is the mechanism of resistance in which a plant shows an ability to grow and reproduce itself or to repair injury to a marked degree in spite of supporting a population approximately equal to that damaging a susceptible host.

Tolerance differed from non-preference and antibiosis in its mechanism. Non-preference and antibiosis require an active insect response or lack of response. Tolerance is more subject to variation as a result of environmental conditions than non-preference and antibiosis. The age or size of the plant and size of insect population strongly influence the degree of tolerance.

Of these three mechanisms, one or a combination of any of the three is present in most cases of resistance. The three mechanisms are interrelated (Fig. 22.5). Such a basic triad of resistance relationships has usually been found to result from independent genetic characters, which are, however, interrelated in their effects. A plant exhibits high tolerance and low antibiosis towards a given insect-pest may show the same level of resistance as one that has a high degree of antibiosis and an average degree of tolerance. The gene(s) for one or more types of resistance may be recombined. This type of recombination in addition to the possible multiple genetic factors in the plant may be the basis of any one of the three characteristics. The genetic expression of these three characters of resistance is frequently modified by various ecological conditions and by other genes.

Methods of breeding for resistance to biotic stresses

The cultivation of resistant varieties has been recognized as the most effective, ideal and economical method of reducing crop losses. The breeding for resistance is generally in no way different than breeding for other traits. However, in resistance breeding the two biological entities, host plant and pathogen (and one physical factor, environment), are involved; whereas in breeding for other traits, the breeder deals with the variability in test (i.e., crop) material only except for abiotic stress (see Chapter 21: Breeding for Resistance to Abiotic Stresses) where both crop and environment are important.

In a traditional approach, the first step in resistance breeding programs is the collection of natural variability followed by the identification of sources of resistance. The next step is to incorporate the resistance gene(s) from the donor parent(s) using various breeding methods, and induced mutations where the susceptible alleles are altered by the use of mutagens. In view of the dynamic nature of pests and pathogens, the resistant gene(s) become ineffective after a few years with the situation exacerbated for single gene-based resistance (e.g., in cereal rusts and other similar systems), as previously explained. Therefore, resistance breeding program is a continuous pipeline and effort for a breeder. An attempt is made in this chapter to present information on breeding strategies including how to manage race specific or race non-specific gene(s)/QTL so that the life span of resistant gene(s) is prolonged, and disease losses are avoided or reduced. In genetic engineering approaches, transgene or gene editing methodologies can be used (see Chapter 27: Molecular Tools in Crop Improvement and Cultivar Development).

Selection for resistance to insect-pests and diseases is relatively easy, but certain host plant genotypes may show differing reactions to races/biotypes. It is desirable to test host genotypes against a wide range of variants of a insect-pest and pathogen before selection is made. This could be done by using variants (biotypes, races, and strains) separately or in known compositions. The use of individual variants is important for studying the genetics of the host–pathogen relationship but is not essential for practical breeding purposes unless specific genes are targeted. The mixture of variants (ideally representative of the environment for which breeding is conducted) can be maintained individually on a range of host genotypes or on culture medium. The breeding material can also be grown at several locations (test sites) where different variants of pathogens are expected to naturally occur, as these consistent prevalence sites can be a good resource for a breeder to establish selection nurseries. Furthermore, coordinated multi-environment testing efforts are also very useful for breeding efforts. The coordinated testing programs of CGIAR institutes the Soybean Uniform Regional Testing coordinated by the USDA, public–private wheat registration tests in Canada, and the coordinated programs of major food crops in India, conducted by the Indian Council of Agricultural Research, are few of the best examples of testing crop varieties under different agroclimatic conditions.

Although similar procedures are adapted in selecting for pest and disease resistance in self- and cross-pollinated crops, there are important differences in handing the two types of crop. In self-pollinated crops, individual plants are selected to form a variety; while in cross-pollinated crops, this is seldom a basis to select for resistance because of self-incompatibility, inbreeding depression, etc. The flower structure of self-pollinated crops poses problems in intermating, and consequently the breeding methods such as recurrent selection are of limited use; however, with the use of genetic male sterility they may be used in autogamous crops too (see Chapter 17: Recurrent Selection in Self-pollinated Crops).

Cross-pollinated crops

Several methods can be used in allogamous crops to improve populations for developing varieties with resistance to insect-pests and diseases. Mass selection/ recurrent selection has been used, while other approaches are shown to be effective as well. The choice of methods comes to the breeder, considering the time, resources, and objectives of the program.

Mass selection/recurrent selection

This is one of the most commonly used breeding methods. Recurrent selection has been used effectively to improve resistance to insect-pests and diseases in forage crops and in other cross-pollinated crop species. It consists of selecting individual plants for resistance from the heterozygous plant population, which is inoculated/infested artificially. The population from the selected plants is reinoculated in the next generation, and susceptible plants are eliminated before intermating to produce seed. For traits with resistance to pests and diseases expressing high heritability, mass selection is useful as a breeding tool. The objective is to achieve a higher proportion of resistant plants in each successive generation (i.e., by recurrent selection), and continually moving the population mean towards better resistance.

Mass selection has been shown to be an efficient method of improving disease resistance in maize populations without adversely affecting yield if adequate population size is maintained (Miles et al., 1980). Estimates of among and within half-sib family variance components for disease score in each of the two maize populations were obtained from artificially inoculated experiments in one, two or three years. These estimates were used to predict the response of mass, half-sib and S_1 selection for resistance to northern corn leaf blight (*Drechslera turcica*); *Diplodia* stalk rot (*Diplodia maydis*), anthracnose leaf blight and stalk rot (*Colletotrichum graminicola*). In addition, among half-sib family various component estimates for yield were obtained from uninoculated experiments for the same families. Genetic correlations between yield in the absence of disease and disease reactions were all near zero and nonsignificant. Correlations among disease reactions were positive and significant.

Recurrent selection over three and four cycles was used to concentrate polygenic resistance in pearl millet (*Pennisetum glaucum*) for ergot (*Claviceps purpurea*) (Chahal et al., 1981). In potato, randomly intermating a group of potato clones, planting seedling tubers in the field test to measure potato leafhopper, *Empoasca fabae* resistance to infestation, and then bulking the seed from the most resistant clones to complete the selection cycle in one year was shown to be effective (Sanford and Ladd, 1983). In five cycles of selection for resistance to infestation, a decrease of 57% in the level of infestation from the original population was noted.

Line breeding

From a population or breeding family, selected plants (based on phenotypic or marker-based tests) are either selfed or interpollinated and the resulting progenies or lines are individually tested for resistance in a progeny row; only the most resistant lines are retained for subsequent breeding. These are then intermated to produce a composite cross. These lines (if sufficiently inbred or DH), can also serve as a parental strain in hybrid production scheme. Inbred lines may be improved for resistance by using the backcross method or convergent improvement.

Polycross

This method consists of selecting resistant plants from a heterozygous population and intercrossing these or inbred lines derived from these, in all possible combinations. The progenies of a polycross can be bulked. Resistant plants are subsequently selected from the bulk population, or the progenies of individual lines can be tested separately.

Synthetic/hybrid varieties

Resistant lines resulting from line breeding or recurrent selection programs can be used to produce hybrid or synthetic varieties. A synthetic variety is produced by intercrossing several selected plants, lines or clones, which have been found to express good general combining ability. Hybrid varieties are produced by controlled pollination between lines. The parental lines ought to be maintained separately so that synthetic or hybrid varieties may be reconstituted as required. Inbred lines may be improved for resistance by using the backcross method or convergent improvement. Rowe and Hill (1981) compared interpopulation improvement methods in alfalfa, and reported that top-cross selection was more effective than polycross selection for resistance to *Colletotrichum trifolii*, when the tester parent had less resistance than the base population, and top-cross selection was less effective when the tester parent was more resistant.

Self-pollinated crops

In self-pollinated crops, pedigree based and backcross methods with certain modifications as per the convenience of breeders are used more frequently and other methods are less frequently used. However, this does not imply that other methods are not effective.

Mass selection

Plants of a similar phenotype are selected for resistance from a population and the progenies are bulked to form the basis of variety. Such varieties are easy to produce. Their heterozygosity might give them certain advantages over pure line varieties for pest and/or disease resistance. The challenges of recurrent selection and potential solutions are presented in Chapter 17, Recurrent Selection in Self-pollinated Crops.

Pure line selection

These are derived from the progeny of a selfed homozygous plant selected from a land variety or commercial variety (if heterozygous). The progeny is retested for resistance and evaluated for other desirable characters in succeeding generations, and, if promising, it is multiplied to produce a new variety. This is a simple and quick procedure; and while common in early 1900s, is uncommon in modern plant breeding era in crops where sufficient prior breeding efforts have been made.

Hybridization

This involves crossing two (or more) pure line varieties with the objectives of transferring resistance from donor parents and of combining characteristics from each parent, that is, forward breeding approach. It enables the breeder to combine resistance to several races or biotypes or to different pests or diseases in a single variety. The F_1 plants from a cross are identical in their genetic constitution in a two parent cross where both parents are homozygous and true breeding (in a three parent cross, or if parents are not homozygous, F_1 generation will show segregation). Segregation occurs in the F_2 generation onwards and homozygosity is regained in succeeding selfed generations. The transgressive segregants for resistance may occur with qualitatively and even quantitatively inherited resistance, and these segregants may be selected in the F_2 and later generations. Selection after hybridization is based on the following methods.

1. *Pedigree method:* if a resistant parent can contribute to improved adaptation, yield, yield components and quality, etc., the pedigree method of handling the segregating generations may be followed. It consists of selecting individual F_2 plants for desirable features, including resistance to insect-pests and diseases. The progenies of these selections are further selected in each succeeding generation until homozygosity is obtained. Artificial epiphytotic conditions are created for insect-pests(s) or disease(s) in the early segregating generations.

   In rice, pedigree method was used to handle the early generation breeding lines where a portion of seeds from plant selections was used to plant the pedigree rows and the rest for screening of disease and insect-pest resistance (Khush, 1971). The dwarf plants from the cross of susceptible (dwarf) and resistant (tall) varieties of rice were selected in the F_3 generation. The progenies of these plants were screened up to later filial generation for pest and disease resistance (Khush and Beachell, 1972). Several selections resistant to common pests and diseases were identified at International Rice Research Institute (IRRI). The pedigree method has been very widely used in breeding for disease and insect-pest resistance and numerous resistant varieties of self-pollinated crops have been developed by this procedure.

2. *Bulk population method*: this method is also suitable for combining characteristics from both parents. In this method, the early segregating generations (F_2–F_6) are bulked together without selection. In later generations, when most plants are homozygous, individual plants are selected for resistance and their progenies are evaluated as in the pedigree method. Artificial epiphytotic conditions are created in the later generations for selecting resistant plants/progenies; however, in a modified scheme, artificial epiphytotic conditions can be imposed in early filial generations to discard susceptible plants so that resistant plant progenies contribute to the bulk population moving to the next generation.
3. *Bulk pedigree method*: some aspects of the pedigree and bulk population methods are combined in this method. Hybrid populations are grown in bulk until a pest or disease attack occurs naturally, when single plants are selected for resistance. At one extreme, bulking may end in the F_2 generation, hence it is no way different from the pedigree method, and if bulking may be continued for many generations in the absence of suitable conditions for selection, it is similar to the bulk population method. The individual plants selected in either case are handled as in the pedigree method.
4. *Backcross method*: if the resistant parent is a wholly unadapted type, the backcross method is appropriate to transfer certain characteristic of a line or variety into an elite variety or breeding material. It is more useful to transfer one gene or a few genes, such as major genes (or QTL) conferring a high degree of resistance, from one genetic background to another. Using this method, genes from wild species have often been transferred to susceptible but otherwise satisfactory varieties. Where resistance is dominant, the F_1 is backcrossed to susceptible parent. The progeny of the first backcross generation is tested for resistance. The resistant plants are backcrossed to the susceptible parent (recurrent parent). After several generations of backcrossing (generally five to six), plants with characters almost identical to the original susceptible variety are obtained. These plants have the added advantage of resistance genes that are transferred from the donor parent. Where resistance is recessive, the progeny of each backcross generation is selfed to obtain homozygous recessive plants. If molecular markers are used, seed or plant selections can be made to chose appropriate parents for crossing in each generation.

This method is generally unsuitable for quantitatively inherited resistance, which are controlled by polygenes except if marker-assisted backcrossing is used to transfer one or two large effect QTL using flanking markers. The backcross method can also be used to transfer resistance to more than one disease and insect-pest resistance trait. Once a variety has been improved for resistance to one trait, it can serve as the recurrent parent for incorporating resistance to the second one. Sometimes it is difficult to recover lines that have a good plant type from single crosses following the pedigree method. Therefore, one or two backcrosses to the improved plant type parent are useful. If needed, for example, if two parents are unable to contribute all the desired characteristics, an F_1 is crossed to a third improved plant type parent to obtain a three-parent cross. The backcross method is also useful for developing isogenic or near-isogenic lines for developing a differential set, and to develop multiline cultivars.

Mutation breeding

The resistance gene(s) sometimes are not available in cultivated types and the wild germplasm is not easily crossable. Often the cultivated variety is higher yielding and widely adapted; however, it is susceptible to a particular disease (race), insect-pest (biotype). If the breeder is interested in bringing novel genes or alleles, mutation breeding for resistance is the obvious choice. Resistance may sometimes be detected under conditions when a disease is so severe that the crop is almost destroyed. Such resistant plants could be due to spontaneous mutations; however, chance out-crossing or seed admixture cannot be ruled out. Orton (1900) selected surviving cotton plants on soil heavily infected with cotton wilt, *Fusarium oxysporum*. Bolley (1905) also obtained flax resistant to *F. oxysporum* f. sp. *lini* under similar conditions. The mutations can also be induced artificially with the use of mutagens. Konzak (1956) presented evidence that disease resistance in higher plants is obtainable via induced mutations. The mutagens could be physical or chemical and have been described in Chapter 14, Mutation Breeding.

An induced mutant, called m1-0 (Favret, 1971a), for resistance to *Erysiphe graminis* f. sp. *hordei* in barley occurs with an unusually high frequency. The mutant is recessive and has pleiotropic effect of producing necrotic flecking in the absence of the pathogen and is associated with a small reduction in yield. Jørgensen (1971) noted that plants carrying m1-0 were not susceptible to any physiological race of mildew in large-scale tests in several parts of the word.

Wallace and Luke (1961) reported that the frequency of obtaining a resistance gene by a mutagenic treatment is extremely low (1 to 35 $\times$ 10^{-8}). Therefore, it may be useful to observe a large number of plants under natural infection. A mutation breeding program was initiated due to the unavailability of sources of resistance in Sri Lanka to blight (*Phytophthora nicotianae var. parasitica)* in sesame (*Sesamum indicum*) (Pathirana, 1992). Seeds of three genotypes were irradiated to six doses of gamma-rays (100 to 750 Gy from Co^{60}). M_1 to M_2 were handled as bulk method. Twenty-one lines from M_3 and M_4 progeny rows of selected M_2 plants were identified as tolerant, yield tested along with MI 3 as a control. The plant survival in selected lines averaged 43.3% compared to 7.2% in MI 3.

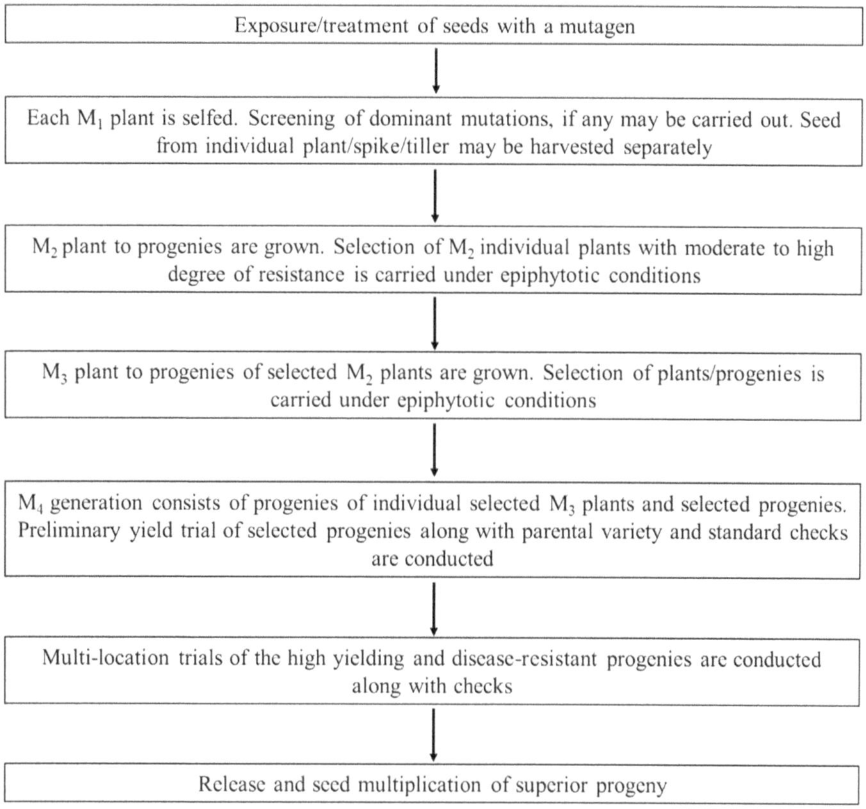

FIGURE 22.6

General outline of a mutation breeding procedure in autogamous crops plants.

The general outline of a mutation breeding in self-pollinated crops is presented in Fig. 22.6. The same procedure (with necessary modifications) may be followed in cross pollinated and vegetatively propagated crops. In cross-pollinated crops, instead of selfing individual plants in M_1 generation, a group of plants are selfed together. In asexually propagated crops, cuttings are used in place of seeds.

Vegetatively propagated crops

Some of the important crops like sugarcane (*Saccharum officinarum* L.), potato (*Solanum tuberosum* L.), strawberry (*Fragaria* L.), raspberry (*Rubus idaeus* L.), pineapple [*Ananas comosus* (L.) Merr.], banana (*Musa* L.), and most of the fruit trees are propagated through vegetative or asexual organs. The plants of these crop species are usually heterozygous. The vegetative propagation maintains this heterozygosity. If the desired level of resistance to pest or pathogen is obtained, it can be maintained easily; however, due to the nature of clonal crop production there is an enhanced threat to production if the clone becomes susceptible. Resistance is sought either by testing mixed populations of clones from different sources of resistance or by selecting the best progenies obtained by hybridizing different clones. Segregation starts in the F_1 itself; therefore, each F_1 plant must be tested separately for resistance. If the breeding program is focused on multiple diseases and/or insect-pest, each clone can be tested concurrently in specialized nurseries for each biotic stress. In crops where grafting is practiced (root and shoot stock), breeders can consider grafting biotic stress tolerant grafts. Grafting has been used as a method of controlling insect-pests, for example, grape aphid (*Phylloxera vitifoliae*).

Management of disease and insect-pest resistance

When a microorganism is able to invade the host cell and multiplies, it becomes a pathogen. The damage to plant cell is essential for the pathogen to access plant's nutrients particularly sugar in the cell; therefore, degradation of the cellular material happens. Plants mount a defense response through the first layer of plant immune system—the pathogen-associated molecular pattern (PAMP)-triggered immunity (PTI) to prevent pathogen's advancement to the cytosol. Plants defense mechanism includes immune receptors found on the plasma membrane that sense the PAMPs (protein encoded by avirulence gene) and the microbial-associated molecular patterns. The detection of PAMPs trigger a physiological change in the cell that is activated by the pattern recognition receptors (PRRs), which commence a cascade response via the recognition of the conserved microbial molecules leading to plant resistance. These receptors monitor conserved pathogen molecules extracellularly such as flagellin and chitin. The second defense system is the effector-triggered immunity (ETI). The ETI comes into play when PTI is unsuccessful to stop the pathogen. At this stage, pathogens release their effectors that continue the invasion of host cells, and these effectors work differently. These pathogenic proteins are recognized in the cytoplasm by the nucleotide-binding and oligomerization domain-like receptors proteins (PRRs similar to Toll-like receptors) and intracellular pathogen effectors, structurally characterized by a nucleotide-binding site leucine-rich repeats (NBS-LRR) proteins of host plants, triggering ETI as a second layer of defense. These generally lead to a hypersensitive response causing apoptosis of the infected host cell. Extracellular and intracellular immune receptors initiate defense responses by directly (physically interacting with pathogen derived elicitors) or indirectly (monitoring modifications of host targets incurred by pathogens).

The ability of plants to defend itself from pathogen depends on the receptors (for pathogen recognition triggering protection response in plants) and effectors (deployed by pathogen to suppress hosts' resistance and permit colonization). The NBS-LRR structure underlie many race specific resistance genes, and these have been shown to have shorter duration of effectiveness compared with race non-specific resistance, which are considered more durable. These types of proteins (NBS-LRR) make for a large family of effector proteins and in plants can be grouped into two major groups by features at their N terminus: one set has a Toll/interleukin-1 receptor-like (TIR) domain and the other a coiled coil domain (leucine zipper region).

The presence of complete resistance exerts a high selection pressure for virulence leading to genes becoming ineffective, making them useless for breeding applications. Therefore, qualitative resistance R genes are used in conjunction with other similar genes (in a gene pyramiding scheme) or with partial resistance sources. Gene pyramiding approaches, particularly enabled by DNA-based marker-assisted selection or backcrossing, is used as a breeding strategy; however, this is not without risk. Although the probability of concurrent mutations in multiple effectors is low, the emergence of a multivirulent pathogen population(s) can render pyramiding approach ineffective.

Another type of resistance with potential usefulness to breeders is the *non-host resistance*, which is the resistance shown by all plants in a species to all genetic variants of a pathogen. Another term in the literature is *basal resistance* which qualitatively speaking has been described for completely effective non-host resistance against pathogens. Quantitatively speaking, basal defense refers to a defense by plants that leads to inhibition of pathogen spread after successful infection and onset of disease. The basal resistance has also been described as quantitative disease resistance. Non-host resistance and basal resistance are postulated to be primarily governed by PTI, and perhaps these can provide as targets for gene editing-based plant protection methods (see Chapter 27: Molecular Tools in Crop Improvement and Cultivar Development).

Depending on the mode of inheritance, several methods have been proposed for a better utilization of the resistance gene(s). Therefore, in the ensuing pages, information is given on how best qualitative/vertical/race-specific and quantitative/horizontal/race non-specific types of resistance genes could be manipulated for the control of diseases and insect-pests. These techniques can prolong the average life span of resistant gene(s) and at the same time reduce the risk of catastrophic losses. In a previous section, the *Sm1* gene example has already been provided.

Recycling and sequential release of resistance gene(s)

This is based on the same principle as the common practice of crop rotations to control certain soil-borne or root-infecting pathogens. Stevens (1949) suggested that a system of variety rotations, similar to crop rotation, should be followed so that inoculums of a particular race(s) do not build up in sufficient quantities over a period of time to create an epidemic. However, the rotated varieties should have different genes for resistance. According to Stevens, where varieties are replaced more frequently, with each rapid increase in the numbers of the "new" race due to new host varieties, there is corresponding decrease of "old" races. After 5 or 10 years of widespread planting of a new host variety, it may be that the formerly well-known races or certain diseases will have become scarce, and that the older host varieties can be replanted with profit if the losses incurred by that disease in the area are very high each year. The utility of the system proposed by Stevens may be of limited value in modern plant breeding era where an on-going breeding program improvements in cultivars are always being made and better yielding and performing varieties are released each year, and planting old varieties is not preferred due to their lower yield. However, if a judicious backcross program is followed keeping the original genotype, with additional genes for resistance added, as new races came up, the system could be balanced, particularly in areas where a biotic stress is extremely damaging and good yield can not be obtained without adequate resistance against this stress. For example, at CIMMYT, a method termed “single backcross-selected bulk” scheme (Singh and Huerta-Espino, 2004) was used in wheat. In this method, elite line is crossed to a group of 8–10 resistance donors, F_1 plants from each cross are backcrossed to obtain a sufficiently large BC_1F_1 population, which is then moved forward under a bulk scheme in a high rust pressure environment.

Sequential release of resistance gene approach was adopted to control stem rust of wheat in Australia between 1938 and 1950 (Watson and Luig, 1963). The premise is to release one gene for resistance and wait until it becomes ineffective; release the second gene and so on. The sequential release strategy was employed for resistance to brown plant-hopper (*Nilaparvata lugens*) at IRRI (Khush, 1979). IR 26 and IR 1561-228-3 varieties of rice with resistant gene *Bph-1* were released in 1973 and 1974, respectively. Towards the end of 1975 and in 1976, these varieties started to show susceptibly at some locations in the Philippines. But by that time multiple disease and insect-pest resistant varieties with *bph-2* resistance gene of BPH were available in new varieties (IR-36 and IR-38), and these were released as replacements for the varieties with *Bph-1*. Later on at IRRI, breeding lines with *Bph-3* and *bph-4* for resistance to BPH were available. A proposal for race prediction and gene rotation to keep development of new disease resistant varieties of rice for farmers ahead of the development of new races of disease organisms was advocated at IRRI (1980) (Fig. 22.7). The system consists of the following steps:

1. Identify the monogene for resistance that is effective against races of the pathogen present in a specific cropping area.
2. Use of monogene (identified in step 1) for resistance to develop varieties for the specific cropping area and release these varieties. This will reduce and possibly eliminate the presently occurring races in the cropping area.
3. Test the resistant monogene at sites remote from the specific cropping area, through international testing program. The race of pathogen that overcomes the monogene in one or more of the remote areas will be representative of the race that will eventually evolve in the cropping area following the introduction of the new varieties containing the monogene (step 2).
4. Identify, based upon results obtained from the program (step 3), a monogene for resistance to the race predicted to occur in the cropping area and use it in the variety development program.

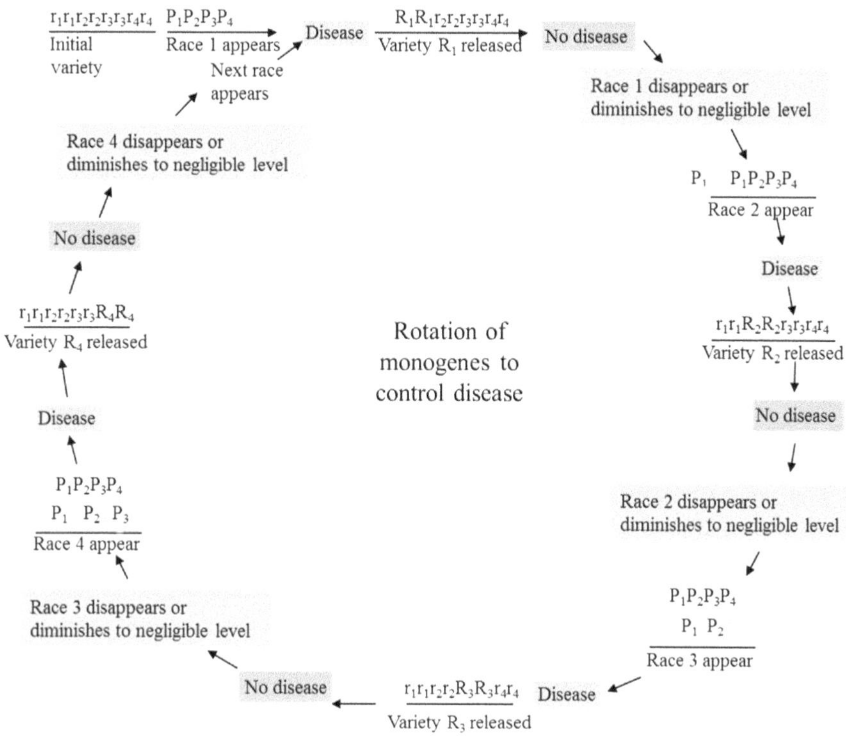

FIGURE 22.7

Scheme for race appearance and resistance gene rotation to develop new varieties with disease resistance that are preemptive of the development of new races of the pathogen. "P" is specific gene for pathogenicity in the pathogen, "p" is the allele of P pathogenic gene that is incapable of causing infection, "R" is the gene for disease resistance, r is the allele of R gene with susceptible reaction, "P" in circle represents gene with reduced prevalence in the pathogen population, that is, reduced frequency.

From IRRI (1980) Int Rice Res Inst, Philippines. Res Highlights 1979, pp 28–29.

The above scheme will likely be of lesser value for a qualitative R gene especially due to complex pathogen population dynamics (varying races active in the geographical area and multiple factors that drive the pathogen race composition and frequencies); however, QTL-based approach may be viable and the durability of APR will be effective for a longer period.

Pyramiding of resistance gene(s)

Simultaneous introduction of diverse genes for resistance into the cultivar was proposed for the first time by Watson and Singh (1952). This consists of introducing two genes, following a system of backcrossing and taking advantage of epistatic reaction types. The system is based on hypothesis that mutation in the pathogen at more than one locus is much rarer than the mutation at one locus; that is, the higher the number of diverse genes introduced into a single variety, the greater would be the longevity of its resistance. The genes should be of nonallelic types, governing resistance to all the virulent races present in a given geographical area or country. Such a variety would offer more than one physiological barrier against the pathogen and prevents a stepwise development of races virulent to varieties possessing different but single genes for resistance. The criticism of this system is that if a race evolves that can attack the combined resistance, then different R genes becomes susceptible simultaneously. Also, there is difficulty in detecting individual genes for resistance in a single genetic stock, because of their common protection against available races of the pathogen. Additionally, if the virulence factors necessary for giving evidence of the corresponding resistance genes are not available, it is rather easy to loose some of the resistance genes preventing the development of variety with maximum number of resistant factors.

Flor and Comstock (1971) developed flax cultivars with multiple rust-resistance conditioning genes. Lines of seed flax (*L. usitatissimum*), each with a different gene for rust resistance, were developed by back-crossing to the cultivar Bison. The monogenic Bison background lines were intercrossed to provide two gene segregates. Intercrossing two gene lines with lines carrying one gene, three gene lines can be developed. The selective pathogenicity of races of *M. lini* was utilized to identify plants that possessed parental resistance genes. At IRRI, pyramiding of two major genes for BPH resistance was also tried as varieties with two genes for resistance to BPH have a longer useful life (Khush, 1979). *Bph-1* and *bph-2* are closely linked. Similarly *Bph-3* and *bph-4* are also linked. However, *bph-1* and *Bph-3*, *Bph-1* and *bph-4*, *bph-2* and *Bph-3*, and *bph-2* and *bph-4* segregate independently of each other and can be combined.

Bacterial blight (BB) caused by *Xanthomonas oryzae* pv. *oryzae* is one of the destructive diseases of rice throughout the world. The over reliance and exploitation of gene *Xa-4* (due to large-scale and long-term cultivation of *Xa-4* carrying varieties) resulted in significant shifts in the pathogen race frequency with large areas shown ineffective *Xa-4* reaction (susceptible reaction). Gene pyramiding using DNA markers was proposed to bring together *Xa-4*, *xa-5*, *xa-13* and *Xa-21* genes (Huang et al., 1997), as it is difficult to pyramid resistance genes by phenotypic selection/conventional breeding because the presence of one major gene obscures the effects of other genes (both in forward and backcross breeding schemes). Gene pyramiding approach is used where research shows that having more than one gene gives a higher level of resistance to pathogen (or pest) than a single gene alone, or the combination of genes can protect against a broader range of pathotypes, or multiple genes can enhance the durability of resistance against the pathogen or pest.

Regional deployment of resistance genes

Resistant varieties with different resistance genes should be developed and recommended for different geographical regions of the country, where the crop covers a sizable area. As pointed out by Nelson (1972), this type of gene deployment is essentially a geographical multiline. A formal plan for regional deployment of genes was put in effect for resistance to crown rust fungus in oats in Iowa (Frey et al., 1973). Reddy and Rao (1979) suggested a strategy for controlling leaf rust, *P. recondita* f. sp. *tritici*, of wheat in India using regional deployment of resistance genes. India could be divided into three regions: (1) central plains, (2) northern Himalayan, and (3) southern Nilgiri and Palani Hills. Stronger genes (*Lr9* or *Lr19* or *Lr10* + *Lr15*) can be deployed in region A, *Lr1* + *Lr10* + *Lr1b* or *Lr3ka* + *Lr10* + *Lr17* in region B, and *Lr3ka* + *Lr10* + *Lr20* or *Lr1* + *Lr10* + *Lr17* in region C. They suggested that a similar work with *Sr* and *Yr* genes can create isogenic lines and will help in the release of stable varieties for all the three prevalent wheat rusts in India. When several genes are in operation, the possibility of build-up of a super-race is minimized, and the pathogen will become stabilized. This approach can be followed for any other crop diseases and insect-pests when enough genes are identified. However, it is important to mention that these approaches will still benefit from integration of APR and race non-specific resistance.

Chromosome or genome substitutions

For breeding resistant varieties, the sources of disease resistance are sought and the resistant gene(s) are incorporated in cultivated varieties. If gene(s) for resistance are not available in the cultivated species, the breeder can transfer resistance from related species/genera. Due to genetic or cytoplasmic or nuclear-cytoplasm genome interactions, chromosomes of different species fail to pair either completely or partially, resulting in sterility of F_1. To facilitate recombination, either the F_1 or the amphidiploid is backcrossed with the recurrent parent. In this way the whole genome or a whole chromosome or a segment of a chromosome of a recurrent parent is replaced by the donor parent. These techniques, along with appropriate examples(s), are discussed in the following subsection and Chapter 6, Wide Hybridization.

The genome of a wild or related species (donor parent) could be incorporated into susceptible cultivar of a cultivated species (recipient) through hybridization and chromosome doubling, using colchicine. In general, the species having higher ploidy level is used as female to avoid nucleocytoplasmic interference. By using colchicine, amphidiploids are produced, which have the full normal diploid genome of both the species. Resistance to clubroot (*Plasmodiophora brassicae*) has been transferred from turnip (*Brassica campestris*) ($2n = 20$, AA) to Swede turnip (*B. napus*) ($2n = 38$, AACC). Johnston (1974) used colchicine to double the F_1 hybrid to produce an amphidiploid, which can be maintained indefinitely and is more easily backcrossed to *B. napus*. Resistant second backcross progenies were virtually identical with the recurrent parent (Fig. 22.8A).

The substitution of a whole chromosome pair of the recurrent parent (susceptible) by one from the donor parent has been used in wheat. The wheat streak mosaic disease is a threat to production of winter wheat in many areas of North

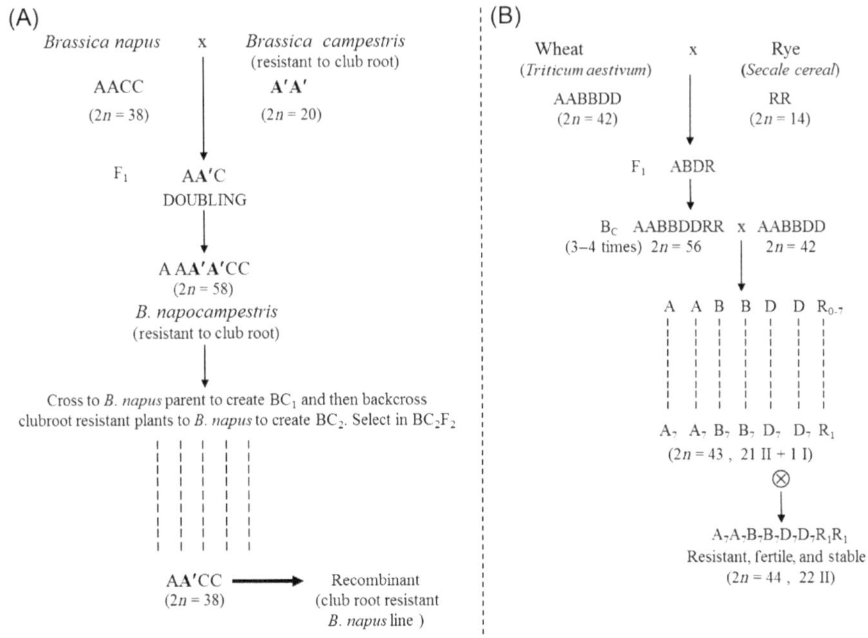

FIGURE 22.8

(A) The substitution of a whole genome for developing club root resistant varieties in turnip. Resistance to clubroot race 3 was successfully transferred from Flemish turnip (*Brassica campestris*) variety to Turnip variety (*B. napus*) by the production of fertile species *B. napocampestris* followed by two backcrosses to the *B. napus* variety. The new variety was resistant to clubroot disease and morphologically similar to the recurrent parent. Similar scheme can be used for other polyploidy crops. *From Johnston, T.D., 1974. Transfer of disease resistance from Brassica campestris L. to rape (B. napus L.). Euphytica. v23: 681–683.* (B) The substitution of a chromosome pair for incorporating resistance from rye to wheat. *From Singh, D.P., Singh, A., 2005. Disease and Insect Resistance in Plants. Sci. Publs., USA and Oxford & IBH, India, 417 pp.*

America. The virus is transmitted by the wheat curl mite (*Aceria teelipae*). Resistance to virus has not been found in *Triticum* species, but immunity and resistance have been found in species of *Agropyron*. Larson and Atkinson (1973) reported a *Triticum* × *Agropyron* line ($2n = 42$) in which chromosomes of 4D, 5D, and 6D of wheat were substituted by homologous chromosomes from *Aceria elongatum* ($2n = 70$, EE), which was found to be immune to wheat streak mosaic resistance. They studied the role of different chromosomes in wheat streak mosaic resistance, and developed different substitution lines for 4D, 5D, and 6D chromosomes of wheat. Substitution of 5D chromosome delayed the development of disease but substitution of 6D chromosome introduced considerable resistance. The procedure for the substitution of a chromosome pair of wheat by rye is presented in Fig. 22.8B. The plants with $AABBDDR_1R_1$ can be crossed with different monosomics of cultivated wheat. The monosomic line, where the loss of a full chromosome pair of cultivated species and addition of a pair from other species do not have any adverse effect (or minimal effect) on agronomic characters, is selected. The addition of the alien chromosome can carry undesirable characters in addition to resistance gene, which can be a major concern in several cases. These lines will only have utility where the damages caused by pathogen or pest outweight any negative effects from the alien chromosome. Also, if mutation (see Chapter 14: Mutation Breeding), gene transformation and gene editing technology (see Chapter 27: Molecular Tools in Crop Improvement and Cultivar Development) are not available or not accepted by consumers.

The transfer of a segment of a chromosome carrying the resistant gene(s) from wild species to a cultivated one is used for developing resistant cultivars. By this method, the chromosomal regions of the wild species are incorporated into the cultivated species, as in the substitution of the whole chromosome. The trisomic plants are treated with X-rays or another source of irradiation prior to meiosis. These plants are used as pollen parent, so that gametes carrying the haploid complement or the gametes, which carry small translocations are more competitive than the gametes carrying the whole extra chromosome. By this method higher frequency of translocations for resistance could be obtained. Such translocations could be terminal, involving small segment or full arm or intercalary one.

Sears (1956) developed the first procedure that used radiation to transfer disease resistance from a chromosome of one species (*Aegilops umbellulata*, $2n = 14$) to a chromosome of another. *A. umbellulata* has good resistance to leaf rust, but its cross with *T. aestivum* does not produce viable seeds. Therefore, Sears crossed *A. umbellulata* with *T. dicoccoides* ($2n = 28$) and produced amphidiploid, which could be crossed with *T. aestivum*. In the latter hybrid, the *A. umbellulata* chromosome did not pair with *T. aestivum* chromosomes. The F_1 plants were partially male fertile and two backcrosses were made to *T. aestivum* cv. Chinese Spring with selection for leaf rust resistance. A resistant plant having an added *A. umbellulata* isochromosome was obtained. Its progeny was irradiated with X-rays prior to meiosis and then used as pollen parents in cross with Chinese Spring. Pollen carrying the *A. umbellulata* isochromosome rarely functioned and was therefore, screened out. However, wheat-*umbellulata* translocations were often less deleterious, and many of the rust-resistant progenies carried translocations. Only one of the translocations proved to be transmitted entirely normally through the gametes. In this line, later named Transfer, the translocation involved wheat chromosome 6B (Sears, 1966). Athwal and Kimber (1972) showed that the *A. umbellulata* chromosome is homeologous, with group 6 chromosomes in wheat. Knott (1971a) made nine backcrosses to add leaf rust resistance of Transfer to the cultivar Thatcher. Although the backcross lines had 1% higher protein content than Thatcher, they were very inferior in baking quality.

Chromosome 1R of rye is a useful source of genes for disease resistance, for example, stem rust (*Sr31*), leaf rust (*Lr26*), stripe rust (*Yr9*), and powdery mildew (*Pm8*) in a gene island block. Subsequently rye chromatin has been incorporated in wheat as substitution lines. In the United States, 15% of commercial varieties carry rye chromatin as per Crespo-Herrera et al. (2017); however, it is hard to confirm the veracity of this number. While traditionally these translocations in substitution lines showed agronomic challenges, but recent efforts have shown success to reduce rye chromatin segments leading to enhanced yield (Howell et al., 2014). More efforts are needed to remove all negative associations due to linkage drag of alien chromosome segment. Ideally the alien chromosome segment should be as small as possible, particularly if the two species show extensive chromosomal differentiation. Otherwise the substitution of an alien chromosome segment for a chromosome segment of the recipient species may result in undesirable duplications, deletions, and linkages of genes. The effect is large in diploids, but is often reduced in polyploids, where the effect of single chromosomes is smaller because of the duplication of genetic material in two or more different genomes. The removal of deleterious genes linked to a desired gene is possible if the chromosomes involved are homologous and crossing over occurs. However, it become increasingly difficult as the degree of differentiation increases, and chromosomes involved are only homeologous and not homologous.

Multiline cultivars

A multiline cultivar is a population of plant that is agronomically uniform but heterogeneous for genes that condition reaction to a disease organism (Frey and Browning, 1971). The use of multiline varieties in self-pollinating crops has been advocated since the late 19th century. The programs of multiline production are based on two radically different philosophies for disease control (Marshall, 1977). In one approach, designated as the "clean crop" approach, each line in the mixture is completely resistant and carries a different single gene resistance and together the multiline mixture would be resistant to all prevalent races of pathogen(s) to be controlled (Jensen, 1952; Borlaug, 1959). The aim of this scheme is to keep the crop as free of disease as possible, and at the same time to reduce the threat of catastrophic disease losses following shifts in the racial composition of the pathogen population. In the second approach, designated as the "dirty crop" approach (Marshall and Pryor, 1979), each line in the mixture has unique resistance gene but are partially resistant; and none of the line is resistant to all known races of the pathogen. Frey et al. (1973, 1975) have argued that such multiline should protect the crop in two ways. (1) They should stabilize the race structure of the pathogen population. This is based on the fact that stabilizing selection against races carrying multiple genes for virulence will ensure that simple races, carrying a single virulence gene, dominate the pathogen population. (2) Since each component of the multiline would be attacked by only one race of the stabilized pathogen population, the remaining lines would act as spore traps, reducing the rate of spread of the disease. In this way, multiline cultivars would have an effect similar to polygenic non-specific or HR in delaying the intercrop build-up of the pathogen population.

The "dirty crop" approach, using partially resistant multilines has a significant potential advantage over the "clean crop" approach, using completely resistant multilines. In "dirty crop" approach multilines and their components confer resistance to only some of the pathogen population. Since moderately susceptible lines are also considered, the breeder is in an advantageous position as they can exercise selection for other characters for example, yield, height, maturity, etc. It would also indefinitely extend the useful life of strong resistance genes, including those that have been ineffective in the past. Hence, it would free the breeder from the difficult task of continually isolating and evaluating new sources of resistance. The only common point between the two approaches is the final aim to produce a variety consisting of a number (6–15) of phenotypically similar lines, which differ in the resistance genes they carry.

The prerequisites of multiline cultivar approach are: proper identification of diverse genetic sources of resistance, adequate race survey, desirable, and commercially acceptable, and, if possible, a widely adapted recurrent parent. By following a conventional or limited backcrossing phenotypically similar but genotypically different isogenic or near isogenic lines, respectively, are developed by crossing the recurrent parent with stocks carrying diverse genes for resistance. All these phenotypcially similar lines are mixed together and distributed to the farmers as a composite variety. If a component line is affected by a new race, it is withdrawn from the composite bulk.

Borlaug (1965) believed that the development of a multilineal hybrid variety will offer the best possible means of incorporating a long-lasting rust resistance into a commercial wheat variety. Once an outstanding single-cross hybrid has been developed, it can be converted into a multilineal hybrid. The first step in this process is the development of many lines phenotypically identical with the male sterile female variety A^s (the modified variety A) through a backcrossing program. Each of these lines, designated male variety A^n, will be genotypically different from the other in stem rust resistance, and also from that of the male sterile female variety A^s, but none of these lines will possess the genes for the restoration of fertility. Therefore, when crossed with the male sterile female variety A^s, the resulting F_1 plants of each cross will be sterile. These male sterile F_1 lines in turn each will be crossed as female with the male variety B^r to provide a heterotic effect, and this second cross will also simultaneously restore fertility to the F_1 commercial hybrid seed formed in the process. The basic commercial variety A and B, both of which are fertile and into which the cytoplasmic sterile and fertility restorer will be incorporated, respectively, should also have different stem rust resistance genes. Each individual F_1 plant of a hybrid produced by this method would be different in its resistance to stem rust and rust epidemic is not likely to develop in such a population.

The first wheat multiline cultivar was Miramer 63, which was released for commercial production from the Colombian program. It was developed by crossing Brazilian wheat Frocor with about 600 varieties/lines. More than 1200 lines phenotyically similar but with resistance from 600 parents were developed.

Miramer 63 was a mechanical mixture of equal seeds of 10 of the best lines with resistance to stripe and stem rust diseases. It yielded more than twice as much as older varieties in some areas. Within 2 years of its release, stem rust was parasitizing on the two component lines; but total losses were always less than the theoretical maximum of 20%. The two stem rust-susceptible lines and two others were dropped from the multiline and four new ones from the reserve of over 600 lines were added to form a new multiline cultivar, Miramer 65.

Emphasis has been placed on the production of multiline varieties in the wheat program in India. Breeders in India have developed multiline varieties of three promising and popular wheat varieties: Kalyan sona; PV-18, and Sonalika (RR 21). These varieties are high-yielding and have good agronomic characters. They were resistant to rusts and other diseases at the time of their release in India, but after 5–6 years of cultivation had fallen susceptible to new races of the pathogen. Therefore, with the aim of replacing these pure line susceptible varieties by their multiline, this program was taken up. By this approach, the life of these varieties may be further prolonged. These three varieties have been used as recurrent parent with several donors as non-recurrent parent. The component lines produced by different breeders all over the country are assembled at the main center of the zone, and the main center grows the components along with their recurrent parents at several locations within the zone, that is, multi-location testing is done and data on several characters such as height, maturity, ear length, disease resistance, etc. are recorded at all the locations. Based on pooled data, component lines that have shown similar behavior at all the locations are assigned to a particular group. Therefore, several groups, each comprising component lines that have proven to be similar on the basis of pooled data at all locations, are made. Component lines belonging to one particular group are mixed in the desired proportions to form the multiline variety, which is tested in varietal trials; and if found promising, can be distributed to the farmers for mini trials. If accepted by the farmers, it can be released for cultivation.

Advantages of multilines: the multiline cultivars have the following advantages when compared with pure line varieties; (1) provide a mechanism of "synthesized HR" which unlike pure line cultivars can utilize several R genes, (2) extend the life of a given R gene(s) and enable a resistance breeding program to be reduced in size while the breeder carries on parallel improvement in the recurrent parent, (3) stabilize the yield to optimize production on a given farm, (4) reduces the risk of homogenizing the pathogen population on a global scale, and (5) hold great promise as a dynamic and natural biological system in effectively balancing the relationship between the host and the pathogen.

Criticisms of the multiline: the multiline approach of management of VR genes for controlling pathogens with great epiphytotic potential such as rusts has been criticized as expensive, agronomically conservative, a breeding ground for new races and possibly even a super-race. This procedure is heavily focused on the R gene(s); however, most breeding programs work on multiple traits compounding the challenges and efforts to build multilines. This method is also dependent on the identification of R genes and is more amenable if linked or gene specific markers are available for backcrossing. Also, backcrossing efforts are required; therefore, genetic gain is compromised compared with forward breeding. Finally, breeding efforts to create these lines are more intensive especially to put them into a complete production ready package of yield, agronomic, disease and quality traits.

Refuge-In-A-Bag

The *Refuge-In-A-Bag* (RIB) approach was popularized after stacking of multiple Bt traits to protect against multiple insect-pests. Plant breeders develop maize hybrids with several *Bt* gene events (see Chapter 27: Molecular Tools in Crop Improvement and Cultivar Development) as described in Chapter 19 (Hybrid varieties). However, the concern that *Bt* gene resistance may become ineffective, motivates crop management approaches to prolong the life of these single gene events (that are stacked, pyramided). The earlier approach for Bt varieties was to plant a structured refuge, consisting of strips of susceptible (non-Bt) hybrids along with Bt hybrid in the field. Although, the stacking of multiple *Bt* genes for insect-pests reduces the development of insect resistance to a single *Bt* gene, due to multiple modes of action for the insect-pest control; the RIB system provides a layer of protection against gene to be rendered ineffective. In the RIB system, seed companies add 5%–10% non-Bt seed (refuge hybrid) in the bag of Bt hybrid that is sold to the farmers. As the percentage of non-Bt seed can be reduced in RIB compared with structured refuge (~20% non-Bt seed in the field), farmers observe a yield increase. Due to the sale of seed in a bag, seed of non-Bt and Bt hybrids are interspersed. The breeder's role is to ensure that the non-Bt hybrid (refuge) is agronomically similar and high yielding with comparable days to maturity and other disease packages. In the absence of this quality control, non-Bt hybrid can lead to disease susceptibility in the Bt hybrid field. In the breeding program, breeders need to be aware of matching hybrids that can serve the role of a non-Bt refuge. Due to the commercial maize breeding programs in the United States maintain breeding efforts on non-genetically modified (non-GM) material and later converting (backcrossing) inbreds with the desired GM traits, it is feasible to pick reasonable non-Bt hybrids as a refuge that match the Bt hybrids.

Breeding for multiple trait resistance

Resistance to two or more diseases has been bred into individual cultivars since Orton (1909) succeeded in combining resistance to *Fusarium* wilt and root knot nematode in cowpea and cotton. Hope and H44 wheat cultivars combined resistance to leaf rust, stem rust and covered smut (Ausemus, 1943). Multiple resistance has long been a breeding achievement in wheat, rice, maize, soybean, and many other crops.

At the IRRI in the Philippines, efforts have been made to eliminate susceptible materials in the early generations. The screening started even in the F_1 generation of a three or more parent cross. The two parent F_1 hybrids could be crossed with other resistant donors to make double or three parent crosses to combine resistance to given diseases and insect-pests. If more donors are available for a pathogen and insect-pest, it is desirable to use good combiners for yield components and plant type, etc. Khush (1977) elaborated the procedure citing an example of a double cross between four parents of which A, B, C, and D are resistant to BB, grassy stunt, brown planthopper, and green leafhopper, respectively. All these traits are monogenic and dominant in their inheritance and are inherited independently. About 400 seeds from the double cross (A/B//C/D) were obtained. These seeds were germinated and inoculated with grassy stunt virus in greenhouse. Approximately 50% of the seedlings were susceptible; and therefore, eliminated. The remaining 200 were transplanted in the field and inoculated with BB. About 50% of these plants were susceptible and rooted out. The remaining 100 plants were harvested separately. Two small seed samples were taken from each, and the progeny were tested for resistance to brown planthopper and green leafhopper. Those carrying the brown planthopper resistance gene (50%) and those carrying green leafhopper resistance gene (50%) were identified. Only progenies carrying both genes (25–30 plants), which segregate for four resistance genes, were grown. The population was subjected to appropriate disease and insect-pest pressures and agronomically desirable plants with multiple resistance were selected in F_3 and F_4 generation to obtain true breeding lines. IR28, IR29, and IR34 are examples of rice varieties with multiple resistance developed at IRRI.

In the scheme to develop cultivars with multiple pest resistance [see Chapter 4: Primer on Population and Quantitative Genetics], it is imperative to work with large population sizes in segregating generations and progeny rows. Due to the culling followed at each generation for several traits, the numbers of "keep" are substantially reduced necessitating a large F_2 generation to begin the breeding population. Also, several breeding populations are needed to be created each year in the program for the objective. Breeders also collaborate closely with pathologists and entomologists to set up screening nurseries (field and/or controlled environment) for making plant and variety selection and characterization.

Another approaches for incorporating the multiple resistance could be the screening for multiple resistance line(s) from germplasm followed by pair wise (or three parent crossing) and screening for more than two diseases and/or insect-pests in the segregating generations. Such breeding pipeline is being followed in Agriculture and Agri-Food Canada, where wheat breeders simultaneously and/or sequentially test for multiple diseases and insect-pests (Fig. 22.9). Numerous varieties have been developed in Canada that have in combination majority of these traits: stem rust, leaf rust, stripe rust, common bunt, loose smut, leaf spot (tan spot, *Septoria*, *Stangnospora*), FHB, wheat midge, and wheat stem sawfly

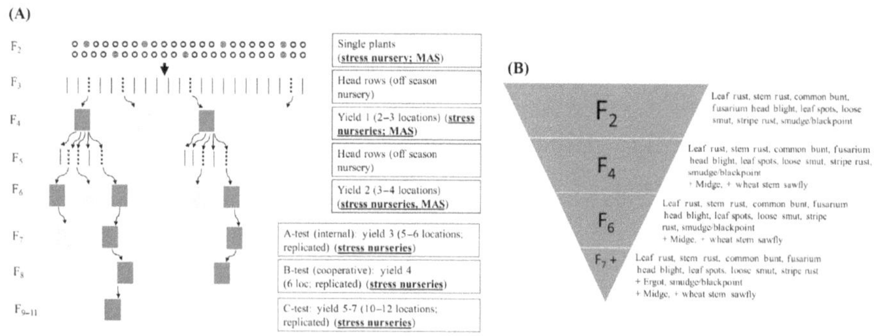

FIGURE 22.9

(A) An example of a usage of specialized stress nurseries in the development of high yield, disease resistant, insect-pest tolerant, and high-quality durum wheat cultivars. (B) Examples of traits that are selected in controlled or field environment conditions. These efforts on trait selection match the importance of a trait in breeding objectives. Population handling from the F_2 generation is shown. MAS, marker-assisted selection.

(Clarke et al. 2009a,b; Ruan et al. 2018, 2019; Singh et al. 2010, 2014, 2015a, 2015b, 2017a,b) (Fig. 22.10). Stress nurseries can be utilized to select in breeding populations, in bulk, or progeny rows, or as a separate nursery independent of yield trial plots. The breeder attempts to select in each generation. For ease, breeder can classify priorities as: (1) highest priority - those diseases and/or insect-pests where cultivar cannot be registered or will not be grown by farmers if resistance is not available in the variety, (2) medium priority - those diseases and/or insect-pests, which do not cause extensive yield, grade, or quality losses or the disease is sporadic and infrequent. Some minimal level of genetic resistance is adequate in the variety, and (3) lower priority - those diseases and/or insect-pests that are currently not high priority or are of very localized importance. Breeders need to be aware about these lower priority pathogens, but no active breeding efforts are made.

These wheat varieties were developed in a forward breeding scheme, with careful choice of parents that complemented the agronomic, quality, disease and insect-pest resistance traits (Fig. 22.10). In the F_2 generation, selection was made for leaf, stem and stripe rust, and leaf spot, followed by selection for FHB and rust resistance in F_4 and F_6 and later generations. F_3 and F_5 generations were grown in a contra season winter nursery and selection against all diseases present was made. DNA-based markers were utilized in early generations for selection of important disease and insect-pest resistance traits, while further characterization and selection for stem rust, leaf rust, stripe rust, common bunt, loose smut, leaf spot (tan spot, *Septoria*, *Stangnospora*), FHB, wheat midge, and wheat stem sawfly resistance was made in F_4 and/or F_6 or later generations. The breeding principle is that breeder should not miss any generation to discard susceptible plants or progeny rows in their program.

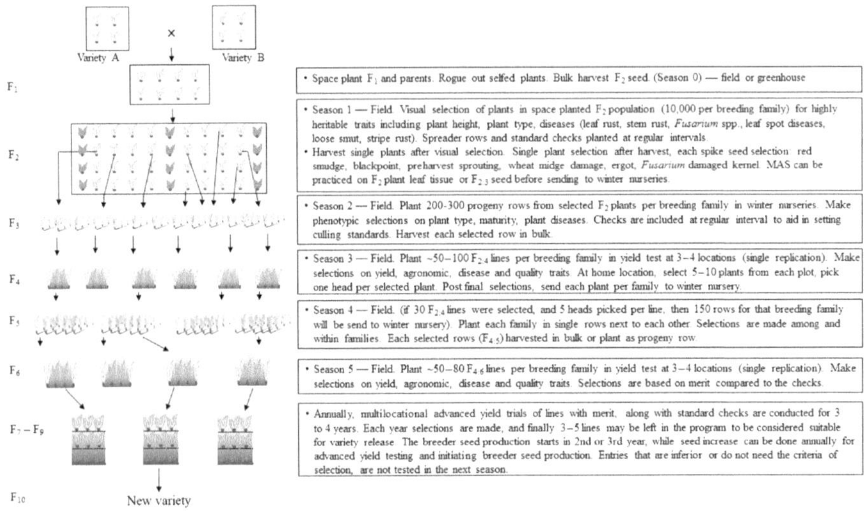

FIGURE 22.10

Scheme of modified pedigree method to develop high yielding, disease resistant, and good quality durum wheat cultivar being followed at Agriculture and Agri-Food Canada. In each segregating season, specialized disease nursery can be utilized as well as opportunistic disease notes can be taken if disease expression is noted and clear distinctions between resistance and resistance checks are noted. DNA marker-based selection can be made in early generation to enrich the population (by selecting homozygous and favorable heterozygote) or fix for the genes of interest. In small grain cereals where large number of F_1 seed can be obtained, a three-parent cross F_1 population can be screened with marker to fix (or enrich) for multiple high priority genes as per the breeding objectives.

If doubled haploid approaches are available to the breeder, seed of the doubled haploid can be simultaneously yield tested and evaluated in specialized disease nurseries (Singh et al. 2012, Singh et al. 2016a). Also, at the haploid stage (prior to chromosome doubling), marker-assisted or genomic selection (see Chapter 27: Molecular Tools in Crop Improvement and Cultivar Development) can be utilized. Irrespective, the effort should be made to use phenotypic or genomic data to make selection in early segregating generations, rather than wait for the line to be developed and then use these molecular and phenotyping tools for characterization. The reason for the earlier generation usage of these tools is to remove the undesirable plants and genotypes leaving more resources to handle more breeding families and with larger population sizes, and conduct more detailed agronomic and disease (and other trait) testing to select suitable varieties to release as cultivars.

CHAPTER

Intellectual property rights and protection 23

Abstract

Innovation is the key for production and processing of knowledge. An ideal regime of intellectual property rights (IPRs) strikes a balance between private incentives for innovators (inventor and/or inventing organization) and the public interest of maximizing access to the fruits of innovation. This chapter describes World Trade Organization's Agreement on Trade-Related Aspects of Intellectual Property Rights (TRIPS) that established intellectual property (IP) rules in the multilateral trading system, allowing more uniformity and conformity across nations on the protection and enforcement of IP rights. Seven IPRs discussed in the TRIPS agreement are described - copyrights, industrial designs, layout designs of integrated circuits, trademarks, geographical indications, trade secrets, and patents. Efforts are made to describe Patents and Plant Breeder's Rights including rights and exemptions, advantages, and disadvantages. An attempt is made to provide information on important topics relevant to a plant breeder particularly cultivar development. This chapter is only for educational purposes and the legal terms, restrictions, rights, limits and all pertinent information should be obtained from official government resources in the respective countries and jurisdictions, as those are the only source of correct legal information.

The earliest record where exclusive rights were given to inventor, dates back to the 5th century and related to the Greek state of Sybaris, South of Italy, which allowed its citizens to obtain a patent for "any new refinement in luxury." In Sybaris, the discoverer of a food recipe was given the exclusive right to use this recipe for one year. This shows the intent from its early inception: to encourage human creativity and enable inventors to reap the benefits of their original ideas. The first law on patents was passed in Venice in 1474, which gave substantive rights to the artisans for their inventions. In 1623 the House of Commons, United Kingdom, passed the act of proprietorship called as Statute of Monopolies. Prior to this, monopolies existed as guilds that controlled imports, production and sales, and control over inventions irrespective of who created them. The Statute of Monopolies provided inventors to retain the rights of their creations, and this would remove monopolies. Inventors were given a 14-year period of exclusive rights on how their inventions were used. In 1710, the Statute of Anne provided

Plant Breeding and Cultivar Development. DOI: https://doi.org/10.1016/B978-0-12-817563-7.00022-2

the inventor 14 years of protection as well as the ability to renew protection for another 14 years.

In the United States, patents are governed by the Patent Act (35 US Code) that established the United States Patent and Trademark Office (USPTO). There are three types of patents issued in the United States—utility, design, and plant. Utility patents are awarded on inventions and cover the usefulness or functionality of a product, design patents protect design elements of an object, while the plant patent is awarded for protection on asexually propagated applicable plants engineered to reproduce without the use of pollen. Plant breeders are able to obtain utility and plant patent on new varieties. Utility patents are enforceable from the day of issuance and valid for 20 years. To obtain protection under the United States law, applicant must submit a patent application to the USPTO, where it will be reviewed by an examiner to determine if the invention is patentable. The United States law grants to patentees the right to exclude others from making, using, or selling the invention.

In India, innovations and novel techniques were retained within families and small social groups. There was no formal system of protection of their rights to benefit from their inventions. In 1856, Government of India introduced an act on protection of inventions, which was based on the British Patent Law (1852). In 1872 the Patterns and Design Protection Act was passed; while the Inventions and Design Act was passed in 1888. The Indian Patents Act (1970), introduced in the parliament in 1965, was modified in 1967 and was passed in 1970. Protection of designs is covered by the Indian Patents and Design Act (1911) with amendment in the act (1978) and amended rules in 1985. A new Design Act, 2000 has been enacted superseding the earlier Designs Act, 1911. The current patents act 2005 became effective in 2005. Trademark protection is in force since 1948 under the 1940 Act; this Act was amended as Indian Trade and Merchandise Marks Act (1958), which came in force in 1959, then came Trade Mark Act, 1999; Copyright Act, 1957 as amended in 1983, 1984, 1992, 1994; International copyright order, 1999; Information Technology Act, 2000; Semiconductor Integrated Circuits Layout Design Act, 2000; and Geographical Indication of Goods (Registration and Protection) Act, 1999. No exclusive legislation exists for the protection of Undisclosed Information (trade secrets) but this is generally covered under Civil Laws.

Inventions are necessary for mankind's progress, and incentivization of innovators while ensuring the preservation of public's interest. This balance is reflected in article 27 of the 1948 Universal Declaration of Human Rights, which recognizes that "*Everyone has the right freely to participate in the cultural life of the community, to enjoy the arts and to share in scientific advancement and its benefits*" *and* "*Everyone has the right to the protection of the moral and material interests resulting from any scientific, literary or artistic production of which he is the author.*" The pertinent question seems to be balancing the interest of the inventor, and that of the society at large in an optimum way. In most organizations, plant breeders are directly or indirectly incentivized through the return of royalties collected from seed sales of commercialized varieties. This model of proportion of royalties shared with inventor (plant breeder) and remaining royalties reinvested in the breeding program ensures a healthy return on investment, and a robust breeding pipeline to continually develop competitive and profitable cultivars.

There are several ways to protect varieties and intellectual properties.

Intellectual property rights

Intellectual property (IP) is the idea, concept, process, or invention, which is the outcome of human mind or human intellect that can be ultimately translated into a useful product, capable of multiplication and use at different times and locations. The main problem with intellectual properties is that they can be copied and used by others, which reduces the benefits that accrue for the original inventors of the properties. IP rights are given to persons over the creations of their minds, giving them an exclusive right over the use of their creation for a certain period of time. These creations are broadly defined—artistic expressions, signs, symbols and names used in commerce, designs, and inventions. In other words, the rights conferred on an individual by the State to derive benefits from the IP for a specified period and to exclude others from doing so is referred to as IP rights. Creator can use their right to negotiate payment in return for others using them. The World Trade Organization (WTO) Agreement on Trade-Related Aspects of Intellectual Property Rights (TRIPS), negotiated during the 1986–94 Uruguay Round, for the first time introduced IP rules into the multilateral trading system. This also allowed more similarity across nations on the protection and enforcement of IP rights. Seven types of Intellectual Property Rights (IPRs) discussed in the TRIPS agreement are as follows:

1. copyrights and related rights,
2. industrial designs,
3. layout designs (topographic) of integrated circuits,
4. trademark,
5. geographical indications (GIs),
6. protection of undisclosed information's (trade secrets), and
7. patents.

The scope of protection of these IPRs has been discussed in Article 9 to 39 of the TRIPs Agreement. Out of these seven IPRs, patents and GIs are of enormous significance for the agriculture sector; while trade secrets are also of significant importance to agricultural companies. In India, there have been laws for the protection of IP rights of four types: patents, copyrights, industrial designs, and trademark. More recently, concern for the protection of IPR has gained prominence compared to the past.

Copyright

Copyright is a form of IP protection granted to protect creative expression of an idea but not fact, idea, system, method of operation, and historical facts. Expression of idea can be in the form of literary, novels, poems, plays, movies, songs, paintings, sculptures, architecture, computer software (protected under the literary works), maps, technical documentation, drawing, and certain other intellectual works. Usually, the copyright act relates to the right of authors to their literary and artistic works. In the US, copyright is a form of protection granted by law for original works of authorship that is fixed in a tangible medium of expression; and these include both published and unpublished works. The copyright is limited both in time and extent, it provides protection for a specified period and duration of copyright is different for different aspects. The copyright prevents anyone from reproducing or unauthorized copyrighting the subject either in entirety or in part. However, it does not prevent another person from using either the idea or the information contained in a copyright material. Copyright differs from a patent as it protects original works of authorship, while a patent protects inventions or discoveries.

Industrial design

Industrial design generally refers to the ornamental or aesthetic aspect of an article rather than its technical features, that is, an industrial design is about how something looks and protection in on a product's unique appearance not what is is made of, how it is made or how it works. Under the TRIPS Agreement, original or new industrial designs must be protected for at least 10 years. As per the Designs Act, 2000, in India the ""design" means only the features of shape, configuration, pattern, ornament or composition of lines or colours applied to any article whether in two dimensional or three dimensional or in both forms, by any industrial process or means, whether manual, mechanical or chemical, separate or combined, which in the finished article appeal to and are judged solely by the eye; but does not include any mode or principle of construction or anything which is in substance a mere mechanical device," and does not include any trade mark, property mark, or any artistic work. Industrial design covers the form and function and how they work together, leading to products that are easy to use, more functional, distinctive and visually appealing.

Layout design of integrated circuits

Under the TRIPS agreement article 35, Member countries are required to protect the layout designs of integrated circuits as per the provisions of the Treaty on Intellectual Property in Respect of Integrated Circuits (IPIC Treaty). As per

WTO, an integrated circuit is an electronic device that incorporates individual electronic components within a single "integrated" platform configured to perform an electronic function. There has to be a component of originality, which implies that these are a result of inventors (creators') own intellectual effort and are not commonplace among creators of layout designs and manufacturers of integrated circuits at the time of their creation. The exclusive rights include the right of reproduction and the right of importation, sale and other distribution for commercial purposes.

Trademark

Trademark is a mark capable of being represented graphically and which is capable of identifying and distinguishing the goods or services of one person (or company) from those of others, and may include any letters, words, names, symbols, designs or devices or any combination, shape of goods, their packaging, and combination of colors. The USPTO describes trademarks as generally a word, phrase, symbol, or design, or a combination of these; and identifies and distinguishes the source of the goods of one party from those of others. In short, a trademark is a brand names and/or designs, which are applied to products or used in connection with services. A service mark is the same as a trademark, except that it identifies and distinguishes the source of service and not goods. A trademark is unique, and protects brand names and logos used on goods and services.

Trademark should identify the product and its origin and should guarantee its quality. Trademark performs the function of advertising the product and creates an image of the product in the minds of consumers and prospective consumers of such good. A good trademark should be distinctive. Distinctiveness may be inherent or acquired and it is preferable to use invented words that are easy to pronounce and remember, consist of single word, easy to spell correctly and write legible, it should be short and appealing to eyes as well as ears. Trademark should satisfy the requirements of registration. Few examples of established trademarks where the brand names have become the most valuable assets of a company include Amazon, Apple, Google, Microsoft, Walmart etc.

As per Article 18 of the TRIPS Agreement, the initial registration and each renewal of registration is for a minimum term of seven years and is renewable indefinitely. This contrasts with other methods of IP protection, such as copyright and patent that have a legal term limit. The owner of copyright simply needs to renew the registration at the expiry of each term and pays the required renewal fees to maintain their trademark.

Advantages of established trademark in business are: (1) increase unit sales, (2) cement customer loyalty, (3) increase revenues and profitability, (4) expand and maintain market share, and (5) differentiate products. Companies can register their mark, and can thereafter use ® symbol. However, if it is not registered with

USPTO, TM for goods and SM for services can be used regardless of filing an application with the USPTO. Different countries may follow different regulations, so it is important for breeding companies to determine the legal requirement in their countries for the use of trademarks.

Geographical indications

GIs are defined as "indications which identify a good as originating in the territory of a Member, or a region or locality in that territory, where a given quality, reputation or other characteristic of the good is essentially attributable to its geographic origin." (Source: Article 22(1), TRIPS, WTO). The GI says where the product comes from, but also identifies the product's special characteristics, which are the result of the product's origins. The TRIPs provisions for GIs have added a new dimension to a good or commodity, which has acquired its reputation from the geographical area in which it is produced. GIs means an indication that identifies good such as agricultural goods, natural goods or manufactured goods or any goods of handicraft or of industry, including food stuff as originating or manufactured in the territory of country, or a region, where a given quality, reputation or other characteristic of such goods is essentially attributable to its geographical origin. In case of manufactured goods, one of the activities of either production or of processing or preparation of the concerned goods takes place in such country, region, or locality. Some of the most known examples include alcoholic drinks (e.g., several types of wines, scotch, tequila), beverages (e.g., Darjeeling tea), cheese (e.g., Roquefort), etc.

As per USPTO, GIs serve the same functions as trademarks, because they are: (1) source-identifiers; (2) guarantees of quality; and (3) valuable business interests. Examples of GIs in the United States include Florida oranges and Idaho potatoes. The United States and most other countries, do not protect generic indications because they are believed to be incapable of identifying a specific business source (or a specifically defined collective producing source). The Geographical Indication of Goods (Registration and Protection) Act, 1999, Rules, 2003, provides for the registration of GIs of Indian goods. In Indian agriculture, there are a number of fruits, vegetables, cereals, and pulses, which shall have significance under the law on GIs. For example, mango, guava, litchi, banana and many other fruits have many varieties associated with GIs. Tea, many vegetables and cereal crops like basmati rice can be considered under the law of GI. Therefore, the TRIPS provisions relating to GIs have widened the scope of protection in agriculture. Inventors can also apply for IP protection through patents, if GI tag is difficult to establish and obtain.For plant breeders, GIs may be important if they are working on varieties or plant types that are covered under such production; however, it is also important if a breeder works on developing a specific plant or variety with characteristics that enable it to be listed in GIs.

Trade secret

When the individual or organization owing an IP does not disclose the property and its associated inventions to any one and keeps it as a closely guarded secret to promote his business interests, it is called trade secret. Thus a trade secret is any sort of information that can be used in business or other concern, provided that such information is sufficiently valuable and confidential, and can provide an actual or potential economic advantage over others. It is a legal term for confidential business information. This information allows a company to compete effectively.

Trade secrets are protected by normal civil laws. In case of trade secrets, there is no concept of registration of term but protection is obtained through legal (nondisclosure) agreement between the owner of the secret and those who have been given access to it. Trade secret may relate to formulae, pattern, device, or compilation of information that is used in one's business and which helps to obtain an advantage over competitors who do not know or use that secret. It can be computer program (e.g., Google search algorithm), formula for a chemical, recipes, process of manufacturing (e.g., Coca-cola), treating or preserving materials, a pattern for a machine or other device, a computer program or a list of customers, design, drawings, or an undisclosed breeding method or cultivar development tool, etc. including other competitively valuable information. Trade secrets offer the following advantages.

- A properly maintained trade secret can remain a secret for a much longer period of time in contrast to temporary protection afforded by patent.
- No strict requirements needed to be satisfied in contrast to patent.
- The cost of filing, contesting, and enforcing a patent is saved.

Trade secrets, however, suffer from the following disadvantages, which often outweigh their advantages.

- Maintaining a trade secret itself is a costly affair.
- It offers no protection from independent invention/innovation that may compromise the innovativeness due to the trade secret.
- Nondisclosure of the invention/innovation does not given others a chance to improve upon the original invention.

In plant breeding industry, the collection of germplasm pool and heterotic pool entries, parents in hybridizations, breeding methods pipeline are some of the examples of trade secret. Employees sign a nondisclosure agreements preventing them to reveal company secrets and at times even a moratorium on the length before they can join a competing company.

Patent

The TRIPS Agreement requires member countries to make patents available for all inventions including products or processes, without discrimination, in all fields of technology. There can be no discrimination for the place of invention and whether products are imported or locally produced (Article 27.1). However, members (governments) can refuse to grant patents on three permissible exceptions: (1) inventions whose commercial exploitation needs to be prevented to protect human, animal or plant life or health or seriously harmful to the environment (under Article 27.2); (2) diagnostic, therapeutic, and surgical methods for treating humans or animals (under Article 27.3a); and (3) certain plant and animal inventions (under Article 27.3b). However, countries excluding plant varieties from patent protection must provide an effective unique system of protection.

A patent is the exclusive right granted by a government to an inventor to prevent others from imitating, manufacturing, using, and selling the invention provided the invention satisfies certain conditions stipulated in the law. That is, exclusivity of right implies that no one can make, use, manufacture, or market the invention without the consent of the patent holder. This right is available only for a specified period of time. A patent in the law is property right and hence, can be gifted, inherited, assigned, sold, or licensed. As the right is conferred by the State, it can be revoked by the State under very special circumstances even if the patent has been sold or licensed or manufactured or marketed in the meantime. A patent is territorial in nature, that is, within the boundaries of a particular country, which has given the patent. The chief requirements for the grant of a patent are as follows:

- novelty,
- inventiveness (non-obviousness), and
- usefulness.

In the United States, utility patent can be obtained that is applicable to all plant varieties (including pure lines and hybrids) as long as it shows; novelty, utility, and obviousness. Plant patents are another type of variety protection in the United States; however, this is limited to vegetatively propagated varieties (Table 23.1). A formal specific government protection for plant varieties in the United States was first implemented in 1930. Called the Plant Patent Act, it was (and still is) limited to clonally (vegetatively) propagated plants. The use of utility patents for plant variety protection was made possible following the United States Supreme Court ruling in 1980 (Diamond vs Chakrabarty ruling on biotech patent). The USPTO extended patent protection to plant varieties in 1985. It is very encompassing and includes plant traits, parts, components, products, cell, breeding methods. The term of a patent is generally 20 years from the date of filling of patent application.

It is important to consider similarities between Plant Variety Protection (PVP) and Utility Patents. Most plant breeders protect their inventions under either plant variety protection or utility patents. Utility patent on crops are generally not granted in other countries (PVP is common there). Breeding companies may protect their inbred lines through trade secret (explained earlier). Major difference between the two protection systems: (1) PVP protected varieties may be used as breeding material by a breeder without requiring the permission of the rights

Table 23.1 Salient points of plant breeder's rights and patent system.

Criteria	UPOV 1978	UPOV 1991	Utility patents	US Plant Patent Act
Protects	Varieties of listed species	Varieties of all species	Plant genotype not normally found in nature	Asexually reproduced nontuberous plants
Requires	Novelty Distinctness Uniformity Stability	Novelty Distinctness Uniformity Stability	Novelty Distinctness Uniformity Enablement	Novelty Distinctness Stability
Disclosures	Full morphological description	Full morphological description	Enabling disclosure (sufficient details of invention) Best mode disclosure (ensuring full and honest disclosure) Deposit of novel material	As complete as possible, photographs and drawings
Claims	–	–	Refer to specific patents	Single varietal claim
Exemptions	Farmer and breeder exemption	Farmer and breeder exemption	Some country patent laws include exemptions	None
Rights	Prevent others from producing for commercial purpose	In addition to UPOV 1978, prevents import and export. Extends to essentially derived varieties	Prevent others from making, using or selling claimed invention	Prevent others from asexually reproducing, selling, or using

From Smith, S. 2008. Intellectual property protection for plant varieties in the 21st century. Crop Sci. 48, 1277–1290.

holder. If the breeder develops a product that is distinct from the parental germplasm and warranting commercialization, it may qualify for protection in its own right. (2) The claims in a utility patent are generally designed to exclude breeding use without express permission from the owner of the property. Breeding companies use utility patents to protect their germplasm base.

The USPTO issues several types of patents that offer different kinds of protection and cover different types of subject matter. The main types are - (1) *Utility patent*: also referred as "patents for invention" are issued for the invention of a new and useful process, machine, manufacture, or composition of matter, or a new and useful improvement thereof. These are the predominant types of patents issued by the USPTO. (2) *Design patent*: these are issued for a new, original, and ornamental design embodied in or applied to an article of manufacture. (3) *Plant patent*: as previously mentioned, it is only for asexually reproduced plant. Plant patents are issued if it is new and distinct, invented or discovered including cultivated sports, mutants, hybrids, and newly found seedlings other than a tuber-propagated plant or a plant found in an uncultivated state.

The Indian Patent Act has been amended three times (1999, 2002, and 2004). The recent amendment of 2004 effective from 2005 extends patent protection to products including process of chemical reactions. This is coupled with the provisions under Plant Variety Protection and Farmers Right Act (PPV & FR Act, 2001) has the potential for scientific research. Most of the inventions relating to disease, weed and insect control, and mechanization of agricultural operations are eligible for protection through patents. Also patentable are chemicals developed for the control of weeds, insect pests, and diseases. In the area of animal husbandry, equipment for diagnosis of disease, and determination of fertility state are eligible for patenting. Vaccines are patentable. In case of biotechnology, various items of molecular biology are patentable. For example, choice of a host cell for bringing together plasmids. Enzymes for clearing and sealing, hybrid plasmids and transformation conditions, and marker assisted selection tools are also patentable. Thus, with a system of protection in the country, scientists will be motivated to come up with new inventions of commercial value.

Novelty

The invention must be new. Novelty is assessed in a global context. An invention is considered novel if it does not form a part of the global state of the art. Information appearing in magazines, technical journals, books, newspaper, etc., will constitute the state of the art. An invention will not be novel if it has been disclosed in the public through any type of publications anywhere in the world before the filing of a patent application. Oral description of the invention in seminar or conference can also spoil novelty. Prior use of the invention in the country of interest before the filing date can also destroy the novelty. Novelty is determined through extensive literature and patent searches. It should be realized that patent search is most crucial parameter for ascertaining novelty as most of the information documents does not get published anywhere else. It is important to note that different countries provide different grace period for prior art. A grace period is the time period after a publication (verbal or written) of the invention, during which a patent application can still validly be filed. In the United States, an inventor has up to 1 year after publishing his invention to validly patent this invention.

Inventiveness (Non-obviousness)

It is invention, and not merely discovery. The invention should not be obvious to a person skilled in the art, and should represent an innovation. The non-obvious requirement refers to the level of difficulty required to invent the technology. If the invention is so obvious that anyone having an ordinary skill would have thought of it, then it does not meet this requirement. The term invention means art, process, method or manner of manufacture, machine, apparatus, or other article or substance produced, which is new and useful.

Usefulness

The subject matter of the patent must have an application or must possess utility, either immediate or in the future, and this application should be useful to the society/nation, for the grant of patent. Inventions are almost always solutions to problems; and therefore, by providing a solution it makes the invention useful. No valid patent can be granted for an invention devoid of utility.

Limits of a patent

There are limits of what can be patented; for example, abstract ideas, laws of nature, and physical phenomena are not patentable subject matter. Additionally, a patent cannot be obtained upon a mere idea or suggestion, and a complete description of the actual machine or subject matter is required for which patent is sought. A patent is limited both in time and space. The two basic limitations of patents are as follows:

1. *Limitation of time*: a patent is valid for a specified period of time from the date of its award. In most countries this period is 15–20 years, this constitutes the limitation of time. The Indian Patent Act (1970) grants protection from 7 to 14 years.
2. *Limitation of space*: a patent is valid only in the country of its award; it is not valid in other country. If an inventor files a patent application in one country, they can file other applications in the rest of the world within 1 year, and the other application is treated as were filed on the date of the first application (in the first country) giving the "right of priority" to inventor. A group of nations may agree to honor the patents awarded by any member country, that is, in European Economic Community. A provisional patent application can be filed (which has a reduced fees and less documentation), and the inventor has one year to file for a non-provisional patent (regular patent). While a provisional patent is not considered a true patent, it protects the inventor's IP for 12 months in the same way a non-provisional patent would. The value of a provisional patent filing is that the filing date of the (prior filed) provisional patents becomes the "effective filing date of the claimed invention" as long as the provisional application sufficiently describes the claimed invention. The

cost of filing a patent is not exorbitant, and in the United States these consist of USPTO filing fees, lawyer fees [if inventor engages the service of lawyer (s)], and drawing fees. Drawings are needed when the nature of the case requires a drawing to understand the invention.

Plant Breeder's Right

Plant Breeder's Rights (PBR) are the rights granted by the government to empower a breeder or an originating institute of a variety to regulate the multiplication and marketing of seed of protected variety for a minimum period of 15–20 years. A person holding PBR title to a variety can authorize other interested persons/organizations to produce and sell the propagating materials of that variety. PBR systems also contain some form of "breeder's exemption" and "farmer's privilege."

The legislations for patenting plants were introduced in many countries, including the United States. The plant patent Act of the United States passed on May 12, 1930 was the first legal protection granted to asexually reproducing plants and was indeed the first patent in the world to benefit plant breeders (see the section on plant patent). Later on, sexually reproducing plants were also made patentable through plant variety protection Act of 1970 in the United States. Initially, the provision of protection was not available to plants because plants and other living organisms being products of nature were not eligible for patent. But the plant patent Act of 1930 recognized that plant breeders create products that are more than mere products of nature. The Act extended protection not only to inventors but even to discoverers, that is, who invents or discovers any distinct and new variety, may obtain patent.

Different forms of such Plant Breeder's Rights were adopted in Europe whereas Netherlands took the lead to adopt PBR by adopting a Plant Breeder's Decree in 1941. However, the issue of PBRs became alive only after an international body, UPOV (*Union pour la protection des Ostentions Vegetales) i.e., International Convention for the Protection of New Varieties* of plants was established in 1961 by eight European countries. Its main objective was to formulate a uniform standard of providing varieties in all the member countries. It has since then been revised in 1972, 1978, and 1991 to make additional provisions. The UPOV convention requires member states to provide the same rights to all other member states. All the member countries are free to decide the minimum levels and form of IP protection, but the member countries have not reached a consensus under which all the member countries were bound to provide certain minimum levels of IP protection including plant varieties. The membership of UPOV is fast increasing and most of the countries have implemented one or the other form of the PBR where breeders or originating industries are granted the exclusive rights to multiply and sell varieties for a specific number of years.

UPOV is an intergovernmental organization that provides for the recognition of plant breeders' rights on an international basis. The mission of UPOV is to "provide and promote an effective system of plant variety protection, with the aim of encouraging the development of new varieties of plants, for the benefit of society." UPOV member countries acknowledge the achievements of breeders of new plant varieties by granting them IP rights called plant breeders' rights. Plant breeders' rights stimulate investment in plant breeding, variety improvement, and commercial propagation of plant material by individuals, private and public breeding organizations, and firms. A grant of rights is country specific. To obtain protection in other countries, the applicant must apply separately to the appropriate authority. Applications originally filed in one country may serve as a basis for claiming priority for an application filed in another UPOV country. With the grant of a Plant Breeders' Rights for a new plant variety, the owner obtains the exclusive rights to produce for sale, and to sell, reproductive material of the variety. The owner is then able to protect the variety from exploitation by others and can take legal action against individuals or companies that are propagating and selling reproductive material without permission. The holder of the rights may also take action to prevent another person or business from using the approved denomination (variety name) of the protected variety when selling propagating material of another variety of the same genus or species.

The original UPOV 1978 act was revised as UPOV 1991 Act, which came into force on April 24, 1998 (See Table 23.1 for comparisons). In Canada, amendments to the PBR Act came into force on February 27, 2015 and included provisions that brought it in line with UPOV 91. The UPOV 91 contained new elements that provide stronger protection for plant breeders than any of the previous conventions. For example, seed conditioning farmers (using farm saved seed and seed treating for sale to other farmers) will need approval from breeders. An effective plant breeders' rights system aims to encourage and support the development of new varieties, increased investment in plant breeding to give farmers increased choices and access to new improved and innovative plant varieties. In the United States, Plant Variety Protection Act provides legal IP rights protection to breeders of new varieties of that are sexually reproduced (by seed) or tuber-propagated. Under the US PVP act, the following conditions have to be met to obtain IP protection:

1. *Variety is new*: that is on the date of PVP application filing, propagating, or harvested material of the variety has not been sold or otherwise disposed off to other persons, by or with the consent of the breeder or breeding organization for purposes of exploitation of the variety in the United States, more than 1 year prior to the date of filing. In any area outside of the United States, the length is more than 4 years prior to the date of filing, except for tuber-propagated and tree or vine (that have longer length of time).
2. *Distinct*: variety should be clearly distinguishable from any other variety that is in existence in public or common knowledge at the time of the filing of the application. A cultivar is distinct if it can be differentiated by one or more identifiable characteristics from all other cultivars currently available or previously developed. Distinctiveness can involve morphological, physiological, molecular, or other characteristics.

3. *Uniform*: a cultivar is uniform if no variation among individuals exists for the distinguishing characteristics that make it distinct from other cultivars. Any variations are describable, predictable, and commercially acceptable.
4. *Stable*: a cultivar is stable if plants remain the same from generation to generation. That is, the variety, upon reproduction, will remain unchanged with regard to the essential and distinctive characteristics of the variety with a reasonable degree of reliability. If there are multiple generations in a cycle of propagation, the plant characteristics should be the same at the end of each such cycle.

Distinct, Uniform and Stable (DUS) test is the main criteria for deciding the novelty of a variety, which in turn is the criteria on the basis of which the Plant Breeder's Rights are granted to a variety. As previously explained, novelty requires that a variety should not have been commercially exploited for more than one year before the grant of PBR protection. Distinctness requires that the new variety must be distinguishable from other varieties by one or more identifiable morphological, physiological, or other characteristics. The new variety must be uniform in appearance under the specified environment of its adaptation to fulfill the criteria of uniformity. Furthermore, the new variety must be stable in appearance and its clonal characteristics over successive generations under the specified environment to satisfy the criterion of stability.

All types of cultivars, including hybrids can be protected including hybrids, as described at the UPOV website. The breeders' rights include exclusive control of the protected variety and all aspects of its propagation. Without the authorization of the breeder, no one is allowed to propagate, sell, or market the plant in any way.

In India, the new plant variety cannot be patented but can now be registered under the PPV & FR Act, 2001. The main objectives of the legislation known as protection of PPV & FR Act, 2001 to protect Plant Breeder's Rights are: (1) to stimulate investments for research and development both in the public and the private sectors for the development of new plant varieties by ensuring appropriate returns on such investments; and (2) to facilitate the growth of seed industry in the country through domestic and foreign investment which will ensure the availability of high quality seeds and planting materials to Indian Farmers. The PPV & FR Act (2001) is similar to UPOV Act (1978) in some respects, has same features of UPOV Act (1991), and is unique in respect of some of its other features. PPV & FR Act (2001) is similar to UPOV Act (1978) in the following respects: (1) protection of varieties of nationally recognized plant species; (2) duration of protection is 15 years or more (18 years in the case of trees and vines); and (3) provision for farmer's privilege. However, PPV & FR (2001) is comparable to UPOV Act (1991) in the following aspects: (1) requirement of novelty, distinctiveness, uniformity, and stability for registration of variety; (2) protection extended to commercial use of all the material of the protected variety; and (3) essentially derived varieties being subject to PBR protection granted to the conceded initial varieties. The PPV & FR Act (2001) in India has certain unique features that are not in the UPOV Acts (1978, 1991): (1) Registration of extant varieties (A variety available in India, which is notified under section 5 of the Seeds Act, 1966, or a farmer's variety, or a variety about which there is common knowledge, or which is in public domain - the DUS criteria are applicable), (2) Registration of farmer's varieties (a variety which has been traditionally cultivated and evolved by the farmers in their fields, or is a wild relative or landrace of a variety about which the farmers possess the common knowledge), and (3) recognition of farmer's right and provision for monetary compensation for these rights from the National Gene Fund. These three classes of plant varieties can be registered under PPV & FR Act (2001) and follow the criteria for their registration. The PPV & FR Act (2001) extends farmer's privilege of UPOV Act (1978) and permits farmers to "exchange, share sell" their farm produce, including seeds, except as branded seed. It also provides protection to farmers from innocent infringement of PBR.

Restrictions to the holders' rights

There are two exceptions to the breeder's right.

- *Research and breeding exemption*: protected varieties may be used for breeding and developing new plant varieties without the permission of the holder of the rights.
- *Farmers' privilege*: farmers may save and plant their own seed of protected varieties on their own land without infringing on the holder's rights (may be country specific).

Breeder's exemption

Protected variety (the initial variety) can be used to create new variation for developing another variety, but not for producing other variety like F_1 hybrid for sale. The PBR for these new varieties will be of the breeder who developed them, and the holder of PBR title of the initial variety will have no claim in it. Under the UPOV 1978 Act, all new varieties evolved using a protected variety were exempted from protection under this provision, But UPOV 1991 Act has somewhat limited the scope of breeder's exemption by bringing "essentially derived" varieties under the cover of PBR protection granted to the initial variety. As per UPOV, variety is considered essentially derived if it is predominantly derived from the initial variety particularly through methods which conserve the essential characteristics of the initial variety. These include the selection of a natural or induced mutation, somaclonal variant, cell fusion, selection of a variant, backcrossing or transformation by genetic engineering, and changes in chromosome numbers. It is considered essentially derived if the new variety conforms to the genotype or the combination of genotypes of the initial variety, except for the differences that result from the method of derivation. Thus a variety produced by, say, mutation or transfer through backcross method/integration by genetic transformation of a single gene will be considered as an "essentially derived variety" and the initial variety will be protected under the PBR title granted to the initial variety. The breeder of such an essentially derived variety will; therefore, be required to obtain permission from the PBR title holder of the initial variety.

Breeders' exemption in one form or the other is available in all PBR systems. The accessibility of protected varieties for use in breeding programs is generally appreciated by plant breeders, as it allows a free gene traffic from one breeder to the other. However, the parental inbreds/lines of a protected hybrid variety are also protected materials and as such they may not be accessible for use as breeding materials.

Farmer's privilege

Farmer's exemption PBR systems generally allow the farmers to use the material of a protected variety produced on their farm for planting of their new crops without any obligation to the PBR-title holder; that is, farmer's privilege applies to the use of seed produced by a farmer for sowing on "own" fields. Under the UPOV 1978 Act, there was an explicit provision for farmer's privilege has been made "optional" and each UPOV member state can either allow or disallow this privilege. However, farmers are not allowed to exchange seeds of protected varieties produced on their farms.

Farmer's privilege is very important for countries (like India) where over 90% of the total cropped area is sown by seeds produced by the farmer's themselves. In addition, a majority of the farmers small landowners will be subject to unjust economic burden if they are forced to pay a royalty on the seed that industry does not allow the option of use of new seed every year.

Farmer's rights

The locally grown varieties or mixture of varieties cultivated by marginal or small land holding farmers are well adapted to poor growing situations. This rich germplasm are selected, assessed, and community shared material are "Farmers Varieties." Often these varieties serve as donor lines of some essential traits needed for the commercially acceptable high yielding new varieties that are bred for wider cultivation areas. The Convention on Biological Diversity (CBD) recognized the role of indigenous peoples, their knowledge and their practices in the conservation and sustainable use of genetic resources. The CBD was the first international treaty that addressed the conservation, sustainable use and equitable sharing of benefits derived from the utilization of biological diversity covering domesticated and undomesticated biodiversity. According to CBD, which came into force in 1993 as the first legal mechanism dealing with biological resources, these plant genetic resources are deemed not to be the heritage of mankind, but are the properties of the countries; and therefore, tradable commodities. This is in contrast to protection of IPRs, which decrees that natural products are God's gift to mankind, and hence cannot be exploited exclusively by any one. In 2001, CBD was followed by the adoption of International Treaty on Plant Genetic Resources for Food and Agriculture. The farmer's rights include compensation for innovations evident from the development of farmer's varieties, compensation for making the genetic resources available and provision of incentives to continuously sustain it, provided support for conservation and utilization activities. There is continuity between the farmers conserved germplasm and plant breeder bred new varieties as one is dependent of the other. Farmers or their communities have a right to keep, use, exchange, share, and market their seeds or planting material, similar to the provisions under the UPOV.

Advantages of PBR

1. It allows breeders to benefit from varieties developed by them, which in turn encourages plant breeding activities.
2. It encourages private companies to come into and invest in plant breeding activities.
3. It enables access to varieties developed in other countries and protected by law of Intellectual Property Rights.
4. Increased competition among various organizations engaged in plant breeding is likely to be beneficial to both farmers and the nation.

Disadvantages of PBR

1. PBR can encourage monopolies in genetic material for specific traits.
2. It may suppress free exchange of genetic material and may encourage unhealthy practices.
3. The holder of PBR title may produce less amount of seed than the demand in order to increase prices for achieving more profit.
4. PBR may result in increased cost of seed, which will increase the burden on poor farmers in developing countries.

CHAPTER

Participatory plant breeding 24

Abstract

Participatory plant breeding (PPB) was proposed in early 1980s as a socio-technological solution to variety development, such that it is complementary to conventional plant breeding. The main participatory research in plant breeding domain is called PPB or client oriented plant breeding (COPB), and participatory varietal selection (PVS). In participatory research, clients (mainly farmers, but other stakeholders too) are intricately included in all major decisions at all stages of a plant breeding program. The difference between PPB and PVS is the demarcation when client participation starts; specifically, clients are engaged from the first stage of breeding pipeline in PPB, while in PVS farmers are involved in the testing of lines developed by plant breeders.

Yearly growing new varieties developed by plant breeders is an attractive option for farmers who can afford seed costs. Formal plant breeding programs are often designed to meet specific requirements of a diverse groups of farmers for different growing environments (edaphic and climatic conditions). This approach is beneficial to farmers in good crop growing environments or those with the capacity to modify growing environments through applying additional inputs (fertilizer and irrigation) to create favorable growing conditions for new varieties. However, for farmers on marginal lands and/or those who are on subsistence farming, there may be a dearth of suitable new varieties. Additionally, mainstream plant breeding where varieties are bred for large agro-ecological zones (with high performance and high stability) can result in a low on-farm varietal diversity, and benefits are reduced where obsolete cultivars are commonly grown.

The complementary approach of participatory plant breeding (PPB) arose from an interdisciplinary effort involving biologists and social scientist. In their seminal paper, Rhodes and Booth (1982) described a case of successful interdisciplinary research (case study: postharvest technology at the international potato center) and suggested a model and guiding principles for teams working toward solving technological problems. This work led to the establishment of participatory research in plant breeding domain to complement plant breeding efforts, rather than as an alternative to plant breeding. Such efforts ensure that the interests and feedback of marginal and economically challenged farmers in diverse agricultural systems lead to the development and expansion of varieties, which

Plant Breeding and Cultivar Development. DOI: https://doi.org/10.1016/B978-0-12-817563-7.00013-1

are selected for specific adaptation in target environments. That is, PPB provides a more market-oriented and farmer-focused approach to help solve diverse problems, as the new varieties are expected to better meet farmer's particular needs for their specific environments, with an expectation of increase in the speed of varietal replacement and often increase the on-farm varietal diversity.

Both PPB and participatory varietal selection (PVS) have been developed and implemented over the past couple of decades as complementary breeding approaches, mainly in developing countries where low resource farmers grow their crop in marginal lands of remote regions. It is practically implemented in areas where the technology transfers or adoption of modern cultivars is low or non-existent. PPB has been widely considered to be more advantageous to use in areas where low yield potential, high stress, and heterogeneous environments exist. Therefore, PPB has emerged to address agricultural problems of poor farmers in developing countries where resources and modern technologies are limited. However, reports of success have been mixed. For example, Sperling et al. (2001) reported that PPB has not been so helpful for those farmers who grow their crop under marginal soil and high stress environmental conditions, necessitating different breeding approaches to reach the poor farmers. Public-sector plant breeders particularly in developing countries have not yet fully used market oriented PPB approaches. However, broadly speaking, the general plant breeding approach of a linear process of research and extension has been almost universally adopted, where breeders first develop, test, and release new varieties, with limited involvement of farmers, and the extension services promote them. In contrast, the private sector has long used market oriented research, where farmers test potential new varieties before their commercialization. This happens at the advanced testing but pre-commercialization stage. Farmers are a part of the testing network, as their fields are used for a large test plot and data obtained is utilized to make decisions on variety commercialization. Research and extension activities overlap because testing creates a demand for the successful new varieties, and farmers field serve as a testbed for the potential performance of the new variety in commercial production.

When public-sector plant breeders use market-oriented approaches of the private sector, they are described as "participatory." The term "participatory" conveys that different people from different organizations including end-users have a significant research role in the major stages of the breeding, evaluation, and selection process, such as setting PPB targets and goals, setting breeding priorities and objectives, selecting genotypes from a heterogeneous population, helping in evaluation and selection of cultivars in farmers' fields and on research stations, releasing and popularizing high performing cultivars, and seed multiplication and distribution. It is important to clarify that private seed companies do not engage farmers in all these listed activities.

The term "collaborative plant breeding" is an alternative to PPB but also describes a particular type of participation (Biggs, 1989). Market-led techniques have been usefully classified by the stage of genetic material involved as farmer-centered plant breeding concerns the entire breeding process (in PPB); whereas farmer-centered PVS is limited to the testing of finished varieties. Both PPB and PVS are a component of participatory crop improvement (PCI). Initially, most PCI projects involved only PVS, in which farmers select among a choice of varieties that they can grow and evaluate on their own farms.

Participatory varietal selection

PVS is an approach where selection of finished or near finished varieties is made by the farmer on her/his own fields and based on their own assessment of what is important and relevant for their farm and production practices. The finished varieties in self-pollinated crops include; advanced non-segregating lines, advanced stage varieties, and released cultivars. However, in cross-pollinated crop they include advanced populations. In PVS, farmers are given a wide range of novel cultivars to test for themselves in their own fields. Both research and extension methods happen in PVS, which help to deploy varieties on farmers' fields; meeting the demand and requirements of different farmers.

PVS works best when it is conducted under "on-farm" conditions, which are very different from on-station testing conditions. Therefore, in the PVS system, evaluation of varieties is done on farmer's fields and not on research stations, which might not precisely represent farmers' fields. One example of this difference is that on-station trials are conduced on optimum conditions targeting yield potential testing, while on-farmer field trials represent yield performance *per se* for the farm conditions. Research focuses also on traits other than just high yield. Farmers evaluate all traits that are important to them and then make trade off amongst all traits. For example, they may accept varieties with lower grain yields but higher grain quality because they consider overall returns more important than just yield. They may also trade lower grain yields for early maturity because they want to have a second cropping. Similarly, they may pick a lower yielding variety with disease resistance instead of a susceptible variety to the same disease. The decision is made because on their farm they can pick varieties that give higher seed yield due to disease resistance, or have better expression of another trait making seed yield a secondary consideration.

A successful PVS program has four stages:

- Participatory rural appraisal (PRA) to identify farmer needs for a cultivar.
- A search for suitable material to test on farmer's condition by farmers.
- Experimentation on the acceptability of the material on farmers' fields.
- Dissemination of the cultivars in accordance with farmers' preference.

Traditional cultivar development approaches differ from PVS. (1) They do not start with a PRA, (2) there are fewer choice of new cultivar and these are limited to those selected after years of formal trials, (3) they tend to involve only a few farmers, if any, and (4) crop management includes a recommended package of practices' that are beyond the resources and risk taking capacity of most resource-poor farmers. On the other hand, PVS is a simple and more direct way of working on multiple traits to assess the value of a variety to farmers. Quality traits (such as milling, cooking and keeping quality, and taste), which are difficult or expensive to evaluate in traditional trials, can be assessed easily in PVS trials. Varietal selection that is client oriented is not controversial, as breeders have long used on-farm trials. However, one challenge in PVS compared to traditional approach is that traditional program developed varieties have been bred for wider adaptation, while PVS approach can be limited to data from the choice of farmers who are participating in evaluation.

All PVS use some form of mother and baby trials, even if they are not named in this way. All the varieties are grown on mother trials in a one field, generally one-replicate design. Typically, there are about 5–6 mother trials. Baby trials are more numerous with each farmer growing a single entry and comparing it to his or her local variety. In the mother trials, qualitative estimates of yield are obtained; but in the baby trials, test coordinators simply collect farmer's perceptions on yield.

PVS has been greatly successful in many crops and countries when used in marginal areas with low-resource farmers. It is also effective in more productive environments where it contributed to increase on-farm varietal diversity and faster varietal replacement. However, for productive environments, on-station trials can represent quite well the situation in farmer's fields so the advantages of PVS in favorable environments, tend to be less than those for marginal areas. One of the great strengths of PVS is that it is both an extension and a research method. For example, PVS trials in upland rice in Ghana resulted in a dramatic spread of new varieties to new villages over a single season. Several international research centers help facilitate substantial networks of national programs with partners. These include CIMMYT initiatives for maize in southern Africa, and wheat in marginal areas in South Asia. International Rice Research Institute and its collaborators under the umbrella of the Consortium for Unfavorable Rice Environments used PVS approaches through national agricultural research and extension systems' rice breeding programs.

The similarities and differences in the national varietal testing networks of India, Philippines, Kenya, Nepal, and Bolivia were reviewed by Witcombe and Virk (1997). In these well-established system varieties were tested in replicated, multi-location, on-station yield trials to identify varieties that should be released and recommended. These trials were examined under conditions when they encountered abiotic stresses such as drought, a common constraint in rainfed rice.

Client oriented plant breeding or participatory plant breeding

Client oriented plant breeding (COPB), often referred as PPB, is a logical upstream extension of PVS and is desirable when the possibilities of PVS have been exhausted. The client driven plant breeding approach involves different participants including breeders, scientists, farmers, along with consumers, extension agents, traders, vendors, farmers' cooperatives, processors, government, and non-government organizations in plant breeding research.

PVS is both the building block for PPB and the means of testing its new products. Farmers accepted cultivars identified by PVS make ideal parents for farmer-centered breeding programs. When these efforts develop potential varieties, farmers can immediately test them. This is one of the greatest advantages of working with farmers in a system that enables research and extension to work in parallel. Typically, several years are saved (from initial cross to adoption) so the rate of return on the investment in plant breeding is considerably increased due to reduced time to cultivar development and seed distribution. However, the scope of PVS is much narrower compared with *client oriented*, *client driven*, *market driven*, or *farmer driven plant breeding* (all terms used analogous to PPB).

Client-oriented or -driven breeding is grouped into two categories:

1. *Formal-led*: where formal breeding programs initiate, manage, and execute plant breeding pipeline and farmers are asked to join plant breeding activities. This can be under the aegis of national programs and international centers.
2. *Farmer-led*: where farmers control, manage, and execute breeding systems and role of scientists and/or development workers is just as a contributing member, although in few instances, scientists (breeders) can support their own varietal selection and seed system.

Table 24.1 provides five models on farmer–breeder participation in the main stages of a breeding program.

Goals of client oriented plant breeding

1. Increase productivity and profitability of crop production through the development and enhanced adoption of suitable, usually improved varieties. This is specially more important in non-commercial crops grown in unpredictable and stress prone environments.
2. Provides benefit to a specific type of user, or to deliberately address the needs of a broader range of users.
3. Build farmer skills to enhance farmer selection and seed production efforts.
4. Enhance biodiversity and accessibility of germplasm to local farmers.

In COPB/PPB programs, the results of PVS are exploited using identified cultivars as parents of crosses. Weaknesses in cultivars are identified in the PVS program and they can be crossed with varieties that have complementary traits to eliminate those weaknesses. PVS is the first step in selecting desirable parents, as it allows local and introduced germplasm to be evaluated using participatory approaches, determines their acceptability to farmers, and identifies important traits. For example, one can cross a high yielding but low-grain quality variety with superior grain characteristics. The COPB approach is ideally suited to the

Table 24.1 The five models that shows connectivity between farmers and breeders during the four main stages of the breeding pipeline.

Model	**Farmer or breeder led**	**Parental selection (source germplasm)**	**Germplasm development (prebreeding)**	**Cultivar development**	**Variety evaluation**
Model A: traditional farmer breeding	Farmer	√	√	√	√
	Breeder	×	×	×	×
Model B: traditional breeder (scientist) breeding	Farmer	×	×	×	×
	Breeder	√	√	√	√
Model C: complete participatory breeding	Farmer Breeder	√	√	√	√
Model D: efficient participatory breeding	Farmer	√	×	×	√
	Breeder		√	√	
Model E: participatory variety selection	Farmer	×	×	×	√
	Breeder	√	√	√	

Adapted from Morris, M.L., Bellon, M.R., 2004. Participatory plant breeding research: opportunities and challenges for the international crop improvement system. Euphytica 136 (1), 21–35.

strategy of selecting parents rigorously, using few crosses and employing large populations. For example, a PPB program in a self-pollinated crop can start with one cross or very few crosses. Even with a low cross-number strategy, the number of crosses covered will gradually increase over time if one, or a few, new crosses are made each year. This helps to maintain the farmer's interests by a supply of novel germplasm and allows a continuing incorporation of new genetic material from the main breeding programs. The fewer breeding populations but larger population sizes allow easier integration with farmer-driven focus and better sharing. Pedigree breeding generates many lines that can only be accommodated with difficulty in a PPB program. However, some form of pedigree breeding have been used, and with success in PPB (e.g., Sthapit et al., 1996). Generally, the most effective participatory methods will keep selection units to a minimum, to provide one or few selection units per farmer. Hence, bulk-population breeding or modified bulk-population breeding, where populations are divided into farmer-selected trait-specific subpopulations are ideal for PPB approaches (Witcombe and Virk, 2001). The subpopulations have characteristics that can be recognized by farmers, and they are able to apply mass selection within these populations.

PPB methods were adopted to take advantages of the strengths of breeders and farmers. A key element of COPB is the collaborative participation of farmers who grow a bulk on their own fields and select among it. The breeder's produce material that is genetically homozygous but highly heterogeneous by advancing the bulk populations from the F_2 to the F_3 generations with minimal selection (Fig. 24.1). This means that breeders share with farmers at a more advanced filial generation when it is expected that there will be a good response to selection between plants, and when segregation in the next generation is no longer a major complicating factor. Using this collaborative breeding, it is possible to replicate selection cost-effectively by giving seed of a particular bulk to many farmers. The selection is thus replicated across physical environments (different farmer's fields) and across farmers (who may have different selection strategies and select for different traits that best meet their needs).

Main stages of client oriented plant breeding

Main stages are as follows. However, stages could vary depending on the type of participation.

1. Identifying and setting breeding objectives.
2. Generating (accessing) genetic variability from local landraces or creating new breeding families or populations.
3. Selecting within variable populations to develop experimental varieties based on the approach (consultative/collaborative), depending on the availability of resources and crop.
4. Evaluating experimental varieties and culling through PVS.
5. Variety release.

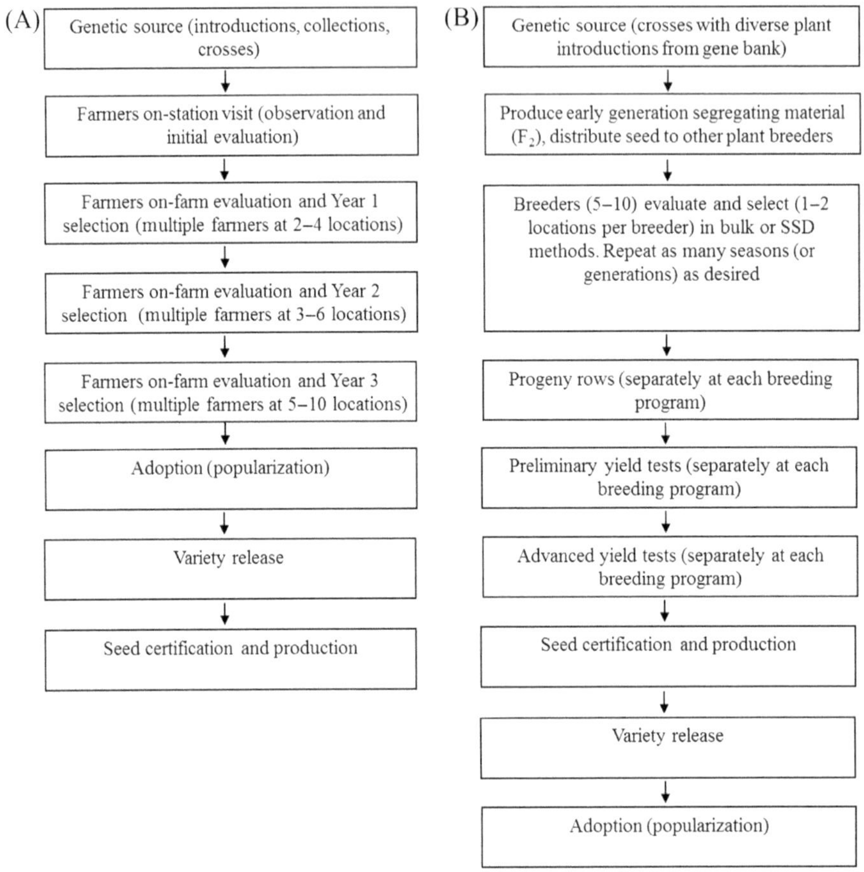

FIGURE 24.1

(A) An example of client oriented plant breeding (also known as participatory plant breeding). Number of farmers can vary as per program requirement and resources and client needs. In each year, the client participation in testing and selection is expanded. (B) An example of breeder collaborated participatory plant breeding. The example presents an approach for better utilization of germplasm collection accessions to create several breeding populations with multiple diverse parents in simple or complex crosses and produce early generation bulks (by a not-for-profit public program), which can be distributed to multiple programs (public and private organizations). Parental source can be other than introductions from gene banks. These breeding programs can cover a broad area of adaptation and diversify program's genetic base. Variety ownership and proper paperwork including material transfer agreements need to be determined, if commercialization is intended. The two schemes (A and B) are not presented in season per season or year per year series, but for a general sequence of steps. Also the methods can be adapted for clonal, OPVs, synthetics, and self-pollinating pure line varieties.

6. Variety adoption - popularization and its management.
7. Collaborating with seed system, that is, seed multiplication and distribution.

Compared to a conventional program (supply driven) where the sequence is selection of new varieties → variety release → seed production → adoption; in COPB (demand driven), the sequence is dissimilar—selection of new varieties → adoption (popularization) → variety release → seed production (Fig. 24.1).

Main types of participation in participatory breeding

- *Conventional*: no farmer participation. In this approach, breeders gauge the requirements of farmers through extension people, their own interactions, farm news and magazines, etc., but there is no formal farmer participation. This is not considered participatory breeding, and is only listed to provide the baseline.
- *Consultative*: farmers are consulted at every PPB stage but the breeder makes decisions. Farmers are consulted in identifying and setting breeding objectives, selection of appropriate parental materials, and traits to focus. In this approach, farmers can participate in making joint selections with a breeder among genotypes in breeders' plots or on-station trials.
- *Collaborative*: breeding decisions are jointly made by farmer and breeder. Extensive two-way communication ensures that farmers and breeders are in sync about selection criteria and research priorities. To revoke or override the joint decision made earlier, both farmer and breeder need to agree on the change(s). Usually this type of participation is effective for self-pollinated crops.
- *Collegial participation*: farmers grow genotypes in their farm fields and make their own plant or genotype selections. Farmers and breeders have organized communications among each other. Farmer decision can be made individually or in a group. Farmers voluntarily supply seeds of selected genotypes to the breeder for further evaluation and seed multiplication.
- *Farmer experimentation*: farmers make their own decision as individuals or in a group about implementation of their research activities with new variety or population without organized communication with breeders. Little or no breeder role, as they do not participate in selection of genotypes or in any farmers' research activities.

One great advantage of PPB is that it can be much faster than conventional breeding. The economic value of this reduction in time can be very large. These types of participation can inform breeders about the research direction based on needs of the specific group of client or targeted farmers. Tailored approaches are created based on physical and economic resource base of the farmers. Farmers can ensure that breeders are assessing trade-offs among traits correctly. Varieties that are taken up (or developed) by farmers produce well under "real life" conditions encountered on the farm, and there is a greater success of adoption of innovation by the farmers. COPB also enhances farmer knowledge and skills so that they can be further integrated in the participation activities and improve their farm management and production.

Justification for client oriented breeding approach and plant breeding considerations

The capacity of any breeding program is limited, and as more crosses are made the populations derived from them must be smaller. However, theoretical considerations provide a strong argument for using large population sizes. In the context of COPB, one possible breeding strategy is to make fewer crosses that are considered most likely to give desirable segregants, and produce large population for meeting client needs. From a viewpoint of generally more traits under selection consideration in COPB, large population sizes are justified too. This strategy of fewer crosses with larger population sizes is well suited to the constraints and advantages of participatory breeding in that:

- PVS aids the selection of most suitable parents. It is effective in identifying locally adapted parental material and in identifying breeding goals, for example, early maturity that assists the selection of complementary parents.
- Large population sizes are easy to manage when grown by farmers. For example, in collaborative breeding a farmer can cost-effectively grow and select from a very large population. There is no additional burden on farmer to grow such population because farmer was in any event going to grow the crop for their consumption and selling. If the PPB material yields adequately, cost (or benefits) are the difference between yield of the COPB population and the yield that the farmer would normally have obtained from their own variety. This also provides breeders with an important consideration, that farmer participation will continue if the breeder is sharing high performing genetic material.
- Selection is carried out in the target environment (minimizing the effects of genotype × environment interactions) and the selection is for traits that farmers consider important. In particular, when breeding for drought-prone environments, conventional multi-location trials are difficult to analyze because trials in the most drought-stressed environments produce many missing values and are often excluded for this reason, despite such trials being the most relevant to farmers. Participatory trials do not suffer from this disadvantage. When a variety fails in a farmer's field, this gives valuable information on the potential acceptability of the variety and is not just a missing value. Although, there should be financial support for farmers to mitigate their catastrophic losses in the event of a crop failure when growing the test population.
- The most common public-sector trial design is a randomized complete block with three or four replications. Even though there are satisfactory statistical techniques of dealing with missing values in this experimental design, it is difficult for these to be accepted in a national varietal trial that involves

entries from several breeders and institutions. For example, during a drought (or another important stress trait) if only a few entries actually produce grain, a common practice is to discard the trial completely. Unfortunately, this means that the missing data are not considered, yet they are not really missing. Instead, they provide valuable information that, under severe drought (or another stress trait) which varieties will fail. Participatory research, in particular PVS, offers a way out of this dilemma. If a new variety fails in a farmer's field but the local variety survives, then this is a valuable datapoint that can easily be accommodated in assessing varietal performance.

- Includes a systematic approach to participatory evaluation such as mother and baby trials system.
- There is a significant reduction in the time needed to breed and test a new variety. As soon as a new product emerges from a PPB program it is included in PVS, and rapid feedback on the desirability of the new material is gained.
- Other aspects of breeder–farmer interactions are: breeders can consult farmers in order to set more realistic goals and choose more appropriate parents; farmers can evaluate material grown on the research station; and farmers can collaborate by growing and selecting breeding material in their own field. The choice will depend on available resources, socio-economic environment, and the extent to which pollination control is needed. Rice, for example, is easy to select because it is a self-pollinating crop. Maize, which is cross-pollinated, would require more effort because isolation distance and pollination control are required.
- In farmer-oriented breeding programs, trials are identified that breeders had not considered important or which they were previously unaware, such as pericarp color or appetite delay in rice (for example, farmers may want varieties that satisfy appetite longer). There are also strict requirements for threshing in rice (for example, number of beatings required) in areas where threshing is done manually.
- Unfortunately, participatory research is not simple and the history of obtaining poor-quality data is a deterrent to its use. Presently, reliable data from participatory trials are needed. For scientists in the formal system, this means having a partnership with a NGO or extension service that has already built relationship with farmers. In addition, staff dedicated to research within the partner organization are important to better align priorities for development activities and for research.

Changes in breeding methodology to maximize farmer-scientist collaboration

Targeting breeding programs more closely to farmer's needs can be done using conventional breeding methods, and such programs have been successful. However, methods can be adapted to maximize the benefits of working closely with farmers. Farmers are willing to grow large populations but will usually find it difficult to test many entries unless assisted or trained by scientists. Hence, in several published examples of PPB, a modified breeding strategy was used that minimized the number of entries any farmers grew but maximized population sizes. In self-pollinating crops, only one or few crosses were made each year but the size of the population derived from them was large. In the cross-pollinating maize crop, a single population per target area was improved by recurrent selection. Theory strongly supports using few crosses with large populations for COPB approaches. However, the choice of parents and cross combinations are critical for success so at least one parent of any cross should be a variety or landrace already popular with farmers or accepted by them in PVS trials.

As previously explained, in self-pollinating crops, bulk-population methods are particularly suited to participatory approaches because they benefit from the opportunity to use large populations and natural selection. Bulk breeding has been very effective when farmers have selected, in their own fields, from heterogeneous bulks of populations of nearly homozygous lines. The evaluation by farmers and breeders of unreplicated nurseries consisting of lines derived from bulk populations (which can be termed "pure line from bulk" breeding) has also been very effective. All lines that proved, over time, to be the best were selected by both farmers and breeders based on the value of using multiple judgments. In cross-pollinating crops such as maize, simple approaches such as mass selection have been effective in PPB.

Comparison of experiment station-based plant breeding and participatory plant breeding

1. Traditional breeding is practiced in areas where crop is produced over a broad, relatively uniform agroecological area(s), whereas PPB is carried out in areas that are not dedicated to large-scale production and are marginal crop production areas where environments are highly variable so selection based on $G \times E$ interaction is not the best selection strategy.
2. Traditional breeding is followed in places where farmers have easier access to agricultural inputs (fertilizers, chemicals, and irrigation); whereas PPB is more suited where agriculture is risk prone, complex (intercropping) and low input, and may include the development of populations or heterogeneous materials.
3. Traditional breeding is practiced where end users are generally similar, whereas PPB is practiced in places where crop end users are diverse and are locally unique.
4. Success of traditional breeding resulted in green revolution in wheat and rice varieties, maize hybrids for temperate and mid-altitude tropical zones, varieties for irrigated, commercial vegetable production, whereas examples of

PPB are pigeon pea and barley breeding in ICRISAT, rice breeding in India and Nepal, maize in India, cassava and beans in Latin America.

5. Traditional breeding programs generally develop many more crosses and breeding families, thereby restricting the size of early generation segregating populations that are evaluated; whereas PPB involves the use of few crosses with large population sizes.
6. In most traditional breeding, farmers participate indirectly in setting breeding goals particularly in programs that aim to produce widely adapted varieties for many farmers which is a model followed for easier and more economical seed production and placement. PPB involves farmers in goal setting in market research. It helps in targeting varieties that will be accepted in defined physical or socio-economic environments. In many cases, early maturity, perhaps even earlier than that of existing landraces or cultivars are found to be as important a trait as higher yield.

Participatory plant breeding turns upside down the delivery phase of a plant breeding program. In a conventional breeding program, the most promising lines are released as varieties, the certified seed is produced and only then farmers decide whether to adopt them or not. In a participatory program the process is driven by the adoption, which takes place during the final stages of selection; and therefore, adoption rates are higher and risks are minimized. The investment in seed production is nearly always paid off by farmer's adoption. These advantages are particularly relevant to developing countries where investments in plant breeding have not resulted in production increases, especially in marginal environments. Adoption rate of improved varieties in marginal environments is lower than in higher producing zones. While the reasons for lower adoption rates are many (such as non-adapted varieties, seed supply, etc.), and the principal causes are related to risk. Resource-poor farmers in marginal environments have few options in crop production, and therefore, the adoption of new technologies is perceived by farmers as risky. This is likely to be the casual factor in farmer preference in continuing to grow unimproved but well known varieties. Participation of farmers in the very initial stages of breeding, when the large genetic variability created by the breeders is virtually untapped, offers farmers the possible to familiarize with new options without exposing the household to any risk. The COPB approaches also exploit the potential gains of breeding for specific adoption through decentralized selection, defined as selection in the target environment, and is the ultimate conceptual consequence of a positive interpretation of genotype $\times$ environment interactions.

CHAPTER

Breeding of crop ideotypes 25

Abstract

Ideotype is a biological model combining morphological and physiological traits. Ideotype breeding is a method of breeding to enhance genetic yield potential (or any other end trait) by modifying individual traits that influence it, and where the breeding goal for each trait is specified. Ideotype breeding is a continuous process because ideotypes are developed to meet the changing needs. Plant breeders can complement traditional breeding methods with ideotype breeding to generate a better understanding on individual traits and to diversify the genetic variation for traits. Ideotype breeding also provides hypotheses on paths to increasing yield or other important end traits, and this approach is also expected to give greater quantity of grains or oil or other useful end product in a specific environment. In this chapter, maize, wheat, rice, common bean, chickpea, lentil, field pea, pigeon pea, mung bean, and urd bean ideotype breeding informations are described.

Ideotype breeding can be used as a method of breeding to enhance genetic yield potential by modifying individual traits where the breeding goal for each trait is specified (Donald, 1968; Rasmussen, 1987). A desirable crop ideotype makes a minimum demand on resources per unit of dry matter produced. The breeding of crop ideotype or breeding of model plants involves defining a crop production environment and designing a plant model combining the morphological and physiological traits that are known to influence performance (i.e., yield in that environment) into one plant type, which is released as a variety to replace an existing cultivar. According to Donald (1968), an *ideotype* is a biological model, which is expected to perform or behave in a predictable manner within a defined environment. More specifically, a crop ideotype is a plant model that is expected to yield a greater quantity or quality of grain, oil or other useful product when developed as a cultivar.

This approach contrasts with traditional plant breeding that involves genetic improvement of specific traits, for example, selection for yield *per se* not necessarily on individual component traits (which is what breeders using crop ideotype methods utilize). The selection for yield *per se* involves the use of diverse desirable parents for crossing with an intent to combine useful traits from each parents into a superior new variety. As described in earlier chapters (on breeding methods), selection of desirable plants/progenies (recombinants) is practiced in

Plant Breeding and Cultivar Development. DOI: https://doi.org/10.1016/B978-0-12-817563-7.00021-0

segregating generations under space planted conditions followed by their testing under microplots (progeny row stage) and larger plots (at preliminary and advanced yield trials) at commercial densities. Most promising varieties are subjected to pre-commercialization testing for their yield potential as well as for their resistance to important biotic and abiotic stresses in different environments and appropriate quality traits. The best line or lines is/are released for commercial cultivation targeted to a particular region/area (environment). Similar concept is applicable for clonal, synthetics, and hybrids. In each stage of traditional breeding, final trait (e.g., seed yield, seed protein, etc.) is measured and selections are made accordingly. While yield *per se* approaches have been most widely used and have worked, theoretical considerations suggest that further improvements can be made if crop ideotypes can be developed based on crop's morphological and physiological traits. For example, in a traditional approach, breeders may sometimes be unaware of the reason for why a cultivar is high yielding. However, crop Ideotype approach addresses that by clearly establishing component (physiological, morphological, biochemical, etc.) traits that govern most important traits for the crop.

The morphological and physiological traits could include: leaf angle, leaf area, photosynthetic rate, stomatal frequency, water utilization, photoperiod, efficient conversion of photosynthate to grain, harvest index (HI), long grain-filling period, slow senescence, plant height, placement of economic part, tolerance to drought, heat and salt, etc. These traits are mostly quantitative in nature and are highly influenced by environmental conditions. Some of these traits are cumbersome to measure in large breeding populations; however, traits such as tolerance to abiotic stresses, leaf shape, size, and its orientation have been used by plant breeders to increase crop productivity. For example, in North America, leaf angle is an important breeding target and maize hybrids have progressively, over time, transitioned to more upright leaf angles in increasing seed planting densities. The ability of recent maize hybrids to produce very high yield (and outperform older hybrids) in dense planting is considered a major reason for continual maize yield increase in the United States and Canada (Duvick and Cassman, 1999). For most cereal crop species, leaf inclination angle (angle of emerging leaf relative to the stem) impacts efficiency of solar radiation interception by the plant canopy.

The economic yield represents the total weight per unit area of a specified plant product (e.g., grains in cereals). The biological yield is defined as the total dry matter produced per plant or per unit area (biomass that is the amount or mass of organic matter in a prescribed area at a given point in time). The biological matter includes material formed above and below ground, although only above ground matter is used due to the complexity to obtain below ground biomass. The HI, calculated as the ratio of economic yield/biological yield, is one of the most important trait impacting seed yield but it is used less frequently by the plant breeders in cultivar development because it is difficult to measure precisely (without large labor and time investments) in many field crops and is highly influenced by environment. Although, with advances in image based phenotyping, this trait is more amenable to be used in breeding [See Chapter 28: Phenomics and Machine Learning in Crop Improvement].

Economic yield is expressed as a product of these four factors; (1) incident solar radiation, (2) the fraction of incident solar radiation absorbed by the crop canopy (also known as canopy absorptance, and correlated with canopy leaf area), (3) overall efficiency in conversion of absorbed energy into crop dry matter, and (4) fraction of the crop dry matter that is allocated to the economic yield (economic part of the crop). This formula integrates these variables from planting date to the harvest date. Therefore, yield depends on biomass and its partitioning and breeders aim to develop cultivars with increased biomass and efficient partitioning of assimilates. High HI indicates that the genotype is more efficient to translocate the assimilates (from leaf and other vegetative parts, i.e., source) to economic parts (i.e., sink) of the plants. For cereals, the HI is the ratio of grain yield to total plant weight (i.e., grain + straw). Theoretically, HI can vary from zero to one and is expressed as a percentage. In cereals, it may be generalized that dwarf type have low biological yield and economic yield. However, semi-dwarf types (e.g., cultivars of wheat and rice) have high economical yield and in turn results in a high HI. For example, more recently in Canada, several high yielding and shorter height wheat varieties have been developed that have become very popular among farmers for their high yield in rainfed conditions, good harvestability, and improved stem lodging resistance.

In Japan, S. Tsunoda, in late 1950s and early 1960s, compared low and high yielding varieties of rice, sweet potato, and soybean. He posited that low nitrogen responders have long, broad, thin, drooping, pale green leaves and tall, weak stems; while high nitrogen responders have erect, short, narrow, thick dark green leaves, and short and sturdy stems. Building on physiological knowledge of crop photosynthesis, he postulated that thick and dark green leaves loose less light through reflection. The reduced leaf size and erect habit permit uniform light penetration to all leaves and also reduce respiration, resulting in increased dry matter production and yield even under low light intensity. The short and sturdy culms minimize lodging, which cause little or no yield loss once panicles are well developed.

Kokubun (1988) reported that soybean breeding will require the development of high yielding varieties adapted to higher planting densities and narrower row widths, and suggested that ideally to increase total dry matter production and partitioning of dry matter to seed in soybean there is a need to improve light intercepting characters in leaves and canopy, improved leaf photosynthesis (higher photosynthetic efficiency), and rapid leaf area index (LAI) increase for better ground coverage. After extensive research, very high planting density does not result in yield increase. Although, early and rapid canopy coverage has shown to be important for yield gains. Other strategies include taller plant height with reduced internode length to give more nodes per plant, and more pods per plant accompanied with more seeds per pod. The other factors helping soybean production have been manipulation of soybean flowering and maturity dates, enabling continual growth of acreage in Northern latitudes.

However, several factors act against the realization of calculated rate of dry matter production. The diurnal changes in solar radiation is strongly related with the rate of photosynthesis of a single fully exposed leaf at normal CO_2 concentration in air, although the rate of photosynthesis does not increase with an increase in light intensity beyond 3000 fc (=foot-candle; defined as one lumen per square foot), as the leaves in the lower region of a crop hardly receive this high light intensity; and therefore, do not function with the same efficiency. Variation in crop surface in relation to interception of solar radiation, thereby assumes significance. LAI is the ratio of leaf surface to land area; and LAI when integrated over time is leaf area duration, which reflects the persistence of leaf area. The rate of crop growth is more closely associated with LAI than with net assimilation rate (NAR). Maximum NAR coincides with maximum LAI for a short period in the life cycle of the crop; and NAR generally decreases with an increase in LAI. LAI is correlated with the rate of dry matter production. In general, in cereals, besides leaves, photosynthesis in leaf sheaths, peduncle, and ear also contributed to grain filling. Plant genotypes differ in photosynthetic efficiency and in patterns of partitioning of the dry matter, which suggested that plant breeders can select efficient cultivars that differ in the pattern of dry matter production.

Ideotypes have been proposed to make full utilization of solar energy as well as higher nutrient supply and moisture availability of soil, and in turn produce greater yield. These ideotypes are aimed to influence positively the sink size, and better partitioning of photosynthate to the sink. For example, in cereals, large panicle with numerous fertile spikelets; and in pulses, large inflorescence/fruiting zone with more pods and large number of seed per pod. In cereals, such as wheat and rice, yield/unit area depends on: plants/unit area × mean number of productive tillers/plant × mean number of grains/ear or panicle × mean grain weight. Tall cultivars differ from dwarf ones in the pattern of dry matter partitioning; therefore, semi-dwarf plant type has been developed by cereal breeders. Cultivars with such a plant type are responsive to fertilizer and irrigation, and have resulted in increased production and productivity. Similarly, yield/unit area in pulse crops depends on: plants/unit area × mean number of pods/plant × mean number of seeds/pod × mean grain weight. Semi-determinate and indeterminate plant types have been developed in pulse crops that are higher yielding than determinate type. An optimum balance among yield traits is an integrated approach of plant type development, as overemphasis on one trait without a balanced approach can give unintended negative outcomes.

Crop ideotypes in several crops are described below.

Maize (*Zea mays* L.)

The ideotype of maize would be one that will perform maximally under a production environment of high fertility, high plant density, narrow row spacing and early planting (Mock and Pearce, 1975). They suggested a maize ideotype for an environment with adequate fertility, high plant densities and narrow row spacings, with early planting dates in the northern plains of United States. They reported that the maize ideotype in this environment will have; (1) stiff and vertically oriented leaves above the ear while the leaves below the ear should be horizontally oriented, (2) maximum photosynthetic efficiency, (3) efficient conversion of photosynthate to grain, (4) short interval between pollen shed and silk emergence, (5) ear-shoot prolificacy, (6) small tassel size, (7) photoperiod insensitivity, (8) cold tolerance in germinating seeds and young seedlings, (9) long grain-filling period, and (10) slow leaf senescence. Adequate moisture and favorable temperatures are also required throughout the growing season, although these are difficult to control. The maize ideotype that should produce optimally when grown under that environment should include the following features.

1. *Leaf orientation*: stiff and vertically oriented leaves (i.e., 80° leaf angle) above the ear and intermediate or horizontal leaf orientation below the ear would be desirable. Lower leaves would check the loss of ground moisture and weeds as well. The vertically oriented leaves above ear would allow better light interception. Pendleton et al. (1968) compared isogenic maize single cross with vertically oriented leaves to its horizontal leaf counterpart at high density. Plants with 80° leaf angle above the ear had higher grain yields than those of the same hybrid with all leaves oriented 80° or with normal leaf orientation at a LAI of four. More recently, several researchers have shown that upright vertically oriented leaves are advantageous for a high input maize production in the United States, and modern maize hybrids have upright vertical leaves.
2. *Maximum photosynthetic efficiency*: plants with high photosynthetic rates are capable of producing high amounts of dry matter; therefore, the maize ideotype should possess maximum and efficient rates of photosynthesis (i.e., high potential to utilize solar energy to dry matter). The races/genotypes of maize were reported to show photosynthetic rates from 21 to 59 mg CO_2/dm^2 leaf area/h at 30°C. Therefore, selection for high photosynthetic rates in maize should be possible.
3. *Efficient conversion of photosynthate to grain*: this includes inherent ability of an plant ideotype to maximize dry matter production and efficiently convert it to grain. Newer maize hybrids are more efficient in distributing photosynthates produced after flowering, and also better reallocate some of the reserved photosynthates produced before flowering to grain. Newer hybrids also maintain their photosynthesis machinery better than older hybrid (i.e., better stay-green).
4. *Short interval between pollen shed and silk emergence*: the male and female parts of maize plants are separately located, hence are likely to lengthen the interval between the initiation of pollen shed and silk emergence at high plant densities. Therefore, it was proposed that an ideotype should be characterized by consistent, rapid silk emergence, and

close coincidence of pollen shed with silk emergence. It is important to note that for production of male and female inbred lines, this becomes quite important because seed production success of inbred lines (prior to their use in hybrid production) requires maximum yield per unit land for better economics.

5. *Ear-shoot prolificacy*: the ability of maize ideotype to produce more than one seed-bearing ear is an important trait. Duvick (1974) increased prolificacy of maize inbred C-103 through backcrossing. When three isogenic selections of the prolific version were compared with original version in hybrid combinations at three planting densities, the prolifics were significantly less barren and higher yielding than the original at the high density (74,100 plants/ha). Modern maize hybrids in North American corn belt are planted in very high planting density (generally around 86,500 plants per hectare or even higher in Iowa) to enable higher grain production per acre. In the production system, minimal branching happens and one corn ear is produced (generally 600–800 seed per ear) compared to two ears with lower total seed per ear, and dense planting with minimal branching and one ear per plant is shown to maximize yield per unit area.
6. *Tassel size*: small tassels should reduce both the competitive ability of tassel and shading of upper leaf layers; and therefore, should be considered an important trait of maize ideotype. The reduced tassel dry weight at pollen shed increased grain yield and reduced the incidence of barren stalks in a diallel set of single crosses among six maize inbreds planted at (98,000 plants/ha).
7. *Cold tolerance of germinating seeds and small developing seedlings*: in certain maize growing areas planting dates coincide with low temperature (10°C); therefore, maize ideotypes should be cold tolerant (i.e., possess the ability to emerge and grow normally when planted in cold and often wet soils). Fortunately, genetic variability is present for this trait in maize populations and in the U.S. corn belt farmers are able to successfully plant in colder soils. In very cold soils, seed won't germinate but as the soil temperature increases it will successfully germinate and emerge.
8. *Photoperiod insensitivity*: photoperiod insensitive genotypes would have ability to develop normally over a great range of photoperiods. Adequate genetic variability for photoperiod response exists in maize to allow selection of insensitivity.
9. *Length of grain-filling period*: grain-filling period of maize ideotype should be sufficiently long to allow maximum production and storage of dry matter but not so long that leaf death occurs before physiologic maturity. The genetic variability for this trait in maize suggested the possibility for selecting materials with short vegetative growth periods (from planting to silking) and long grain-filling periods (silking to physiologic maturity).

While proposing the above ideotype in maize, Mock and Pearce (1975) assumed that various morphological and physiological traits could be combined through breeding into one plant type. Cairns et al. (2012) tested a large set of tropical and subtropical maize inbreds and single cross hybrids under reproductive stage drought stress and well-watered conditions. Patterns of biomass production, senescence, and plant water status were measured throughout the crop cycle. Under drought stress early biomass production prior to anthesis was important for inbred yield, while delayed senescence was important for hybrid yield. Under well-watered conditions, the ability to maintain a high biomass throughout the growing cycle was crucial for inbred yield, while a stay-green pattern was important for hybrid yield. Duvick (2005) reported the following traits changed in conjunction with genetic gain in modern era U.S. maize hybrids (on a per plant basis): (1) plant and ear height (lower ear height placement over time); (2) leaf angle (more upright, greatest in leaves over the ear); (3) tassel size (tassel weight and branch number have seen a decline over time); (4) LAI (may vary, but generally has increased over time); (5) leaf rolling (increased in newer hybrids); (6) stay-green (delayed leaf senescenece has increased over time); (7) tillers (decreased); (8) silk emergence (date generally is earlier, accentuated due to high plant density); (9) anthesis-silking interval (days between anthesis and silking) has reduced; (10) barren plants (less barren plants in newer hybrids, giving higher numbers of ears per plant per unit area); (11) grain filling period (days of grain filling has increased in newer hybrids), and (12) HI (little to no change).

Wheat (*Triticum aestivum* L.)

Farrer, developer of wheat variety Federation in Australia in early 1900's, was the first to visualize an ideotype of wheat, with narrow leaves, few tillers, and early maturity, coupled with an ability to thrive on dry soils. In 1930s, Norin 10 a short-statured wheat with stiff straw, high tillering, resistance to lodging, and yellow rust disease was grown in Japan. It had good response to nitrogen. Norin 10 was brought into the United States from Japan by S.D. Salmon, and this led to the breeding of semidwarf winter wheat variety Gaines in the U.S. by Orville Vogel in late 1950s. Gaines was the product of Norin 10 × Brevor cross and had red kernel. The same material was used by Norman Borlaug in Mexico at CIMMYT (Centro International de Mejoramiento de Maiz Y Trigo-International Centre for Maize and Wheat Improvement) to develop semidwarf spring wheat. Borlaug introduced shuttle breeding to speed generations and shorten cultivar development time (i.e., growing of alternate generations under two diverse environments).

These locations (Near Toluca and Ciudad Obregon, both in Mexico) differed in soil type, temperature, rainfall, and photoperiod. The shuttle breeding led to the development of strains with relative insensitivity to photoperiod as well as a broad spectrum of stem rust resistance. It also helped to reduce the considerable time needed to breed new cultivar. As a result, several high yielding semidwarf

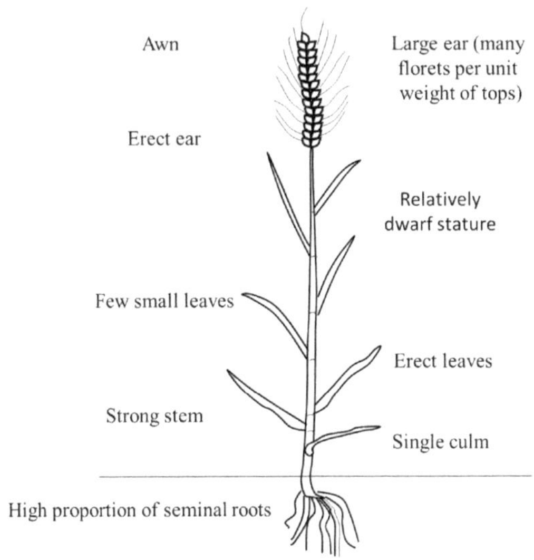

FIGURE 25.1

A basic wheat ideotype, designed to give a high grain yield as a crop community.
(From Donald, C.M., 1968. The breeding of crop ideotypes. Euphytica 17, 385403.).

cultivars of spring wheat were successfully developed. These cultivars in late 1960s, revolutionized wheat production by increasing crop productivity under good soil fertility and irrigation managements in several developing countries, including India and Pakistan. This progress in wheat production and productivity was called green revolution (the term coined by William Gaud in 1968). The farmers who were harvesting 1–2 tons of wheat yield per ha started harvesting >5 tons/ha.

Donald (1968) proposed a wheat ideotype suited to well fertilized, well-watered lands or irrigated areas of northern Europe. The morphological attributes of this ideotype are based on physiological considerations that would enable it to give an optimum partitioning of dry matter according to the purpose for which cultivars will be used. The model was expected to yield greater than the available cultivars. The important features of the model (Fig. 25.1) are described as follows.

1. *A short and strong stem*: a relatively short and stout stem reduces the risk of lodging at higher fertility and moisture, and increased weight of ear.
2. *Erect leaves*: plants with vertical leaves can be crowded with less mutual competition, hence would permit a greater population of ears per unit area.
3. *Few small leaves*: fewer small and erect leaves per culm would accommodate high density of culms, which are sufficient to intercept all light and yield better than cultivars that have high number of longer leaves per culm. Leaf area is known to be negatively correlated with the rate of photosynthesis, and stomatal frequency is positively correlated with photosynthetic rate.

4. *A large and erect spike*: there are indirect evidences that the wheat spike (ear) is normally a limiting sink for photosynthates; therefore, large spike with many fertile florets would be desirable. An erect ear would facilitate illumination of all sides of all the ears in a community of erect ears. However, these factors should not be looked in the absence of other factors. For example, a large and erect spikes without sufficient number of spikes per unit land, and high number of seed per spike cannot produce a high yield on farmer's field.
5. *The presence of awns*: there is evidence that additional surface provided by awns will contribute significantly to photosynthesis (contribute $>10\%$ of the total grain dry weight). However, more recently high yielding awnless varieties have been developed.
6. *A single culm*: a community of wheat plants with a single culm (only main stem and no tillers) will be a unidirectional organization toward ear and grain formation, hence will give greater production/unit area than is given by a variety that tillers freely or even sparsely. This happens because there would be no internal competition between the developing ears and younger tillers for nutrients.

The above ideotype with its strong single culm, few small erect leaves, and large spike can be bred from available materials and would give greater production if grown at high planting density, high fertility, well-watered soil, good weed management, and with maximum utilization of light. At lower fertility/density level and limited moisture supply, such an ideotype would not be useful because it would be of low competitive ability relative to its neighbors (which are also weak competitors).

More recently, researchers have proposed that in a changing climate (shifts in weather patterns and increase in climatic variability, with more regular occurrence of extreme weather events) wheat ideotype in Europe needs improvement in light conversion efficiency and longer grain-filling duration resulting in increased HI (Semenov and Stratonovitch, 2013); while in Australia early anthesis and longer grain-filling period are required with higher radiation use efficiency, maximizing grain size, and rate of grain fill (Wang et al., 2019). As previously explained, an ideotype is dependent on the growing environment including crop management practices.

Rice (*Oryza sativa* L.)

In case of rice, a program for incorporating genes for fertilizer response from *japonica* rice into *indica* cultivars was sponsored by Food and Agriculture Organization in 1950s. This program led to the development of two varieties: ADT 27 in Tamil Nadu (India) and Mahsuri in Malaysia. Between 1959 and 1962, Japanese scientists explained the association of yielding ability with plant type. These workers classified highly fertilized responsive varieties into two categories: those with increased panicle number (suitable for transplanting), and those with increased panicle size (suitable for direct seeding). Murata et al. (1957) expressed photosynthetic production (P) as equal to the product of total leaf area (A) $\times$ the rate of photosynthesis per leaf unit (P_0)

× the efficiency of light utilization (*f*) to give $P = AfP_0$. Lowering of light utilization (*f*) is due to mutual leaf shading in upper leaves and is related to low penetration of light to the lower leaves. The fertilizer responsive varieties grown at higher fertility level permit a high light penetration through the plants, and have a high efficiency of light utilization. At low nitrogen level and/or wide spacing, such varieties showed markedly reduced yield due to greatly reduced total leaf area. In contrast, varieties with low response to fertility when heavily fertilized showed low penetration of light and utilization, increased respiration hence reduced yield; however, under low nitrogen and/or wide spacing, they exhibit higher light penetration and utilization, and show increased yield as well.

Jennings (1964) suggested that in rice an ideal or model plant type would consists of semidwarf stature, high tillering capacity, and short, erect and thick leaves. This concept was utilized at the International Rice Research Institute (IRRI) at the Philippines, where Dee-geo-woo-gen, a Taiwanese variety of short stature and erect leaf habit was crossed with Peta, a tall, lax and locally adapted cultivar. This cross resulted in the development of IR 8, which was released for cultivation in 1966. IR 8 had a combination of desirable traits, such as prolific tillering, dark green and erect leaves, and strong stems. It had a better response to nitrogen fertilizer with a biomass of about 18 tons/ha with 8–9 tons/ha of grain yield. This was a significant improvement in HI from previously grown tall rice varieties (HI of ~0.3, about 4 tons/ha grain yield). Another Taiwanese semidwarf *indica* variety, Taichung (Native) 1, IR 8 and Jaya were frequently used as donors for nitrogen responsiveness, as a result several high yielding cultivars were produced. IR 8, Jaya and their derivative cultivars with improved plant type were responsible for a quantum jump in yield of rice in tropical Asia in the late 1960s and early 1970s. Since then only marginal increments in rice yield have occurred, which were due to incorporation of genes for tolerance to biotic and abiotic stresses, and required emphasis on quality traits. Also, semidwarf varieties can produce large number of unproductive tillers and excessive leaf area, mutual leaf shading, reduce canopy photosynthesis, and sink size especially under direct seeding.

To achieve larger gains in rice yields, IRRI scientists have proposed characteristics of ideotype which is called as new plant type (NPT). The major components of NPT included: low tillering capacity (3–4 tillers when direct seeded, and 9–10 tillers when transplanted), no unproductive tillers, 200–250 spikelets per panicle, very sturdy stems, dark green, thick and erect leaves, vigorous root system, and increased HI. Donors for these traits have been identified in rice germplasm, mainly bulus or *javanica* type from Indonesia which are now referred to as tropical *japonicas*. The work with NPT in rice is underway with an expectation of release of new cultivars.

Bean (*Phaseolus vulgaris* L.)

Adams (1973) proposed the development of plant architecture using unique bean germplasm that would morphologically and physiologically contribute to improved yield. This ideotype was based on 11 characteristics (six morphological and five physiological traits), which are given as follows.

1. *Stem*: single sturdy stem or with minimum erect branches; numerous nodes and medium long internodes.
2. *Leaves*: numerous small leaves with vertical orientation and numerous small mesophyll cells and high stomatal index.
3. *Racemes*: axillary at each node, many flowered, short pedicle, overall length not long.
4. *Pods*: long and many seeded, thin walled at maturity.
5. *Seeds*: as large as possible with acceptable limits of the commercial class to which variety belongs.
6. *Growth habit*: determinate, narrow, erect profile. Although, Adams (1982) made a change leading to indeterminate growth habit. In later research, it was noticed that high yielding determinate type cultivars were less stable over a broad range of environments compared to indeterminate types.
7. *Growth rate*: rapid accumulation of optimum leaf area.
8. *Growth duration*: early and sequential establishment of nutritional units and long periods from flowering to maturity.
9. *Minimum uptake and transport*: sufficient for all requirements, generally high.
10. *Photosynthesis rate (net CO_2 exchange)*: high and durable for all leaves.
11. *Translocation rate*: movement of photosynthate from leaf to sink high.

The above plant type was based on the need to diversify and improve genetic base of both architectural and yield traits of North American bean cultivars. However, genetic variation was limited for physiological traits in the plant germplasm. Also, it was difficult to determine such traits with accuracy that were often associated with specific growth habit and/or seed size classes. Due to these reasons as well as due to negative correlations found among the first-order yield components, plant types in beans was revised by Adams (1982) which was called architype. Such an ideotype can be grown at higher plant densities (500,000 plants/ha); and indeterminate (Type II) growth habit cultivars had higher yield, and yield stability in less favorable environments. The variability for many morphological traits was considerable in the bean germplasm. Therefore, the major differences between original and revised architypes was the transition from determinate to a more upright indeterminate Type II growth habit.

The expected characteristics of a high-yielding architectural plant type in beans are - tall, with main stem nodes numbering 12–15, moderate number of basal branches (3–5), indeterminate growth, large plant size but not with extended vine growth, upper internodes longer and more numerous than basal internodes, thick stem diameter, narrow plant profile, high value of first-order yield components in keeping with commercial class requirements, and LAI near 4 at flowering. Additional characteristics which are also of merit includes; small leaves capable of light induced orientation, full season leaf area duration, higher specific leaf dry weight, stem and root storage and remobilization during pod and seed filling, high rate of seed filling, and longer duration of linear phase of seed filling.

Beans are grown very widely in different cultural conditions in tropical and subtropical climate across the globe. All variability can be classified into four discrete growth habits known as Type I (strictly determinate referred as bush bean; have few nodes and short internodes), Type II (upright short vine), Type III (vine type), and Type IV (pole bean; are indeterminate but differ in vine extension, stem strength, and branch number and angle). Within each of these types there is a great variation for leaf size, internode length, seed number, seed size, seed color as well as for pod texture. Hence, no single ideotype could be suitable for all situations. It would be desirable to design architypes (or ideotypes) to meet different cultural and environmental conditions in a range of seed types and sizes.

According to Kelly (2001), the plant architecture in beans has been changed dramatically in number of seed and pod types through breeding efforts. The development of Bush Blue Lake determinate snap beans from traditional pole type cultivars helped retain a vital industry in the Pacific Northwest. The transition from prostrate vine Type III navy beans to determinate bush beans revitalized the dry bean industry in Michigan. The change from determinate to upright short vine Type II has resulted in record yields and changes in the production system among Michigan bean growers. Architectural changes in pinto and great northern beans have been among the most dramatic. Through recurrent selection, unique architectural traits were introduced into these medium-seed-sized classes. Type II architecture was not present in pinto beans prior to the development of the pinto ideotype. Changes in architecture associated with indeterminate growth habit in Andean dry beans have resulted in improved yield and yield stability for more variable tropical conditions.

Chickpea (*Cicer arietinum* L.)

The *desi* and *kabuli* types are two races of chickpea and are also referred to a *microsperma* and *macrosperma*, respectively. These types also differ in plant architecture, and within these types there is also variation with respect to yield components. These two races are easily crossable and produce fertile hybrids. The important differences are given in Table 25.1.

Singh (1997) suggested that the crosses may be made between *desi* and *kabuli* types and selection may be made for either type in view of the environmental requirements. The hybridization of tall Russian *kabuli* types that have podding restricted to the top one-fourth of the plant with *desi* types could yield segregants for erect habit combined with an increase in fruiting zone, and hence increased number of pods per plant. Plants with few primary branches and more apical secondary branches may be preferred during selection. Varieties with a 1000 seed weight of 180–200 g are observed to be higher-yielding under optimum date of sowing than large seeded ones (Singh D.P., Unpublished data). According to Singh (1997) an ideotype for late sown rainfed conditions of north India would require early seedling vigor, a plant height of 50–60 cm, 2–3 primary branches, 9–10 secondary branches,

Table 25.1 Characteristics of *desi* and *kabuli* chickpea.

***Desi* chickpea**	***Kabuli* chickpea**
Plants are relatively short with smaller leaflets, sometimes prostrate (bushier) habit.	Plants are relatively tall, generally erect to semi-spreading with more primary branches.
Anthocyanin pigmentation is present on flowers and stem.	Flowers are white, have no anthocyanin pigmentation on flower as well on stem.
More pods per plant.	Less pods per plant.
More seeds per pod.	Less seeds per pod.
Smaller seeds (150–200 g/1000 seeds) which are wrinkled (of irregular shape) of various colors.	Large seeds (>250 g/1000 seeds) which are rounded (sheep headed) less wrinkled with salmon white color.
Seed coat is thick with a generally tight adherence to the cotyledons.	Seed coat is very thin, yet adheres well to the cotyledons.

60–70 pods per plant, 2 seed per pod, and a 1000 seed weight of 200–250 g. Additionally, the cultivar should have resistance to one or two major diseases and/or pests in the area.

Siddique et al. (1984) reported that at low density (23 plants/m^2) plants grew as isolated units for most of their early life and interfered less with each other than at the higher density of 50 plants/m^2. Each branch was relatively more efficient in producing seeds at low rather than at higher density, and the HI was fairly uniform among branches. They proposed that a higher yielding ideotype for short seasons comprising plants with no more than two branches suitable for sowing at high density may be developed.

Sedgley et al. (1990) proposed an ideotype for Mediterranean environments that are cool and wet in winter followed by rapid warming in spring, culminating in terminal drought. Generally rapid onset of spring drought dictates that early pod set should be a prime strategy for avoiding drought stresses. Chickpea requires mean daily temperatures (>15°C) for successful pod set of early flowers; therefore, in addition to early flowering, tolerance of suboptimal spring temperatures is also required. Additionally, under high input systems and winter sowing in the Mediterranean-type environments, an erect and limited branching habit sown as high seed density will result in higher HI and biological yield.

Lentil (*Lens culinaris* Medikus)

Lentil is considered to be the oldest and most widely grown pulse crop across the globe. The *macrosperma* (large seeded) types are mainly grown in the Mediterranean region and in the New World. The cultivars have seeds of 6–9 mm in diameter and yellow cotyledons. There is little or no pigmentation in flowers or vegetative parts. The *microsperma* (small seeded) are found mainly in the Indian subcontinent and parts of the Near East. The cultivars have small seeds of 2–6 mm in diameter and red orange or yellow cotyledons. These are generally shorter and more pigmented and have smaller leaves, leaflets, and pods than *macrosperma* types.

Erskine (1983) reported positive and significant phenotypic correlations between seed and straw yields with values of $r = 0.53$ and 0.34 for *microsperma* and *macrosperma* accessions, respectively. Mean seed and straw yields across locations were 1222 and 1366 kg/ha, respectively; and HI varied from 0.19 to 0.41. The genetic correlation between seed and straw yield was 0.76. This showed that selection for increased seed yield would not adversely affect straw yield, and selection for a high straw yield could result in cultivars suitable for forage production.

Correlations were studied in *macrosperma* (exotic materials) and *microsperma* (indigenous materials) types in winter (dry season) in India (Table 25.2). Plant height and number of pods per plant had significant and positive correlations with yield per plant in both types (i.e., in large and small seeded types). Therefore, for yield enhancement it is would be desirable to increase pods per plant by increasing pod bearing length. To achieve this, hybridization between indigenous small seeded semi-spreading (that produces lateral branches from the base of the plant) widely adapted cultivars were crossed with exotic large seeded, early maturing with erect habit (Fig. 25.2). Such crosses produced transgressive segregants with an increase in pod bearing length of the branches.

Table 25.2 Correlation coefficients of yield per plant with important primary yield traits in lentil.

	Types of lines and cultivars		
	Exotic lines	Indigenous lines	
Yield traits	a	b	c
Days to flower	−0.302**	−0.220**	0.336**
Plant height (cm)	0.386**	0.342**	0.391**
Primary branches per plant	–	–	0.247**
Pods per plant	0.909**	0.878**	0.870**
Seeds per pod	–	–	0.076
Hundred seed weight (g)	−0.009	−0.054	0.049

Notes: *a = 30 lines and cultivars having 1000 seed weight of > 30 g* (macrosperma *type).*
b and c = 45 and 1300 lines and cultivars, respectively, having 1000 seed weight of < 30 g (i.e., microsperma *type).*
Adapted from Singh, D.P., 1997. Tailoring the plant type in pulse crops. Plant Breed. Abstr. 67 (9), 1213–1220.
*** denotes significance at P-value of 0.01.*

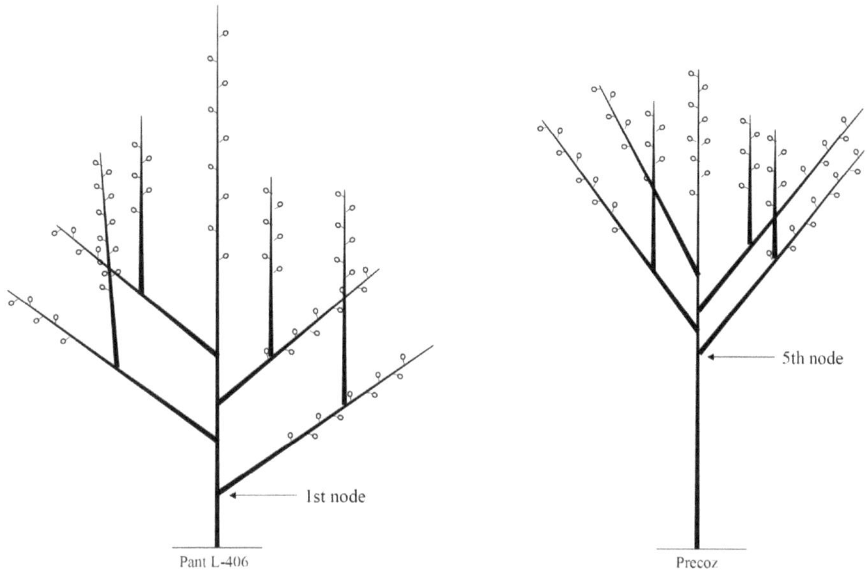

FIGURE 25.2

Distinct plant types in lentil, showing the main stem and nodal position of primary branches, secondary branches, and pods. Pant L-406 is a spreading type and Precoz is an erect type.

(From Singh, D.P., 1997. Tailoring the plant type in pulse crops. Plant Breed. Abstr. 67 (9), 12131220.).

Field pea (*Pisum sativum* L.)

Pea thrives well in places with a cool climate and is grown almost in all the temperate regions of the world. Two contrasting plant types exist: the tall types are indeterminate in growth and the dwarf types are determinate to semi-determinate. The tall types, in general, are superior in yielding ability but are inferior in standing ability.

The pea breeders have utilized three genes for developing different plant types. The gene "*af*" converts all leaflets to tendrils, "*st*" is responsible for reduced stipule size, and "*tl*" gene converts all tendrils into leaflets. A homozygous recessive allele at any one of the three loci reduced plant productivity with *stst* having the most pronounced effect (Snoad et al., 1985). Plants with *afaf stst* have been referred to as leafless type. Such types have increased standing ability due to greater number of tendrils; however, are inadequate in seed yield due to very small stipule size. Stelling (1994) studied the performance of isogenic lines and cultivars differing both for *af* and *st* locus in experiments, which included plants either supported by wire netting to prevent lodging or unsupported to permit natural lodging. The traditional leaf type (*AfAf StSt*) exhibited the highest yield potential, but also the largest yield reduction due to lodging. The yield potential of semi-leafless plant type (*afaf StST*) was less than that of the leafed type but due to its better standing ability, the yield of unsupported semi-leafless crops was less reduced by lodging. The additional reduction of stipule size by "*st*" (i.e., the plants with *afaf stst* genotype) resulted in decreased yield, which was too large to be compensated by the positive effects of standing ability.

During the past 25–30 years, several varieties of field peas with leafless (*afaf stst*) or semi-leafless (*afaf StSt*) plants have been developed in India and released for commercial cultivation. These cultivars can be grown at high population density (44 plants/m^2, as compared to 33 plants/m^2). All of these cultivars are resistant to powdery mildew (*Erysiphe polygoni*), however, many of these are susceptible to rust (*Uromyces fabae*). In rust susceptible cultivars, tendrils dry at a faster rate leading to the loss of standing ability resulting in yield losses during harvesting. Pea stem is hollow and fragile; therefore, lodging is a serious drawback. Singh (1997) suggested that stem of peas needs to be made more sturdy. There is a need to search entire pea germplasm for this trait to incorporate sturdiness in elite cultivars. In case no variability is found in *Pisum*, then the gene for stem sturdiness could be transferred from the other genera, like *Cicer* or *Vicia* through the use of biotechnological tools. The sturdier stem would also provide tolerance to shoot borers. Researchers have been studying quantitative trait loci for stem thickness and lodging (and their relationship), and height (*Le*) and *Af* gene have been reported for explaining lodging (Smitchger and Weeden, 2019), although relationship between stem thickness and lodging is not clearly established.

Pigeon pea (*Cajanus cajan* (L.) Millspaugh)

Pigeon pea is grown in the tropical and subtropical regions of the world and about 90% of the world production is from the Indian subcontinent. It is grown under a wide range of cropping systems. Traditionally, long-duration (180–260 days) crop was grown in a mixture with cereals. However, higher yielding and early maturing (<160 days) cultivars have been developed in the 1970s in India.

Short duration pigeon peas have medium plant height, larger leaf area and higher relative growth rate and NAR values during the reproductive phase compared to traditional long-duration types. Short duration genotypes are relatively insensitive to photoperiod and temperature interactions. Productivity is optimal at high plant densities (25–35 plants/m^2) in non-traditional as well as in traditional areas of pigeon pea cultivation (Ariyanayagam and Singh, 1994). Chauhan and Johansen (1996) proposed plant types of short duration pigeon pea for subtropical and tropical environments. For subtropical environments, yield maximizing traits would include photoperiod insensitivity, annual growth habit, short height (<150 cm), moderate biomass (13 t/ha), higher partitioning, reduced length of reproductive period, and water logging tolerance. For tropical environments, moderate photoperiod sensitivity, perenniality, high rate of biomass production, drought, and water logging tolerance are important characters.

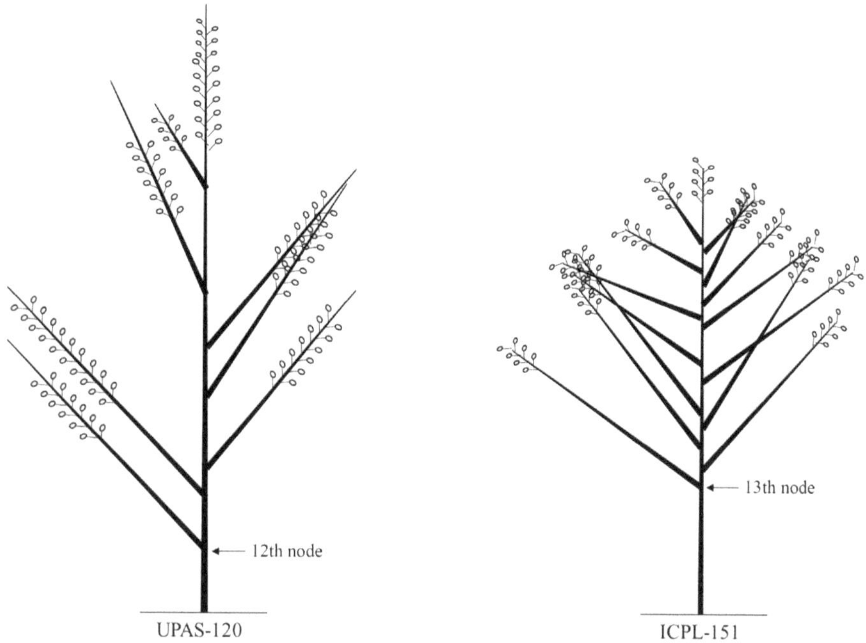

FIGURE 25.3

Distinct plant types in pigeon pea, showing the nodal position of primary branches and pods. UPAS-120 is an early indeterminate type, and ICPL-151 is an early determinate type.

(From Singh, D.P., 1997. Tailoring the plant type in pulse crops. Plant Breed. Abstr. 67 (9), 12131220.).

Within the short duration group two distinct plant types; determinate and indeterminate are available (Fig. 25.3). High yielding early types could be produced by making crosses between determinate and indeterminate types and selecting for large seeds and longer pod bearing length with more primary branches. In late types, emphasis on more branches and seeds per pod may be desirable in addition to increased pod bearing length and seed size (Singh).

Mung bean (*Vigna radiata* (L.) Wilczek)

Mung bean is extensively cultivated in South East Asia. It is also cultivated in Australia, the United States and in local areas of many Caribbean, African, and Middle Eastern Countries. In India, mung bean is grown in wet season, largely in mixture with cereals or in rotation, and rarely as pure crop. It is grown in between two major cereal crops (i.e., rice and wheat) in northern India in the spring/summer season, or in winter season in southern parts of India in rice fallows as pure crop.

Table 25.3 Correlation coefficients of important yield traits with yield per plant in mung bean.

	Season			
	Spring	Rainy		
Character		a	b	c
Days to flower	−0.074	0.444**	0.074**	0.174**
Plant height (cm)	0.636**	0.813**	0.363**	0.488**
Primary branches per plant	–	–	0.557**	0.464**
Bunches per plant	0.771**	0.678**	0.676**	0.682**
Pods per plant	0.881**	0.816**	0.793**	0.719**
Pods per bunch	–	–	0.337**	0.271**
Seeds per pod	0.578**	0.659**	0.348**	0.307**
100 seed weight (g)	0.093**	0.603**	0.283**	0.160**
Number of lines	**39**	**39**	**420**	**420**

** denotes significance at *P*-value of 0.01.
Note: *Where a, b, and c are three different years.*
(From Singh, D.P., 1997. Tailoring the plant type in pulse crops. Plant Breed. Abstr 67 (9), 12131220.).

The correlation studies were conducted by D.P. Singh and his coworkers in rainy season (rainfed) as well as in dry season/spring season (irrigated) crops (Table 25.3). Results indicated that across the seasons, number of pods per plant had the highest positive and significant correlations with yield per plant followed by bunches per plant, plant height and seed per pod; suggesting the need of increasing number of pods, bunches, seeds, and plant height. In the rainy season, correlation between number of primary branches per plant and pods per bunch were lower than the correlation between bunches per plant and yield per plant. This suggested that in the rainy season, an increase in the number of bunches per plant may be a priority that would ultimately increase pods per plant. Moreover, an increase in branch number would lead to a bushy plant type and increased pods per bunch may lead to flower and/or pod drop. The correlations between days to flower and yield per plant, and 100 seed weight and yield per plant were positive and significant in rainy season and were of no consequence in dry season. This showed that in rainy season cultivars, duration of crop as well as seed size could be increased further.

Singh (1997) suggested that for the Indian subcontinent production zones there is a need to develop semi-determinate plants of 70–75 cm plant height, a large inflorescence having more bunches per plant, 3–4 branches, increased average pod length with pods containing 8–10 seeds of average size (30–40 g/1000 seeds). The selection of cultivars for dry season may include the fast initial growth [i.e., early seedling vigor, determinate growth with shorter maturity duration (60–65 days), longer pods containing >10 seeds per pod, and a 1000 seed weight of 35–40 g]. Tolerance to shattering and moderate seed dormancy is desirable for both seasons.

Black gram (*Vigna mungo* (L.) Hepper)

Black gram (Urd bean) is cultivated in the Indian subcontinent primarily as intercrop with cereals, pigeon pea etc., during the rainy season and as pure crop on residual moisture in winter (in rice fallows) and in spring/summer season (i.e., irrigated crop) in between two main crops.

Two growth types, erect (determinate) and semi-spreading or spreading (indeterminate) are present. No single plant type will be optimum for all the production situations of black gram. Plants which produce more pods per plant with more seeds per pod would be desirable. Pods per plant has the highest positive and significant correlation with yield. The number of pods per plant could be increased by increasing branches per plant, bunches per plant or pods per bunch. Singh (1997) proposed that a large inflorescence combined with more bunches per plant will give better pod set. This could be achieved by selecting plants with more bunches per plant with 2–4 primary branches per plant. The number of seeds per pod and seed weight also show significant positive correlation with yield per plant. Plants with pods of 6–8 seeds and 1000 seed weight of 40 g may be preferred. At Pantnagar University, India D.P. Singh developed Pant Urd-31 of black gram utilizing the above approach. This variety has semi-determinate plant type, and is grown in all black gram producing states of India, and presently occupying highest acreage attesting to farmer adoption of this variety. The farmers are realizing high yield up to 2 tons/ha. It is resistant to all foliar diseases of black gram; therefore, provides stable yield across seasons and locations. It has early maturity (70–75 days) and reduced plant height of about 70 cm in wet season. Black gram variety, Pant Urd-31 fits well in different cropping systems as it is photo-thermo insensitive, hence is suitable for different seasons as well. It has average seed size (38.6 g/1000 seeds) with black color, which is acceptable to consumers.

Considerations for ideotype breeding

The crops are grown in a wide range of environmental conditions for their end product(s). There is a great variation in their morphological, physiological, and economic traits. Therefore, ideal plant type would be different for each crop for a specific environment and interest in the end product. The ideotype breeding is a continuous process because ideotypes are developed to meet changing needs. In addition, ideotype is beneficial as a complementary tool to traditional breeding primarily because ideotype breeding provides better understanding of the target component or individual or combination of traits important for the target phenotype (e.g., seed yield in row crops). The ideotype breeding assumes that enough genetic variation is present for the morphological and physiological traits in the germplasm and the genes can be transferred and recombined by hybridization. By integrating ideotypes (e.g., component traits producing yield) in traditional breeding, breeders can maintain a larger genetic diversity in their breeding program for each of the individual traits for a long-term success and viability/competitiveness of the program. In the plant breeding world, "breeder's eye" term has been used to classify successful breeders who can make visual selection in early generation and progeny rows before yield data are available, and this can be attributed consciously or inadvertently to ideotyping (traits) by the breeder to decide "keep" and "discard." However, it takes numerous years to develop such knowledge and ability to integrate ideotype in a visual cue, rather than measuring individual component traits of the ideotype. With the advent of machine-based phenotyping methods (see Chapter 28: Phenomics and Machine Learning in Crop Improvement), ideotype approach has become more feasible because trait information can be extracted numerically in a deeper and larger throughput.

CHAPTER

Field plot designs in plant breeding

26

Abstract

Plant breeders need to consider the choice of proper experimental design appropriate for the crop and filial generation under testing. For example, in early filial generations (e.g., progeny rows or preliminary yield trials or first stage of advanced yield trials) where the size of breeding funnel is wide, unreplicated tests are preferred and are often necessary due to seed limitation. Therefore, experimental designs suitable for such testing need to be utilized. As the width of breeding funnel becomes narrower, in advanced yield trials alpha-lattice or latinized row-column type of designs are deployed. In the final pre-commercial adaptation testing where fewer candidate varieties remain, higher replications in an incomplete block designs or randomized complete block designs are used. Breeding team should attempt to remove or minimize manageable sources of error in an experiment and create homogenous units or blocks. Each data point should be considered in the context of cost per data point, with the intention to reduce inefficiency and wastefulness of resources. Breeders need to consider that good data quality is extremely important, and paramount in the implementation of protocols including good field testing, data collection, and appropriate analyses. In this chapter, major replicated and unreplicated designs applicable in plant breeding trials are described.

Plant breeders use various field plot experimental designs to measure the performance of clones, pure lines, inbred lines, and hybrids during both generation advancement and precommercial testing. The necessity of proper experimentation comes from the requirement to make valid and meaningful comparisons between genotypes for selection, and the interaction of genotypes to various treatments (e.g., fertilizer, seeding rate, or other factors if they are relevant for the crop production), as well as to determine the performance of genotypes across different environmental sites (locations, year, locations × years) also known as genotype × environment interaction (i.e., stability analysis).

Plant Breeding and Cultivar Development. DOI: https://doi.org/10.1016/B978-0-12-817563-7.00027-1

Fundamentals of experimental designs

In the simplest scenario of a plant breeders' interest to compare the performance of two distinct genotypes (clones, pure lines, inbred lines, hybrids), they can grow both genotypes next to each other in identical plot dimensions. However, is it valid to grow two genotypes side by side and make comparisons, while there may be field variation due to soil, moisture, nutrients, weeds? How would a plant breeder know that the difference in yield of genotype#1 and genotype#2 is due to genetic reasons (procedure to separate varietal difference from other sources of variation)? In order to proceed with the comparison of two genotypes, there is a need to find yield difference between plots planted to different varieties, and yield difference between plots planted to the same variety. The difference among experimental plots treated alike, that is, more than one plot of the same genotype say genotype#1 is the experimental error. Therefore, replication (where the same genotype is grown more than once in the same test) helps to provide an estimate of experimental error. Experimental error is a measure of variation that exists among observations on experimental units treated similarly. The sources of variation causing experimental error includes; variability in the field, dissimilar seed source of same genotype, inadequate experimental technique, and poor emergence etc. Replication also improves the precision of an experiment by reducing the standard error of the difference between treatment means. Therefore, plant breeder strives to have as many replications as possible. However, due to resource constraints this is not generally feasible. As replication increases, the estimate of the mean becomes more precise.

The main reason for determining optimum number of replications is to manage Type II error. Type I (α) error is rejecting the null hypothesis when true, and Type II (β) error is accepting (i.e., fail to reject) when false. Power of the test $(1-\beta)$ is the probability of rejecting a null hypothesis when false. This is the correct outcome, and it is important because the failure to detect treatment differences could be caused by insufficient replications. Researchers have generated information on optimum number of replications in selection experiments (Bos, 1983; Gauch and Zobel, 1996). Statistical approaches have been shared to explore optimum number of replications in different scenarios (Casler, 2015). Therefore, sufficient replications are needed for user acceptable power of the test, as well as manage resource availability. It is important to note that the advantage(s) of replications are gone if improper field/lab protocols/techniques are used that biases observations. Additionally, plant breeders try to increase the scope of inference about the performance of a genotype by growing it over more environments with fewer replications per environment rather than more replications per environment.

Analysis of variance (ANOVA) is a partitioning of variance into recognized sources of variation. There are two basic assumptions of variance analysis: (1) treatment and environmental effects are additive, and (2) experimental errors are random, independent, and normally distributed about a zero mean and with a common variance. If these assumptions are not met, there will be heterogeneity of error; therefore, false significance levels for certain comparisons will be declared.

When a plant breeder puts a field test with genotypes of interest (experimental material along with appropriate checks) in a replicated test, after testing of these assumptions and performing influential statistics (to detect potential outliers), the interest is to see if the genotypes are a significant source of variation (say *P*-value < 0.05; assuming that is the Type I error rate breeder is comfortable to make). If the *P*-value is lower than the accepted Type I error, it is declared that genotypes (their individual mean) are significantly different from each other for the trait under study. In this scenario, planned (experimental material compared to the checks) or unplanned comparisons (compare means to all genotypes) are made using *t*-test, which compares mean between two treatments (genotypes, populations, etc.). Means comparison for *planned* comparisons can be made using least significant difference (LSD); while means comparison for *unplanned* comparisons are made using multiple comparison tests such as protected- or unprotected-LSD, Tukeys, Scheffe, and Bonferroni. Multiple comparison tests reduce experiment wise error rate (probability of making at least one error in an experiment when there are no differences between treatments) that falsely declare a difference between two means. Therefore, as more comparisons are made the experiment wise error rate goes up. For example, if there are 10 genotypes in a test, number of comparisons to be made $[(10) \times (10 - 1)]/2 = 45$. If Type I error of 5% is used, the probability of accepting H_0 for an individual comparison $= (1 - 0.05) = 0.95$. The probability of experiment wise error rate over all 45 possible comparisons will be $(1 - 0.05)^{45} = 0.099$. Therefore, the probability of falsely declaring a difference increased from 0.05 (single comparison) to $(1 - 0.099) = 0.90$ (with all possible 45 comparisons are made). This implies that there is a 90% chance of observing at least one significant result, even when all of the tests are actually not significant. This shows the importance of accounting for unplanned comparisons before declaring significance between treatments (i.e., genotypes) in plant breeding trials.

Common replicated experimental designs in plant breeding and cultivar development

Completely randomized design

Observations of treatment effects are not grouped into blocks. Treatment effects are randomly applied to the experimental units (i.e., experimental design is completely random - observation from one treatment has nothing in common with an observation from a different treatment).

$$\text{Model is: Trait} = \text{mean} + \text{Treatment} + \text{Error}$$

where, trait will be the response variable measured, for example, seed yield. Treatment is genotypes or entries. Completely randomized design (CRD) is appropriate for experiments with homogenous experimental units (e.g., lab experiments). In field, there is large variation among experimental plots, and therefore, CRD is less powerful to detect true differences between means. This design is uncommon in breeding trials.

Randomized complete block design

In a randomized complete block design (RCBD), sources of variation in a single environment test includes genotype (i.e., treatment), blocks (i.e., complete block), and error. Treatments (e.g., genotypes = experimental lines and checks) are randomized within a block, and the randomization (each randomization is unique) is repeated for each block. Randomization removes systematic bias and helps to improve the accuracy of treatment means. Experimental units in a block should be as uniform as possible, so that observed differences are due to treatments. Variation among blocks is maximized, and variation within blocks is minimized.

Blocking is an arrangement of experimental units (plots) into homogeneous subgroups, and it is done to help reduce the effects of field variability. It is less effective if variation changes in small intervals. Ideally, researchers need to understand the factors (soil, nutrient, elevation, cultivation, etc.) prior to the planting of test in field, so that blocking can be done against the direction of gradient; however, this is generally not known. Therefore, RCB designs are generally not suitable where large and unknown variations are expected in the field. An incomplete block design or spatial adjustments, explained later in this chapter, are more appropriate in these circumstances. With blocking, the principle to use is that all treatments within a block are experiencing uniform field variability. For example, if the direction of slope is going higher to lower elevation from south to north, blocking is done in east to west direction ensuring that within a block all genotypes have similar levels of field gradient.

RCB designs group one complete replicate in each block. They are useful when the number of lines in the trial is not large, so that there is less variability within than among blocks. The variability among blocks is thus removed from the plot residuals. Additionally, with large number of entries, RCBD may be inappropriate if blocking does not create homogenous blocks. Incomplete block designs are more appropriate with large number of entries, say more than 30–50 entries depending on the size of plots in the yield trial. In a different scenario with smaller plot sizes, more plots can be accommodated in the same area. In this situation, more entries can be put in each block because the variation experienced within each block is homogenous. The ability to detect existing differences among treatments increases as the size of the experimental error decreases. Blocking allows variation among blocks to be measured and removed from experimental error. However, incomplete block designs separate within and among block variation, allowing for better treatment comparison due to lower experimental error. In RCBD, replication = block, and in literature both terms are used interchangeably. However, in incomplete designs, replications ≠ block.

Replications (or blocks) and randomization cannot overcome human introduced factors such as poor field procedures and protocols; for example, uneven seed depth while planting, planting in too dry or too wet soils, poor weed management, uneven plot lengths, non-uniform seed to seed distance, lack of disease and insect-pest notes, poor harvest methods. Breeders need to use appropriate and properly calibrated machinery to avoid introducing biases. The disease and pest management are debated topics as two differing viewpoints exist. Some breeders consider the presence of disease beneficial as they can make selections if disease is present; while other breeders consider that a random and non-uniform disease or insect-pest presence helps select for the disease or insect-pest and not yield potential of genotype. Ultimately, the decision needs to be made by the breeder considering the objective of the trial. However, good notes will help a breeder on selection decisions. Additionally, proper checks and controls are beneficial to determine if the entire test is experiencing a stress (biotic and abiotic) enabling more precise and accurate selection (when there is a uniform stress presence in field) or to make a decision how to select from sub-sections of field (when there is a non-uniform stress presence in field). Checks should be multifaceted, for example, check can serve multiple purpose - seed yield, standability, maturity, disease and insect-pest resistance traits. Good record keeping is needed for all aspects of lab and seed operations, in addition to breeder notes. Examples of CRD, RCBD, and Latin square [which is one type of Incomplete block design (IBD)] are presented in Fig. 26.1.

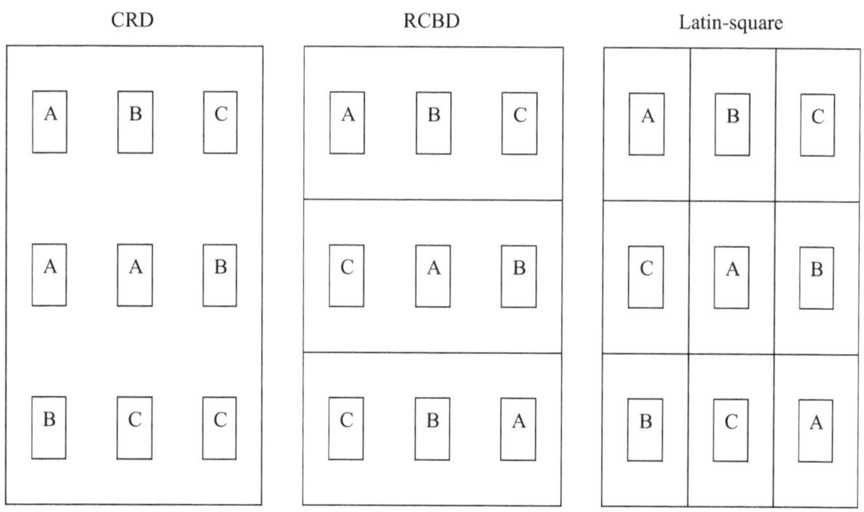

FIGURE 26.1

Examples of CRD, RCBD, and Latin square for a hypothetical example with three genotypes, and three replications per genotype. *CRD*, Completely random design; *RCBD*, randomized complete block design.

In figure 26.1, in CRD, genotypes A, B, and C are replicated three times without any blocking. In RCBD, each genotype is present once per block, and blocking is done in one direction only. In Latin square design, each of the three genotypes is blocked in two directions (N-S and E-W). Latin square is advantageous over RCBD, but due to blocking in both directions, the test size grows very rapidly as number of genotypes increase. For example, for 10 genotypes replicated three times, 30 plots will be needed in an RCBD; while 100 plots will be needed in a Latin square design. Another type of plant breeding trials are row-column designs, where a control is replicated in both row and columns to capture as much as variation as possible. A latinized alpha design (resolvable incomplete block design; number of entries in a test is a multiple of block size) can be effectively used in plant breeding, and is easier to deploy even with larger number of entries compared to a latin square. In this design, breeder assembles an alpha design that is first arranged with rows as incomplete blocks and then order of entries is rearranged to balance entries present across the columns.

Incomplete block designs

When the number of genotypes to be tested are large, there is likely greater field variability even within the block. This variability may be partly controlled by grouping plots within large replicates or complete blocks into much smaller *incomplete blocks*. The partitioning of complete blocks into smaller more homogeneous incomplete blocks permits more of the residual variation among plots to be removed from estimates of genotype means, which gives incomplete block design an advantage over RCB design where more genotypes are included in the trial. The most used incomplete block experimental designs in plant breeding tests are *alpha-lattices for replicated trials* and *augmented designs (AD) or gridding for unreplicated trials.*

Lattice designs, first proposed by Yates (1936a,b), are a type of IBD. Lattice designs are used when large number of entries need to be tested; therefore, it is not possible to have homogeneity among the experimental units within a complete block. Lattice designs overcome this issue by breaking complete block into smaller incomplete blocks. Blocks are grouped so that each group of blocks constitutes one complete replication of the treatment (resolvable). Earliest documentation of analysis are provided by Yates (1936a,b) and Goulden (1937).

Few types of IBD are:

- *Balanced or square lattice:* each entry occurs together in the same block with every other entry with an equal number of times in the test. All pairs are compared with the same precision, which is a major advantage over other types of IBDs. However, this design is impractical if large number of entries are included in a test. Some conditions on the setting up of the balanced lattice are:
  - Blocks per replicate (b) and plots per block (k) are equal ($b = k$).
  - Number of treatments must be a perfect square ($t = k^2$).
  - For complete balance, number of replicates (r) $= k + 1$.

- *Partially balanced incomplete block designs*: different entry pairs occur in the same blocks with an unequal number of times; therefore, not all pairs are compared with the same precision. That is, mean comparisons of entries within a block compared to entries not in the same incomplete block are compared with differing levels of precision. However, unlike balanced lattice, any number of replications can be used.
- *Rectangular lattice designs:* it is a form of resolvable incomplete block design. Square lattice designs have experimental and statistical limitations for limited numbers of treatments and blocks that are available. Harshbarger (1946, 1947, 1949) extended the square lattice principle to simple and triple rectangular lattices with $t = b \times (b - 1)$ and $k = b - 1$. For example, a 5×6 lattice has 5 plots per block, 6 blocks per replicate, and 30 treatments. In the rectangular lattice, the number of varieties is the product of two consecutive integers, and the number of replications of each variety is either 2 (simple rectangular lattice) or 3 (triple rectangular lattice) and their multiples.
- *Alpha-lattice*: alpha-lattice designs are replicated designs that divide the replicate into incomplete blocks that contain a fraction of the total number of entries. Genotypes are distributed among the blocks so that all pairs occur in the same incomplete block in nearly equal frequency. They do not need $k \times k$ or $k \times (k + 1)$ number of entries (where k = units per block).
  - Number of treatments $(t) = b \times k$.

Alpha-lattice provides flexibility in choice of b and k; and therefore, is extensively used by plant breeders (over other complete or incomplete block designs); although, they are less efficient in treatment separation compared to square and rectangular designs.

Below a test layout of an alpha-lattice is provided:

- Two-rep alpha-lattice test: $t = 120$ (# of genotypes), $r = 2$ (# of reps), $b = 15$ (# of incomplete blocks per rep), and $k = 8$ (# of genotypes per incomplete block).
- Condition of alpha-lattice $t = b \times k$ ($120 = 15 \times 8$) is meet in this example.
- Variation is partitioned into four sources (from a single location, replicated trial): replication, entries, incomplete block error (i.e., difference among incomplete blocks within a replication), and experimental error.

In Fig. 26.2, a four-row plot seeder is depicted to plant a breeding trial (going up a pass or column, and then coming down in the next pass or column). In this test, there are 5 ranges (also known as, rows) per rep (or 10 ranges in total) and 24 passes. Note that some programs use row as synonym for a range, and column as a synonym for a pass. Plots 1–8, 9–16, 17–24, 25–32, and so on are the incomplete blocks (15 incomplete blocks per replication). Replication 1 consists of plot #1–120, and rep#2 consists of plot#121–240. Entries are present in the incomplete block in the first replication but not present together in the second replication incomplete blocks. In this example, 120 entries will include experimental material and checks. ANOVA generated best linear unbiased means

Direction of 4-row plot seeder

	Rep1					Rep2					
-> -> -> ->	1	25	49	73	97	121	145	169	193	217	-> -> -> ->
<- <- <- <-	2	26	50	74	98	122	146	170	194	218	<- <- <- <-
-> -> -> ->	3	27	51	75	99	123	147	171	195	219	-> -> -> ->
<- <- <- <-	4	28	52	76	100	124	148	172	196	220	<- <- <- <-
-> -> -> ->	5	29	53	77	101	125	149	173	197	221	-> -> -> ->
<- <- <- <-	6	30	54	78	102	126	150	174	198	222	<- <- <- <-
-> -> -> ->	7	31	55	79	103	127	151	175	199	223	-> -> -> ->
<- <- <- <-	8	32	56	80	104	128	152	176	200	224	<- <- <- <-
-> -> -> ->	9	33	57	81	105	129	153	177	201	225	-> -> -> ->
<- <- <- <-	10	34	58	82	106	130	154	178	202	226	<- <- <- <-
-> -> -> ->	11	35	59	83	107	131	155	179	203	227	-> -> -> ->
<- <- <- <-	12	36	60	84	108	132	156	180	204	228	<- <- <- <-
-> -> -> ->	13	37	61	85	109	133	157	181	205	229	-> -> -> ->
<- <- <- <-	14	38	62	86	110	134	158	182	206	230	<- <- <- <-
-> -> -> ->	15	39	63	87	111	135	159	183	207	231	-> -> -> ->
<- <- <- <-	16	40	64	88	112	136	160	184	208	232	<- <- <- <-
-> -> -> ->	17	41	65	89	113	137	161	185	209	233	-> -> -> ->
<- <- <- <-	18	42	66	90	114	138	162	186	210	234	<- <- <- <-
-> -> -> ->	19	43	67	91	115	139	163	187	211	235	-> -> -> ->
<- <- <- <-	20	44	68	92	116	140	164	188	212	236	<- <- <- <-
-> -> -> ->	21	45	69	93	117	141	165	189	213	237	-> -> -> ->
<- <- <- <-	22	46	70	94	118	142	166	190	214	238	<- <- <- <-
-> -> -> ->	23	47	71	95	119	143	167	191	215	239	-> -> -> ->
<- <- <- <-	24	48	72	96	120	144	168	192	216	240	<- <- <- <-

FIGURE 26.2

Example of an alpha-lattice layout for a breeding test with $t = 120$ (# of genotypes), $r = 2$ (# of reps), $b = 15$ (# of incomplete blocks per rep), $k = 8$ (# of genotypes per incomplete block).

(BLUE) from a mixed model (with genotypes as fixed effect factor, and remaining factors as random effect), or best linear unbiased predictor (BLUP) from a completely random model can be generated; and the experimental entries are compared to the checks or amongst each other. Decision can be made by the breeder to either retain experimental entries better than the checks or can be based on advancing a portion (%) of entries for the next round of testing.

In some situations, particularly with larger field variation, breeders may still find it useful to perform spatial adjustment analysis. This can be done on each rep within a test (see sections on Spatial Adjustments, Nearest Neighbor, and Moving Mean Adjustments) to determine if adjusted yield should be used in the experimental design model testing. However, this needs to be used with caution because spatial adjustments work best when genotypes are more homogenous units.

Common unreplicated designs in plant breeding and cultivar development

Plant breeders developing pure line or inbred lines will use unreplicated tests (at one or more locations) in the earlier generations, generally starting in progeny row testing and sometimes 1–2 more generations. It is important to note that unreplicated designs do not substitute a properly planned and laid out replicated design, but these designs or analysis tools are invaluable to breeders in early generations when limited seed is available, and larger number of entries need to be tested. Few common approaches in unreplicated trials are described below.

Check plot method

Breeders use control or check plots for partial corrections for site heterogeneity. Control and/or check plots play an important role in experimentation. Breeder can utilize a systematic arrangement of check plots within the unrandomized test plots of breeding lines. The check plots are repeated regularly throughout the unreplicated trial; therefore, they can be used to indicate major fertility variations and to provide a crude estimate of experimental error between plots. A plot of check variety is grown adjacent to every plot or in repeating pattern of breeding material. The yield of each experimental plot or entry is expressed in terms of the yield of its adjacent control. This criterion may be used as a selection index. Any difference in yield between check and the test line will be more due to the genetic difference than to the difference in soil fertility. The frequency of check plots, their arrangement, and the method of adjusting test plot values can be varied. Kempton (1984), described the evaluation of F_5 lines wheat lines that were planted as drilled plots in a nursery incorporating check plots of two varieties, every sixth column of plots being occupied by two alternating check varieties. Checks may not exceed one in five unless heritability of test lines for the trait of interest is expected to be low. The same principle could be used for comparisons among individual plants. The check plot system has the disadvantage of requiring many additional plots; however, the system has advantages that there is no limit to the number of entries, only test plots of interest to be harvested, and replications are not required. While checks serve as a benchmark for selection decisions, control genotypes can be grown to ensure proper seeding and plot arrangement. A genotype with an established morphological trait or trait expression can be used as a control to ensure field planting accuracy as well as a gauge for important traits such as the presence of disease and days to maturity. This method is not commonly used, as it requires more resources, and also works reasonably well only in very uniform fields.

Grid method

The blocking principle proposed by Fisher (which is the best measure of local control) has been used by Gardner (1961), who suggested selection within grids as a means of reducing the effect of environmental variation. The contiguous control and the blocking principles serve the same purpose; that is, to ensure more comparable growing conditions by reducing the environmental effects. Though he used this method for the improvement of maize, it can also be used in self-pollinated crops. The procedure consists of dividing the field into grids of 40 plants each and selecting within each grid the four highest yielding plants corresponding to a 10% intensity of selection. This was to reduce environmental variation among the compared plants so that selection would be more effective. In this method, a fixed blocking principle of local control is used. An excellent linear response of 3% per generation from mass selection for grain yield was obtained.

Gridding is a simple way to control residual variation, and can be used in replicated or unreplicated nurseries and does not require repeated checks. The nursery is divided into a set of incomplete blocks. Selection is done within blocks. For example, see Fig. 26.3, where a nursery or unreplicated trial of 1000

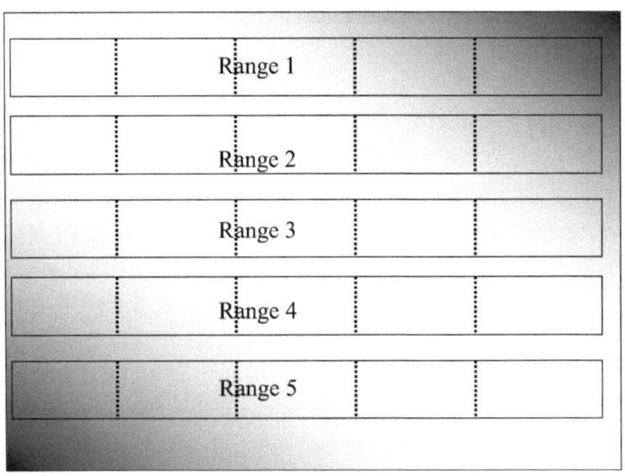

FIGURE 26.3

Example of a gridding layout of 1000 lines planted, divided into five ranges of 200 lines each.

lines may be planted in the field, divided into 5 ranges of 200 lines each. The field has a gradient, such that top right and bottom left have more water while the lighter shaded has moisture deficit. If breeder does not impose any sort of blocking or gridding on the field and attempt to select the best 20% of the lines on the basis of phenotypic acceptability, she/he will select mainly from the ends of ranges one and five, which have ample water (but not water logging) providing higher yield. To avoid this problem, breeder can divide each range into five smaller, more homogeneous blocks of 40 plots each (represented by the dotted lines in Fig. 26.3) and select the best eight lines in each block. Breeder will select the best 20% (or any other selection intensity) mainly on the basis of differences in genotypic value rather than water availability.

Honeycomb method

The honeycomb designs were developed by Fasoulas during 1970s in Greece for testing individual plant yield. The field layout of numerous and genetically distinct individual plants in the honeycomb designs is hexagonal; and every plant, like the one marked by the arrow, is surrounded by plants forming concentric hexagons of various sizes (Fig. 26.4). Plants are selected only if they outyield all other plants within the moving hexagon including checks (solid circles). The size of the chosen hexagon, that is, the number of plants within hexagon determines the intensity of selection. Thus, one plant out of 7, 19, and 37 plants in the three hexagons (in ascending order) give corresponding intensities of selection 14.3, 5.3, and 2.7%. Missing hills present no problems as it would not affect the efficiency of selection due to many plants included in the moving hexagon. The honeycomb method utilizes the moving block principle (see Moving average method) for local control. The progress through selection increases, as the intensity of selection increases.

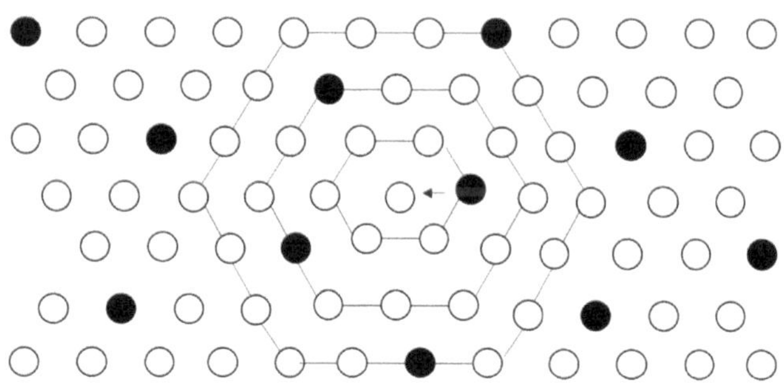

FIGURE 26.4

The field layout of the honeycomb method (Fasoulas, 1973). The test plants are represented by open circles and checks by solid circles. The arrow represents the candidate variety.

Moving average method

An alternative to using check plots is to use a completely randomized layout without controls. Richey (1924) suggested a method based on the regression of yield on a moving average as means of corrected soil heterogeneity. At the simplest level, the mean of a certain number of test plots on either side of a central test plot is calculated, and the central yield test is expressed relative to, or as a percentage of the mean of those test plots. The method is then applied to the next and to successive plots; providing a set of adjusted test plot values, adjusted for soil and/or environment variation in the immediate vicinity of each test plot. The method has the advantage that effort and trial space are not expended on check plots, but has the disadvantage that the average genetic background with which the individual test plots are being compared will not be the same; therefore, breeders try to create test with similar genetic background and days to maturity (e.g., one or more breeding populations with common and/or similar parents). While these stratifications can help minimize, they do not eliminate soil heterogeneity and genetic variation, and these together may contribute to the adjustments made. In order to provide the moving mean (MM) information, all plots must be harvested; although it is possible that majority of these plots were discard on trait information collected during the growing season. Also, all plots are harvested to come up with better spatial adjustments. Generally, at least eight of neighboring plots should be used in determining the moving average.

The performance of a genotype is expressed as a difference between it and the adjacent control plots or moving averages or as percentage of the control plots or moving averages. One of the assumptions required for such adjustment is the existence of a high correlation (>0.50) between check and test plots, or moving averages provide an inadequate representation of environmental differences (Mak et al., 1978). To preclude such error, analysis of covariance was suggested in which the check plots or moving averages are used as the independent covariate. The regression of the dependent variable on the independent variable after elimination of replication and entry variation is linear, so the change in yield is represented by the linear regression coefficient, b. Wherever there is no association with the yield of neighboring plots, the covariance method will give b value of zero and over, and adjustments will then be avoided. The moving average method has advantages that the number of entries is not limited, replication is not necessary, and one does not have to devote 25%–50% of the land to check cultivars as it is done in the check plot method.

Spatial analysis or covariance adjustment

Variance is a measure of the dispersion from the mean (single variable), while covariance is the measure of change of two variables together (i.e., strength of correlation between two variables). Analysis of covariance (ANCOVA) is a statistical technique that combines the methods of the analysis of variance (ANOVA) and regression analysis, also called as "ANOVA with covariates." Both blocking and ANCOVA can be used to control the experimental error as they are complementary to each other. Blocking can be used only to cope with sources of variation that are known or predictable (*a priori*; known prior to starting the experiment). In contrast, ANCOVA can handle unpredictable sources of variation that occur during the experiment (*a posteriori*) so long as there is one covariate, or more, that represents the heterogeneity between plots and is quantitatively measurable. Thus, ANCOVA is a useful supplementary procedure to further reduce errors that cannot be removed by blocking.

Spatial analysis is not an experimental design. In traditional analysis, it is assumed that entries within a block are experiencing minimal spatial heterogeneity (e.g., soil fertility, effect of crop rotation, previous seasons test layout, factorial tests, and cultivator effect). This is generally not always true: RCB, IBD may not always be effective in reducing the experimental error when blocking cannot cope with plot-to-plot variation (within blocks) that are generated by competition between varieties, soil variation and fertility, or weather-related conditions such as rainfall and snow cover.

Spatial analysis exploits the fact that neighboring experimental plots in the field should be more alike (i.e., similar) one another than distant plots. In nearest neighbor adjustment (NNA), plot performance is adjusted for spatial variability by using information from the immediate neighboring plots (i.e., use information from the neighboring plots in order to reduce or remove the undesirable effect of heterogeneity and hence improve the estimation of model parameters). Important condition for spatial analysis is that treatments (i.e., genotypes) must be similar (e.g., material from the same population or similar pedigree, adaptation, maturities, *et cetera*). If experiment is planted as an RCB or IB, covariance adjustment [NNA or MM] can be used by breeding programs (especially, if there is a lot of spatial variation not accounted by blocking) to generate the plot yield prior to analysis as an RCB. In the example described, experimental layout (Fig. 26.5) and moving means analysis are presented (Table 26.1). Researchers use different

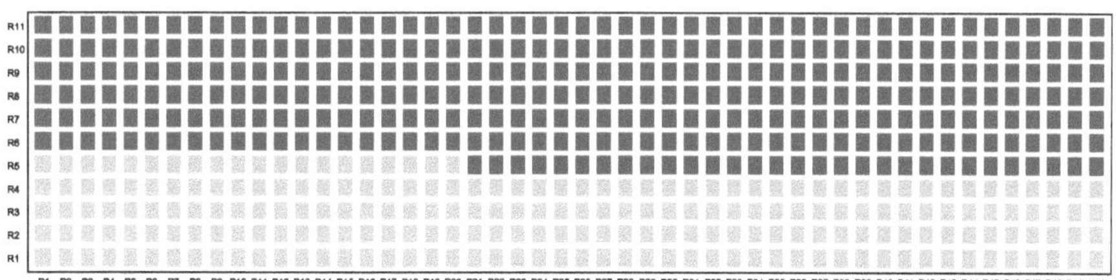

FIGURE 26.5

Layout of two tests with 220 plots (including experimental entries and checks) (light shaded) and 330 plots (including experimental entries and checks) (darker shaded) in field. There are 11 ranges and 50 passes for the example of two tests. Each test consisted of entries with similar maturity and from the same breeding population, although the tests can have multiple populations as long as they have similar maturities and genetic background.

Table 26.1 Calculation of moving means adjustment in a hypothetical test with 220 plots with no replication of breeding lines.

Plot	Genotype ID	Plot yield (g)	NN mean (g)	NN plot mean calculation	Adjusted yield (g)	Calculation of adjusted plot yield	Yield trend (g)	Calculation of yield trend	Adjusted yield (% over the mean of all checks)	Calculation of adjusted yield reflected as a % of the mean of all checks
1	Geno#84	1142.3								
2	Geno#2	1275.5	1121.2	Calculated as average of six nearest neighbors = (1142.3, 1155.7, 1085.9, 1126.1, 1066.1, 1151)	1303.3	Calculated as: [(plot yield − NN mean) + test grand mean] = [(1275.5 − 1121.2) + 1149.0]	− 27.8	Calculated as: (plot yield − ayield) = (1275.5 − 1303.3)	119.1	Calculated as: (ayield/mean of the checks) = 1303.3/1093.0
3	Geno#112	1155.7								
4	Geno#202	1085.9								
5	Geno#102	1126.1								
6	Geno#168	1066.1								
7	Geno#203	1151.0								
8	Geno#23	1140.9								
–										
–										
–										
220	Geno#39	1280.8								
Average plot yield (all entries), that is, test grand mean		1149.0								
Mean of the checks		1093.0								

Note: In this example, an unreplicated test with 220 plots (11 checks, 2 parents, 207 experimental entries were tested at one location. NN calculations are shown for plot #2, but calculations are done for all entries based on six nearest neighbors (e.g., for plot#1), the NN mean will include plot yields of #2 to 7 (six plots). This is done for all plots.

numbers of nearest neighbors to include in the adjustment as too few or too large a number of neighbors will bias the adjusted plot yield. As previously stated, it is important that entries in a test are not too dissimilar. The main steps are: (1) calculate nearest neighbor (NN) yield, (2) calculate adjusted plot yield, (3) calculate yield trend, and (4) calculate adjusted yield as a % of check mean.

Augmented designs and its variations

Augmented designs (ADs) were proposed by Federer (1956, 1960, 1961) for situations where entries are not replicated, but controls (or check lines) are available to account for field error. He proposed that a replicated blocked design can be used for control, and blocks can be augmented with unreplicated entries. Therefore, AD have incomplete blocks, but experimental entries are not replicated; and it uses incomplete blocks (IBs) to remove some field variation from the plot residuals. In an AD, a large set of experimental lines are divided into small IB (against one or two directions of field variation). In each IB, a set of controls are included; therefore, every check occurs in each incomplete block but the experimental lines are unreplicated. Design is used to compare unreplicated genotypes to each other and/or with controls (also known as checks). Because the design is unreplicated, the repeated checks are used to estimate the error mean square, and sometimes the block effect (subplot error; by including a few additional controls in each incomplete block, i.e., modified AD). The block effect is estimated from the repeated check means, and then removed from the means of the test varieties. This reduces error and increases precision. Block effects could also be estimated from the means of the test varieties in each block or by replication of few entries (partial-rep design).

Modified AD is a derivative of systematic control design. In some IBs, there is a systematic placement of check with more than one control. AD has advantages over unreplicated nurseries in which block effects are estimated without repeated checks (gridding) due to blocking. AD provides flexibility in # of entries, # of controls, # blocks, #entries/blocks, which are important considerations for a plant breeder. The inclusion of repeated checks allows breeders to generate an estimate of standard error, allowing comparison of experimental entries (genotypes) for selection as well as experimental genotypes versus check cultivar(s).

However, there are few disadvantages with AD. Repeated checks add a substantial number of plots to the trial leading to less space for experimental entries (e.g., in a 4×4 incomplete block with one control = 1/16 = 6.25% fewer entries). Due to the design layout, relatively few degrees of freedom are available for experimental error, which reduces the power to detect differences among treatments.

An example of plot layout of an AD is shown in Fig. 26.6. In this design, blocking can be done in one direction. Information on the specific presented example,

- 2 checks ($c = 2$).
- 30 blocks ($r = 30$).

1->	Border row	Border row	Border row	Border row	Border row	Border row	Border row	Border row	Border row	Border row	1->
2->	1	67	133	199	265	331	397	463	529	595	2->
1 3->	Check#1	68	Check#1	200	Check#2	332	398	464	530	596	3->
4->	3	69	135	201	267	333	399	Check#2	531	597	4->
4<-	4	70	136	202	Check#1	334	400	466	532	598	4<-
3<-	5	Check#1	137	Check#2	269	335	401	467	533	599	3<-
2 2<-	6	72	138	204	270	336	402	468	534	600	2<-
1<-	7	73	139	205	271	337	403	469	535	601	1<-
1->	8	74	140	206	272	Check#2	404	470	536	602	1->
2->	9	75	141	207	273	339	405	471	537	603	2->
3 3->	10	76	142	208	274	340	406	472	Check#1	604	3->
4->	11	77	143	209	275	341	407	473	539	Check#1	4->
4<-	12	78	144	210	276	342	408	474	Check#2	606	4<-
3<-	13	79	Check#2	211	277	343	409	475	541	607	3<-
4 2<-	14	80	146	212	278	Check#1	410	476	542	608	2<-
1<-	15	Check#2	147	213	279	345	411	477	543	609	1<-
1->	16	82	148	214	280	346	Check#2	478	544	610	1->
2->	Check#2	83	149	215	281	347	413	479	545	Check#2	2->
5 3->	18	84	150	216	282	348	414	480	546	612	3->
4->	19	85	151	Check#1	283	349	415	481	547	613	4->
4<-	20	86	152	218	284	350	416	482	548	614	4<-
3<-	21	87	153	219	285	351	Check#1	483	549	615	3<-
6 2<-	22	88	154	220	286	352	418	Check#1	550	616	2<-
1<-	23	89	155	221	Check#1	353	419	485	551	617	1<-
1->	24	90	156	222	288	354	420	486	552	618	1->
2->	25	91	157	223	289	355	421	487	553	619	2->
7 3->	26	92	158	224	290	356	422	488	554	620	3->
4->	27	93	159	225	291	357	Check#2	489	Check#1	Check#1	4->
4<-	28	94	160	226	292	358	424	490	556	622	4<-
3<-	29	95	Check#2	227	293	359	425	491	557	623	3<-
8 2<-	30	96	162	228	294	360	426	492	558	624	2<-
1<-	31	97	163	229	Check#2	Check#1	427	493	559	625	1<-
1->	32	Check#1	164	230	296	362	Check#1	494	560	626	1->
2->	33	99	165	231	297	363	429	495	Check#2	627	2->
9 3->	34	100	166	232	298	Check#2	430	496	562	Check#2	3->
4->	35	101	Check#1	233	299	365	431	Check#2	563	629	4->
4<-	36	102	168	Check#1	300	366	432	498	564	630	4<-
3<-	37	103	169	235	301	367	433	499	565	631	3<-
10 2<-	38	104	170	236	302	368	434	Check#1	566	632	2<-
1<-	39	105	171	237	303	369	435	501	567	633	1<-
1->	Check#1	106	172	238	304	370	436	502	568	634	1->
2->	41	107	173	239	305	371	437	503	569	635	2->
11 3->	Check#2	108	174	Check#2	306	372	438	504	570	636	3->
4->	43	109	175	241	307	373	439	505	571	637	4->
4<-	44	Check#2	176	242	308	374	440	506	572	638	4<-
3<-	45	111	Check#2	243	309	375	441	507	573	639	3<-
12 2<-	Check#2	Check#2	178	244	310	376	Check#1	Check#2	Check#1	640	2<-
1<-	47	113	179	245	311	377	443	509	Check#2	641	1<-
1->	48	114	180	246	Check#1	378	444	510	576	642	1->
2->	49	115	181	247	313	379	445	511	577	643	2->
13 3->	50	116	182	248	314	380	446	512	578	644	3->
4->	51	117	183	249	315	381	447	513	579	Check#2	4->
4<-	52	118	184	250	316	382	448	514	580	Check#1	4<-
3<-	53	119	185	251	317	383	449	515	581	647	3<-
14 2<-	54	120	186	252	318	384	450	516	582	648	2<-
1<-	55	121	187	253	319	385	451	517	583	649	1<-
1->	56	122	188	254	320	386	452	518	584	650	1->
2->	57	123	189	255	321	387	Check#2	519	585	651	2->
15 3->	58	124	190	256	322	388	454	520	586	652	3->
4->	59	125	191	257	Check#2	389	455	521	587	653	4->
4<-	60	126	192	258	324	390	456	522	588	654	4<-
3<-	Check#1	127	193	259	325	391	457	523	589	655	3<-
16 2<-	62	128	194	Check#1	326	392	458	524	590	656	2<-
1<-	63	129	Check#1	Check#2	327	Check#1	459	Check#1	591	657	1<-
1->	64	Check#1	196	262	328	394	460	526	592	658	1->
2->	65	131	197	263	329	395	461	527	593	659	2->
17 3->	66	132	198	264	330	Check#2	462	528	594	660	3->
4->	Border row	Border row	Border row	Border row	Border row	Border row	Border row	Border row	Border row	Border row	4->

FIGURE 26.6

Example of plot layout of an augmented design.

- Error degrees of freedom = $(r - 1)(c - 1) = 29$ (Experimental error is estimated by treating the checks as if they were treatments in a RCBD).
- 600 new entries ($n = 600$).
- Number of plots = $n + r \times c = 600 + 30 \times 2 = 660$.

- Plots per block = 660/30 = 22; (c = 2, n = 20), if plots per block is not a whole number based on the formula, blocks can be unbalanced (e.g., the last block can have fewer or more entries).
  Number of replications of a check should be the square root of the number of new entries in the trial or higher (since fewer replication will reduce precision; reduces the power to detect differences among treatments).

In this example, a four-row planter was used but for planting single rows. Each plot, except checks (i.e., control), is an entry (which is unreplicated) as shown by the number in each cell. Check #1 and #2 are present once (assigned randomly) in each incomplete block. A plant breeder can use one or more of these repeated checks; however, the use of more than one check can provide a broader scope of inference for the cut-offs. Repeating the same check more than once in the same block can also be used to develop error variance for the among and within block. However, a plant breeder must decide whether to use the resources to include more checks or experimental entries. Generally, in an AD, entries will not be statistically significantly different from the checks; therefore, researchers will make selection using a cut-off and retain a certain percent of entries.

P-rep designs

The partial-replicated or p-rep designs involve the use of replicated plots for a percentage "p" of the test lines; therefore, called as "p-rep" designs (Cullis et al., 2006). These designs have incomplete block, but a small proportion of entries are replicated, and its variations are called as the partial-replicated design. The partial-replicated design is akin to AD, except portion of experimental entries are replicated in each trial providing connectivity, genotypic relevance, and an ability to identify genotypes meeting the criteria. The p-rep designs are built on an AD layout but the plots of standard varieties (also sometimes referred as controls) are replaced by additional plots of test lines. The number of replicated plots is calculated as a percentage of the test lines (p), and this is the basis of name "p-rep" design. In the basic version, control plots can be replaced by the test lines that need to be replicated. Other variations include the use of few check lines and a percentage of test lines to be replicated. If *a priori* information is available, the choice of test lines to be replicated can be fixed; however, the choice of test lines to be replicated can be made based on seed availability or a breeding objective (e.g., seed or plant morphological trait). If a $p = 20\%$ is used, a test of 200 entries may have 40 entries replicated twice (total plots = 80), while the remainder 160 lines are not replicated. These can be a part of early generation testing, where seed quantity is low, but breeder has an interest to make selection across a range of environments. In the above hypothetical example, the subset of lines can be tested at four different locations (i.e., environments): ten lines will be replicated at each of the four locations. If there are "n" locations, breeder can replicate $(1/n)^{th}$ of entries at a location so that each entry is tested with two replicates in one of the n locations, giving a wider inference space. Williams et al. (2014) presented improvements in the method for generation of p-rep designs based on α-arrays that allows for a much broader class of designs to be constructed, thereby providing more flexibility to breeders in its usage.

Williams et al. (2011) proposed augmented p-rep design, which is a combination of augmented- and p-rep designs. In the augmented p-rep design, controls are replaced with partially replicated entries. At each location, breeder identifies the subset of entries that are partially replicated on two plots each and determines a resolvable incomplete block design with two replicates for this subset. Therefore, the blocks of this design are augmented by the unreplicated entries. In their work, Williams et al. (2011) showed the advantages of this design over the either of the two designs that it is built on.

CHAPTER

Molecular tools in crop improvement and cultivar development 27

Abstract

Molecular marker and biotechnology techniques are an integral part of plant breeding methodologies in major crops. Worldwide, in major crops, adequate molecular and biotechnology tools that have been developed to assist breeders to make selection decisions, trait characterizations, and genetic manipulation. The motivation for these tools was driven by the need to remove or minimize extensive phenotyping, which could be costly or time consuming. Molecular tools covered in this chapter include molecular markers, marker assisted selection, marker assisted backcrossing, and marker assisted recurrent selection. Other molecular marker driven tools covered in this chapter include genome wide association studies and genomic selection. The two main biotechnology tools described in the chapter are plant transformation to create genetically modified organisms, and genome editing.

In the earlier stages of molecular markers, restriction fragment length polymorphism, random amplified polymorphic DNA, amplified fragment length polymorphism, and simple sequence repeat (SSR) or microsatellite, were common in plant breeding applications. However, now single-nucleotide polymorphism (SNP) is the preferred marker platform, if available in a crop. While initially SSR were preferred due their multi-allelic and codominant nature, as well as the ease of genotyping; these assays were not high throughput and some crops had limited molecular diversity for such repeats. The move toward SNP platforms is due their high-throughput nature, genome abundance, automation, and cost efficiency (number of markers per unit price). With the development of next-generation sequencing, genomes of multiple crops were rapidly sequenced in a high throughput low-cost manner, setting the stage for a marker revolution in plant breeding application. With these advances, plant breeders moved away from one trait marker assisted selection (MAS) to multiple trait selection in a high-throughput manner especially through genomic selection (GS).

The earlier stages of marker technology involved genetic diversity studies and genetic mapping of monogenic traits, and later involved quantitative trait loci (QTL)

Plant Breeding and Cultivar Development. DOI: https://doi.org/10.1016/B978-0-12-817563-7.00014-3

mapping. Linkage disequilibrium (LD) between markers and causal variants is needed to detect QTL. The earliest QTL mapping methods were driven by geneticists and used F_2 or $F_{2:3}$, recombinant inbred lines, doubled haploid (DH) lines, and backcross generations obtained from biparental crosses. Marker trait linkages were computed, and marker in high LD with gene or QTL of interest was considered as potentially useful for making MAS. Moreover, geneticists and breeders would validate the usefulness of marker(s) in multiple populations and ensure a tight linkage to ensure false positives (marker present, but gene or QTL of interest is not) are minimized. However, these approaches invariably were limited due to non-closely linked or non-gene specific marker, mainly due to sparse coverage on the genome. More recently, sequencing based markers give a dense coverage, so tightly (or even gene specific) linked markers can be identified, thereby helping breeders to effectively use MAS. In this chapter, some of the main molecular marker tools available to plant breeders, and examples of their usage are provided.

Identification of molecular markers linked to gene or Quantitative Trait Loci of interest

While in some references, the utility of QTL and MAS of QTL is questioned, breeders have successfully used markers linked to QTL for plant breeding decisions and outcomes. The main steps in the development of molecular markers for breeding applications are described in Fig. 27.1.

For a wider use of QTL in MAS, robust and repeatable markers are needed. Ideally, the marker is in the gene itself (causal or non-causal); that is, perfect marker. A perfect or diagnostic marker has no recombination with the gene or QTL of interest, and is on the gene sequence itself. This type of marker has a perfect probability of selection under MAS. Independent mapping populations need to be used for the linked (even perfect) markers' validation. The availability of diagnostic markers that are useable in multiple genetic backgrounds is useful for a breeding perspective. Breeders strive to ensure that relevant germplasm in the program is subjected to marker validation, prior to its usage. Even if the marker was developed in a different lab, breeder has to check that marker is working in their program's germplasm pool by performing phenotyping and marker assessment on population(s) that are representative of germplasm in the breeding program. Instead of a single linked marker, flanking markers (marker linked on both side of the gene) can be used for an increased certainty that the QTL region has been inherited in the progeny.

The major breeding use of molecular markers for targeting gene(s) or QTL has been to select plants (in pedigree, bulk, single seed descent, DH, and backcross method) or families (in pedigree and backcross method) in breeding populations using MAS. Assays can be multiplexed where possible, to increase the inference scope for traits and genes and/or QTL per trait. In MAS, allele associated with

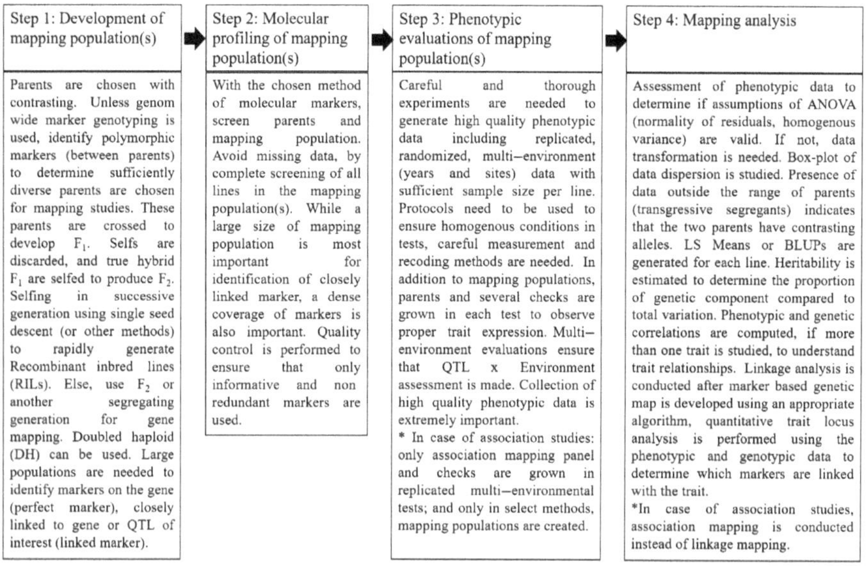

FIGURE 27.1

An overview of identification of molecular markers linked to gene or QTL of trait of interest (i.e., linkage mapping). In case of association studies, a large panel of diverse genotypes is assembled and association mapping is conducted. *QTL*, Quantitative Trait Loci.

positive outcome are selected and genetic material with the undesirable allele can be discarded (if progenies have homozygous alleles) or can be used for gene enrichment (if homozygous and heterozygous progenies are present). The MAS technique is especially desirable where phenotyping is expensive or time consuming. Difficulty in phenotyping can be due to complexity in establishment of specialized nurseries for trait selection in field or greenhouse (e.g., disease and insect–pest resistance). Other scenarios can include quantitative traits with lower heritability; therefore, breeder may have an objective to select for QTL in progeny in earlier generation, followed by more thorough phenotypic evaluation (more replications, locations) in later generations when fewer genotypes (but having the gene or QTL present) are remaining. MAS is also useful when phenotypic evaluation is expensive and/or time consuming; for example, milling yield in wheat. Other advantages of markers can be in situations where trait expression happens at maturity, but the use of MAS can reduce the number of progenies to test at the end of season. These may include traits such as days to maturity and seed traits. The ability to reduce population size is advantageous: (1) ability to work with larger starting population sizes per breeding population (see Chapter 4: Primer on Population and Quantitative Genetics), (2) cost savings, as undesirable plants can be discarded at seed or seedling stage, and (3) time saving, as more rapid cycling can be practiced in a backcross breeding scheme.

It is important to note that markers do not change the reproductive biology and type of cultivar developed; therefore, markers do not necessarily save time to develop cultivars. For example, in a pure line variety, marker do not reduce the time to develop a variety in a forward breeding scheme (using pedigree, bulk, SSD, DH methods, or their variants) but only in a backcross scheme (see Chapter 13: Backcross Method) as markers can remove the need for an extra season if progeny testing is needed. Other advantages of MAS are deciphering trait control, which can help strategize breeding scheme and methods. Markers are also very effective in helping identifying recombinants, for example, if two genes (controlling same traits or different traits) are linked in repulsion. In this situation, a large breeding population size with dense coverage can help identify progeny with recombination, wherein it contains positive allele for the two linked genes. Examples of repulsion linkage are common, for example, disease resistance loci in wheat. Molecular markers are also useful in case pleiotropy is present (see Chapter 4: Primer on Population and Quantitative Genetics); particularly if selection for one gene is insufficient for trait phenotype, which is dependent on interaction of more than one gene to get the desired trait expression (Singh et al., 2013a,b; 2014).

While QTL have been successfully used for the improvement of many crops, such as in rice, maize, wheat, soybean, there are several challenges that need to be considered. One of the major limitations of QTL targeting MAS is that in the absence of common marker set among breeding groups and institutions, markers identified in one lab may not be universally effective. This can happen due to various reasons; difference in recombination frequencies, genetic background, lack of polymorphism, marker robustness, etc. Furthermore, identification of usable markers linked to QTL can be time consuming and expensive; consequently, a coordinated effort is beneficial for a wider applicability across different backgrounds. While for perfect markers, MAS can be simpler; MAS for QTL is prone to errors as the selection is for a region not specific gene. Inadvertently, undesirable genes may be selected. However, this can be overcome to some extent by the use of tight flanking markers (i.e., closely linked markers on both 3′ and 5′ ends). Haplotype mapping is important, where set of alleles at linked loci are present on the chromosome and are inherited together. It is important to generate marker profile to cover all haplotype blocks [Haplotype is the combination of alleles in a single gene, or alleles across multiple genes, or SNPs that are not in the gene region]. These haplotype blocks create the framework for genome wide association mapping. For example, the international HapMap project developed a haplotype map of the human genome, and haploblock structure allowed the identification and focus on a subset of SNPs that describe most of the trait variation.

Genome wide association studies

Association mapping was proposed as a complementary strategy to QTL linkage mapping, and is used to identify associations between genotype (or genotype classes) and phenotype. In the most basic statistical analysis, once phenotypic data is available on a panel of lines that have been genotyped with molecular markers (preferably, SNPs), a series of single-locus statistic tests are conducted. Each test independently investigates every SNP for association to the phenotype. When the studies span the entire genome (SNPs for a genome level association study), it is referred as genome wide association studies (GWAS).

GWAS are a powerful tool that overcomes the limitations of biparental populations, such as limited inference scope (i.e., only alleles present in the parents can be studied) and lower mapping resolution. Using GWAS, complex traits can be dissected at high mapping resolution. The association mapping approach is based on LD to detect correlation between genotype (or genotype class) and phenotype in a sample of unrelated individuals, while linkage mapping is based on physical linkage or recombination frequency in the mapping population of biparental cross. In contrast, association mapping uses a diverse population and high-density SNP; therefore, samples multiple alleles and creates maps at higher resolution. The extent of LD determines mapping resolution, and higher mapping resolution is obtained when LD declines rapidly with increasing physical distance (Fig. 27.2). LD is the non-random association of alleles at different loci, and it is calculated as the deviation of the observed frequency of a haplotype from its expected value in each population. Assuming allelic frequencies at locus 1 (P_A, P_a) and locus 2 (P_B, P_b), LD coefficient D (one measure of LD) can be computed for AB as:

$$D_{AB} = P_{AB} - P_A P_B$$

where, P_{AB} is gamete frequency, and P_A and P_B are allele frequencies. D equal to 0, indicates linkage equilibrium; while D not equal to 0, indicates linkage disequilibrium. LD is measured with r^2 (correlation between these loci), which is $r^2 = D^2_{AB}/P_A P_a P_B P_b$, where $P_a = 1 - P_A$, and $P_b = 1 - P_B$. Perfect LD is when $r^2 = 1$ (there are only two of the four possible haplotypes, AB, Ab, aB, and ab, and have the same allele frequencies).

In crops with rapid LD decay (allogamous crops), a denser marker coverage is needed compared to self-pollinating (autogamous crops) because these have higher recombination rates.

Another major consideration for GWAS is to account for population structure. GWAS utilize a large collection of diverse genotypes. While a large panel of genotypes are desired, these need to be sampled aiming to avoid redundancies or overlaps. For example, plant breeders tend to use lines from their program for GWAS, which is sub-optimal if genetic diversity is limited. It must be noted that population structure can cause spurious associations and declare false SNP to be significantly associated with the phenotype. This happens due to selection and coancestry. Therefore, population substructure is an important covariate that is accounted in the model to ensure high fidelity results. In response, software programs have been developed or accessible through R library to account for population stratification. Bayesian Information Criteria and other metrics are used to determine the need to involve

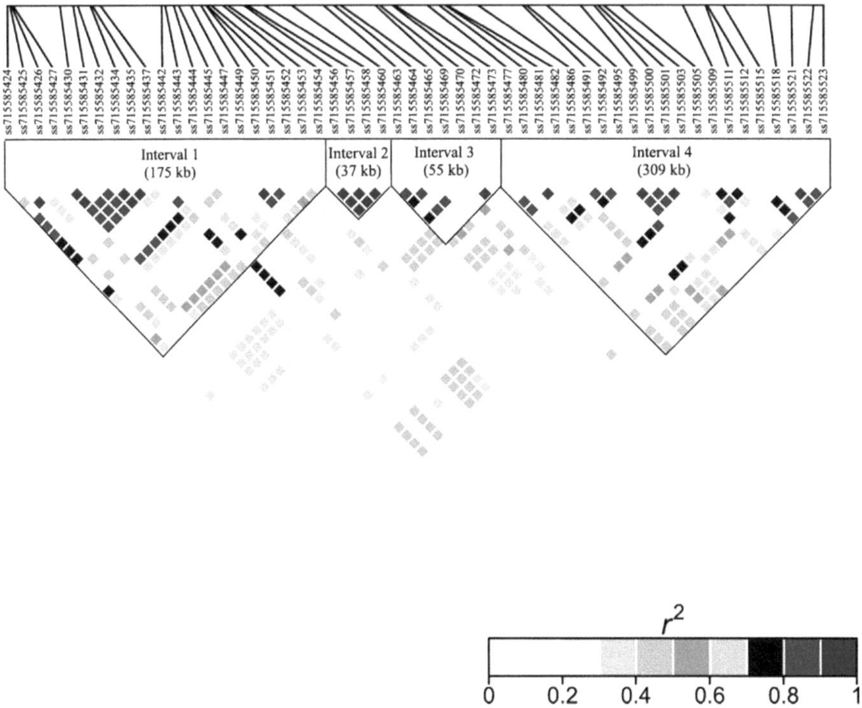

FIGURE 27.2

Example of linkage disequilibrium heat map of an iron deficiency chlorosis QTL region on Gm03 chromosome [linkage disequilibrium (r^2) of 57 SNPs within a 576 kb chromosomal region]. This shows the presence of four separate regions in a previously considered large single QTL controlling the trait. *QTL*, Quantitative Trait Loci.

From Assefa, T., Zhang, J., Chowda-Reddy, R.V., Lauter, A.N.M., Singh, A., O'Rourke, J.A., et al., 2020. Deconstructing the genetic architecture of iron deficiency chlorosis in soybean using genome wide approaches. BMC Plant Biol. 20 (1), 42.

population structure in the model. Principal component analysis and structure analysis are popular ways to quantify and adjust population stratification. There are programs that measure ancestry of each sample in the genotype panel using SNP data. Researchers can use these results to remove redundant genotypes, and also use principle components as covariates to adjust for ancestry effects (Fig. 27.3).

Depending on the genetic control of traits, i.e., qualitative versus quantitative, different methods for phenotypic trait analyses are used. The most common approach for quantitative traits uses a generalized linear model, similar to linear regression. The assumptions of ANOVA are implied (see Chapter 26: Field Plot Designs in Plant Breeding), and if these assumptions are not valid, data transformation may be needed. If the studies are focused on qualitative traits, logistic regression is the preferred method, while contingency table methods can also be

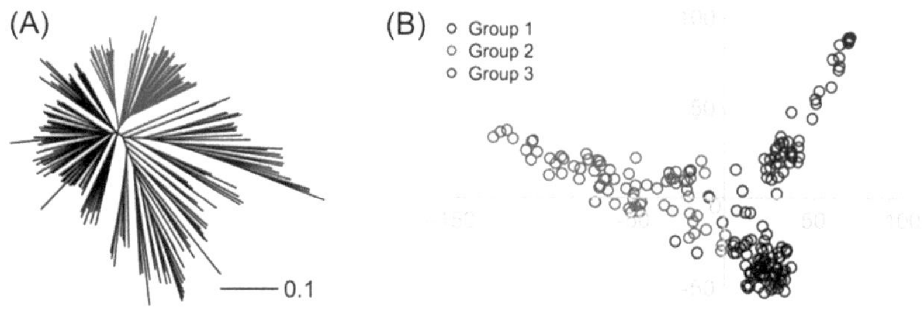

FIGURE 27.3

Population structure of a 214 genotypes soybean association panel used to study sudden death syndrome. (A) Neighbor-joining tree of the 214 soybean genotypes in GWAS panel. The three trees are in gray scales. (B) PCA plot—first two principal components of the panel including subgroups as defined by the neighbor-joining analysis. The PCA results were used to account for the population structure in this study, and the three subgroups were genotypes belonging to geographical subgroups (China, Japan, and South Korea). *GWAS*, Genome wide association studies; *PCA*, principal component analysis.

From Zhang, J., Singh, A., Mueller, D.S., Singh, A.K., 2015. Genome-wide association and epistasis studies unravel the genetic architecture of sudden death syndrome resistance in soybean. Plant J. 84 (6), 1124–1136.

used. These methods are most relevant if the phenotype is measured on a category scale, for example, disease rating on a 1–5 or similar scales. In this scenario, an ordinal logistic regression (measure of effect size can be produced by an adjusted odds ratios) with proportional odds assumption can be used to obtain best linear unbiased predictor (BLUP) for each accession prior to commencing genome wide association analysis. Logistic regression can also adjust for covariates (and other factors). Contingency table tests use chi-square tests (or Fisher's exact test) to test the null hypothesis that there is no phenotype–genotype association. Genotype-based association tests the association between single locus and the phenotype, while haplotype-based association tests the association between haploblock (or SNP block) and the phenotype. One of the major considerations in GWAS is to adjust for multiple tests (each SNP being tested for association).

Depending on the significant level used by a researcher (1% or 5% are common) as each SNP is tested, error rate (comparison wise and experiment wise error rate) increases. Therefore, researchers try to determine multiple testing correction threshold. The comparison wise error rate is the probability of a type I error which is set by the experimenter for evaluating each comparison, while experiment wise error rate is the probability of making at least one type I error when performing the whole set of comparisons. For GWAS, researcher has to balance both type I error (rejecting the null hypothesis when it is true) and power (probability that we make the right decision when the null hypothesis is incorrect, i.e., correctly rejecting the null hypothesis) of the test. The main methods to correct for these multiple tests are:

1. Bonferroni correction: it is generally the most conservative test. If the type I error rate of 1% is used, with Bonferroni correction type I error rate (type I error rate/number of independent markers tested) will be 0.0000002 if 50K markers (e.g., 50K SNP marker chip in soybean) are used. The assumption for Bonferroni correction is that each association test of the 50K marker is independent of all other tests; however, this is rarely the case because of high LD in self-pollinating crops and in most cross-pollinating crops.
2. False discovery rate (FDR): an estimate of the proportion of significant results that are false positives; therefore, FDR controls the expected proportion of false positives among the rejected null hypotheses. Hence, it is generally less conservative method compared with Bonferroni correction. One of the challenges with FDR method is that false negatives are identified if SNP are in high LD. FDR need to be carefully considered for crop species, for example, cross pollinating crops (with lower LD) should have more stringent FDR than self-pollinating crops (with higher LD).
3. Permutation test: it is a computationally intensive method but more straight forward approach to determine significant threshold. In studies with large number of markers, it becomes challenging to perform permutations tests, but approximation algorithms have been developed to ease computation. Permutation tests generate the empirical distribution of test statistics in a study dataset when the null hypothesis is true. The premise of the method is that genotype–phenotype relationship in the data is broken by randomly reassigning phenotypes of each individual to another individual. Researchers have used empirical significance level $P < .001$, determined by 1000 permutations (where $N = 1000$ is used with number of genotypes in the study).

The results of GWAS are most commonly presented as a Manhattan plot, where significant SNPs are declared based on the choice of cutoff method (Fig. 27.4).

Association studies lead to the identification of SNP associated with the trait in two ways: (1) direct - SNP is in the genic region; therefore, it can directly be responsible for the trait phenotype, and the SNP is referred as functional SNP; although SNP in the non-genic regions can also be functional and contribute to trait phenotpyic variation (Ashikari et al., 2005; Salvi et al., 2007), and (2) indirect - SNP is in a high LD with the gene or QTL, and this leads to the association. For a plant breeder, GWAS can be useful for the study of traits and genes with major effects. For traits controlled by quantitative inheritance, rarely are breeding targets (loci) identified that will have an immediate outcome and applicability in cultivar development due to their smaller effect on the phenotype. However, GWAS is still a useful tool for trait studies, and dissection of genetic control providing candidate genes, which requires extensive validation and in some situations (indirect SNP) fine mapping, before they can be used in a cultivar development

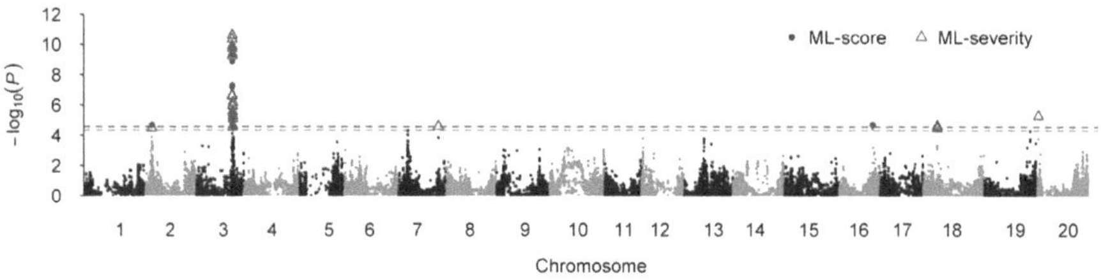

FIGURE 27.4

GWAS results presented as a Manhattan plot for IDC in soybean using computer vision and machine learning algorithms to obtain phenotypic (trait) raw data from image data. Negative log10-transformed *P* values from a genome wide scan for IDC resistance score (circle) and severity (triangle) by using a mixed linear model are plotted against positions on 20 chromosomes. The dash lines indicate the significance level of FDR = 0.05 for ML-score and ML-severity, respectively. The significant SNP are those which are greater than the FDR cutoff. The GWAS was done using ML generated phenotype abiotic stress score and severity. *FDR*, False discovery rate; *GWAS*, genome wide association studies; *IDC*, iron deficiency chlorosis; *ML*, machine learning; *SNP*, single-nucleotide polymorphism.

From Zhang, J., Naik, H.S., Assefa, T., Sarkar, S., Chowda-Reddy, R.V., Singh, A., et al., 2017. Computer vision and machine learning for robust phenotyping in genome-wide studies. Sci. Rep. 7, 44048. Available from: https://doi.org/10.1038/srep44048.

scheme (Fig. 27.5). As previously explained, functional validation of candidate genes is required to validate the findings, and phenotypic differences must be sufficiently large and consistent for application in cultivar development.

Marker assisted backcrossing

In Chapter 13, backcross method is described in detail. Marker assisted backcrossing (MABC, see Collard and Mackill, 2008) provides substantial advantages to backcross method, as phenotypic selection may not be needed or minimized when moving one or few genes from a donor parent to an elite background. If the gene to transfer is recessive (*aa*), the BC_1F_1 segregates as *Aa* and *aa*; therefore, progenies are selfed for one generation to determine the *Aa* versus *AA* types. The *AA* type is discarded before making the next backcross. With the application of molecular markers, this extra step has become redundant, and F_1 plants can be grown and DNA extracted from seed or young plant tissue to determine *Aa* and *AA* types. AA types can be removed, and crosses can be made with *Aa* types. Additionally, another advantage of markers in a backcross breeding program is if the gene to be moved comes from an unadapted or related species. In this scenario, the breeder has to be careful to avoid inadvertently bringing in undesirable genes linked to the desirable target gene (termed linkage drag). Very large population sizes are needed to be tested and phenotyped to identify recombinants, and even then in certain situations

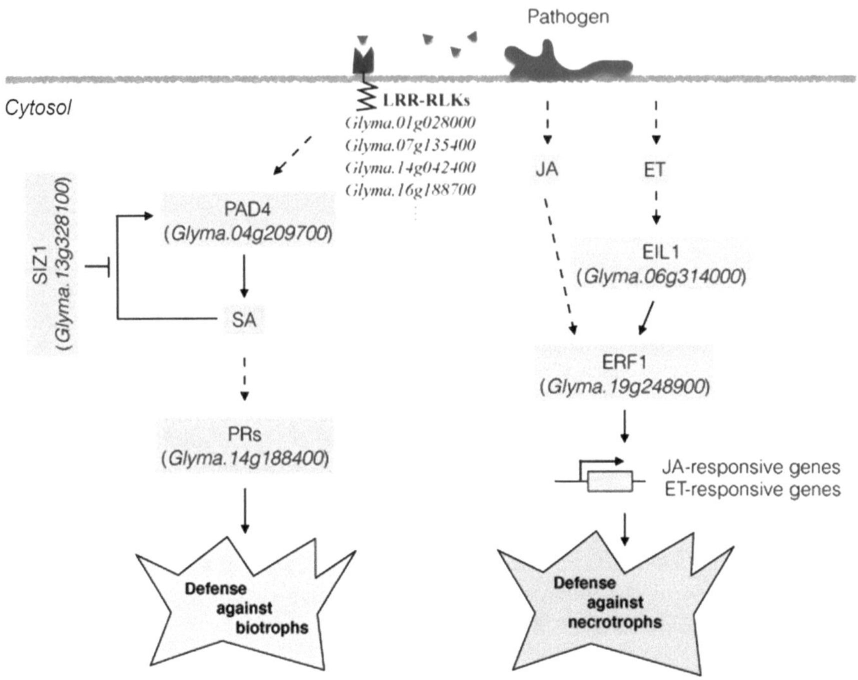

FIGURE 27.5

Example of a GWAS main and epistatic interactions on *Sclerotinia sclerotiorum* resistance in soybean. The study revealed multiple modes of resistance by identifying several candidate genes. The presence of candidate genes in salicylic acid, jasmonic acid, and ethylene pathways suggested a pathogen–host response that includes defenses against biotrophs and necrotrophs indicating a hemibiotrophic nature of the pathogen. *EIL1*, Ethylene-insensitive 3-like 1; *ERF1*, ethylene response factor 1; *ET*, ethylene; *GWAS*, genome wide association studies; *JA*, jasmonic acid; *LRR-RLKs*, leucine-rich repeat receptor-like protein kinases; *PAD4*, phytoalexin-dependent 4; *PRs*, pathogenesis-related proteins; *SA*, salicylic acid; *SIZ1*, small ubiquitin-related modifier E3 ligase (SUMO).

From Moellers, T.C., Singh, A., Zhang, J., Brungardt, J., Kabbage, M., Mueller, D.S., et al., 2017. Main and epistatic loci studies in soybean for Sclerotinia sclerotiorum resistance reveal multiple modes of resistance in multi-environments. Sci. Rep. 7 (1), 3554.

phenotype based selection does not overcome problems associated with the lack of ability to discern the effect of linked undesirable gene as its expression may not be sufficiently contrasting. MABC [and marker assisted recurrent selection (MARS)] reduces the need for large population sizes.

There are two main stages to MABC:

(1a) Foreground selection (stage I): markers may be used to screen for the target trait, which may be useful for traits that have laborious phenotypic

screening procedures or recessive alleles. BC_1F_1 plants are genotyped with markers tightly linked or causal variant to the target gene or QTL.
(1b) Recombinant selection: ideally a functional or causal variant marker (i.e., perfect marker) should be used; however, in the absence of such markers flanking markers can be used to avoid erroneous (allele) selection because of recombination between the marker and the target gene. This ensures that the target favorable allele is selected, and linkage drag is minimized. Furthermore, homozygous and heterozygous genotypes can be selected for gene enrichment; and reduces the problems of small population size, if only the favorable homozygotes are retained.

The reliability of selection (i.e., reliability of marker to predict phenotype, assuming no crossover interference; Collard and Mackill, 2008) is equal to (1 - recombination frequency between target locus and marker).

If flanking markers are available, the reliability of selection is

$$= 1 - (2 \times \textit{recombination frequency between the target locus and flanking marker A} \\ \times \textit{recombination frequency between the target locus and flanking marker B})$$

Therefore, if the flanking marker X on one side of gene A is 8 cM from the target locus, reliability of selection will be 92%, if the flanking marker Y on the other side of gene A is 5 cM from the target locus, reliability of selection will be 95%. However, if both flanking markers are available, the reliability of selection using both markers will be 99.2%. This shows the value of using flanking markers to increase the ability to accurately select the target locus.
(2) Background selection (stage II): selecting backcross progeny (that have already been selected in 1a and 1b for the target trait) with background markers. In other words, markers can be used to select against the donor genome to accelerate the recovery of the recurrent parent genome. This ensures identification of BC_1F_1's that carry maximum proportion of recurrent parent genome. This is accomplished by genotyping backcross progenies (selected in 1a and 1b) using genome wide markers. These two stages are carried out in each round of backcrossing until a final product is developed with the presence of favorable allele(s) in a higher proportion of recurrent parent, than will be achieved with no MABC approach in same number of backcrossing generations.

Hospital (2001) explored the relationship between backcross generations, donor segment length and population size, and showed that more number of backcrosses are beneficial to reduce the cost of backcross program and handle smaller donor segment length (cM). In comparison with no MAS, in only two backcross generations, markers located at $\sim$10 cM from the introgressed gene permits a reasonable reduction of donor segment length at affordable cost. However, if narrow donor segments are breeding outcome objectives, more generation of backcross and larger populations are needed. Background selection is of high importance in both scenarios but fewer number of backcrosses are used compared to non-background selection method. More backcross generation can also lead to a reduced cost due to the need of smaller population sizes; however, it will lead to longer period to develop new breeding line or variety.

Marker assisted recurrent selection

Recurrent selection (see Chapter 16: Population Improvement, and Chapter 17: Recurrent Selection in Self-pollinated Crops) is a cyclic breeding system for the improvement of complex traits, where superior genotypes are identified and intermated to produce the next generation. In a broader view, each breeding program runs a recurrent selection program because the cycle starts with the choice (picking) of parents that are hybridized to produce segregating progenies leading to superior (than checks) pure lines or inbreds or clones, which are then crossed to start the next cycle. However, it should be noted that recurrent selection in cross pollinated crops has shorter time duration per cycle. One of the most important objectives of recurrent selection is to focus on a particular quantitative character(s) in a population, without a marked loss of genetic variability. Consequently, recurrent selection provides a means of cyclical selection through which favorable genes are increased in improved population (new population) and it is superior to the original population (base or source population) in mean values for the trait(s), and genetic variability is maintained. This contrasts to a pure line, inbred, DH, or clone cultivar development program, where the entire breeding program is seen as a population and crossing is between a selected few parents at the beginning of each cycle (contrasting with intermating), although in both cases genetic variability is attempted to be expanded with continual infusion and maintenance of favorable genetic variability.

In MARS scheme, a large number of favorable alleles are accumulated using QTL-linked markers to identify desirable genotypes for use as parents in crossing in every selection cycle. Therefore, compared with a phenotype-based approach, favorable QTL alleles from diverse sources are accumulated more rapidly. An earlier example of MARS in wheat was presented by Ribaut et al. (2010) (also see Fig. 27.6). They reported that using MARS scheme, breeders can utilize QTL information generated on their populations of interest (or using from other programs as long as these QTL are validated in own program) to develop superior lines with an optimum combination of favorable alleles originating from both parents and including multiple traits. The three main stages are: (1) population development, (2) recombination, and (3) line development. If markers linked to QTL are unavailable, an additional step between (1) and (2) is followed for QTL detection. Population development can include multiple breeding populations.

MARS method aims to rapidly accumulate favorable alleles from several genomic regions within a single population; and therefore, as a prerequisite, targeted genes or QTL need to have fully validated (fine mapped and tested in

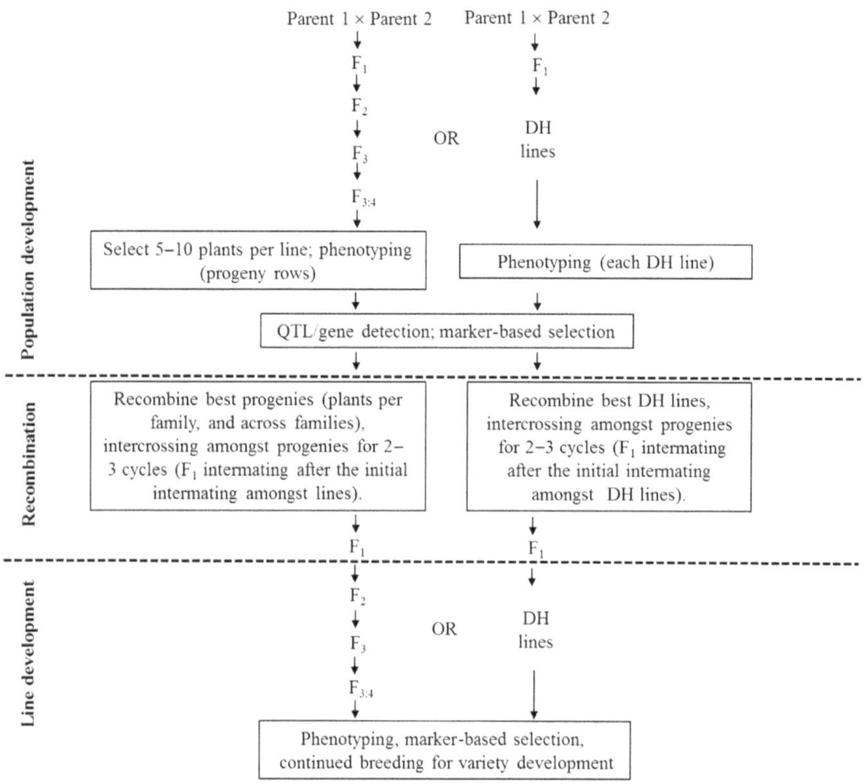

FIGURE 27.6

An example of MARS scheme in non-DH (left) and DH (right) systems. The three main stages of MARS are population development, recombination, and line development. "QTL detection" will involve either marker development and then selection or MAS with previously identified markers. In population development, number of plants per line will depend on number of traits and QTL or gene targeted for selection; with larger family size needed if higher number of QTL and genes are involved. In the population development stage, if markers are already available, breeder can utilize complex crosses or work with multiple populations rather than a single population. Phenotyping is done (multi-location) to obtain trait data for marker identification and development. If cultivar development is the main intent, phenotyping in population development step allows breeders' visual selections. The MARS scheme ensures that all favorable QTL (and/or genes) can be combined in a single line. In recombination, intermating can be done sequentially amongst lines or in pairwise non-sequential manner. Breeders attempt that lines are used in crossing scheme to ensure all QTL have equal chances of representation. *DH*, Doubled haploid; *MARS*, marker assisted recurrent selection; *QTL*, Quantitative Trait Loci.

different genetic backgrounds). This focus to move multiple genes or QTL, sets it apart from MABC where the aim is to move one or two genes (and generally from non-elite background) into an elite variety. Both require linked or gene specific markers, but the scale is different. Similar to MABC, breeders concurrently may have multiple MARS populations, since each MARS population have similar or completely dissimilar objectives and target traits. In Chapter 4: Primer on Population and Quantitative Genetics, the complexity of working with multiple traits was presented. The chance to accumulate favorable genotypic class decreases progressively with more number of genes; and with 10 genes, desirable genotype quotient (DGQ) was 0.0563 in F_2. If we assume favorable allele frequency of 0.5, the probability of recovering favorable genotype with all ten favorable alleles will be 0.000977 ($=0.5^{10}$), which is one in thousand. If pursuing MARS, the favorable allele frequency can be increased at each locus, and if this frequency is 0.90 at each locus (which is feasible with MAS), the probability of recovering favorable genotype with all ten favorable alleles will be 0.349 ($=0.9^{10}$), which is one in three. This is a significant advantage for a breeding program working on building a multiple trait package, as an increased allele frequency at each locus will lead to positive increase in the mean performance of the population for the selected traits. It is important to note that if allele frequency equals 0.5 (say either A or a), genotype frequency with the desirable allele is 0.75 (Aa, AA; or Aa, aa), which means that genetic enrichment can happen at an even higher percentage.

Eathington et al. (2007) compared a set of lines in a population derived after one year of MARS method against lines selected through conventional breeding schemes from the same population. Variables were considered as uniform as possible, to avoid any confounding results. For North American and European corn breeding programs, 248 breeding populations were compared for MARS and conventional approaches. They reported that MARS-derived lines performed better than conventionally selected lines. In soybean, comparison across 43 breeding populations showed a higher seed yield using MARS than the conventional method. On the flip side, when different selection models were compared in a study including 23 maize breeding populations from eight different breeding programs, it was noted that multiple trait index performed poorly compared with a grain yield model. This is not unexpected, because the multiple trait index model used dissimilar weights for seed yield in comparison to the yield alone model (yield model had 62% more weight on yield traits compared with the multiple trait index model). Also, MARS scheme works best for loci with larger effects and selection for yield QTL will be more challenging in this scheme. However, utilization of MARS scheme can allow breeding pipeline activities as well as identification of marker trait association and linkages. The other major limitation of MARS is the extensive cross-specific genotyping required to identify progeny for recombination, and genotyping is limited to few QTL. Genomic selection offers better prospects to combine multiple QTL and traits, compared with MARS.

A *breeder SNP chip*, a multiplexed chip, that is built to include 20–50 or more markers that a breeder is most interested in their program can be a good solution for this problem. Additionally, if a breeder SNP chip is to be constructed, it is advisable to include specific markers that target important QTL and gene, and also include a good genome coverage for their potential utilization in genome wide selection (and in this scenario, 200–500 SNP markers will be absolute minimum, and number of SNP markers used in a chip will be crop and its genome dependent).

Genomic selection

Genome wide prediction and selection is the current state of the art for molecular marker applications in cultivar development. Meuwissen et al. (2001) first proposed methods that could predict phenotype based on genotype information and termed it genomic selection, i.e., "GS." GS enables the estimation of link between the marker (genotype) and trait (phenotype), which is built on two groups of individuals: (1) training or reference population (all individuals are genotyped and phenotyped), and (2) breeding population (all individuals are genotyped but not phenotyped). After the marker effects are estimated, these can be deployed in the population to predict breeding values (see Chapter 4: Primer on Population and Quantitative Genetics, for breeding value explanation). The reference population should represent breeding programs genetic diversity using a high-density and -throughput marker system. In plant breeding, selection of important traits is based on economically important traits, which are generally quantitatively inherited. Estimated breeding values, based on phenotypic data can be calculated by BLUP. A high correlation coefficient is observed between true breeding value and estimated breeding values with the usage of genomic BLUP (gBLUP; VanRaden, 2008; Meuwissen et al., 2001). The BLUPs reflect a higher selection accuracy when selecting for marker based breeding value estimates. In GS, genome wide markers are used in genetic evaluation so that all QTL (of all expression sizes) are in LD with at least one molecular marker (commonly used marker technique is SNP), to estimate genome estimated breeding value (GEBV). The GEBV are estimates of breeding value that are based on genome wide dense molecular markers (for each individual GEBV is the summation of marker effects). This contrasts with conventional methods where selections are based on phenotypic values, and under some methods varying degrees of family pedigrees. The most common method is gBLUP, although Bayesian and least squares based methods are also used. gBLUP differs from traditional pedigree BLUP as a genomic relationship matrix is used in place of pedigree relationship matrix, and utilizes information from LD, co-segregation, and additive genetic relationships (Habier et al., 2013).

GS has quickly gained wide applicability in major crops and particularly in large private and public organizations that are capable to handle large populations and genotype data. For a successful GS pipeline, i.e., GS model development;

phenotyping component needs to be robust and done in environments representative of the target area of adaptation. Traits that are phenotyped include those considered difficult or expensive to measure. The GS strategy is most suitable for implementation in large and well-funded programs with in-house molecular marker capabilities and the ability to obtain marker data and analytics in narrow timelines. This genotyping and phenotyping requirement comes from the optimal balance of cost and outcomes. An easier to phenotype trait can be adequately improved without the need of GS. Perhaps the most important advantage of GS is a more rapid cycling of breeding lines as parent in the program, which significantly and positively impacts genetic gain, if done properly and accurately. Additionally, GS can provide an estimate of breeding value before they reach the growth stage when phenotype is expressed and measured; even in seed (using its DNA) before putting it in the ground to complete the seed cycle. This affords a huge advantage to the breeder because they can work with large populations to combine multiple traits and genes, but then only grow a select group that are predicted by GS to have the performance meeting trait cutoffs.

The GS method allows capturing QTL with small effects better than any other genotype based selection method described earlier (MAS, MABC, and MARS), and can substantially reduce the phenotyping footprint. Overall, GS brings economic advantages to the program because resources can be optimally invested in growing and phenotyping individuals that will be better performer. However, for effectiveness of GS and its application in a breeding program, breeders need to ensure that training or reference population is sufficiently large and includes a sample of the entire program's genetic diversity. This is important because model fidelity can be compromised if inaccurate association between the genotype and phenotypes are estimated. Also, the same problem persists if the training set and breeding population are genetically dissimilar.

GS is a cyclic process, and the GS models needs to be continually updated so that previously unseen individuals are included in the training or reference population including new parents (from within and outside organization) and new lines that are developed. Breeders strive to retain a portion of discarded individuals so that model is able to better differentiate between keeps and discards. This is applicable for all traits of interest. This updating of model is also driven to account for recombination and mutation that can bias the estimate of association between the SNP and genes that governs the phenotype. Seed yield is a major trait in row crops; and because it is a resource intensive phenotype to collect, it remains as one of the major target to predict the field performance of a genotype (prior to actual field testing). However, other traits are also important and GS models can be developed for those as well and concurrently several traits can be selected.

The main steps in the GS model development are presented in Fig. 27.7.

1. *Assemble the original population* (training-validation set): assemble the original training-validation set, which will be used to develop GS models. The original population is composed of genotypes covering all of the important parents, checks

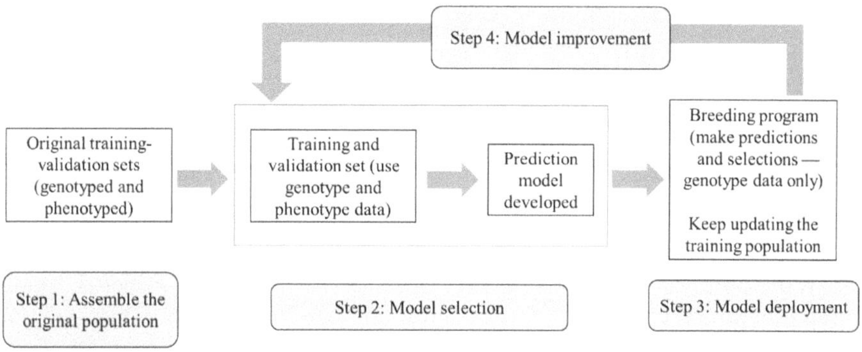

FIGURE 27.7

Genomic selection steps for cultivar development.

and experimental genotypes of the breeding program (it is advisable to include experimental lines from a complementary breeding program if their genetics is regularly or semi-regularly used in one's program). The entire training-validation set is genotyped with SNP markers and phenotyped for traits of interest (for which genomic prediction model needs to be developed). Generally, GS is used on complex traits with lower heritability (for simpler and higher heritability traits, MAS or MARS may be more effective in forward breeding).

2. *Model selection*: develop a GS "prediction model" that integrates the genotypic and phenotypic data from the training population. The original population consists of the training (model generation), validation (model validation) and may include testing (model testing) sets. Some programs may not include the testing set, as training and validation sets can help in model development and testing its accuracy. However, some programs may decide to keep a testing set, which is composed of lines that are never used in training and validation sets, to give an additional model check (for accuracy) prior to the deployment in the program. The relative sizes of these sets can be 80:20 for the training and validation set, respectively, for a fivefold cross validation. Alternatively 80:10:10 for the training, validation and testing sets, respectively, for *k*-fold cross validation and the ability to test on an independent set. The model generates a GEBV for genotypes for which genotypic information is available. GEBVs represent the overall value of a line to be used as a potential parent for crossing and/or for selection as a superior genotype (to advance for further testing). GEBV informs the breeder on the expectation of field performance (for the trait of interest) without field testing.
3. *Model deployment*: after the accuracy of the GS model is deemed satisfactory, it can be implemented in the breeding program. Breeder can make predictions and selection in the breeding program on individuals that have only been genotyped and not phenotyped.

4. *Model improvement*: in the breeding program, after selection with GS is done in early stages (filial generations), in later generations phenotyping is also done. Since the line is breeding true (or is a doubled haploid line), and now has field phenotypic data and marker genotypic data, this information is used in model improvement. Update the training population (in start of step 2), and continually repeat steps 2–4. This step is essential to ensure that a robust model is built on the current genetic material.

Since the method was first proposed by Meuwissen et al. (2001), with further research efforts, GS models have become more robust with better handle on optimum marker density, trait heritability, genetic relationship (kinship and population structure), training population size, and the effective population size of the breeding program in determining prediction accuracy. Researchers have identified problems of mixed phenotypes (binary, ordinal, continuous) that are common in plant breeding efforts, and explored deep learning genomic-based prediction model for multiple traits (Montesinos-López et al., 2018). Their research explored comparison of univariate models, which precludes exploitation of trait correlations, and multiple trait deep learning with mixed phenotypes models; and reported modest gains in continuous traits in the multiple trait deep learning model while little improvements were noted for binary and ordinal traits. Other researchers have noted that deep learning approaches were not better for genomic prediction except if non-additive variance was large (Abdollahi-Arpanahi et al., 2020). Gradient boosting was reported to be better than parametric models when additive, dominance, and epistatic interactions were present. With continual advancements and refinements of machine learning methods geared towards genomics data, it will be interested to determine the usefulness of such methods on GS.

The effectiveness of GS models is determined by prediction accuracy, which is the correlation between predicted genotypic value and the underlying true genotypic value. The prediction accuracy is a function of the training population size, marker density, and trait heritability (on entry mean basis). Researchers have also reported the usefulness of identify by descent from marker data to allow for a rapid identification of lines carrying specific alleles, which can increase the accuracy of genomic relatedness and diversity estimates, improving genomic prediction (Shook et al. 2020). The prediction accuracy generally increases with increased training population size and more markers; however, in structured population, increasing the number of markers or the number of individuals in training population may not increase prediction accuracy. Therefore, when plant breeders are looking to apply GS, they start-off by assembling the training set, determine training population size, required marker density in their crop (i.e., number of markers) and then calculate GEBV. Subsequently, in addition to prediction accuracy, there is a need to obtain high predictive ability, which is the correlation between predicted breeding values and the observed phenotypic values (because true breeding values are generally unknown in real plant breeding datasets). Robust models need to have high predictability (correlation between phenotype and GEBV) and high accuracy (correlation between genotypic value and GEBV). Also, plant breeding cycle involves multiple, usually competing, objectives; and genomic breeding strategies and framework using multi-objective optimization principles have been proposed to help in these situations (Akdemir et al. 2018).

Most of the successes of GS have been demonstrated in diploid organisms. For polyploids, there are current challenges in genome assembly, mapping of sequence reads, genotyping for multiple alleles and heterozygotes (particularly in autopolyploids), distinguishing homologues (from same subgenome) and homoeologues (from different subgenomes), linkage mapping and association studies, and GS. With advances in genotyping technologies and bioinformatics, some of these challenges will likely be addressed. One of the proposed solution (Bevan et al., 2017) includes sequencing populations of progenitors and wild relatives to help identify ancestral haplotigs (haplotype-specific contigs). The markers defining haplotypes can be used in breeding program enabling GS.

There are numerous ways GS methods are advancing the breeding pipelines: (1) reducing cycle time as potential parents are identified earlier than with phenotypic method alone, (2) increased selection intensity as lines can be selected with more accuracy if a robust GS model is available, (3) genetic variance in the breeding program and in breeding families can be maintained or increased, and (4) with the ability to discard earlier, larger population sizes can be carried, if desired, without additional costs. There are other applications including genomic prediction models for genebank scans to identify useful accessions for the identification of useful sources of resistance (Peixoto et al., 2017).

Other examples of molecular marker application in plant breeding

While common methods (MAS, MABC, MARS, GWAS, and GS) have been explained previously, other application of molecular markers includes usage in plant variety protection, calculation of genetic similarity, study of heterotic pools, and cytoplasmic male sterility system for hybrid production using MABC. The Plant Variety Protection Office of the US Department of Agriculture (USDA) allows the use of biochemical and molecular markers to describe the variety. Similarly, the Canadian Food Inspection Agency allows the use of molecular markers for distinguishing new variety with other varieties. Researchers routinely use molecular markers to study genetic diversity; for example, if they are interested in exploring genetic similarity to determine suitability for use as a parent to enhance genetic variability. However, genetic diversity study should be accompanied with phenotypic evaluation so wide variability is not confused with low or undesirable performance characteristics. Similar to studies on genetic diversity, heterotic pools can be investigated and/or established with the use of molecular markers. White et al. (2020) studied ex-PVP inbreds, and by combining genotypic and phenotypic data were able to generate information on the usefulness of these inbreds as a breeding resource and enhance the understanding of heterotic grouping within North American proprietary maize germplasm.

Practical consideration of marker applications in a breeding program

Molecular markers are an invaluable tool for a plant breeder with multifaceted applications. Plant breeders continually assess the role of molecular markers and the problems this technology can solve in the breeding pipeline. One of the challenges that breeding programs face is the timeliness or data turnaround in the breeding pipeline. For example, if there is a need for a rapid turnaround between field harvest and planting in contra-season or greenhouse, breeders need to have a well-established marker lab (centralized or outsourced), so that the marker screening results are timely available. These data need to come in an easily interpretable manner, and integrated with phenotypic and other data. Also, regular controls must be used in each run (replicated and randomized) to ensure that marker results are correct. Careful cost-benefit analysis must be conducted so that resources are optimally invested in marker technology and its use. It will be beneficial to involve principles (and experts) of *operations research* (OR) to help make better decisions that are optimal or near-optimal solutions for complex decision-making problems (Cameron et al., 2017). The OR principles are not just important for marker applications but for multiple facets of the program. Careful implementation of OR methods can help improve economic or profit genetic gain; because ultimately genetic gain in isolation is sub-optimal, and higher profitability is the driver for the farmers and breeding programs. For farming operations, this may come from better usage of genetics and resources and marketing decisions, while for a plant breeding program it (higher economic or profit genetic gain) implies that more competitive cultivars can be developed with fewer resources and costs.

Plant transformation for crop improvement

In major world crops used for non-food and oil applications, plant transformation technology has been used to make crop improvements. Plant transformation can be used to make transgenic plants, that is, insert the DNA from another organism into the genome of a crop species of interest. This differs from a cis-transformation, where DNA (cis-genes) from the same species is inserted or moved (using transformation or traditional methods). The cis-transformation is not common in plant breeding approaches, where traditional breeding method can easily achieve the same outcome. In transformation, a transgene (inserted DNA) from another species is moved creating a transgenic plant or genetically modified organism (GMO). Transformation can be done in unicellular or multicellular organisms. Transformation is the primary method for large scale production on medically important compounds, such as insulin. To produce insulin for diabetics, the insulin gene from humans is inserted into a plasmid (a genetic structure in a cell that can replicate independently of the chromosomes, typically a small loop of bacterial DNA in the cytoplasm of a bacterium), which is returned to the bacteria and allowed to produce insulin in a fermentation tank from where it (i.e., insulin) is harvested. In 1982, the Food and Drug Administration (FDA) on the United States approved the use of human insulin making it the first GMO medical product. Other vaccines have been developed using GMO technology. GMO-derived chymosins are used for cheese production.

The realization of GMO crops came from a discovery that a natural gene transfer mechanism operated between *Agrobacterium tumefaciens* and host plants, where the bacterium injected DNA (from its plasmid) which carried genes that produced a tumor in the plant. Independent works of Marc Van Montagu and Jeff Schell, Mary Dell Chilton, and Robert T. Fraley was foundational for the GMO success. Van Montagu and Schell worked on crown gall disease in plants. In 1974, they were the first to discover that *A. tumefaciens* (soil microbe that induces plant tumor) carries a circular molecule of DNA called Ti plasmid, which is responsible for the formation of plant tumor. Subsequent work by them and independently by Chilton established that T-DNA, a segment of Ti plasmid defined by left and right border sequences, is copied and transferred into the genome of the infected plant cell where it becomes part of the host plant genome. Chilton and her team are credited for the development of first transgenic tobacco. Fraley and his colleagues developed protocols to insert a foreign gene into petunia and tobacco plants using *A. tumefaciens*. Interestingly, Van Montagu and his colleagues, Chilton and her colleagues and Fraley and his colleagues, independently presented their findings at the 1983 Miami Biochemistry Winter Symposium: Advances in Gene Technology. Van Montagu, Chilton and Fraley were awarded the 2013 World Food Prize, as they with their colleagues and teams heralded the agricultural biotechnology era with GMO.

In 1996, the first cultivation of staple transgenic crop was initiated with the successful introduction of genetically engineered soybeans that were resistant to the herbicide glyphosate (commercially called as Roundup). Before the availability of herbicide tolerance traits, farmers had to intensively manage weed control through herbicide tank mixes, and multiple applications. This was done to account for different weed plant species, and crop growth and development stages of both crop and weed species. However, with genetically modified crops with herbicide tolerant trait, farmers were able to spray their fields with glyphosate without killing their crop while eliminating all narrow- and broad-leaved weeds. It must be noted that eventual over reliance on limited herbicide chemistry has created herbicide tolerant weeds, which are increasingly becoming more difficult to chemically control. As a prevention, companies developed stacked (multiple herbicide chemistry) traits in new cultivars; however, it remains to be seen if this approach works in the long term as the fear of "super weed" persists.

Briefly, the stages in plant transformation for new transgene event (i.e., integration of foreign gene into a given locus) are: (1) delivery of the DNA into a single cell, and (2) regeneration into full fertile plants. While *Agrobacterium* is the main method for gene delivery in plants, particle guns are also used. For *Agrobacterium* mediated method, the tumor forming genes are replaced with the transgene (or cis-gene) of interest. Wounded plant tissue explants, for example; shoots, leaves, stem, etc., are inoculated with *Agrobacterium* that contain the genes of interest, facilitating the transfer of the gene of interest into the DNA of the plant tissue (the explant used). Thereafter, the plant tissue is cultured on media containing antibiotics (or another selection marker) that will kill plants except those with gene construct (with gene of interest and selective marker gene). In this artificial selection environment (e.g., antibiotics), only cells with gene construct (with transgene, i.e., foreign DNA) survive on the media (with the antibiotic, or another selection agent used in the gene construct). The explant is cultured on tissue media with plant hormones to induce plant regeneration forming callus. The calli are moved to another media first for shoot and then for root induction through hormones. Once a plantlet (regenerated plant) is ready, it is moved to soil media. If a single T-DNA locus is inserted, these seedlings will be hemizygous for the transgene in T_0 generation. Upon selfing these T_0 generation plants, the progeny will segregate in 1 homozygous for transgene: 2 heterozygous for transgene: 1 homozygous for null (i.e., negative for transgene). The final step in plant transformation is the confirmation of transgene event through assay of choice and testing of functionality. It must be noted that microspore regeneration will produce a homozygous outcome after chromosome doubling; however, majority of transformation are on in-planta transformation, floral dip and callus regeneration methods due to higher efficiency of transformation in these methods. After selfing (ensuring diploid and homozygosity) and identification on transgenic event, scientists assess number of copies that have been inserted, as well as the position of transgene to ensure that there is no yield drag or other unintended consequence of the transgenic event. Once all established criteria are met, line can be commercialized or used as a parental strain with necessary regulatory approvals.

In maize breeding, companies keep their inbred lines in native state (no-GMO) while development and testing (crossing with tester, then evaluating the hybrid) are on-going. Promising inbred lines are converted for GM trait using backcrossing, while testing is going on. Seed production is done on converted lines. It is important to realize that breeders do not transform each line; instead, they use backcrossing to convert lines prior to hybrid production. In a high value crop such as hybrid maize, it is more prudent to breed with non-GM material and convert lines before hybrid seed production. It keeps the gene pool clean and amenable to convert with any of the available gene event, depending on targeted market. It also enables quicker integration of new transgene, and getting rid of transgenes no longer desired or required by the breeding organization.

In self-pollinating crops, once a line is genetically modified using transformation, it can serve as a gene donor source (or set of lines in different maturity groups), and breeding is done with the transgene in the genome of the lines. That is, originally transformed lines can be used as a parent in generating breeding crosses and primarily as donors in backcross to convert elite lines into the GM trait line. As the breeding cycles advance, backcrossing is no longer needed as new parents (new lines developed each year) are already converted lines. The biggest challenges in self-pollinating crops (such as soybean) come when the breeding program decides to remove a current trait with a new transgene. In this scenario, few elite parents are converted by removing the old transgene and inserting new genes. However, breeding program needs to be cautious of genetic bottleneck if a very small portion of lines are converted and used in developing new breeding populations. A massive effort is undertaken each time a self-pollinating crop breeding program moves to a new transgene and removes previously marketed transgene.

The genetic engineering technology allowed the development of varieties with unique traits that did not exist in nature. While providing numerous advantages, it must be noted that the cost of development and commercializing a new GM event is an expensive endeavor with numerous regulatory checks and a lengthy time process in each country it needs to be released. When a transgene has been developed by an organization it undergoes extensive testing, and applications to obtain approvals from multiple agencies. For example, in the United States, approvals are required from the USDA, the Food and Drug Administration, and the Environmental Protection Agency before GMOs can be grown on a farmers' fields and made available to consumers.

In crop breeding, in addition to herbicide tolerance, the other traits include insect resistant (primarily through *Bt* gene), drought tolerance (e.g., maize), virus resistance (e.g., papaya). Within herbicide tolerance, several different types of herbicide tolerant GM varieties are available with single or stacked protection. In the United States, currently grown GM crops are cotton, sugar beet, soybean, maize, canola, papaya, alfalfa, and summer squash. The research on newer genes and traits is a continuous activity, as researchers try to address on-going production challenges or meet farmers and consumers requirements. These include increased abiotic and biotic stress tolerance, improved nutrition, removal of antinutritional factors, and many more. In addition to breeding applications, plant transformation has been used for research on gene function and related studies.

The main technical challenges of plant transformation-based genetic engineering are poor transformation efficiency in some crops, and non-random transgene insertion and varying copy number insertions, which can cause unintended and undesirable consequences including lower yield or negatively impact other traits. This lead researchers to explore better precision through genome editing (GE) technologies: (1) zinc finger nucleases, (2) meganucleases, (3) transcription activator-like effector nucleases, and (4) clustered regularly interspaced short palindromic repeats (CRISPR)/CRISPR-associated protein 9 (Cas9) or CRISPR from *Prevotella* and *Francisella* 1 (Cpf1), also known as the CRISPR/Cas9 and CRISPR/Cpf1 system, respectively. These technologies have enabled plant scientists to make targeted deletions, point mutations, and sequence insertion.

Genome editing technology for crop improvement

More recently, GE technology called as CRISPR/Cas9 system has rapidly gained acceptance as a more effective technology to change the DNA of an organism. This change can be addition, deletion or altering nucleotides at specific locations in the genome. The reason for the excitement about CRISPR-Cas9 (or similar systems) system for GE is lower cost, faster, and higher efficiency than other existing GE methods.

The CRISPR-Cas9 system is an outcome of independent work in CRISPR biology and genome engineering that merged with the work of J. Doudna and Emmanuelle Charpentier and their groups (Jinek et al., 2012). Their work was built on earlier work by numerous researchers who worked on genome editing and CRISPR biology leading to the 2012 seminal paper by Doudna and Charpentier groups (Jinek et al., 2012). They reported that microbial immunity mechanism can be utilized for programmable genome editing, and were also responsible for the biochemical characterization of guide RNA (gRNA) and Cas9 enzyme-mediated DNA cleavage. Their work was utilized to show the application of CRISPR in eukaryotic cells by Feng Zhang group (Cong et al., 2013) and other researchers the same year.

In nature, a naturally occurring GE system existed in bacteria as they try to protect themselves from invading viruses, where the bacteria captures portions of DNA from invading viruses to create DNA segments called CRISPR arrays or spacers. The CRISPR arrays serves as imprints for bacteria to "recall" the previous attack by the virus or its closely related types. In case, the virus attacks again, bacteria uses the CRISPR array to produce RNA segments to target virus DNA followed by the usage of Cas9 attack enzyme (that is produced by bacteria when attacked) to chop the DNA apart neutralizing the threat by disabling the virus.

In the native (bacteria) system: the Cas9 protein is guided by CRISPR RNA (i.e., crRNA) and a trans-activating CRISPR RNA (tracrRNA). crRNA contains a 20-nucleotide segment (defined by the spacer sequences) that gives target specificity, and tracrRNA stabilizes the structure and activates Cas9 to cleave the target DNA (protospacer). The presence of a protospacer-adjacent motif (PAM) is a prerequisite for DNA cleavage by Cas9, where PAM is generally an NGG sequence directly downstream from the target DNA. This is a three-component process, but the GE revolution came by converting this three component system into a simpler two step system in eukaryotic cells. This was achieved by combining the targeting specificity of crRNA with structural properties of tracrRNA, as a gRNA - a chimera created by fusing the 3′ end of the crRNA to the 5′ end of the tracrRNA (see Bortesi and Fischer, 2015).

The diagrammatic representation of the CRISPR/Cas9 system is shown in Fig. 27.8A–C (Bortesi and Fischer, 2015; Manghwar et al., 2019; Zhang et al.,

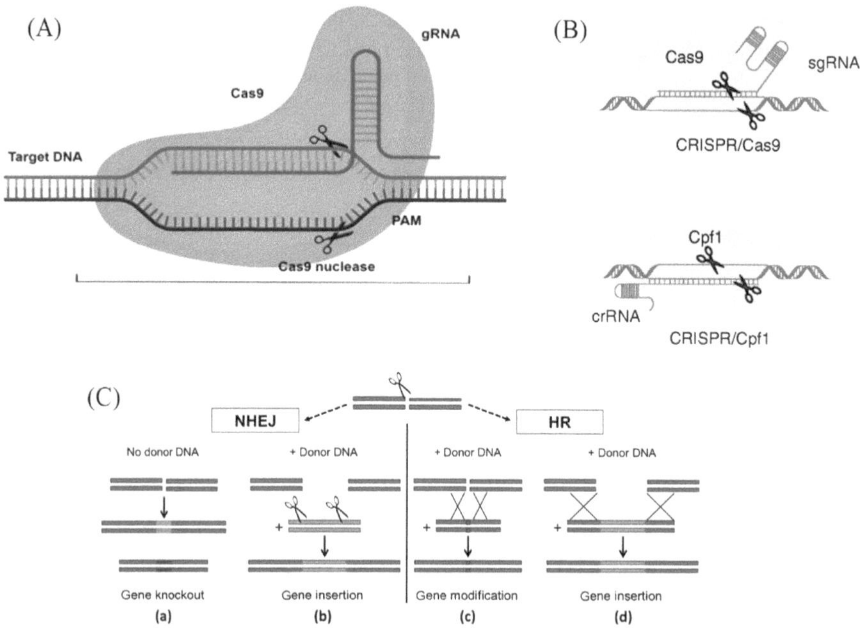

FIGURE 27.8

Diagrammatic representation of the CRISPR/Cas9 system and DNA repair mechanisms. (A) The crRNA guides the Cas9 protein. The tracrRNA stabilizes the structure and activates Cas9 to cleave the target DNA. The sgRNA with binding specificity through 20 bases, identifies the target gene and the Cas9 protein cleaves a double-stranded break in the target DNA, through the help of PAM sequence. (B) Cas9 and Cpf1 systems. (C) Double-stranded breaks are repaired by NHEJ or HR also known as HDR in CRISPR/Cas9-based gene editing. *Cas9*, CRISPR-associated protein 9; *CRISPR*, clustered regularly interspaced short palindromic repeats; *crRNA*, CRISPR RNA; *HDR*, homology directed repair; *HR*, homologous recombination; *NHEJ*, nonhomologous end joining; *PAM*, protospacer-adjacent motif; *sgRNA*, single-guide RNA; *tracrRNA*, trans-activating CRISPR RNA.

(A) From Manghwar, H., Lindsey, K., Zhang, X., Jin, S., 2019. CRISPR/Cas system: recent advances and future prospects for genome editing. Trends Plant Sci. 24 (12), 1102–1125. (B) From Zhang, Y., Massel, K., Godwin, I.D., Gao, C., 2018. Applications and potential of genome editing in crop improvement. Genome Biol. 19 (1), 210. Published correction appears in Genome Biol. 20 (1), 13. (C) From Bortesi, L., Fischer, R., 2015. The CRISPR/Cas9 system for plant genome editing and beyond. Biotechnol. Adv. 33 (1), 41–52.

2018). In plants, the steps for gene or DNA editing (Bortesi and Fischer, 2015) includes: (1) create a genetic sequence called *"guide" RNA* that matches the DNA sequence we want to modify (using guide sequence), (2) the guide sequence as a part of gRNA is added to a cell along with Cas9 protein, (3) the gRNA finds the target DNA sequence and Cas9 cleaves the DNA, and (4) once the DNA is cut, gRNA and Cas9 leave and cell's own DNA repair machinery add or delete nucleotide base pair or insert customized DNA sequence at the cut site. In other words, the *single guide RNA (sgRNA)* directs Cas9 to bind genomic DNA through 20-nucleotide sequence for gene specificity, and introduce a double-stranded break (DSB; Fig. 27.8A). While CRISPR/Cas9 is a more common used term; other proteins with higher efficiency and capabilities exist. For example, CRISPR from *Prevotella* and *Francisella* 1 (*Cpf1*) is more efficient than Cas9 as it requires only *crRNA* while *tracrRNA* is not needed (to cut DNA) (Fig. 27.8B). Also, Cpf1 is smaller sized than Cas9 with a smaller sgRNA molecule. Cpf1 cuts DNA at different places and has more choices than Cas9 (PAM sequence is NGG), as Cpf1 PAM is TTTN, where N can be A, C, and G. Therefore, genome editing options are broadened including in the AT-rich or T-rich regions, thereby more flexibility exists for editing sites compared with Cas9. Cpf1 creates sticky ends compared with blunt ends created by Cas9; therefore, more efficiency and specificity is possible to incorporate new sequences.

There are two main mechanisms for DNA repair (Fig. 27.8C): (1) nonhomologous end joining (NHEJ) or (2) homology directed repair (HDR). In most eukaryotic cells, NHEJ pathway generates insertions and deletions during DSB repair, and repair by NHEJ are generally more error prone and usually results in insertion (if the donor DNA is available, which is simultaneously cut and has compatible overhangs) or deletion (gene knock out). NHEJ-break ends can be ligated without a homologous template, but is more efficient repair mechanism and in native state more active in the cell than HDR. In the presence of a DNA template with homology to the sequences flanking the DSB location, HDR can repair the DSB in an error-free manner. Therefore, HDR with a donor DNA introduces precise nucleotide substitution, or donor DNA (gene) insertion. Although, in most cells both repair pathways are active, HDR is generally less efficient that NHEJ in the absence of a homologous template. Therefore, it is important to determine the optimal HDR conditions of crop-CRISPR system for better success.

There are exciting technological advancements underway in GE methods, and this is expected to continue in the future. For example, Anzalone et al. (2019) described prime editing, which is a GE method that directly writes new genetic information into a specified DNA site using a Cas9 endonuclease fused to a prime-editing guide RNA (pegRNA) allowing target site specification and encoding the desired edit. This system includes a prime editor, which consists of a Cas9 nickase fused to an optimized reverse transcriptase (RT), and two RNA molecules (pegRNA and sgRNA). The pegRNA drives the Cas9 nickase to a specific DNA position and acts as a donor and a template for the RT, and forms a single strand nick at a precise DNA position. The sgRNA forms a second nick on the opposite (non-edited) DNA strand to increase prime-editing efficiency. This allows for repair and insertion of the desired edit on both DNA strands and removes the need for DSB in current CRISPR systems. This new system works for small modifications, and is unable to create large insertions and deletions.

The success of CRISPR/Cas9 and CRISPR/Cpf1 have applicability for plant breeders in multiple ways; gene knockout (deletion), gene insertion, gene modification (substitution), gene knock-in, and site-specific mutation.

The following two methods are used for genome editing:

1. Transgenic methods—Cas9 and gRNA are cloned into a single plant transformation vector, and *Agrobacterium*-mediated transformation is done on the crop variety of choice. This inserts the GE components (Cas9 and gRNA) into the genome, creating a transgenic plant. Cas9 and gRNA expressing transgenic plants are identified using molecular analysis. To confirm GE success, sequencing is done and plants with successful GE in the target gene are identified. At this step, crossing followed by selfing/backcrossing and selection is done to segregate out the transgenic components (Cas9 and gRNA).
2. Alternatively, non-transgenic method can be used. These include guided nuclease based expression system, Cas9-encoding mRNA and gRNA, ribonucleoprotein complexes delivered via particle bombardment, transgene free protoplasts, virus-based expression vectors, transient expression-based genome editing system, etc. These methods have been developed to reduce off-target mutation and avoid transgene integration (allowing it to be non-transgenic). The biolistic delivery method (for plant transformation) of CRISPR/Cas9 ribonucleotide allows its rapid degradation reducing off-target mutation and provides a non-transgenic T_0 plant.

Researchers have used GE (CRISPR based) for numerous traits. Zhang et al. (2018) in their review covered several success by researchers - for example, canker resistance (grapefruit and oranges), bacterial disease resistance (rice and tomato), decreased polyunsaturated fatty acids (*Camelina sativa*), drought tolerance (maize), earlier maturity (tomato), yield component traits (rice), herbicide resistance in multiple crops (acetolactate synthase and 5-enolpyruvylshikimate-3-phosphate synthase), amylose and amylopectin content (maize, potato, and rice), grain weight and protein (wheat), powdery mildew resistance (tomato and wheat), and virus resistance (cucumber). There are numerous opportunities to use GE (CRISPR based) to improve plant traits: disease resistance, insect resistance, adaptation traits, nutritional traits, preharvest sprouting, drought tolerance, salt stress tolerance, cold tolerance, nutrient deficiency tolerance, herbicide tolerance, and quality traits. The list is long and continually expanding. Other possible targets include removing allergens from food crops such as peanuts. The applications can be endless, but from a plant breeding perspective there are three main advantages: (1) quicker and cheaper than traditional and transgenic approaches, (2) more precise in editing if target gene and genome sequence is known, and (3) ability to produce non-transgenic plants that have been gene edited.

In the United States, in 2018, the USDA decided to not regulate plants that could otherwise be developed through traditional breeding techniques provided they are not plant pests or developed using plant pests. This included GE approaches, where transgene material is bred out as explained earlier in this section. In 2020, this policy was updated and the US regulations now mandate that the oversight is on product rather than process. However, some countries regulate the process to create as well; therefore, in these countries, GE in plants will still be regulated and considered genetically modified. This implies that if GE is used to design a plant that can be created using conventional plant breeding approaches, that plant will be exempt from regulation including nucleotide and amino acid changes and deletions. This suggests that gene cassettes can be created if they include endogenous genes, paving way to assemble genetic information with more precision. However, it indicates that transgenes transfer or creating new genes interfering or affecting metabolic pathways will still need regulatory reviews and approvals. These exceptions have the potential to bring a rapid technical innovation with the usage of GE approaches. Genome sequencing and gene annotations are needed to identify more gene targets in multiple crop species.

CHAPTER

Phenomics and machine learning in crop improvement 28

Abstract

The advances in phenotyping, that is, phenomics, through the development of hardware and software solutions are bringing new opportunities in cultivar development. The phenotyping systems in field, including aerial and ground based semi- or fully-automated systems, and controlled environment growing units are all observing the infusion of sensors. Together, the combination of advanced sensors and data analytics has unlocked previously unattainable breeding approaches and information. These comprise high throughput phenotyping (HTP), sensors and phenotyping platforms, and machine learning (ML) methods for handling high dimensional data and large datasets. This chapter provides an overview of HTP, sensors, ML methods, deep learning and smart breeding, with an emphasis on the application of these tools and methods for cultivar development.

Visual phenotyping has been a cornerstone of plant breeding activities, since the beginning of this profession, as most tasks were done with minimal mechanization. Thereafter, with mechanization, new machines were created to generate more reliable plot data, which increased the footprint of breeding pipelines leading to crop improvement. These included tractors, planters, harvesters and combines, and similar other equipment that allowed more work to be done in shorter times. The next wave of technology came through computerization and its integration with various instruments and as stand-alone units. For example, the use of computers for statistical analysis of plant breeding data, scanners, barcodes, semiautomation in packaging and planting, global positioning system enabled planting and chemical applications, etc. With the advances in genomics-assisted breeding (Chapter 27: Molecular Tools in Crop Improvement and Cultivar Development), plant breeding and cultivar development saw significant success and increased genetic gain. More recently, the infusion of advanced sensors and data analytics has unlocked previously unimaginable outcomes. These include high throughput phenotyping (HTP) (sensors and platforms) and machine learning (ML) methods for both phenotyping and genotyping applications.

The advances in trait phenotyping come from a two-pronged approach: (1) ability to collect data previously considered untenable, difficult, or impossible,

Plant Breeding and Cultivar Development. DOI: https://doi.org/10.1016/B978-0-12-817563-7.00011-8

and (2) large data analytics through advances in computing power and use of ML algorithms. The traits may include economic yield, yield components, agronomic, morphological, anatomical, disease, insect, pests, other plant health parameters, abiotic factors including water stress (drought or flooding), salt tolerance, temperature (heat, cold, and freezing), and quality factors.

Each crop has its own unique set of targeted traits, although different crops may have quite dissimilar and/or differing numbered traits under breeding consideration. Due to programmatic constraints (time, costs), breeders are unable to phenotype traits with sufficient information depth, temporal frequency, and in a time series (i.e., temporal resolution) that are optimal; for example, plant growth and development stages, multiple disease ratings, yield components, etc. Additionally, trait ratings and evaluations require experienced researchers who spend an inordinate amount of time in breeding nurseries and plots. While it is essential for breeders and staff to spend as much time as possible in field to observe and collect meaningful information in breeding populations and trials; not all tasks are equally important or time efficient, nor can the same person be at multiple places at the same time. Furthermore, to increase the amount of genetic variation and to recover desirable recombinants, the number and size of breeding families needs to be increased, which are additional phenotyping constraints.

These factors lead to the proposition of HTP and its application in plant breeding. *HTP can be defined as the use of sensors in conjunction with aerial, ground, or controlled environment system to rapidly phenotype large number of plants with some level of automation*. Particularly, image-based phenotyping is generally integral to HTP applications. The choice of deploying HTP system is governed by balancing requirements - information gained, cost, ease of use, measurement accuracy, and correlation with the traits of interest. With appropriate data analytics, HTP can empower plant breeding programs to increase efficiency and success in cultivar development and research, and provide synchronous data collection on multiple phenotypes (and simultaneously on multiple genotypes). This enables monitoring and detection of dynamic events for important traits, such as growth stages. Some of the examples of HTP systems applications include plant phenotype data collection, time series data collection, large-scale data collection, timely data collection, large-scale genomic studies due to corresponding phenotype data, and selection decisions. HTP can reduce the operating costs in the long term, and it can help to reduce measurement errors while ensuring timely data collection. The use of HTP is a critical component of phenomics (see section "Phenomics" for details).

A breeding program should strive to minimize all errors in the programs. It includes many aspects, for example, in seed packaging, planting sequence, seed planting depth, plot length uniformity, weed-free fields, timely and accurate trait data collection, data transfer and storage, metadata, harvesting, data analysis and many more. Many of these can be handled with a well-trained and prepared work team, development of easy to follow and scientifically sound protocols, adherence to established and agreed upon protocols, and well maintained and serviced machines. Meanwhile, infusion of HTP and ML can significantly influence trait

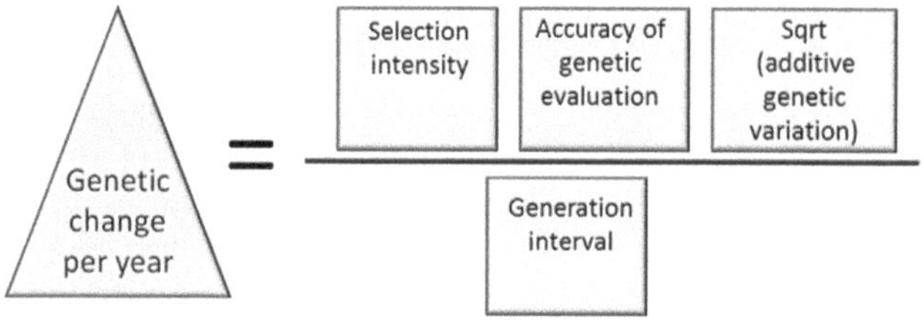

FIGURE 28.1

The genetic gain equation. Advances in phenotyping (e.g., high throughput phenotyping) and data analytics (e.g., machine learning) can assist in increasing all terms in the numerator, positively increasing the genetic gain.

data collection and data analytics, respectively; and allow working with increased population size, improve trait measurement accuracy, and better use of selection intensity leading to an improved response to selection (Fig. 28.1).

Phenomics

The word genotype and phenotype were introduced by Johannsen in the early 20th century (1909, 1911). These concepts came from Johannsen's work on pure line breeding experiments (1903). As per Johannsen, organism's phenotype is determined by its genotype through the process of development, under the influence of the environment. Phenotype refers to the observable physical properties of an organism. These include appearance, development, and behavior. Phenotype word came from Greek word "phaínein" with root Latin word "phaeno" and root Greek word "phaino", which means "to show", "to appear", "to shine", or "visible." The root Greek word "typos" and Latin word "typus", which means "a type", "a form", or "a kind."

A plant phenotype is the set of structural, physiological, and performance related traits, measured on a qualitative or quantitative basis, of a genotype in a given environment. The term "phenomics" is also motivated by the concept of phene. In plant sciences, in some instances researchers do not strictly refer to traits when referring to a phenotypic variable; however, they may use the term "trait" to define a characteristic such as a canopy shape or characteristic of vegetation (Pieruschka and Poorter, 2012). Therefore, the relationship between genetic information and phenotypic expression becomes a characteristic of a mixed number of individuals, with unclear relationship between the genotype and phenotype in heterogenous populations (different ecotypes, species). This motivated the description of "phene" as an alternative to trait, making "phene" a phenotype equivalent to what "gene" is to genotype. Subsequently, the "gene-phene" terminology implies a broader concept of gene–trait association without issues listed earlier with mixed populations. However, the usage of phene terminology is still not prevalent.

After the rapid advances post-genomics breakthrough, particularly in genome sequencing, researchers realized the limitations in phenotyping to match the genomics revolution. Houle et al. (2010) advocated for large-scale phenotyping as necessary to connect and understand genomic variants and how they affect phenotypes. They defined phenomics as the acquisition of high-dimensional phenotypic data on an organism-wide scale. Phenomics can significantly advance biological sciences, and in conjunction with technologies that enable HTP, is becoming mainstream. These include both platforms (field, controlled environment, ground and aerial systems) and sensors (digital, thermal, infrared, etc.). These advances have come with the matching and increased analytics capabilities to use high-dimensional data, particularly, through the use of ML methods.

Genomics and phenomics complement each other, and one can be limiting without the other in most cases, particularly because phenotypic data are generally the outcome traits, for example, yield, biotic and abiotic stress resistance, morphological, physiological, anatomical, and quality traits. These advances are particularly useful for genotype–phenotype mapping including linkage and association mapping, and establishment of causality and explanation at the phenotype level. However, phenomics leads to a data deluge situation as it collects data at both spatio-temporal scales at multiple organization levels in large genotype panels (Dhondt et al., 2013).

Phenome is the aggregate of all the expressed traits of an individual, and is a counterpart equivalent of genome that encompasses all genes present in an individual (Pieruschka and Poorter, 2012). Plant phenomics is the study of phenomes of multiple genomic expression states, in all possible environmental conditions, at multiple organizational levels (cell, tissue, organ, whole plant, canopy), in a dynamic time series (Dhondt et al., 2013). For a given genotype, its phenome consists of all possible phenotypes; hence, phenomics is the study of phenomes of multiple genotypes. Phenomics is used to represent the broader context of a higher dimensional, throughput, and resolution data. These aspects are accrued particularly in plant breeding because breeders handle large populations (in different filial generations, and stages of testing) on multiple traits at differing scales in a range of environments (Fig. 28.2). Furthermore, several traits require a time series data collection, for example, developmental stages, disease progression, etc., to parse out biological and physiological processes.

The need for resolution comes from breeder's interest to extract meaningful trait features from sensors. When referring to spatial resolution, these may imply a higher number of pixels per cm^2 in image based phenotyping, and also refer to the need to have separation of scales (tissue, organ, plant, and canopy) to enable better parsing of plant or canopy parts in field trials. The temporal resolution implies the length of time between phenotyping (data acquisition), for example imaging, of the same field and/or plant organ or parts. Therefore, when referring to temporal resolution it also means the time needed to acquire an image

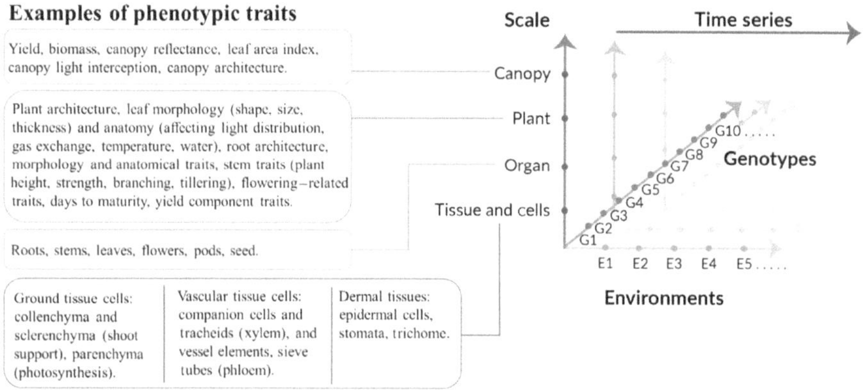

FIGURE 28.2

Phenomics representing higher dimensional, throughput, and resolution data collected in plant breeding programs, which measure traits at multiple organization scales, use large population sizes (filial generations, and stages of testing) that are tested in a number of environments. Regularly, time series data collection is needed. A high throughput phenotyping system (with sensors, and ground and aerial based phenotyping) is integral to the phenomics application in plant breeding.

(frequency of imaging and data acquisition or another data type) with sufficient depth to collect important information that can be valuable for selection in plant breeding. For example, temporal resolution for reflectance data may imply the number of time points to collect these data using a sensor without missing the desired biological state (growth and development stages), or disease progression measured through image-based phenotyping without missing pertinent time point (s). Overall, the complexity of types and streams of sensors, spatial and temporal data, multitude of genotypes, and numerous environments create large dimensional data, which requires special data handling methods, particularly ML.

In this chapter, field-based phenotyping and associated sensors are described; however, general principles for field and lab have common elements.

Phenotyping systems

The two main types of field-based phenotyping systems based on their deployment are: (1) aerial based, and (2) ground based (Fig. 28.3).

Aerial based field phenotyping systems

The aerial-based phenotyping systems include Unmanned Aerial Vehicles (UAV) or drones. The UAVs differs from Unmanned Aerial System (UAS), as UAV

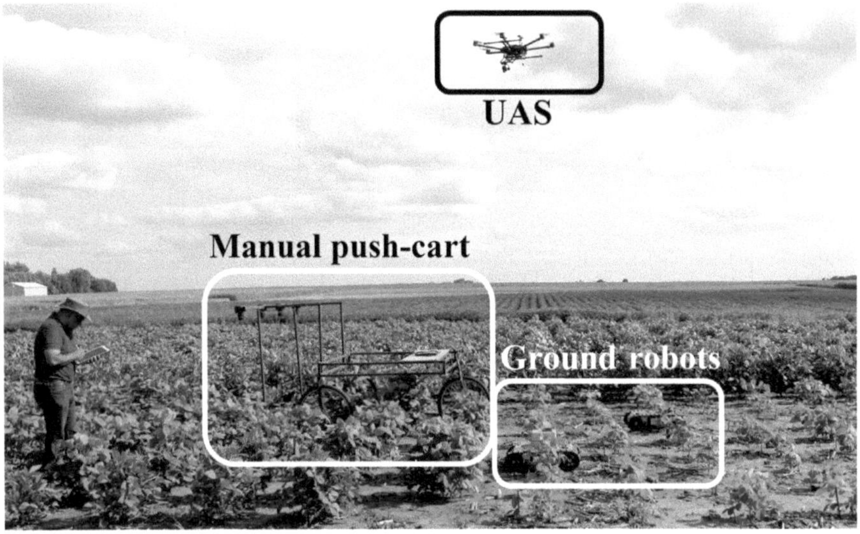

FIGURE 28.3

Breeder note taking and trait collection pipeline involving manual rating by the breeder, ground robots, manual pushcart, and drones, each with mounted sensors.

refers to an aircraft that is unmanned (without a human pilot on-board) and is a component UAS. UAS includes UAV, a ground-based controller, and a system of communications between the vehicle and controller. On the basis of types of rotors, drones can be fixed wing and multi-rotor (single-rotor, multi-rotor).

The single-rotor drone (unmanned helicopter) were among the earliest units in aerial plant phenotyping. The source of energy was based on electric or combustion motor. The combustion engine enable longer flight times and higher payload; however, the costs were high and maintenance and operability were more complex. The next innovation was multi-rotor drones that are currently the more popular choice among researchers in plant phenotyping applications. These types of drones are quite versatile and are easier to operate, with flexibility on take-off and landing from fields. The cost of drones was generally affordable for most researchers interested to enter aerial based phenotyping, and reasonable sensor payload can be carried onboard. In addition, the barrier to entry for pilots' (ground controller) training and operation was lower, further propagating the usage of these drones. However, due to the workload on rotors that provide the lift off from ground, the battery life is severely limited (15–35 minutes per flights depending on the type of batteries and type/number of sensors on-board). Due to lower flight time, large fields cannot be phenotyped in one flight, necessitating battery replacement (rechargeable) at frequent intervals. In addition, flights can't be flown at very low heights (say, <10 m) to avoid downward wind draft-based canopy movement (which can create ghosting in canopy images); consequently, high resolution cameras at higher flight heights may be needed for certain applications requiring higher resolution data.

Compared with multi-rotor drones, fixed-wing drones provide an advantage on the larger area covered per flight between battery changes. The drone wings provide "lift", therefore the motor conserves energy during lift-off and flight, enabling power to be effectively utilized for GPS and sensors. Unlike, multi-rotor drones, fixed-wing drones cannot "hover" over one spot. However, for some plant breeding programs with very large field sizes, fixed wing aircraft may sometimes be advantageous; although, for majority plant breeding fields, multi-rotor may be more practical to use. In addition, fixed wing drones are more stable in higher winds compared with multi-rotor drones. Fixed-wing drones are generally more expensive, and require more space for take-off, landing, and making turns in the air. Fixed-wing drones, in addition to longer flight time, faster speed of flight, and larger area covered, can accommodate more sensors (e.g., multiple types of cameras) on the same flight. However, sensors coregistration is needed among sensors so data from multiple sensors can be correctly assigned to individual plots. This is particularly important if breeders are interested to assimilate information from multiple sensors on the same plot at the same time. Drone speed and sensor capture rate need to be aligned so that the desired resolution is achieved.

Some of the salient points for application of drone-based plant phenotyping include several phases: (1) planning, (2) testing, (3) data collection, (4) data analytics, and (5) data storage, followed by actionable outcomes including selection decisions. The details of each of these phases are explained in Fig. 28.4. Additionally, few important considerations include:

1. GPS for accurate positioning to generate geo-tagged data of field plots, thereby connecting sensors data with individual plots on which inference will be made. This is done at the real-time kinematic (RTK) level accuracy, and is spread across the field to be phenotyped per run ensuring that variation in elevation and terrain features are considered. Generally, 4–5 (or more, if larger field sizes are flown through the drone) ground control points (GCP) are sufficient. The lack of GCP established at the beginning of the season (prior to drone flight), creates issues with orthomosaic maps. The orthomosaic map is created by stitching photos from the drone flights (sensors on cameras take continual images or measurement at a regular interval), and is adjusted for topography, lens distortion, camera tilt, etc., to give an orthorectified (geometrically corrected) map. The GCPs become even more important if time series data are collected (on the same field). The stitching of images also reduces the data amount and the entire field can be researched (trait data extraction) at the same time.
2. Plot extraction and image processing (through ML or Machine Vision methods) is conducted. If the purpose of the drone flight is the establishment of methods to extract traits from sensors data, ground truthing is done. For ground truthing, sufficient and high quality data are manually collected. Ground truth can also be collected in some automated or semiautomated

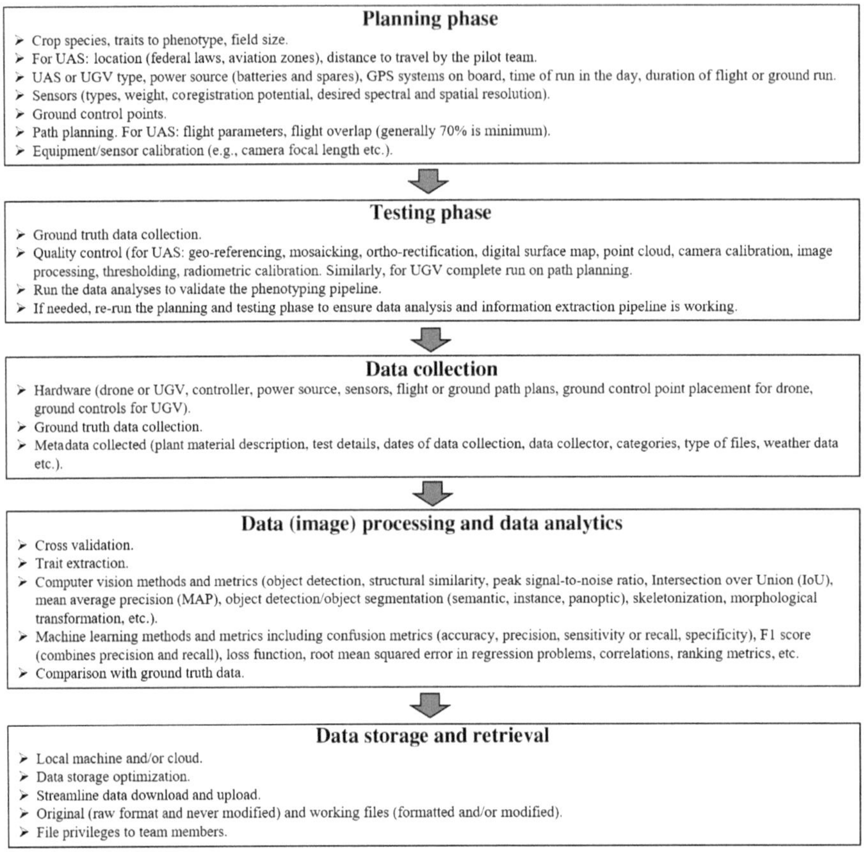

FIGURE 28.4

The main phases in the deployment of phenotyping systems and sensors in a breeding program. Under each phase, salient aspects and/or steps are included, which are important considerations for a breeding team.

manner, if the fidelity of these equipments are satisfactory from multiple prior research. If the purpose of the aerial phenotyping is trait extraction for actionable outcomes (selection decisions), and if ML method or image analysis has been satisfactorily established, no or limited ground truth measurements are required. If a breeder is uncomfortable in the outcomes of image processing trait data and utilizes extensive ground measurements with each flight, it adds more work with each flight; and therefore, somewhat defeats the purpose of aerial based phenotyping. In this case, breeder should consider improving the image processing and/or ML methods so that they can be applied with a high confidence in the breeding program. Sufficient statistical analytic and engineering capabilities need to be in place for a proper execution of plans.

3. Types of sensors (see section "Sensors" for full details): sensors should be chosen carefully so useful data are obtained. Depending on the choice of sensor, appropriate trait data are collected. Common types of sensors including digital imagery (tri-band), multi and hyperspectral cameras. These sensors covering a wide wavelength, are utilized to obtain numerous traits such as abiotic and biotic stress. Other traits include morphological and physiological traits that relate to plant architecture and health, respectively. Using multiple images, 3D point clouds can also be generated through structure from motion methods.
4. Drone based phenotyping while streamlines the data collection pipeline, still requires a manual operator (drone pilot) and other expenses (battery, on-board sensors, maintenance costs, and staff time). Hence, appropriate sensors that can generate information desired to make breeding selections should be deployed. More sensors on the drone can sometimes make it more stable in flight, but reduces the flight time so an optima needs to be developed.
5. Drone flight policies in the jurisdiction of its operation. Breeder needs to be aware of policies and guidelines, so they are compliant to country laws. In addition, field sites need to be planned so drone flights are not excluded due to federal aviation flight zones.

Ground based field phenotyping systems

While aerial phenotyping is faster and more streamlined, and there are several advantages to remote sensing; some applications require a closer investigation and data collection on proximal sensing. Particularly, for plant breeding applications where genotypes are grown in plots (dissimilar genotypes next to each other in adjacent plots), not all traits can be collected from aerial based remote sensing. For example, traits that are collected under the top canopy such as stem diseases, lower canopy diseases, several plant organs that are hidden from the top view imagery, etc. These requirements motivated an early effort to convert tractors and small sprayers to carry sensors for non-invasive and non-destructive phenotyping. These phenotyping vehicles are driven in between rows in a rapid manner, simultaneously carrying multiple sensors. These ground vehicles are driven by liquid fuel; consequently, power is not a constraint for power usage and length of phenotyping runs. However, these large machines cause soil compaction and can be disruptive to the canopy causing plant damage (depending on the size and design of the ground vehicle). Without changing tires (narrow width), tractors and sprayers are generally not amenable for driving through rows after canopy closure in row crops as they can cause plant disruption and damage in rows of movement.

These types of larger phenotyping unit gave way to smaller more mobile units that are capable to drive in between rows collecting high resolution data, due to closer proximity of the sensor to the plant. The ground vehicles are motorized (generally battery powered), and semi- or fully autonomous to run in between rows. These units came with different width for usage in different crops. While off the shelf units are available, technically competent groups modify/assemble their own customized ground vehicles (Gao et al. 2018). These ground vehicles are capable of all-terrain movement and work in a range of conditions for enhanced applicability. These are sufficiently lightweight to avoid issues of soil compaction, if repeated runs are made. These units carry multiple sensors and give RTK level accuracy GPS (for autonomous mode) to avoid collision with plant material enabling improved navigation under varying field conditions and crop stages. The improvement of such ground units is an active area of research, but not at the same pace as improvements of sensors and aerial system functionality. Overall, ground-based phenotyping is slower in data collection than aerial imagery due to lower speed of ground vehicles; however, they can overcome the issues of top and side view data collection associated with drones. For example, researchers have shown the utility of ground robots to estimate pod count in soybean from field plots using deep learning methods (Riera et al., 2020). This is just one example of the utility of ground robots coupled with ML methods for obtaining complex data.

In addition to ground vehicles, stationary tower and gantry system are also ground based phenotyping systems. Gantry system can provide super high temporal and spatial resolution data on plants grown on the same field. In these phenotyping structures, sensors are mounted on a movable system fully loaded with multiple and sophisticated sensors (although the gantry is not amenable to move between fields at other locations). This unit moves up and down the field to capture data. The sensors are generally enclosed in protective manner; therefore, phenotyping can be conducted in all environmental conditions. However, the rate of data collection is not high throughput compared to aerial systems and even small ground vehicles. While the gantry system is a useful research tool, but it is not scalable for full scale plant breeding and cultivar development applications. Stationary towers can include an affixed tower that is placed at the beginning of the season at a set coordinate, and it carries multiple sensors. These can provide information from field plots in the field of view of sensor(s). However, the resolution will vary based on distance; and as a result, stationary towers are generally not practical for mid- to large-sized plant breeding programs.

For small- to mid-sized plant breeding programs, a cheaper but effective solution is a manual pushcart with a suite of sensors mounted on it. The construction of such lightweight units with wheels and turning maneuverability is fairly simple, and can carry multiple sensors and laptops; including a battery pack for additional power, if needed, for phenotyping. The pushcart can be constructed to carry 2–6 sensor row units that phenotype each row independently. This creates a significant improvement in throughput compared to one-person phenotyping, at a minimal cost. Due to the narrow wheels, a manual pushcart can continue to collect data in row crops even after canopy closure with minimal crop disturbance. Depending on the row spacing and crop species, the manual cart can be constructed in a customized manner for wheel spacing and height of row units carrying sensors. If desired, geo-referencing can be done using a GPS unit. However, for mid- to large-sized plant breeding programs, a manual pushcart is not ideal as the system is not truly high throughput, is non-automated, and also prone to operator fatigue and errors thereby compromising data quality.

Controlled environment phenotyping

While most cultivar development programs are primarily field based, there are several uses of controlled environment units in indoor settings. For a plant breeding and related application these includes - generation advancements, research projects, regulated experiments, seed germination, disease and other stress screenings, vernalization treatment, insect rearing, cold and incubation, drying rooms, etc. These controlled environment units can be complemented with multiple sensors and accomplish phenotyping in semi- or fully-automated manner.

Greenhouse rooms and growth cabinets (walk-in, and reach-in) are examples of controlled environments units. Walk-in are larger units compared to reach-in chambers, and normally have a higher clearance to grow crop plants that reach a taller height. Growth cabinets are modular, can be generally custom manufactured (dimensions) by the manufacturer as per the order. These give a wide array of options for lightning, temperature, humidity, water control options, and adjustable racking and shelving options. Greenhouses have a larger physical footprint than growth cabinets. Greenhouses are primarily detached, and ridge and furrow style (with connection at the eave by a common gutter) types. Greenhouses are constructed with different materials, although metal are more sturdy and longer lasting. These are covered by glass, fiberglass, polyethylene, and polycarbonate material. Costs can vary depending on the location and types of construction material. Greenhouse rooms are larger and have multiple rooms, that can be individually controlled (temperature, water, nutrient, humidity, and light). Greenhouses can be large enough to conduct a single seed descent or modified bulk method for a small grain cereal program and obtain four generations in a year, with the final cycle that leads to plant pull. These can be grown in progeny rows in the crop season shortening the time to develop cultivars. The possibility to use these greenhouse based generation advancement for trait phenotyping, for example, temperature, water, biotic stresses, and phenotypic selection using sensors or platforms described earlier is attractive. This usage can be for static or dynamic (for plasticity) response of genotype or breeding families.

More recently, automated (no human intervention and control) and/or semi-automated (with some level of human intervention and/or control) phenotyping systems have been developed. These are useful for some plant breeding related research, including study of biochemical compounds, transgenic research, some physiological research. However, due to a lower correlation of growth cabinet and greenhouse to field performance, these are not ideal for cultivar development programs focussed on performance traits, including seed yield (Fig. 28.5). Although, genomic assisted selection can be effectively deployed in large greenhouses where no field performance phenotyping data is collected. Screenings for race-specific

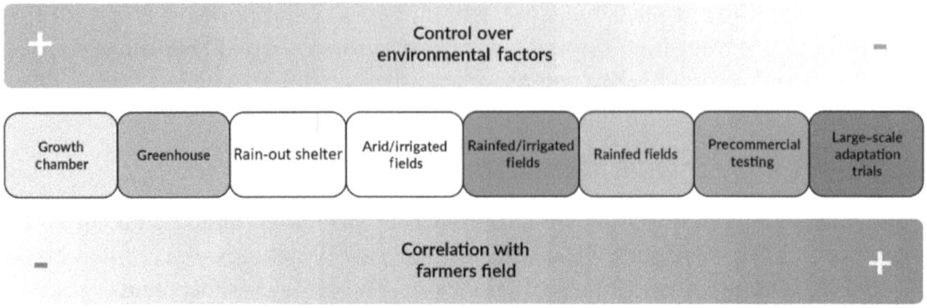

FIGURE 28.5

Examples of phenotyping levels and correspondence with cultivar performance in farmer's field. *Adapted from José L Araus, Jill E. Cairns. 2014. Field high-throughput phenotyping: the new crop breeding frontier. Trends in Plant Science. 19(1): 52-61.*

resistance can be done in controlled environmental for traits that have a good correlation between indoor screening and field expression/performance.

Sensors

Sensor is a device that detects or measures physical, chemical, or biological property. For a plant breeder, these include a broad range of device/equipment/machine that can generate information on yield and its component traits, canopy biomass, growth and development, physiological, morphological, anatomical traits, abiotic and biotic stresses, sugar, proteins, metabolites, high value chemicals, etc. The choice of appropriate sensor in breeding programs is very important. An ideal sensor needs to be affordable, simple to use, mobile, and with lower technical complexity to operate and more streamlined data extraction and analytics. These ensures that breeders will be able to maximize the usage of sensor tools to obtain useful phenotyping information for application in the breeding pipeline. Lately, the interest in sensors have focused on imaging methodologies and continual advances are ongoing. The speed and capabilities of sensors are continually increasing, while the costs are reducing. These image-based sensors involve data collection on quantitative traits with complex features, such as stresses (biotic and abiotic) and yield component traits. These sensors include simple digital camera (tri-band Red Green Blue camera), spectroscopy (multispectral and hyperspectral cameras), thermal infrared camera, fluorescence cameras, stereo vision and laser scanners for 3D imaging, and tomographic imaging (that performs sectioning using penetrating waves, e.g., computed tomography, magnetic resonance imaging).

Electromagnetic (EM) spectrum is the range of all types of EM radiation. Wavelength is the distance between one peak or crest of a wave of light, heat, or other energy type and the next peak or crest. Electromagnetic spectrum is all the wavelengths of visible and invisible light (Fig. 28.6). In crop applications, we

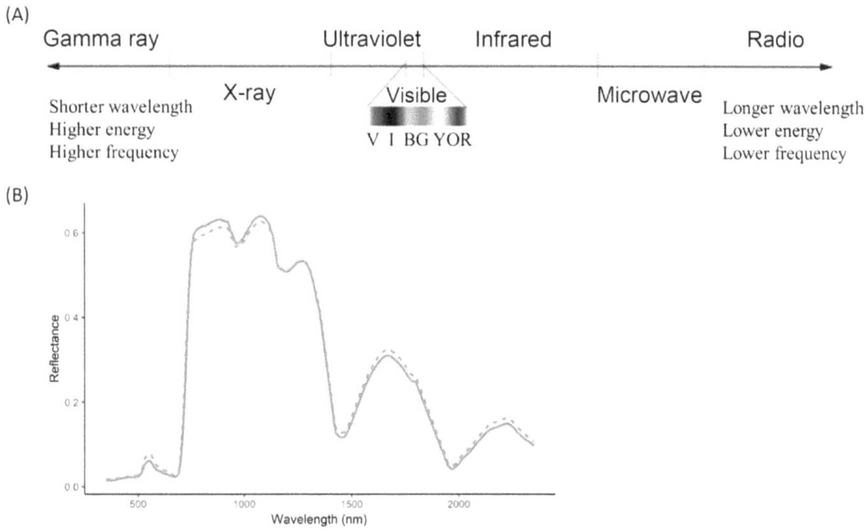

FIGURE 28.6

(A) Electromagnetic spectrum including visible and infrared (near infrared, shortwave infrared, midwave infrared, infrared, longwave infrared) wavelengths. (B) Plot of reflectance data from soybean elite varieties (*solid line*) and landraces (*dashed line*) using a spectroradiometer. This reflectance data is useful to observe difference between accession and to determine differentiating wavebands.

primarily rely on energy from the sunlight (electromagnetic energy). Electromagnetic radiation can be expressed in terms of energy (measured in electron volts), frequency (measured in cycles per second, or Hertz), or wavelength [measured in meters, Angstrom (Å) (1 Angstrom = 10^{-10} Meters = 0.1 Nanometer), micrometers (μm) or nanometers (nm)].

Each sensor has its own utility depending on the breeding objective and target trait focus. However, lately there has been a lot of interest in remote sensing using sensors that span a wide range of electromagnetic spectrum: visible (VIS: 400–700 nm), near infrared (NIR: 700–1100 nm), shortwave infrared (SWIR: 1100–3000 nm), midwave infrared (MIR: 3000–5000 nm), infrared (IR: 6000–7000 nm), and thermal or long-wave infrared (TIR or LWIR: 8000–15,000 nm) wavelength regions.

These wavelengths have different uses and applications for plant science and breeding. For example, the visible wavelength includes any trait that human eyes can discern, and VIS region imaging can do the same. These can include traits including stress symptoms on leaves, as reflectance is mainly a result of absorption by the photosynthetic pigments. Plants reflect near infrared light but healthy plants reflect more than stressed plants; consequently, NIR can be used to monitor plant health and make selections of better breeding materials. SWIR wavelength helps to estimate the amount of water present in plants (and soil); and thus, are useful for water use efficiency measurements, particularly drought. The MIR wavelength shows emitted thermal radiation. In plants, nutrient stress, pathogen infection, insect infestation, and senescence are accompanied by reduced reflectance in the MIR wavelengths compared with healthy and green foliage. Scientists use IR region to observe water vapor in the atmosphere, and hence it is integral to weather observations and forecasts. The TIR wavelengths can gauge water and land surface temperatures. Since the actively growing plants cool the air above them by releasing water through evapotranspiration, TIR can inform on the amount of water plants are using.

In remote sensing, the two types of sensors are: (1) active, and (2) passive. Active sensors provide their own source of energy to illuminate the objects they observe. This is done by sensor emitting radiation on the plant or canopy that is under study; detecting and measuring the radiation that is reflected from the plant/canopy. On the other hand, passive sensors detect natural energy (radiation) that is emitted or reflected by the plant or canopy under study. Due to a lack of its own source of energy, passive sensors generally use the sunlight radiation.

The common types of active sensors include LiDAR (Light Detection And Ranging), which uses light in the form of a pulsed LASER (light amplification by stimulated emission of radiation) to generate precise 3D information about the shape of the phenotyped area; for example, to create dense 3D point clouds that can be used to obtain object heights and shapes. A LiDAR instrument mainly consists of a laser, scanner, and GPS receiver. LiDAR and other ultrasonic sensors can generate 3D point clouds for canopy architecture, plant height, canopy biomass. Terrestrial Laser Scanners (TLS) are active sensors that can be used to acquire detailed 3D data of objects including plant canopies. TLS are the ground based LiDAR system and work on the same principles; that is, emit laser pulses, measure the time of return of the LASER, and calculate sensor-to-surface distances to generate a high resolution 3D surface map. The resolution from TLS can be in mm to cm scale; however, scanning is done from multiple perspectives (e.g., each of the four sides of the field), and the perspectives for each test area are coregistered. The TLS type of approaches are fairly fast but not high throughput. It is a cheap way to produce 3D models, although the test area that can be phenotyped at a time is generally in sub-acre area.

Passive sensors include different types of radiometers and spectrometers in VIS, NIR, SWIR, MIR, IR, and TIR wavelenghts. Passive sensors include hyperspectral, multispectral, NIR and tri-band digital camera, fluorescence, and thermal infrared cameras. The hyperspectral sensors (for example, spectroradiometers) contains the most wavebands as it detects hundreds of very narrow spectral bands (can be even 1−3 nm or more) in the VIS, NIR, SWIR, MIR wavebands of the electromagnetic spectrum. This provides a very fine resolution and separation (spectral resolution) to differentiate the test genotypes and computation of numerous vegetation indices and physiological traits. Hyperspectral imaging combines imaging and spectroscopy, thereby generating high spectral resolution data (although copious). Hyperspectral datasets are generally composed of about

100–200 spectral bands of narrow bandwidths (5–10 nm or lower). On the other hand, multispectral datasets are usually composed of 4–10 bands of large bandwidths ($>$ 50 nm). Hyperspectral images are represented as a 3D datacube with spatial information collected in the $X–Y$ plane, and spectral information represented in the Z-axis. Hyperspectral data are advantageous because these can help determine specific bands, from the entire continuous spectrum of wavebands; that are most informative for the phenotype detection and its prediction.

On the basis of acquisition modes, the two main types of hyperspectral imaging are: (1) line scanning (or push broom), and (2) snapshot (or single shot). In the push broom mode, the detectors are arranged in a linear array at the focal plane of the image that is formed by lens in the camera. Either the camera or the platform (with plant sample on it) are moved to record each pixel individually. Each individual detector in the camera measures the energy for a single ground resolution cell (part of the sample), thus the size and field of view of the detectors defines the spatial resolution of the system. The forward motion allows pixel line to be recorded. The imaging plant (in field or on a platform in indoor settings) are perpendicular to the camera. A separate linear array measures each spectral band. The lens is "pushed" along in the flight track direction (i.e., along track) across the plant object being phenotyped. While the pushbroom technology is effective for other applications; in plant breeding settings the spectral resolution for in-field resolution is inadequate due to wind-induced plant organ movement, making it ineffective for the granularity of scale that may be desired by the breeder. In contrast, snapshot camera simultaneously acquires the spatial and spectral data, and all bands are captured at once. However, this is an evolving technology and challenges include balancing spectral-spatial resolution, power requirement for deployment on phenotyping systems, and calibration in varying field conditions (illumination, weather, etc). Finally, the data analytics pipeline for information extraction is still not routine. The results from the curve-based hyperspectral spectroradiometer and hyperspectral imaging cameras can allow customization of wavebands in multispectral cameras, to only include informative wavebands and remove non-informative wavebands.

Multispectral cameras measure the intensity of electromagnetic radiation in a smaller range (but multiple wavebands) of the spectrum than a hyperspectral camera. For example, multispectral cameras can include red, green, blue, near infrared and red edge waveband and generate spectral signatures for numerous plant traits and vegetation indices. More recently, multispectral cameras with dynamically selectable bands have become available. Customizable multispectral cameras allow breeders to choose the wavebands that they want to include on their camera. This is analogous to a breeder using a single nucleotide polymorphism (SNP) chip, where breeder can decide which SNPs to include to screen their materials. These SNPs can be linked/associated with traits and genes/QTL of interest. Similarly, breeder can customize multispectral cameras that can include precise wavebands that cover the traits of interest (traits can be extracted from the combination of wavebands). These hyperspectral and multispectral cameras can be imaging (pixel) or non-imaging based (curves). Multispectral cameras are cheaper than hyperspectral cameras.

The NIR and tri-band (RGB) cameras can provide information on plant height, leaf area and shape, canopy shape and cover, stand count, plant organ detection and count, plant maturity, biomass estimation, biotic stress and abiotic stress identification and quantification, and weed detection, etc. These cameras are generally the least expensive, and easier to mount on ground and aerial systems; and trait extraction methods can be less cumbersome. These are also low weight, while generating high resolution images; and for this reason, are very attractive for both aerial and ground based phenotyping. In addition, these types of cameras do not normally create large data amount, a problem common with hyperspectral cameras.

For water stress phenotyping, thermal imaging cameras are useful. These types of sensors can be used to monitor canopy temperature and crop water deficit under water stress conditions, which causes leaf temperature increase. As a result, thermal imaging cameras allow researchers to measure crop stress, canopy temperature, drought tolerance, and water use efficiency. These sensors serve as an indirect measurement of water stress, because plants not getting sufficient water have a reduced transpiration rate that increase leaf temperature. Furthermore, with a combination of visible and thermal imaging, crop water stress index can be computed more effectively to determine crop water status and stomatal conductance in plants (Möller et al., 2007). While tractor mounted or manual pushcart can be utilized to support thermal imaging cameras, UAS can provide a higher throughput (Fig. 28.3). Thermal imaging sensors are prone to give erroneous outputs with variation in the height of the UAS flights, and variable weather conditions, for example, cloud cover and wind speeds.

Remote (or distal) and proximal sensing are common techniques for data acquisition from plant canopies at organ or tissue levels with or without disruption and physical contact with the object. Remote sensing is more commonly associated with mini-satellite and UAS based phenotyping systems, while proximal is done with handheld, manual pushcart, or ground robot units where sensors are close to the plant canopy and organs. Most sensors are based on imaging; therefore, pixels are extremely important for imaging technology and methodologies. Researchers continue to attempt to obtain more information per unit pixel, as this provides better trait measurement. Calculation of cm/pixel, that is, ground sampling distance (GSD) is as follows:

Sensor width (in mm) = W_S

Focal length (in mm) = F_L

Flight height of the drone (m) = Ht

Image width (in pixels) = I_W

Image height (in pixels) = I_H

Ground sampling distance (in cm/pixel) = $\frac{(W_S \times Ht \times 100)}{(F_L \times I_W)}$

Image distance width (width footprint of a single image on the ground, in meters) = $\frac{(I_W \times GSD)}{100}$

Image distance height (height footprint of a single image on the ground, in meters) $= \frac{(I_H \times GSD)}{100}$

Area covered by a single image (in m^2) $= I_W \times I_H$

Using a hypothetical example, with two scenarios:

1. $W_S = 20$, $F_L = 50$, $Ht = 25$, $I_W = 4000$, $I_H = 3000$
   GSD = 0.25, Image distance width = 10, Image distance height = 8, and Area covered by single image = 75.
2. $W_S = 20$, $F_L = 50$, $Ht = 50$, $I_W = 3000$, $I_H = 1500$

   GSD = 0.67, Image distance width = 20, Image distance height = 10, and Area covered by single image = 200.

This example is provided for educational purpose only. With increasing drone height, GSD and area covered by a single image will increase if the same camera is used. In future, super-resolution analysis will become more prominent as it attempts to infer subpixel information from data and maps between low- and high-resolution images (thereby connecting ground-based high resolution, with UAS based, and ultimately satellite-based data with lower resolution but increased spatial coverage). This, in conjunction with multimodal fusion in image processing, can advance data processing (spatial, altitude, and temporal) for actionable outcomes in plant breeding programs. In future, there is an expectation of empowerment of phenotyping to even non-expert phenotypers with the development of smartphone and handheld applications for trait estimations and measurements.

For a plant breeding application and integration in the pipeline, several factors are important as the sensor usefulness depends on flexibility, accuracy to trait correlation, affordability, and throughput. While the cost of sensors is continually coming down, it is still a substantial investment and may require equipment insurance for automated or semiautomated utilization. Additionally, there are maintenance and depreciation costs, and the technology can become "old" in a relatively short timeframe. Multifunctionality in a sensor may be useful, for example, spectroradiometers that can calculate important vegetation indices correlated to several traits. Plant breeders can also assess the size of their program, resources, and objectives to determine suitable phenotyping systems and sensors. However, investments in phenotyping method are generally upfront compared to molecular marker based methods (see chapter 27) where operational costs are generally in each filial generation as genotyping happens to make breeding decisions.

Data analytics

The data from majority of aerial and ground based sensors requires a combination of data analytics approaches including ML, Deep Learning (DL), and traditional statistical analyses. ML is widely applicable to diverse data types and objectives within plant breeding and agricultural sciences; with successful application for high-dimensional data from genomics, transcriptomics, proteomics, and metabolomics, and phenomics. In this section, the focus is on the ML and its subset, DL, methods in relation to plant breeding activities particularly for plant phenotyping. Phenomics generally creates higher dimensional cube than genomics (Dhondt et al., 2013), necessitating ML algorithm more frequently.

Machine learning

Arthur Lee Samuel defined ML as the "field of study that gives computers the ability to learn without being explicitly programmed." ML refers to computer algorithms that can learn patterns from the data to make automatic decisions without explicit programming (Singh et al., 2016). The ML data analytics takes a multidisciplinary approach that draws principles of probability theory, statistics, decision theory, visualization, and optimization. The ML methods became more common in the plant breeding domain with the advances in phenomics (and genomics) including high throughput methods, as these create large amounts and high-dimensional data relating inputs (e.g., image data) to output of interest (e.g., organ identification, stress quantification, growth and development date prediction, disease prediction, disease movement and spread, yield prediction, drought and flooding stress, chemical application date and effects, weed detection and identification, etc.).

ML requires an input (training dataset) to create a desired output, and hence most applications requires a customized dataset for each ML algorithm that needs to be developed, unless a researcher can leverage previously created dataset and augment it with data that lacked in the previous dataset. This ensures that a representative sample space is covered, and ML algorithm will be relevant and accurate to solve the problem in real world scenario. Therefore, collection of high quality and large quantity data are important in the development of robust ML models. Rather than a more simplistic study of one variable at a time, the ML methods enable querying large datasets to discover patterns and insights by concurrently looking at a combination of variables (e.g., canopy and plant features, weather variables, genetic factors, etc.) (Shook et al., 2020). This enables linking the inputs to the outputs when it is too complex to model them using simple mathematics. Of course, advanced ML is not needed in all situations and can be an overkill in several instances (e.g., if simple linear models are sufficient), and inadequate in other (e.g., if less data are available to build the ML model).

Due to the rapid increase of the ML application in multiple domains; scalable, robust, and flexible software tools have been created to develop ML models. These tools use different programming languages such as Python, R, C++, C#, Java, etc. Examples of these software development platforms include (listed alphabetically): Accord.net, Amazon Machine Learning, Apache (Mahout, MLLIB), Azure, Google Cloud AutoML, Keras, Knime, Pylearn2, Pytorch, RapidMiner, Scikit-Learn, TensorFlow, Weka, etc. Most of these are open source, and some are more user friendly than others; for example, RapidMiner is suitable for non-programmers due to its drag and drop menu. Some of the cloud-based services provide pretrained models so that users can utilize in various applications. These run on Graphics Processing Unit (GPU), Central Processing Unit (CPU), and mobile computing platforms. GPUs are better suited for image analysis (photos and videos). The differences between these two GPU (many cores, high throughput, parallel computing, higher computing capacity - thousands of operations at a time; more efficient for DL) and CPU (fewer cores, low latency, good for serial processing not parallel processing, and more limited numbers of operations at a time), govern which one to use.

There are numerous ML methods to choose from for a given problem. As a result, it is important for plant breeder or practitioner to make a thoughtful choice on the specific ML method to deploy for an exact problem. For example, Regression based ML methods include: Artificial Neural Network Regression, Elastic Net Regression, Gaussian Process Regression, Linear Regression, Logistic Regression, Partial Least Squares Regression, Piecewise Regression, Ridge Regression, Stepwise Regression, Tree Regression, etc. Classification type ML methods include: Artificial Neural Networks, Decision Trees, Extreme Learning Machine, K-means clustering, K-Nearest Neighbor, Linear Discriminant Analysis, Naive Bayes Classifier, Random Forests, Self-Organizing Map, Support Vector Machines, Quadratic Discriminant Analysis, etc. Common DL methods include: Artificial Neural Network, Convolutional Neural Network, Recurrent Neural Network, Stochastic Gradient Descent, and Long Short Term Memory. The applications of these ML software's range from data manipulation, data mining, dimensionality reduction, classification, regression, clustering, pattern recognition, object detection, computer vision, image analysis, image classification, filtering, etc.

The choice of ML method can be made by considering the type and amount of data and intended outcomes, and appropriateness of ML method for specific type of problem. Plant breeders can maximize the benefit of ML methods if expert programmers are part of their team and stronger collaborations exist between plant breeder (domain scientist), engineers (sensors, phenotyping platforms), and data analyst/computer scientist (ML, DL methods).

An overview of the machine learning approach

The choice of ML method is done from two different viewpoints (Singh et al., 2016): (1) learning process (how the features are learnt), and (2) modeling objectives (what is being learnt). Based on learning process, two main types of approaches include supervised and unsupervised. In a supervised approach, data labels are provided to the ML model, for example, if model development objective is to automate identification of three different diseases (disease A, disease B, and disease C) in soybean. In a supervised approach, each image (with one of three disease) will be labeled by the disease phenotyping expert with a label on the disease visible in each image. In this situation, the learning process is "supervised." On the other hand, if no label was given to the images, the learning process is "unsupervised."

In the supervised learning process, the model tries to learn a map between the input dataset and the corresponding output labels. In the three-disease example, it implies that a supervised ML model learns to map all three soybean diseases to the disease label we had given. In a good performing model, it will show a high accuracy, precision, sensitivity, and specificity. In the unsupervised learning approach (examples include clustering and dimensionality reduction techniques), no specific output labels are associated with input images hence the model identifies features present in the images. For example, presence of specific symptoms associated with each disease. It is also possible that the features identified in the unsupervised method is not intuitive or meaningful to the plant breeder. In a third variant, "semi-supervised" learning, ML models are developed by using partly labeled data, that is, ML algorithm is trained by only labeling a partial set of the training data. Semisupervised learning saves on manpower (for manual annotation) over the supervised learning methods. Both supervised and semisupervised ML approaches require manual input (i.e., labeling, feature extraction); thus, human intervention and expertise are required, and models can be error prone if insufficient or poor feature extraction and labeling is done.

Unsupervised classification of pixels in an image, separate pixel from each other without prior knowledge from class of pixels, and clustering algorithms is used to determine in which class each pixel fall into. This creates groups composed of samples with similar features in comparison to the dataset as a whole. Visualization is done using cluster maps, and the identity of clusters is not known *a priori* of the analysis. However, the class identified by the clustering methods can be subsequently used as class labels for supervised ML models. Unsupervised ML approaches can help overcome inter-rater variation (variation between two different people rating the same object) and intra-rater variation (variation between the ratings of the same person on the same object at different times). Unsupervised ML can help in object classification using some criteria that are not apparent to human raters, and this could potentially lead to more accurate classification and quantification performance.

The modeling objective relates to model being trained to distinguish between two different data patterns or to be able to learn and generate similar patterns synthetically (Singh et al., 2016). Building on the three-disease example, if the diseases are being distinguished based on disease A having yellowing leaves, disease B having round lesions on leaves, and disease C shows leaf necrosis: in the discriminative model, the model can find these differences, and the model does not learn other features of the object (disease images); therefore, many supervised models are also discriminative. In a supervised discriminative model, the model can learn features that are relevant for maximizing the performance of the task. However, discriminative models will not be able to differentiate disease A with new disease D in another crop species with yellowing leaves. Consequently, supervised discriminative models are suitable for a specific predetermined task. The generative model on the other hand, captures the overall data pattern and is able to generate synthetic images, and hence it is useful for many decision tasks at the same time. If the discriminative model is developed with a large amount of data, it usually performs better for classification tasks than generative models, for example, classification of disease (images) into their own group. In most cases, if lower data volume is present, generative models may outperform discriminative models, and are less prone to model overfitting issues (i.e., a model learns training data very well but performs very poorly for unseen test data). ML models can belong to cross classification categories of supervised-unsupervised and discriminative-generative. Singh et al. (2020) reviewed the growing capabilities of ML methods together with image-based phenotyping, which can extract new insights from curated, annotated, and high-dimensional datasets across numerous crops and stresses. They proposed an overarching strategy to utilize ML methods for methodical application in plant stress phenotyping at different scales across multiple stresses, program goals, and environments.

The ML approach identifies a hierarchy of features and generalized trends from available data (can be of different format and type), and can integrate disparate and often redundant data to draw coherent (even non-intuitive) patterns for outcomes, including classification task (e.g., if symptoms are from physiological stress or physical stress), object detection (e.g., insect-pest detection), object tracking (e.g., plant growth and development or organ tracking from images), and segmentation (e.g., multiple disease detection on the same canopy etc.). The ML model training is done using a large fraction of the dataset, known as training dataset or training population, and is supposed to represent the entire population. The remaining dataset is called testing dataset or testing population, and it is used to test the developed model. Once the model is developed using the training dataset, a validation is done (with data from the same or independent population dataset); this step is called model validation. The training set consists of labeled images (in supervised learning approach) of each disease or stress class (for example, disease A, disease B, and disease C). The training set is used to train a classifier to learn the relevant features of each class for classification. Model is trained and validated, and model evaluation is done to test the quality of the classifier using the testing dataset by using the trained model to predict the labels from unseen images from test data. Testing set is important because it gives unbiased estimation of model success. Model performance can be measured using various metrics, such as accuracy, precision, sensitivity or recall, specificity, F1 score, etc. Calculation of these metrics is explained below:

Consider a simple problem where the task is to classify images into diseased and healthy, that is, we are training a ML model to automate this classification. There will be four classes of outcomes:

- True positive (TP): prediction is "Diseased" and object (information from the image) is diseased. This means is a correct classification.

- True negative (TN): prediction is "Healthy" and object (information from the image) is healthy. This means it is a correct classification.
- False positive (FP): prediction is "Diseased" and object (information from the image) is healthy. This means it is an incorrect classification.
- False negative (FN): prediction is "Healthy" and object (information from the image) is diseased. This means it is an incorrect classification.

Depending on the objective of the project, FP and FN can be equally bad or may have differing value. In the above example, FP may imply researcher apply fungicide (if the disease is a fungal disease), although the variety (in that image) is healthy. This is a waste of chemical, time, machine usage where fungicide application was not needed. On the other hand, FN implies that model indicated a healthy variety although it is diseased. This will mean yield losses due to disease and no application of fungicide.

Accuracy is the ratio of the correctly labeled diseased (+ve) and healthy (−ve) objects (images) over the whole pool of objects (images).

$$\text{Accuracy} = \frac{(TP + TN)}{(TP + TN + FP + FN)}$$

Precision is the ratio of the correctly diseased (+ve) labeled objects (images) by the model to all diseased (+ve) predicted objects (images). This provides information on how many diseased images prediction is actually diseased.

$$\text{Precision} = \frac{(TP)}{(TP + FP)}$$

Sensitivity (or Recall) is the ratio of the correctly diseased (+ve) labeled objects by the model to all actual diseased objects (images). This provides information on; of all objects (images) that are diseased, how many were correctly predicted to be diseased.

$$\text{Sensitivity} = \frac{(TP)}{(TP + FN)}$$

Specificity is the ratio of correctly healthy (−ve) labeled objects by the model over all the objects that are actually healthy. This gives information on how many healthy images prediction were correctly predicted.

$$\text{Specificity} = \frac{(TN)}{(TN + FP)}$$

F1-score (or *F*-Measure) combines both precision and recall into one value. This metric is used if both these values are important and a balance is desired. A higher F1-score value indicates that both precision and recall values are high and equal.

$$\text{F1-Score} = \frac{2 \times (Precision \times Recall)}{(Precision + Recall)}$$

Deep learning

There are several limitations of traditional ML approaches, specifically in handling high-volume and high-dimensional multisensor data (Singh et al., 2018). The traditional ML methods require "handcrafted features" (i.e., *a priori* user-identified features), or in other words, handcrafted methods or rules. For clarity, handcrafting refers to the task of choosing appropriate input features (e.g., a color channel from a digital image) or designing transformations (e.g., algorithm such as scale invariant feature transform (SIFT) that are used to detect and describe local features in a digital image) that are applied to the raw datasets prior to model training. This is done to enhance the ML model performance. These feature extraction methods are primarily heuristic; and thereby, involves time-consuming trial-and-error steps. The success in developing useful model may depend on the level of experience and domain knowledge of the data scientist for this important step. Therefore, it is required that plant breeders work closely with data scientists in their teams. In the traditional ML, we specify the computer on what to do, for example, SIFT transformation followed by Support Vector Machine to perform classification. This differs from more recent subset of ML methods called DL, which do not require any hand crafting of features, and we let the model to learn the most appropriate representations for the task.

Among ML methods, DL is the most promising for their application to complex plant phenotyping tasks. DL is an ML technique that teaches computers (i.e., what to do). The DL models are generally trained by using a very large set of labeled data and neural network architectures that contain many layers. One of the popular architecture for image classification is convolutional neural networks (CNNs). DL algorithms excel in hierarchical feature learning through first extracting simple patterns, followed by more complex non-linear patterns (features of features); as a result, DL has emerged as the primary ML tool of choice for image-based phenotyping problems that can alleviate some of the challenges associated with other ML methods. DL based methods allow learning of complex, non-linear, and hierarchical features from a variety of sensors, for example, tri-band, multi and hyperspectral cameras. DL models typically work well with raw data (such as image data), without the need for the trial-and-error based hand-crafted feature extraction process (Fig. 28.7). In comparison to ML models, automatic feature extraction happens in DL; additionally, the image representations learnt used by DL can be many and highly complex. This is more important for more complex sensors, such as hyperspectral cameras and can be quite effective for phenotype extraction at more depth than possible with visual phenotyping. Like ML, training of DL models can be either supervised or unsupervised. Although, the advantage of DL methods is that no feature identification or feature engineering is needed; the automatically learned features are difficult to interpret from the domain (scientific) perspective.

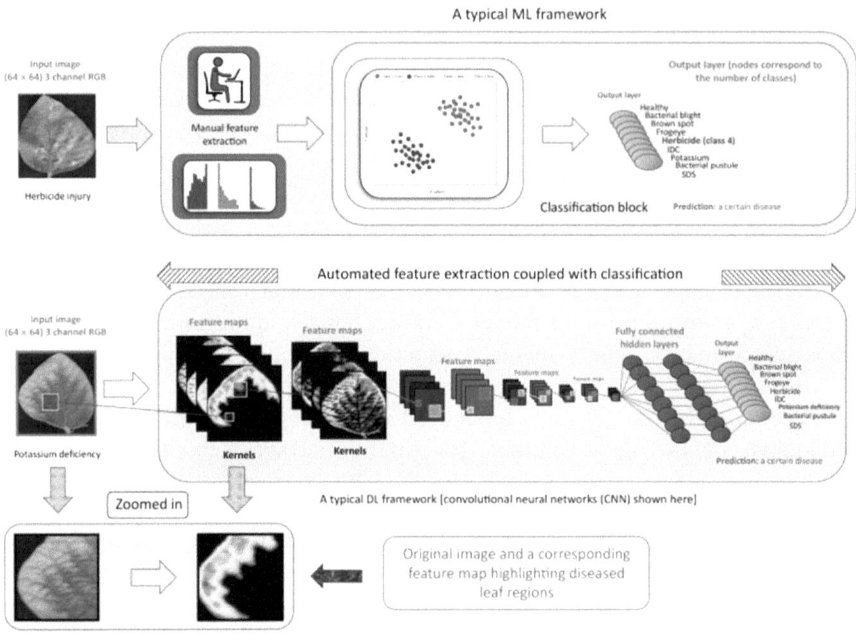

FIGURE 28.7

Key Differences between ML and DL models. The ML pipeline typically consists of two stages (1) human user identifies key features to extract from image data (heuristic–intuition, domain expertise, or scientific hypothesis). Image processing algorithms are used to extract these features from the images. (2) The feature set is subsequently used in ML application, for example, disease identification analysis. On the other hand, in a DL pipeline raw image data are fed in the model, and during model training, features are automatically extracted. This leads to image classification to the most likely class, and the model performance metrics are used to determine if the user is satisfied with the results. *ML*, Machine learning; *DL*, deep learning.

From Singh, A.K., Ganapathysubramanian, B., Sarkar, S., Singh, A., 2018. Deep learning for plant stress phenotyping: trends and future perspectives. Trends in Plant Sci. 23 (10), 883–898. doi:https://doi.org/10.1016/j.tplants.20.

Typically, DL models are built using many layers of neurons in which each layer is connected to subsequent layer using weights and bias parameters. In the case of image processing using DL, the input image (e.g., diseased leaf) is transformed using the hierarchical non-linear transformations, leading to the final decision (identification, classification, quantification, etc.) at the output layer. Compared with traditional ML methods where feature handcrafting has been the norm using limited data, DL approaches are significantly more scalable and robust. Since unsupervised DL methods remove the need for feature engineering, that is, human annotations, or feature engineering by handcrafted rules/

methods and/or human selected features are not needed; these save breeding resources. In plant breeding and sciences, annotations is a large and complex task requiring substantial time and resourses even if large data are available. Furthermore, DL typically outperforms human-engineered features in most ML tasks. On the other side, DL approaches typically need huge amount of training data to learn a very large number of model parameters. Specifically, in a supervised setting, the training process requires sufficiently large amount of annotated data that can be difficult and expensive to obtain in many circumstances.

For that reason, weak supervision and active learning approaches typically become useful in training DL models. Active learning methods reduce the amount of labeling needed by DL models while achieving good model performance (Nagasubramanian et al., 2020). Another important method is transfer learning. Transfer learning can leverage preexisting trained networks (for image processing tasks popular models' examples include, GoogLeNet, ResNet, VGG, MobileNet, DenseNet), while minimizing or avoiding the expensive and resource intensive data-labeling. The supervised DL models require large amount of labeled data, which is expensive, resource, and time intensive. Transfer learning is more applicable if large amount of labeled data are unavailable, while active learning is more useful if large amount of unlabeled data are available and labeling is not feasible (due to time, cost, resources).

DL approaches are also generally black-box models and suffer from a lack of interpretability or explainability. For this reason, until interpretability methods were proposed, plant breeders were skeptical of DL methods and their outcomes/decisions. These new methods have started to remove bottlenecks for wide adoption of such analyses; for example, to rapidly and accurately phenotype biotic and abiotic stresses in soybean (Ghosal et al., 2018; Nagasubramanian et al., 2020). However, ML and DL methods do not circumvent aerial and ground based high throughput automated phenotyping problems due to the complexity of sensors (sophistication, capability), environments (spatial, temporal), and biology (textural, morphological, and physiological) attributes that presents complex and continually evolving challenges for data processing. In addition, these methods are not being applied in real-time. For example, an operator will upload drone or ground based sensor data on local machine and servers (in their labs/organization). These analyses are done in a batch-style that can take hours or days depending on data size, type, machine computing ability. This generates lagging indicators, hence after-the-fact decisions are made. In a comparative scenario, a plant breeding program may obtain harvest data from a mechanical plot combine. These data are sent to the organization's statistical support unit personnel who will perform the analysis, and breeder will use these to make breeding decisions. These then cascades to multiple subsequent decisions including advancement and marketing decisions; for example, which genotypes will advance to the next generation or season, and varieties that will be promoted for seed production and sales. These can also be utilized for breeding method applications on plant pulls etc. This same analogy can be used for aerial and ground phenotyping systems that can provide yield prediction from anatomical, morphological, physiological traits such as yield predictors and yield organ estimators (spikes, pods, siliques, etc.). Another example can be an opportunistic disease phenotyping in plant breeding programs.

With the combination of RTK level position for plots and plants, and on-board analytics disease quantification can be completed informing selection/discarding decisions prior to the disappearance of differentiating symptoms between resistant and susceptible classes. The on-board real-time analytics can allow breeders to react without delay, if such situation exist; for example, plot maintenance actions, as well as breeding method decisions, and selections. These real-time analytics utilizes Edge computing, which is distributed or parallel computing, where computation and data storage is done closer to the location where it is needed; for example, phenotyping systems. Currently, DL models remain less frequently used for on-board processing in phenotyping systems with high-dimensional data (such as hyperspectral cameras) where there are stringent battery power constraints. Therefore, energy storage and on-board (GPU) capabilities needs to be improved to realize the vision of edge computing in plant breeding applications. For most scenarios, off-line batch analytics meets the current needs of a plant breeder, but this is expected to change as breeder try to develop cultivars meeting the ever changing climatic conditions.

Phenomics and machine learning applications in plant breeding

There are numerous avenues for the use of phenomics and ML for plant breeding applications. These include the understanding of underlying physiological mechanisms involved in the observed phenotype that can provide genetic drivers of traits for cultivar development. A properly implemented phenomics–ML system on high-dimensional time series data (physiological, morphological, etc.) on numerous genotypes can enable an efficient framework for breeding traits. These can be accomplished in a non-destructive manner that was previously challenging. Additionally, phenomics data can be used to predict (and/or measure) seed and biomass yield in the growing season, canopy closure, canopy area, leaf area index, dry matter accumulation, genotype plasticity in differing environments, canopy temperature, canopy reflectance, vegetation indices (regular and optimized), plant height, and many other traits. With the use of ML, optimized wavebands can be developed in addition to established vegetation indices (Parmley et al., 2019a,b). They also showed the potential of HTP of canopy to capture useful data on genotype-specific differences in water status, heat tolerance, chlorophyll concentration, and canopy size. These are useful in genotype selection in a breeding program at a cheaper cost and high rank correlations as high as 0.6–0.8, establishing their application in cultivar development. Researchers can also perform genetic studies to identify and validate important quantitative trait loci (QTL) and genes.

The phenomics–ML approach can also help to develop better crop simulation models. Crop models are comprehensive dynamic mathematical descriptions of crop behavior and its interaction with the soil and environment (crop management conditions, climatic and soil parameters), and used to predict genotype performance under diverse environmental conditions. These models use daily weather data, soil data, crop parameters (pertaining to canopy traits, growth and development) and genetic parameters (e.g., photosynthesis, respiration, canopy height, etc.) and output the final trait, for example, seed yield. These crop simulation models are primarily statistical (describe output with some empirical function) and mechanistic (explain output using underlying physiological processes). For successful models (accurate predictions), vigilant parameterization of genetic coefficients and weather and soil environment descriptions are needed. These crop growth models can be quite useful, and plant breeders need to examine their usefulness more critically, as these models can help define more informed hypothesis as well as assist in ideotype-based breeding strategies. However, these models are generally more generic, and currently not created to predict yield performance in a breeding program where thousand to million unique plants and genotypes are present at a given time. Therefore, researchers can utilize phenomics to create more inclusive and diverse crop models, while ML approaches can provide trait prediction (either with supervised or unsupervised methods) with an eventual merging of crop models with ML based real time data analytics.

Another obvious application of phenomics–ML approaches is in breeding of stress traits. Although visual rating is the primary way of recording genotype response to abiotic and biotic stresses, these are problematic due to inter- and intra-rater variability, and is not easily scalable to large population. Visual assessment varies in rater accuracy, speed, and precision. HTP methods using multiple sensors types in conjunction with ML can identify and quantify symptoms associated with nutrient deficiencies and toxicity, bacterial-fungal-viral diseases, insect-pest infestations, weeds, and chemical injury, across multiple nurseries and environments. These improve accuracy and speed of phenotyping. These methods can go beyond identification and quantification, and provide control with drone or ground robot based mitigation strategy (mechanical, chemical, etc.) implemented in precise spot-control improving environmental sustainability and reducing costs. The phenotype or trait information generated using computer vision and ML have other applications. For example, ML methods were used to translate RGB images (iron deficiency chlorosis stress) of a diverse set of soybean canopies into a numerical rating scale and used for genome wide association studies based QTL discovery (Zhang et al., 2017). Naik et al. (2017) combined image capture, image processing, and classification workflow into a real-time phenotyping smartphone app for iron deficiency chlorosis severity rating in the field. Such tools can be eventually used by farmers and researchers for rapid data collection for scouting and breeding/research, respectively. Hyperspectral camera and associated ML based image analysis can also be useful for "early" stress detection, before visible to human rater (Nagasubramanian et al., 2018, 2019). The early identification of stress symptoms in farmer's field can help early decision making and stress mitigation strategies, and also useful for breeding selections.

While majority of plant breeding efforts have focused on above ground traits, below ground traits (relating to root traits) have lagged behind. This is not surprising because the study of root traits is resource intensive (labor, cost, and time); however, considering that roots provide the basis for nutrient uptake, water acquisition and structural integrity, more efforts are needed. Particularly, environmentally resilient cultivars that are needed to meet the production requirement in a changing climate, cannot be built without improving both above and below ground traits. Root phenotyping experiments have used field trials, controlled environment units (using special paper towel etc.), rhizotrons, and transparent soil. While more approaches exist with controlled environment approaches (e.g., germination paper, agar plates, transparent soil, or other media, aeroponics, hydroponics, etc.). However, these lack correlation with field performance (Fig. 28.5). On the other hand, indoor studies allow for controlled experiments with more genotypes, higher throughput, automation, and varying environmental variable to study genotype responses.

Earlier field based root studies included trench digging to expose the plant root zones, soil cores and mini-rhizotron, where durable and transparent tubes are installed in the ground with a light-weight cylindrical imaging device lowered into the tubes to obtain images and track root growth over season(s). These transparent subterranean tubes are buried at an angle in the soil, and require roots to intersect with the tube allowing for non-invasive and repeated imaging. Other methods have included manual root sample extraction followed by image-based methods, ground-penetrating radar, and X-ray CT scanning. Several imaging pipelines and softwares are available to obtain root traits (generally in a destructive manner), except where plants are grown in a manner that facilitates repeated time series data collection (e.g., in controlled conditions, Falk et al., 2020a,b). Recently, researchers have shown successful usage of ML to develop an automated soybean nodule acquisition pipeline (Jubery et al. 2020). Due to the complexity of roots, ML is especially exciting because a vast amount of data points can be collected in a short time compared to manual efforts. There is still a need to develop non-destructive HTP approaches for field-based root excavation/sample acquisition, imaging, and trait extraction. In addition to digital imagery, 3D imaging approaches such as X-ray computed tomography (CT), magnetic resonance imaging (MRI), and positron emission tomography (PET) are also being used but these are not high throughput for breeding applications. Root studies are observing continual hardware and software improvements, and there is a need for sustained effort in the study of root traits.

The integration of phenomics–ML with genomics is opening new doors for plant breeders. HTP can provide high quality phenotypic data for genomic prediction (GP) model development and improvement. Most GP models incorporate marker and/or pedigree information only, but not information on secondary traits. In this sense, the integration of secondary traits related to the target trait into GP models has shown improved accuracy in wheat (Juliana et al., 2019) and soybean (Zhang et al., 2017). Zhang et al. (2017) reported improved prediction accuracy when GP alone was compared with GP-secondary trait phenotype value as a covariate. In wheat, researchers showed that a GP + HTP traits as covariate showed equal or better prediction accuracy than the GP model alone (Juliana et al., 2019).

Advances in aerial and ground-based phenotyping, sensors (remote and proximal), phenomics and ML are positively impacting plant breeding pipelines. In Figs. 28.8 and 28.9, examples of the application of these methods are provided for cultivar development.

While these are non-exhaustive examples of application in cultivar development, the intent is to show that phenomics, HTP, ML methods can really

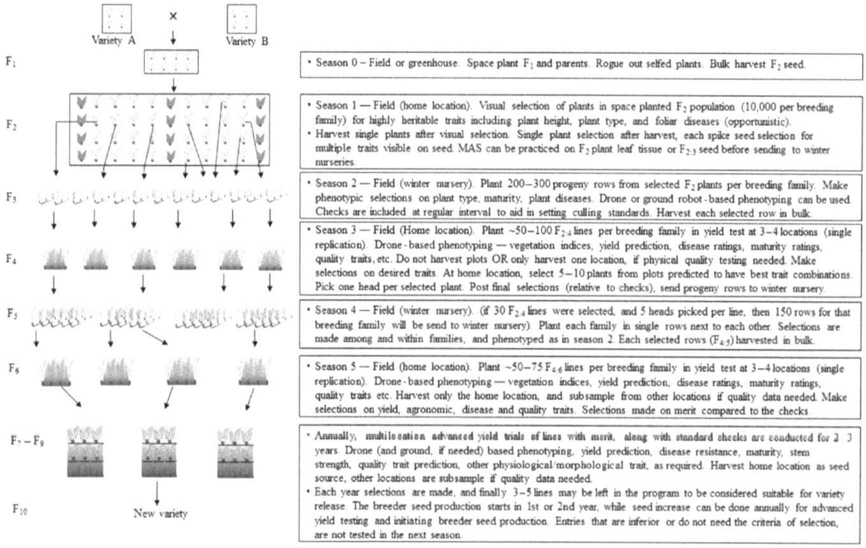

FIGURE 28.8

The use of phenomics, high throughput phenotyping, and machine learning algorithms in a modified pedigree method. Population sizes that are higher than listed in the figure can be handled. Due to the combination of mechanization and advanced data analytics, not all locations need to be harvested, which allows a significant cost and time savings. The scientific literature contains example of the use of these methods for yield prediction, crop health, stress trait phenotyping, crop maturity ratings, and stem strength, etc.While not included, genomic selection can be used for individual seed, plant or progeny row selection prior to planting. The combination of genomic and phenomic selection will give the maximum advantage, provided accurate models are developed and used for both.

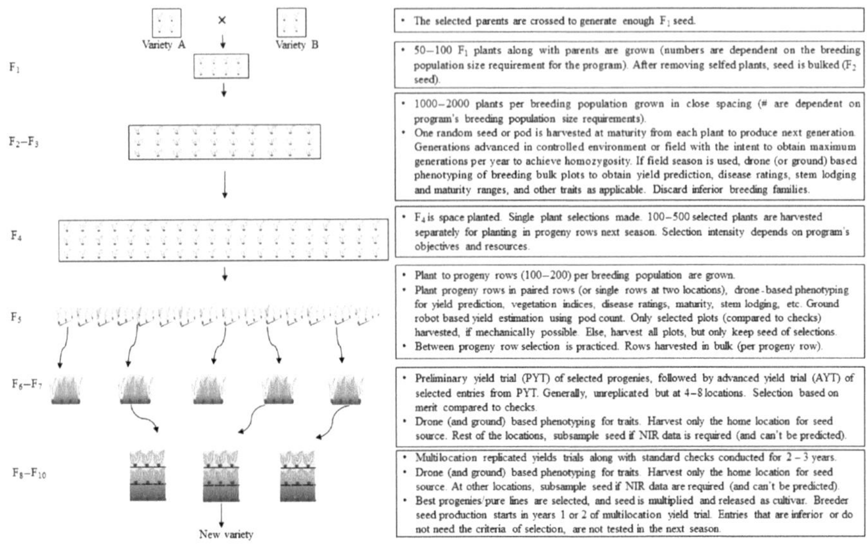

FIGURE 28.9

The use of phenomics, high throughput phenotyping, and machine learning algorithms in a modified single pod or seed descent method. Due to the combination of mechanization and advanced data analytics, not all locations need to be harvested, which allows a significant cost and time savings. The scientific literature contains example of the use of these methods for yield prediction, crop health, stress trait phenotyping, crop maturity ratings, and stem strength, etc. In the doubled haploid method, these tools can be applied at the progeny row testing stage. While not included, genomic selection can be used for individual seed, plant or progeny row selection prior to planting. The combination of genomic and phenomic selection will give the maximum advantage, provided accurate models are developed and used for both.

streamline the pipeline that can increase the size of the program (number of populations and its sizes), increased trait under consideration, reduce cost to the program, and improve the accuracy of selection (see Fig. 28.1).

Smart breeding

All the methods explained in this chapter help to create a "smart" breeding scenario. The concept of smart breeding includes extensive utilization of technology and analytics that empowers breeders to make informed selection, and strategize pipeline decisions (Fig. 28.10). The principles of cyber-physical systems motivate the creation of cyber-agricultural systems based smart breeding program. The cyber part consists of (1) data collection (high-throughput), computing (data

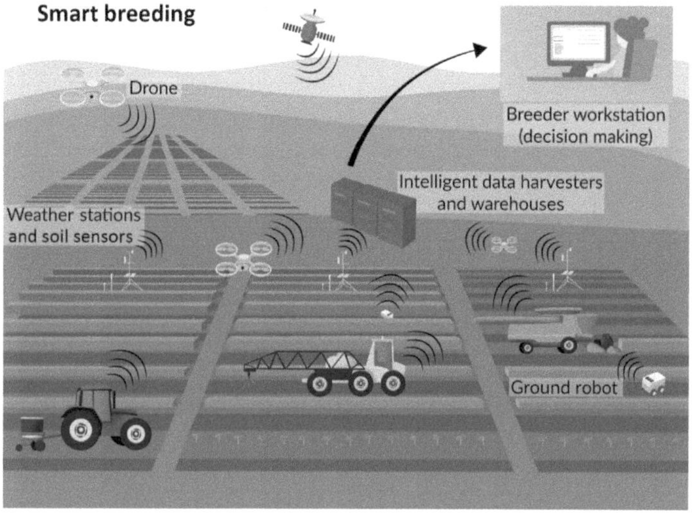

FIGURE 28.10

An overview of cyber-agricultural systems based smart breeding field operation. Multiple sensors (imaging, soil, weather) and phenotyping systems (drone, ground robots, satellite) in conjunction with farm operation and machinery to generate breeding relevant data, which are stored on machines or cloud, data analytics are performed, and results are accessible to plant breeder to make selection decisions. Other decisions include the controlling of operations and coordination of various phenotyping tools to obtain pertinent data. Genomic tools can be integrated in a smart breeding program to complement phenomics tools.

analyses, e.g., simple statistics and ML), data curation and storage (i.e., data organization and storage for easier transmission between breeding groups such as breeder, field staff, analysts, etc., and retrieval from hard drives and/or private cloud servers, if privacy is a concern), and (2) communication between the physical elements; sensors/controller and actuator (e.g., breeder made selection decisions). In a smart breeding process, multiple data streams are created (e.g., phenotypic and genomic, soil sensors, cameras, weather stations), and these data are transferred to data harvester/storage, for data analytics. Plant breeder makes the selection decision, which informs the staff and phenotyping systems operators on next tasks. For example, more data to be collected or selection/discard decisions to be made.

The smart breeding farm, although may seem costly at a cursory glance, is quite cost effective in situation where the size of the program or breeding organization is large and multi-dimensional with moving parts between crops, groups, etc. Smart breeding does not remove the need for traditional breeding; in fact, it can be damaging to the long-term success of the program if computer-based breeding is hastily implemented. Smart breeding can only be as successful as the quality of the measurement and prediction models that are built (developed, validated and continually improved, and span the required inference space for good predictions); and breeder insights and decision making are paramount. Smart breeding supplements the breeder's toolbox, by providing enhanced information to make better decisions in addition to existing tools that are routinely used. It allows more flexibility to the breeder in their pipeline, and gives them more time in the field and lab to observe and improve the entire pipeline, and make less breeding/selection errors. Smart breeding is not a substitute for hard work, scientific rigor, interdisciplinary collaboration, and vision, to be successful in cultivar development efforts.

Bibliography

References

Abdollahi-Arpanahi, R., Gianola, D., Peñagaricano, F., 2020. Deep learning versus parametric and ensemble methods for genomic prediction of complex phenotypes. Genet. Sel. Evol. 52 (1), 12.

Adams, M.W., 1973. Plant architecture and physiological efficiency in the field bean. Potential of Field Beans and other food legumes in Latin America. Centro Internacional de Agricultura Tropical, Cali, DC, pp. 226–278.

Adams, M.W., 1982. Plant architecture and yield breeding in *Phaseolus vulgaris*. Iowa State J. Res 56, 225–254.

Akdemir, D., Beavis, W., Fritsche-Neto, R., et al., 2019. Multi-objective optimized genomic breeding strategies for sustainable food improvement. Heredity 122, 672–683. Available from: https://doi.org/10.1038/s41437-018-0147-1.

Akerman, A., Mac Key J., 1948. The breeding of self fertilized plants by crossing. Some experience during 60 years of breeding at the Swedish Seed Association. In: Akerman, A., et al. (Ed.), Svalof 1886–1946, pp. 46–71.

Allan, R.E., 1987. Wheat. In: Fehr, W.R. (Ed.), Principles of Cultivar Development, Vol 2. Crop Species, Macmillan Publ. Co., New York, pp. 699–748.

Allard, R.W., 1960. Principles of Plant Breeding. Toppan Co. Ltd, Tokya, Japan, 485 p.

Anzalone, A.V., Randolph, P.B., Davis, J.R., Sousa, A.A., Koblan, L.W., Levy, J.M., et al., 2019. Search-and-replace genome editing without double-strand breaks or donor DNA. Nature 576 (7785), 149–157.

Ariyanayagam, R.P., Singh, N.B., 1994. Pigeonpea breeding- accomplishments and challenges. Plant Breed. 64, 773–782. Abstr.

Asaga, K., Yoshimura, S., 1969. Field resistance of sister lines of rice crosses to blast diseases. Ann Phytopatho. Soc. Jpn. 35, 100 (Abs).

Ashikari, M., Sakakibara, H., Lin, S., Yamamoto, T., Takashi, T., Nishimura, A., et al., 2005. Cytokinin oxidase regulates rice grain production. Science 309, 741–745.

Asker, S., 1979. Progress in apomixis research. Hereditas 91, 231–240.

Assefa, T., Zhang, J., Chowda-Reddy, R.V., Lauter, A.N.M., Singh, A., O'Rourke, J.A., et al., 2020. Deconstructing the genetic architecture of iron deficiency chlorosis in soybean using genome-wide approaches. BMC Plant Biol. 20 (1), 42.

Athwal, R.S., Kimber, G., 1972. The pairing of an alien chromosome with homoeologous chromosomes of wheat. Can. J. Genet. Cytol. 14, 325–333.

Ausemus, E.R., 1943. Breeding for disease resistance in wheat, oats, barley, and flax. Bot. Rev. 9, 207–260.

Baker, R.J., 1971. Theoretical variance of response to modified pedigree selection. Can. J. Plant Sci. 51, 463–468.

Barrus, M.F., 1911. Variation of varieties of beans in their susceptibility to anthracnose. Phytopathology 1, 190–195.

Bashaw, E.C., 1962. Apomixis and sexuality in buffel grass. Crop Sci. 2, 412–415.

Bateson, W., 1909. Mendel's Principles of Heredity. Cambridge University Press, Cambridge.

Bateson W., Punnett R., 1906. Experimental studies in the physiology of heredity. Reports to the Evolution Committee of the Royal Society, 3, pp. 1–53.

Beeck, C.P., Wroth, J.M., Falk, D.E., Khan, T., Cowling, W.A., 2008. Two cycles of recurrent selection lead to simultaneous improvement in black spot resistance and stem strength in field pea. Crop Sci. 48, 2235–2244.

Bellon, M.R., Brar, D.S., Lu, B.R., Pham, J.L., 1998. Rice genetic resources. In: Dowling, N.G., Greenfield, S.M. (Eds.), Sustainability of rice in the Global Food System. Pacific Basin Study Center; IRRI, Davis, CA; Manila, PH, pp. 251–283.

Bernardo, R., 2014. Essentials of Plant Breeding. Stemma Press.

Bevan, M.W., Uauy, C., Wulff, B.B., Zhou, J., Krasileva, K., Clark, M.D., 2017. Genomic innovation for crop improvement. Nature. 543 (7645), 346–354.

Biffen, R.H., 1905. Mendel's laws of inheritance and wheat breeding. J. Agric. Sci. 1, 4–48.

Biggs, S.D., 1989. Resource-poor farmer participation in research: a synthesis of experiences from nine national agricultural research systems. OFCOR Comparative Study Paper No. 3. The Hague: International Service for National Agricultural Research.

Blakeslee, A.F., Belling, J., Farnham, M.E., Bergner, A.D., 1922. A haploid mutant in the Jimson weed, Datura stramonium. Science 55 (1433), 646–647.

Blixt, S., 1975. Why didn't Gregor Mendel find linkage? Nature 256, 206.

Blum, A., 1987. Methods of plant breeding for drought resistance. In: Monti, L., Porceddu, E. (Eds.), Drought Resistance in Plants. Commission of the European Communities, Luxembourg, pp. 23S-254.

Bolley, L.H., 1905. Breeding for resistance or immunity to disease. Proc. Am. Breed. Assoc. 1, 131–135.

Borlaug, N.E., 1959. The use of multilineal or composite varieties to control air borne epidemic diseases of self pollinated crop plants. First Int. Wheat Genet. Symp. Winnipeg (1958), pp. 12–27.

Borlaug, N.E., 1965. Wheat, rust, and people. Phytopathology 55, 1088–1098.

Bortesi, L., Fischer, R., 2015. The CRISPR/Cas9 system for plant genome editing and beyond. Biotechnol. Adv. 33 (1), 41–52.

Bos, I., 1983. About the efficiency of grid selection. Euphytica 32, 885–893.

Bresler, E., McNeal, B.L., Caner, D.L., 1982. Saline and Sodic Soils. Principles-Dynamics-Modeling. Springer-Verlag Berlin Heidelberg.

Briggs, F.N., Allard, R.W., 1953. The current status of the backcross method of plant breeding. Agron. J. 45 (4), 131–138.

Briggs, F.N., Knowles, P.F., 1967. Introduction to Plant Breeding. Reinhold Publishing Corporation, Davis, CA, 426 p.

Brim, C.A., 1966. A modified pedigree method of selection in soybeans. Crop Sci. 6, 220.

Brim, C.A., Burton, J.W., 1979. Recurrent selection in soybeans. II. Selection for increased percent protein in seeds. Crop Sci. 19, 494–498.

Brim, C.A., Cockerham, C.C., 1961. The inheritance of quantitative characters in soybeans. Crop Sci. 1, 187–190.

Brim, C.A., Stuber, C.W., 1973. Application of genetic male sterility to recurrent selection schemes in soybeans. Crop Sci. 13, 528–530.

Brim, C.A., Young, M.F., 1971. Inheritance of a male sterile character in soybean. Crop Sci. 11, 564–566.

Brown, C.M., Forsberg, R.A., 1987. Oat. In: Fehr, W.R. (Ed.), Principles of Cultivar Development, Vol.2. Crop Species, Macmillon Publ, Co., New York, pp. 295–345.

Brown, H.B., Ware, J.O., 1958. Cotton. Mc Graw-Hill, New York, 566 p.
Bruce, A.B., 1910. The Mendelian theory of heredity and the augmentation of vigor. Science 32, 627–628.
Burton, J.W., Brim, C.A., 1978. Recurrent mass and within half-sib family selection for high percent oil in soybeans. Agron. Abstr. 48 p.
Burton, J.W., Carter Jr., T.E., Wilson, R.F., 1999. Registration of 'Prolina' soybean. Crop Sci. 39, 294–295.
Cairns, J.E., Sanchez, C., Vargas, M., Ordoñez, R., Araus, J.L., 2012. Dissecting maize productivity: ideotypes associated with grain yield under drought stress and well-watered conditions. J. Integr. Plant Biol. 54 (12), 1007–1020.
Cameron, J.N., Han, Y., Wang, L., Beavis, W.D., 2017. Systematic design for trait introgression projects. Theor. Appl. Genet. 130, 1993–2004.
Casler, M.D., 2015. Fundamentals of experimental design: guidelines for designing successful experiments. Agron. J. 107, 692–705.
Chahal, S.S., Gill, K.S., Phul, P.S., Singh, N.B., 1981. Effectiveness of recurrent selection for generating ergot resistance in pearl millet. SABRAO J. 13, 184–186.
Charlesworth, D., Willis, J.H., 2009. The genetics of inbreeding depression. Nat. Rev. Genet. 10 (11), 783–796.
Chase, S.S., 1949. Monoploid frequencies in a commercial double cross hybrid maize and its component single cross hybrids and inbred lines. Genetics 34, 328–332.
Chase, S.S., 1952. Production of homozygous diploids of maize from monoploids. Agron. J. 44 (5), 263–267.
Chauhan, Y.S., Johansen, C., 1996. Ideotypes of pigeonpea (*Cajanus cajan* (L.) Millsp.). Abstracts, 2nd International Crop Science Congress, New Delhi, India, 17–24 November 1996, 242.
Clarke, J.M., Knox, R.E., DePauw, R.M., Clarke, F.R., Fernandez, M.R., McCaig, T.N., et al., 2009a. Brigade durum wheat. Can. J. Plant Sci. 89, 505–509.
Clarke, J.M., Knox, R.E., DePauw, R.M., Clarke, F.R., McCaig, T.N., Fernandez, M.R., et al., 2009b. Eurostar durum wheat. Can. J. Plant Sci. 89, 317–320.
Clausen, J.D., Keck, D., Hiesey, W.H., 1945. Experimental Studies on the Nature of Plant Species. II. Plant Evolution through Amphidiploidy and Autopolyploidy with Examples from the Madiinae, 546. Carnegie Inst. Wash. Publ, pp. 1–174.
Collard, B.C., Mackill, D.J., 2008. Marker-assisted selection: an approach for precision plant breeding in the twenty-first century. Philos. Trans. R Soc. Lond. B Biol. Sci. 363 (1491), 557–572.
Collins, G.N., 1921. Dominance and vigor of first generation hybrids. Amer. Nat. 55, 116–133.
Compton, W.A., 1968. Recurrent selection in self-pollinated crops without extensive crossing. Crop Sci. 8, 773.
Compton, W.A., Comstock, R.E., 1976. More on modified ear-to-row selection in corn. Crop Sci. 16, 122.
Comstock, R.E., Robinson, H.F., Harvey, P.H., 1949. A breeding procedure designed to make maximum use of both general and specific combining ability. Agron. J. 41, 360–367.
Cong, L., Ran, F.A., Cox, D., Lin, S., Barretto, R., Habib, N., et al., 2013. Multiplex genome engineering using CRISPR/Cas systems. Science 339 (6121), 819–823.
Correns, C., 1900. G. Mendel's Regel Uber das Verhalten der Nachkommenschaft der Rassenbastarde. Berichte der Deutschen Botanischen Gessellschaft 18, 158–168.

Crespo-Herrera, L.A., Garkava-Gustavsson, L., Åhman, I., 2017. A systematic review of rye (*Secale cereale* L.) as a source of resistance to pathogens and pests in wheat (*Triticum aestivum* L.). Hereditas 154, 14.

Crow, J.F., 1947. A consequence of the dominance hypothesis of hybrid vigor. Records Genet. Soc. Amer. 16, 28.

Crow, J.F., 1948. Alternative hypotheses of hybrid vigor. Genetics 33, 477–487.

Cullis, B.R., Smith, A., Coombes, N., 2006. On the design of early generation variety trials with correlated data. J. Agric. Biol. Environ. Stat. 11, 381–393.

Cuthbert, R.D., DePauw, R.M., Knox, R.E., Singh, A.K., McCaig, T.N., Fetch, T., et al., 2016. AAC Elie hard red spring wheat. Can. J. Plant Sci. 96 (5), 919–927.

Cuthbert, R.D., DePauw, R.M., Knox, R.E., Singh, A.K., McCaig, T.N., McCallum, B., et al., 2017. AAC brandon hard red spring wheat. Can. J. Plant Sci. 97 (2), 393–401.

Davenport, C.B., 1908. Degeneration, albinism and inbreeding. Science 28, 454–455.

DePauw, R.M., Shebeski, L.H., 1973. An evaluation of an early generation yield testing procedure in *Triticum aestivum*. Can. J. Plant Sci. 53, 465–470.

DePauw, R.M., Knox, R.E., Clarke, F.R., Wang, H., Fernandez, M.R., Clarke, J.M., et al., 2007. Shifting undesirable correlations. Euphytica 157, 409–415.

DePauw, R.M., Knox, R.E., Humphreys, D.G., Thomas, J.B., Fox, S.L., Brown, P.D., et al., 2011. New breeding tools impact Canadian commercial farmer fields. Czech. J. Genet. Plant Breed. 47, S28–S34 (Special Issue).

d'Erfurth, I., Jolivet, S., Froger, N., Catrice, O., Novatchkova, M., Mercier, R., 2009. Turning meiosis into mitosis. PLoS Biol. 7 (6), e1000124.

Dhondt, S., Wuyts, N., Inzé, D., 2013. Cell to whole-plant phenotyping: the best is yet to come. Trends Plant Sci. 18 (8), 428–439.

Doebley, J., Gaut, B.S., Smith, B.D., 2006. The molecular genetics of crop domestication. Cell. 127, 1309–1321.

Doggett, H., 1965. Disruptive selection in crop improvement. Nature 206, 279–280.

Donald, C.M., 1968. The breeding of crop ideotypes. Euphytica 17, 385–403.

Dorrell, D.G., Downey, R.K., 1964. The inheritance of erucic acid content in rapeseed (*Brassica campestris*). Can. J. Plant Sci. 44, 499–504.

Downey, R.K., Harvey, B.L., 1963. Methods of breeding for oil quality in rape. Can. J. Plant Sci. 43, 271–275.

Downey, R.K., Rakow, G.F.W., 1987. Rapeseed and mustard. In: Fehr, W.R. (Ed.), Principles of Cultivar Development, Vol. 2. Crop Species, Macmillan Publ. Co., New York, pp. 437–486.

Driscoll, C.J., 1972. XYZ system of producing hybrid wheat. Crop Sci. 12, 516–517.

Driscoll, C.S., Jensen, N.F., 1964. Characteristics of leaf rust resistance transferred from rye to wheat. Crop Sci. 4, 372–374.

Dudley, J.W., 2007. From means to QTL: the Illinois long term selection experiment as a case study in quantitative genetics. Crop Sci. 47, S20–S31.

Duvick, D.N., 1974. Continuous backcrossing to transfer prolificacy to a single-eared inbred line of maize. Crop Sci. 14, 69–71.

Duvick, D.N., Cassman, K.G., 1999. Post-green revolution trends in yield potential of temperate maize in the north-central United States. Crop Sci. 39, 1622–1630.

Duvick, D.N., 2005. Genetic progress in yield of United States maize (Zea mays L.). Maydica 50, 193–200.

East, E.M., 1908. Inbreeding in corn. Rept. Connecticut Agric. Exp. Sta 1907, 419–428.

East, E.M., 1936. Heterosis. Genetics 21, 375–397.
East, E.M., Mangelsdorf, A.J., 1925. A new interpretation of the hereditary behavior of self-sterile plants. Proc. Nat. Acad. Sci. 11, 116–183.
Eathington, S.R., Crosbie, T.M., Edwards, M.D., Reiter, R.D., Bull, J.K., 2007. Molecular markers in a commercial breeding program. Crop Sci. 47, S154–S163.
Eberhart, S.A., 1970. Factors affecting efficiencies of breeding methods. Afr. Soils 15, 669–680.
Eberhart, S.A., Russell, W.A., 1966. Stability parameters for comparing varieties. Crop Sci. 6, 36–40.
Eenink, A.H. 1977. Genetics of host-parasite relationships and stability of resistance. In: Proc. FAO/IAEA Internat. Symp. on the use of induced mutations for improving disease resistance in crop plants. Vienna-SM-215/55, pp. 45–47.
Emsweller, S.L., Brierley, P., Lumsolen, D.V., Mulford, F.L., 1937. Improvement of flowers by bdreeding. U.S. Dept. Agric. Yearbook 890–998.
Erskine, W., 1983. Relationship between the yield of seed and straw in lentil. Field Crops Res. 7, 115–121.
ERS-USDA, 2019. Recent trends in GE adoption. <https://www.ers.usda.gov/data-products/adoption-of-genetically-engineered-crops-in-the-us/recent-trends-in-ge-adoption.aspx> (accessed 05.07. 20).
Falconer, D.S., 1981. Introduction to Quantitative Genetics., 2nd Edition Longman Group Ltd, London, pp. 1–133.
Falconer, D.S., Mackay, T.F.C., 1996. Introduction to Quantitative Genetics, 4th Edition Longmans Green, Harlow, Essex, UK.
Falk, K.G., Jubery, T., Mirnezami, S.V., Parmley, K.A., Sarkar, S., Singh, A., et al., 2020a. Computer vision and machine learning enabled soybean root phenotyping pipeline. BMC Plant Methods. v16, Article number: 5.
Falk, K.G., Jubery, T.Z., O'Rourke, J.A., Singh, A., Sarkar, S., Ganapathysubramanian, B., et al., 2020b. Soybean root system architecture traits study through genotypic, phenotypic and shape based clusters. Plant Phenom. Available from: https://doi.org/10.34133/2020/1925495. Article ID: 1925495.
FAO, 1999. Women: users, preservers and managers of agrobiodiversity. <http://www.fao.org/3/x0171e/x0171e03.htm#P181_22270> (accessed 05.07.20).
FAO, 2004. Building on gender, agrobiodiversity and local knowledge. <http://www.fao.org/3/y5609e/y5609e00.htm#Contents> (accessed 05.07.20).
FAO/IAEA, 2018. In: Spencer-Lopes, M.M., Forster, B.P., Jankuloski, L. (Eds.), Manual on Mutation Breeding, Third edition Food and Agriculture Organization of the United Nations, Rome, Italy, 301 pp.
FAOSTAT. 2020. <http://www.fao.org/faostat/en/#data/QC> (accessed 05.07.20).
Farmer, J.B., Moore, J.E.S., 1905. On the meiotic phase (reduction divisions) in animals and plants. Quart. J. Microscopical Sci 48, 489–557.
Fasoulas, A. 1973. A new approach to breeding superior yielding varieties. Arist. Univ., Thessaloniki, Publ. 3, pp. 42.
Favret, E.A., 1971. Different categories of mutations for disease reaction in the host organism. In Mutation breeding for disease resistance, IAEA, Vienna, pp. 107–116.
Federer, W.T., 1956. Augmented (or Hoonuiaku) designs. Hawaiian Planters' Record 55, 191–208.

Federer, W.T., 1960. Augmented designs with two-, three-, and higher way elimination of heterogeneity. Abstr. Biometrics 17, 166.
Federer, W.T., 1961. Augmented designs with one-way elimination of heterogeneity. Biometrics 17, 447–473.
Finlay, K.W., Wilkinson, G.N., 1963. The analysis of adaptation in a plant breeding programme. Aus. J. Agric. Res. 14, 742–754.
Flor, H.H., Comstock, V.E., 1971. Flax cultivars with multiple rust-conditioning genes. Crop Sci. 11, 64–66.
Flor, H.H., 1942. Inheritance of pathogenicity in *Melampsora lini*. Phytopathology 32, 653–669.
Flor, H.H., 1956. The complementary genic systems in flax and flax rust. Ad. Genet 8, 29–54.
Florell, V.H., 1929. Bulked-population method of handling cereal hybrids. J. Am. Soc. Agron 21, 718–724.
Francis, T.R., Kannenberg, L.W., 1978. Yield stability studies in short season-maize. I. A descriptive method for grouping genotypes. Can. J. Plant Sci 58, 1029–1034.
Frei, T.L., 2020. Breeding maize maternal haploid inducers. Plants (Basel) 9 (5), 614.
Freisleben, R.A., Lein, A., 1942. Über die Auffindung einer mehltauresistenten Mutante nach Röntgenbestrahlung einer anfälligen reinen Linie von Sommergerste. Naturwissenschaften. 30, 608.
Frey, K.J., 1967. Mass selection for seed width in oat populations. Euphytica. 16, 341–349.
Frey, K.J., Browning, J.A., 1971. Breeding crop plants for disease resistance. Int. Atomic Energy Agency, Vienna, Proc. of a Panel, Vienna, IAEA/FAO, pp. 45–54.
Frey, K.J., Browning, J.A., Simons, M.D., 1973. Management of host resistance genes to control diseases. Z. Pflanzenkrankh Pflanzensch 80, 160–180.
Frey, K.J., Browning, J.A., Simons, M.D., 1975. Multiline cultivars of autogamous crop plants. SABRAO J. 7, 113–123.
Gabriel, K.R., 1971. The biplot graphic display of matrices with application to principal component analysis. Biometrika 58, 453–467.
Gao, T., Emadi, H., Saha, H., Zhang, J., Lofquist, A., Singh, A., et al., 2018. A novel multirobot system for plant phenotyping. Robotics 7, 61.
Gardner, C.O., 1961. An evaluation of effects of mass selection and seed irradiation with thermal neutrons on yield of corn. Crop Sci. 1, 241–245.
Gardner, C.O., 1963. Estimates of genetic parameters in cross-fertilizing plants and their implications in plant breeding. In: Hanson, W.D., Robinson, H.F. (Eds.), Statistical Genetics and Plant Breeding. Natl. Acad. Sciences, Natl. Res. Council, Washington, DC, pp. 225–252.
Gardner, C.O., 1978. Population improvement in maize. In: Weldon, D.B. (Ed.), Maize Breeding and Genetics. John-Wiley and Sons, New York, USA.
Gartner, C.F., 1849. Versuche and Beobachtungen fiber die Bastarderzeugung im Pflanzenreich. Stuttgart. 790pp.
Gauch, H.G., Zobel, R.W., 1996. Optimal replication in selection experiments. Crop Sci. 36, 838–843.
Gaynor, R.C., Gorjanc, G., Bentley, A.R., et al., 2017. A two-part strategy for using genomic selection to develop inbred lines. Crop Sci. 57, 2372–2386.
Gayon, J., Zallen, D.R., 1998. The Role of the Vilmorin Company in the Promotion and Diffusion of the Experimental Science of Heredity in France. 1840–1920. J. History Biol. 31, 241–262.

Genter, C.F., 1976. Mass-selection in a composite of intercross of Mexican races of maize. Crop Sci. 16, 556–558.

Ghoshal, S., Blystone, D., Singh, A.K., Ganapathysubramanian, B., Singh, A., Sarkar, S., 2018. Bringing consistency to plant stress phenotyping through an explainable deep machine vision framework. Proc. Natl. Acad. Sci. U S A 115 (18), 4613–4618.

Ghosal, S., Blystone, D., Singh, A.K., Ganapathysubramanian, B., Singh, A., Sarkar, S., 2018. An explainable deep machine vision framework for plant stress phenotyping. PNAS. 115 (18), 4613–4618.

Gill, B.S., Raupp, W.J., 1987. Direct genetic transfer from *Aegilops squarrosa* L. to hexaploid wheat. Crop Sci. 27, 445–450.

Goulden, C.H., 1937. Modern methods for testing a large number of varieties. Dominion of Canada, Dept. Agr., Tech. Bul. 9.

Goulden, C.H., 1941. Problems in plant selection. In: Proceedings 7th International Genetics Congress, 1939, pp. 132–133.

Grafius, J.E., 1965. Short cuts in plant breeding. Crop Sci. 5, 377.

Grant, V., 1982. Periodicities in the chromosome numbers of the angiosperms. Bot. Gaz. 143 (3), 379–389.

Griffing, B., 1956. A generalized treatment of diallel crosses in quantitative inheritance. Heredity 10, 31–50.

Guha, S., Maheshwari, S., 1964. *In vitro* production of embryos from anthers of datura. Nature 204, 497.

Habier, D., Fernando, R.L., Garrick, D.J., 2013. Genomic BLUP decoded: a look into the black box of genomic prediction. Genetics 194 (3), 597–607.

Hallauer, A.R., 1967a. Development of single-cross hybrids from two-eared maize populations. Crop Sci 7, 192–195.

Hallauer, A.R., 1967b. Performance of single-cross hybrids from two-eared maize populations. Ann. Hybrid Corn Industry Res. Conf. Proc. 22, 74–81.

Hallauer, A.R., 1981. Selection and breeding methods. In: Frey, K.J. (Ed.), Plant Breeding II. Iowa State University Press, Ames, IA, pp. 3–55.

Hallauer, A.R., Sears, J.H., 1969. Mass selection for yield in two varieties of maize. Crop Sci. 9, 47–50.

Hallauer, A.R., Sears, J.H., 1972. Integrating exotic germplasm into corn belt maize breeding programs. Crop Sci. 12, 203–206.

Hallauer, A.R., Eberhart, S.A., 1970. Reciprocal full-sib selection. Crop Sci. 10, 315–316.

Hallauer, A.R., Miranda, J.B., 1981. Quantitative Genetics in Maize Breeding. Iowa state University Press, Ames, pp. 159–269.

Hanna, W.W., Bashaw, E.C., 1987. Apomixis: its identification and use in plant breeding. Crop Sci. 27, 1136–1139.

Hardy, G.H., 1908. Mendelian proportions in a mixed populations. Science 28 (706), 49–50.

Harlan, H.V., Pope, M.N., 1922. The use and value of back-crosses in small-grain breeding. J. Hered 13, 319–322.

Harlan, H.V., Martini, L.M., 1938. The effect of natural selection in a mixture of barley varieties. J. Agric. Res. 57, 189–199.

Harlan, H.V., Martini, M.L., Stebbins, H., 1940. A Study of Methods in Barley Breeding. U.S. Dept. Agri. Tech. Bul 720, 26 p.

Harlan, H.V., Martini, M.L., Stebbins, H., 1940. A study of methods in barley breeding. U.S. Dept. Agri. Tech. Bul. 720, 26.

Harlan, J.R., 1971. Agriculture origins: centers and non-centers. Science 174, 468–474.

Harlan, J.R., 1975a. Geographic patterns of variation in some cultivated plants. J. Hered. 66, 182–191.

Harlan, J.R., 1975b. Our vanishing genetic resources. Science 188, 618–621.

Harlan, J.R., de Wet, J.M.J., 1971. Towards a rational classification of cultivated plants. Taxon 20, 509–517.

Harrington, J.B. 1952. Cereal breeding procedures. Plant Production and Protection Series No. 14 Agric. Dev. Paper No. 28, FAO, Rome: 122 p.

Harshbarger, B., 1946. Preliminary report on the rectangular lattices. Biometrics Bull 2, 115–119.

Harshbarger, B. 1947. Rectangular lattices. Virginia Agricultural Experiment Station, Memoir 1.

Harshbarger, B., 1949. Triple rectangular lattices. Biometrics 5, 1–13.

Hartl, D.L., 1988. A Primer of Population Genetics, 2nd edition Sinauer Associates, Inc, Sunderland, MA.

Hartwig, E.E., Kilen, T.C., Young, L.D., Edwards Jr., C.J., 1982. Effects of natural selection in segregating soybean populations exposed to phytophthora rot or soybean cyst nematodes. Crop Sci. 22 (3), 588–590.

Hayes, H.K., Rinke, E.H., Tsiang, Y.S., 1944. The development of synthetic varieties of corn from inbred lines. J. Amer. Soc. Agron. 36, 998–1000.

Hays, H.K., Garber, R.J., 1919. Synthetic production of high protein corn in relation to breeding. J. Am. Soc. Agron. 11, 309–319.

Holbrook, C.C., Burton, J.W., Carter Jr., T.E., 1989. Evaluation of recurrent restricted index selection for increasing yield while holding seed protein constant in soybean. Crop Sci. 29, 324–329.

Hooker, A.L., 1972. Southern leaf blight of corn-Present status and future prospects. J. Environ. Qual 1, 244–249.

Hopkins, C.G., 1899. Improvement in the chemical composition of the corn kernel. Ill. Agric. Exp. Stn. Bull. 55, 205–240.

Hospital, F., 2001. Size of donor chromosome segments around introgressed loci and reduction of linkage drag in marker-assisted backcross programs. Genetics 158 (3), 1363–1379.

Houle, D., Govindaraju, D.R., Omholt, S., 2010. Phenomics: the next challenge. Nat. Rev. Genet. 11 (12), 855–866.

Howell, T., Hale, I., Jankuloski, L., Bonafede, M., Gilbert, M., Dubcovsky, J., 2014. Mapping a region within the 1RS.1BL translocation in common wheat affecting grain yield and canopy water status. Theor. Appl. Genet. 127, 2695–2709.

Huang, N., Angeles, E., Domingo, J., Magpantay, G., Singh, S., Zhang, G., et al., 1997. Pyramiding of bacterial blight resistance genes in rice: marker-assisted selection using RFLP and PCR. Theor. Appl. Genet. 95, 313–320.

Hull, F.H., 1945. Recurrent selection and specific combining ability in corn. J. Amer. Soc. Agron. 37, 134–145.

Hull, F.H., 1946. Overdominance and corn breeding where hybrid seed is not feasible. J. Amer. Soc. Agron. 38, 1100–1103.

Humphreys, D.G., Knox, R.E., 2015. Doubled haploid breeding in cereals. In: Al-Khayri, J.M., Jain, S.M., Johnson, D.V. (Eds.), Advances in Plant Breeding Strategies: Breeding, Biotechnology and Molecular Tools. Springer, New York, pp. 241–290.

Hutchinson, J., 1965. Essays on Crop Plant Evolution. University Press, London.

IRRI, 1980. Highlights 1979. IRRI, Philippines, pp. 28–29.

IRRI, 1980. Int Rice Res Inst. Philippines. Res. Highlights 1979, 28–29.

ISAAA, 2016. Brief 52: Global Status of Commercialized Biotech/GM Crops: 2016.
ISAAA, 2018. Brief 54: Global Status of Commercialized Biotech/GM Crops: 2018.
Jan, V.V.S., de Macedo, C.C., Kinet, J.-M., Bouharmont, J., 1997. Selection of Al-resistant plants from a sensitive rice cultivar, using somaclonal variation, *in vitro* and hydroponic cultures. Euphytica 97, 303.
Jenkins, M.T., 1934. Methods of estimating the performance of double crosses in corn. J Am. Soc. Agron. 26, 199–204.
Jenkins, M.T., 1940. Recurrent selection as a means of modifying combining ability in corn. Agron. J. 43, 311–315.
Jennings, P.R., 1964. Plant type as a rice breeding objective. CropSci. 4, 13–14.
Jensen, N.F., 1952. Intra-varietal diversification in oat breeding. Agron. J. 57, 621–624.
Jensen, N.F., 1970. A diallel selective mating system for cereal breeding. CropSci. 10, 629–635.
Jensen, N.F., 1988. Plant Breeding Methodology. Wiley Inter Science Publications.
Jinek, M., Chylinski, K., Fonfara, I., Hauer, M., Doudna, J.A., Charpentier, E., 2012. A programmable dual-RNA-guided DNA endonuclease in adaptive bacterial immunity. Science 337 (6096), 816–821.
Johannsen, W.L., 1903. Ueber Erblichkeit in Populationen und reinen Linien. Eine Beitrag zur Beleuchtung schwebender Selektionsfragen. Gustav Fischer Verlag, Jena.
Johannsen, W.L., 1909. Elemente der exakten Erblichkeitslehre. Gustav Fischer, Jena.
Johannsen, W.L., 1911. The genotype conception of heredity. Amer. Naturalist. 45 (531), 129–159.
Johannsen, W.L., 1926. Elements der exacten Erblichkeitslehre. Gustav Fischer, Jena.
Johnson, H.W., Bernard, R.L., 1962. Soybean genetics and breeding. Adv. Agron. 14, 149–211.
Johnston, T.D., 1974. Transfer of disease resistance from Brassica campestris L. to rape (*B. napus* L.). Euphytica 23, 681–683.
Jones, D.F., 1917. Dominance of linked factors as a means of accounting for heterosis. Genetics 2, 466–479.
Jones, D.F., 1918. The effect of inbreeding and cross breeding upon development. Conn. Expt. Sta. Bull 207.
Jones, D.F., 1939. Continued inbreeding in maize. Genetics 24, 462–473.
Jones, D.F., 1945. Heterosis resulting from degenerative changes. Genetics 30 (6), 527–542.
Jones, H.A., Clarke, A.E., 1943. Inheritance of male sterility in the onion and the production of hybrid seed. Proc. Ames. Soc. Hort. Sci. 43, 189–194.
Jubery, T.Z., Carley, C.N., Singh, A., Sarkar, S., Ganapathysubramanian, B., Singh, A.K., 2020. Using machine learning to develop a fully automated soybean nodule acquisition pipeline (SNAP). bioRxiv 2020.10.12.336156.
Jørgensen, J.H., 1971. Comparison of induced mutant genes with spontaneous genes in barley conditioning resistence to powdery mildew. In: Mutation Breeding for Disease Resistance. Proceedings, IAEA and FAO, pp. 117–124.
Juliana, P., Montesinos-López, O.A., Crossa, J., Mondal, S., Pérez, L.G., Poland, J., et al., 2019. Integrating genomic-enabled prediction and high-throughput phenotyping in breeding for climate-resilient bread wheat. Theor. Appl. Genet. 132, 177–194.
Kao, K.N., Kasha, K.J., 1969. Haploid from interspecific crosses with tetraploid barley. Barley Genet 2, 82–87.

Karpechenko, G.D., 1928. Polyploid hybrids of *Raphanus sativus* L. X *Brassica oleracea* L. Z. indukt. Abstamm. Vererb. Lehre 48, 1–85.

Kasha, K.J., Kao, K.N., 1970. High frequency haploid production in barley (*Hordeum vulgare* L.). Nature 225, 874–876.

Kaufmann, M.L., 1961. A proposed method of oat breeding for Central Alberta. Cereal News 6, 15–18.

Kaufmann, M.L., 1971. The random method of oat breeding for productivity. Can. J. Plant Sci. 51, 13–16.

Kawai, T., Amano, E., 1991. Mutation breeding in Japan. Plant Mutation Breeding for Crop Improvement. IAEA, Vienna, pp. 47–66, Khalaburdin, A.P. 1991.

Keeble, F., Pellew, C., 1910. The mode of inheritance of stature and the time of flowering in peas. J. of Genet. 1, 47–56.

Kelly, J.D., 2001. Remaking bean plant architecture for efficient production. Ad. Agron. 71, 111–143.

Kempton, R.A., 1984. The design and analysis of unreplicated trials. Vortage fur Pflanzenzuchtung 7, 219–242.

Khush, G.S., 1971. Rice breeding for disease and insect resistance at IRRI. Oryza 8, 111–119.

Khush, G.S., 1977. Disease and insect resistance in rice. Adv. Agron 29, 265–341.

Khush, G.S., 1979. Genetics of and breeding for resistance to the brown planthopper threat to rice production in Asia. International Rice Research Institute, Philippines, pp. 321–332.

Khush, G.S., Beachell, H.M., 1972. Breeding for disease and insect resistance at IRRI. Rice Breeding. International Rice Research Institute, Los Banos, PH, pp. 309–322.

Kihara, H., Ono, T., 1926. Chromosomenzahlen Und Systematische Gruppierung Der Rumex-Arten. Zeitschrift Für Zellforschung Und Mikroskopische Anatomie 4 (3), 475–481.

Kirk, J.T.O., Hurlstone, C.J., 1983. Variation and inheritance of erucic acid content in *Brassica juncea*. Z. Pflanzenzuchtg 90, 331–338.

Knott, D.R., 1961. The inheritance of rust resistance. VI. The transfer of stem rust resistance from *Agropyron elongatum* to common wheat. Can. J. Plant Sci 41, 109–123.

Knott, D.R., 1971. The transfer of genes for disease resistance from alien species to wheat by induced translocations. Mutation Breeding for Disease Resistance. IAEA, Vienna, pp. 67–77.

Knox, R.B., 1967. Apomixis: seasonal and population differences in a grass. Science 157, 325–326.

Kokubun, M., 1988. Design and examination of soybean ideotypes. JARQ 21 (4), 237–243.

Konzak, C.F., 1956. Induction of mutations for disease resistance in cereals. Brookhaven Symposia in Biology 9, 157–176.

Ladizinsky, G., Abbo, S., 2015. The Annual Species of the Cicer Genus. In: Ladizinsky, G., Abbo, S. (Eds.), The search for wild relatives of cool season legumes. Springer, Cham, Switzerland, pp. 29–53. Springer Briefs in Plant Science.

Lamkey, K.R., Edwards J.W., 1999. Quantitative genetics of heterosis. In: Coors, J.G., Pandey, S. (Eds.), The Genetics and Exploitation of Heterosis in Crops. ASA, CSSA and SSSA WI, pp. 31–48.

Larkin, P.J., Scowcroft, W.R., 1981. Somaclonal variation—a novel source of variability from cell cultures for plant improvement. Theor. Appl. Genet 60, 197–214.

Larson, R.I., Atkinson, T.G., 1973. Wheat-Agropyron chromosome substitution lines as sources of resistance to wheat streak mosaic virus and its vector, *Aceria tulipae*. In: Sears, E.R., Sears, L.M.S. (Eds.), Proc 4th Int/Wheat Genet Symp, Columbia, MO, pp. 173–177.

Lawrence, P.K., Frey, K.J., 1975. Backcross variability for grain yield in oat species crosses (*Avena sativa* L. x *A. sterilis* L.). Euphytica 24, 77–85.

Le Couteur, J., 1836. On the varieties, properties, and classification of wheat. Shearsmith, London.

Levine, M.N., Stakman, E.C., 1918. A third biologic form of *Puccinia graminis* on wheat. J. Agric. Res. 13, 651–654.

Levitt, J., 1980a. Response of plants to environmental stresses. Vol. 1, Chilling, Freezing, and High Temperature Stresses, second edition. A.P., New York.

Levitt, J., 1980b. Response of Plants to Environmental Stresses. Vol. 2, Water, Radiation, Salt and other Stresses, second edition. A. P., New York.

Lin, S.C., Yuan, L.P., 1980. Hybrid rice breeding in China. Innovative Approaches to Rice Breeding. International Rice Research Institute, Philippines, pp. 35–51.

Lonnquist, J.H., 1964. A modification of ear-to-row procedures for the improvement of maize populations. Crop Sci. 4, 227–228.

Lonnquist, J.H., 1967. Mass selection for prolificacy in maize. Zuchter 37, 185–188.

Lonnquist, J.H., Williams, N.E., 1967. Development of maize hybrids through selection among full sib families. Crop Sci. 7, 369–370.

Love, H.H., 1927. A program for selecting and testing small grains in successive generations following hybridization. Am. Soc. Agron. J. 19, 705–712.

Lupton, F.G.H., Whitehouse, R.N.H., 1957. Studies on the breeding of self-pollinating cereals. I. Selection methods in breeding for yield. Euphytica 6, 169–184.

Mac Key, J., 1962. The 75 years development of Swedish plant breeding. Hodowla Roslin Aklim. i Nasiennictwo 6, 437–467.

Mak, C., Harvey, B.L., Berdahl, J.D., 1978. An evaluation of control plots and moving means for error control in barley nurseries. Crop Sci. 18, 870–873.

Maluszynski, M., Nichterlein, K., van Zanten, L., Ahloowalia, B.S., 2000. Officially released mutant varieties-the FAO/IAEA database. Mut. Breed. Rev. 12, 1–84.

Manghwar, H., Lindsey, K., Zhang, X., Jin, S., 2019. CRISPR/Cas system: recent advances and future prospects for genome editing. Trends Plant Sci 24 (12), 1102–1125.

Marshall, D.R., 1977. The advantages and hazards of genetic homogeneity. The Genetic Basis of Epidemics in Agriculture. 287, (1), pp. 1–20.

Marshall, D.R., Brown, A.H.D., 1975. Optimum sampling strategies in genetic conservation. In: Frankel, O.H., Brown, J.G. (Eds.), Crop Genetic Resources for Today and Tomorrow. IBP Handbook.

Marshall, D.R., Pryor, A.J., 1979. Multiline varieties and disease control. 2. The "dirty-crop" approach with components carrying two or more genes for resistance. Euphytica 28, 145–159.

Mather, K., 1943. Polygenic inheritance and natural selection. Biol. Rev. 18, 32–64.

Mather, K., 1950. The genetical architecture of heterostyly in *Primula sinensis*. Evolution 26, 103–112.

Mather, K., 1953. The genetical structure of populations. Symposia of the Society for Experimental Biology. 7, pp. 66–95.

Matzinger, D.F., Cockerham, C.C., Wernsman, E.A., 1977. Single character and index mass selection with random mating in a naturally self-fertilizing species. In: Pollak, E.,

Kempthorne, O., Bailey, T.B. (Eds.), Proc. Intern. Conf. on Quant. Genet. The Iowa State University Press, Ames, pp. 503–518.
McMullen, M., Jones, R., Gallenberg, D., 1997. Scab of wheat and barley: A re-emerging disease of devastating impact. Plant Dis. 81, 1340–1348.
Mendel, G. 1866. Versuche über Plflanzen-hybriden. Verhandlungen des naturforschenden Ver-eines in Brünn, Bd. IV für das Jahr 1865, Abhand-lungen, 3–47.
Merezhko, A., 1998. Impact of plant genetic resources on wheat breeding. Euphytica 100, 295–303.
Merezhka, A.F., 1998. Impact of plant genetic resources on wheat breeding. In: Braun, H. J., coworkers (Eds.), Wheat: Prospects for Global Improvement. Kluwer Academic Publs, The Netherlands, pp. 361–369.
Mesterházy, Á., Bartók, T., Kászonyi, G., Varga, M., Tóth, B., Varga, J., 2005. Common resistance to different Fusarium spp. causing Fusarium head blight in wheat. Eur J Plant Pathol. 112, 267–281.
Meuwissen, T.H., Hayes, B.J., Goddard, M.E., 2001. Prediction of total genetic value using genome-wide dense marker maps. Genetics 157 (4), 1819–1829.
Miles, J.W., Dudley, J.W., White, D.G., Lambert, R.J., 1980. Improving corn population for grain yield and resistance to leaf blight and stalk rot. Crop Sci. 20, 247–251.
Miller, P., 1768. The gardeners dictionary. John and Francis Rivington.
Miller, P.A., Rawlings, J.O., 1967. Selection for increased lint yield and correlated responses in upland cotton, *Gossypium hirsutum* L. Crop Sci. 7, 637–640.
Mock, J.J., Pearce, P.B., 1975. An ideotype of maize. Euphytica 24, 613–623.
Moellers, T.C., Singh, A., Zhang, J., Brungardt, J., Kabbage, M., Mueller, D.S., et al., 2017. Main and epistatic loci studies in soybean for *Sclerotinia sclerotiorum* resistance reveal multiple modes of resistance in multi-environments. Sci. Rep. 7 (1), 3554.
Moll, R.H., Stuber, C.W., 1971. Comparisons of response to alternative selection procedures initiated with two populations of maize (*Zea mays* L.). Crop Sci. 11, 706–711.
Möller, M., Alchanatis, V., Cohen, Y., Meron, M., Tsipris, J., Naor, A., et al., 2007. Use of thermal and visible imagery for estimating crop water status of irrigated grapevine. J. Exp. Bot. 58 (4), 827–838.
Montesinos-López, O.A., Montesinos-López, A., Gianola, D., Crossa, J., Hernández-Suárez, C.M., Martín-Vallejo, J., 2018. Multi-trait, multi-environment deep learning modeling for genomic-enabled prediction of plant. G3 8 (12), 3829–3840.
Morris, M.L., Bellon, M.R., 2004. Participatory plant breeding research: opportunities and challenges for the international crop improvement system. Euphytica 136 (1), 21–35.
Mujeeb-Kazi, A., Rajaram, S., 2002. Transferring alien genes from related species and genera for wheat improvement. In: Curtis, B.C., Rajaram, S., Macpherson, H.G. (Eds.), Bread Wheat Improvement and Production. FAO Plant Production and Protection Series. No. 30.
Muller, H.J., 1927. Artificial transmutation of the gene. Science 66, 84–87.
Murata, Y., Osada, A., Iyama, J., Yamada, N., 1957. Photosynthesis of rice plants (IV). Plants factors constituting photosynthetic ability of the rice plants growing on paddy field. Proc. Crop Sci. Soc. 25 (B), 133–137. Japan.
Murray, H.J., 1969. Successful use of irradiation breeding to obtain Verticillium-resistant strains of peppermint, Heneha piperiea L. Induced Mutations in Plants. IAEA, Vienna, pp. 345–370.
NAAS, 2013. Role of Millets in Nutritional Security of India. Policy Paper No.66. National Academy of Agriculture Sciences, New Delhi, 16.

Nabors, N.W., Daniels, A., Nadolny, L., Brown, C., 1975. Sodium chloride tolerant lines of tobacco cells. Plant Sci. Lett. 4, 155–159.

Nagasubramanian, K., Jones, S., Sarkar, S., Singh, A.K., Singh, A., Ganapathysubramanian, B., 2018. Hyperspectral band selection using genetic algorithm and support vector machines for early identification of charcoal rot disease in soybean stems. Plant Methods 14, 86. Available from: https://doi.org/10.1186/s13007-018-0349-9.

Nagasubramanian, K., Jones, S., Singh, A.K., Sarkar, S., Singh, A., Ganapathysubramanian, B., 2019. Plant disease identification using explainable 3D deep learning on hyperspectral images. BMC Plant Methods 15, 98. Available from: https://doi.org/10.1186/s13007-019-0479-8.

Nagasubramanian, K., Singh, A.K., Singh, A., Sarkar, S., Ganapathysubramanian, B., 2020. Usefulness of interpretability methods to explain deep learning based plant stress phenotyping. arXiv:2007.05729.

Nagasubramanian, K., Jubery, T.Z., Ardakani, F.F., Mirnezami, S.V., Singh, A.K., Singh, A., et al., 2020. How useful is active learning for image-based plant phenotyping? arXiv:2006.04255.

Naik, H.S., Zhang, J., Lofquist, A., Assefa, T., Sarkar, S., Ackerman, D., et al., 2017. A real-time phenotyping framework using machine learning for plant stress severity rating in soybean. Plant Methods. 13, 23. Available from: https://doi.org/10.1186/s13007-017-0173-7).

Nanda, J.S., Sharma, S.D. (Eds.), 2003. Monograph on Genus Oryza. Sam Publishers, Plymouth, UK, pp. 283–289.

Nelson, R.R., 1972. Stabilizing racial populations of plant pathogens by use of resistance genes. J. Environ. Qual 1, 220–227.

Nelson, R.R., 1973. Breeding Plants for Disease Resistance. Pennsylvania State University Press, p. 401.

Nelson, R.R., 1978. Genetics of horizontal resistance to plant diseases. Annu. Rev. Phytopathol. 16, 359–378.

Newman, L.H., 1912. Plant Breeding in Scandinavia. Canadian Seed Growers Association. Ottawa.

Nicolaisen, W., 1934. Die Grundlagen del' Immunitatszuchtung gegen *Ustilago avenae* (Pers.) Jens. Z. Zucht, A XIX, I-56.

Nilsen, K.T., N'Diaye, A., MacLachlan, P.R., Clarke, J.M., Ruan, Y., Cuthbert, R.D., et al., 2017. High density mapping and haplotype analysis of the major stem-solidness locus SSt1 in durum and common wheat. PLoS ONE 12 (4), e0175285.

Nilsson-Ehle, H., 1909. Nagot on nuvarande principer vid hostvete foradlingen pa. Svalof (Present principles in breeding winter wheat at Svalof). Sveriges Utsadesf. tidskr 18, 165–170.

Orton, W.A., 1900. The wilt disease of cotton and its control. Bull. Div. Veg. Physiol. Path. U.S. Dep. Agric 27.

Orton, W.A., 1908. The development of farm crops resistant to disease. U. S. Dept. Agr. Yearbook 1908, 453–464.

Padmanabhan, S.Y., 1973. The great Bengal famine. Ann. Rev. Phytopathol. 11, 11–26.

Painter, R.H., 1951. Insect Resistance in Crop Plants. Macmillan, New York.

Parmley, K.A., Higgins, R.H., Ganapathysubramanian, B., Sarkar, S., Singh, A.K., 2019a. Machine learning approaches for prescriptive plant breeding. Scientific Reports 9, 17132.

Parmley, K., Nagasubramanian, K., Sarkar, S., Ganapathysubramanian, B., Singh, A.K., 2019b. Development of optimized phenomic predictors for efficient plant breeding decisions using phenomics assisted selection in soybean plant phenomics. Plant Phenomics 2019, Article ID 5809404.

Pathirana, R., 1992. Gamma ray induced field tolerance to Phytophthora blight in sesame. PlantBreed. 108, 314−319.

Payne, T.S., Stuthman, D.D., McGraw, R.L., Bregitzer, P.P., 1986. Physiological response to three cycles of recurrent selection for grain yield improvement in oats. Crop Sci. 26, 734−736.

Peixoto, L.A., Moellers, T.C., Zhang, J., Lorenz, A.J., Bhering, L.L., Beavis, W.D., et al., 2017. Leveraging genomic prediction to scan germplasm collection for crop improvement. PLoS One. 12 (6), e0179191.

Pendleton, J.W., Smith, G.E., Winter, R.S., Johnston, T.J., 1968. Field investigations of the relationships of leaf angle in corn (*Zea mays* L.) to grain yield and apparent photosynthesis. Agron. J. 60, 422−424.

Pesek, J., Baker, R.J., 1969. Desired improvement in relation to selection indices. Can. J. Plant Sci. 49, 803−804.

Pieruschka, R., Poorter, H., 2012. Phenotyping plants: genes, phenes and machines. Funct. Plant Biol. 39 (11), 813−820.

Rasmussen, D.C., 1987. An evaluation of ideotype breeding. Crop Sci. 27, 1140−1146.

Rasmusson, J., 1934. A contribution to the theory of quantitative inheritance. Hereditas 18, 245−261.

Reddy, M.S.S., Rao, M.V., 1979. Resistance genes and their deployment for control of leaf rust of wheat. Indian J. Genet. Plant Breed. 39, 359−365.

Reyniers, F.N., Truong-Binh, L.J., Nicou, R., 1982. Breeding for drought resistance in dryland rice. Drought Resistance in Crops with Emphasis on Rice. International Rice Research Institute, Los Banos, Philippines, pp. 273−292.

Rhodes, R., Booth, R., 1982. Farmer-back-to-farmer: a model for generating acceptable agricultural technology. Social Science Department Working Paper 982: Lima, Peru: Centro Internacional de la Papa.

Ribaut, J.M., de Vicente, M.C., Delannay, X., 2010. Molecular breeding in developing countries: challenges and perspectives. Curr. Opin. Plant Biol. 13 (2), 213−218.

Richards, L.A., 1954. Diagnosis and Improvement of Saline and Alkali Soils. USDA Handbook 60, United States Dept. of Agriculture.

Richey, F.D., 1924. Adjusting yields to their regression on a moving average, as a means of correcting for soil heterogeneity. J. Agric. Res. 27, 79−90.

Richey, F.D., 1927. The convergent improvement of selfed lines of corn. Am. Nat. 61, 430−449.

Riera, L.G., Carroll, M.E., Zhang, Z., Shook, J.M., Ghosal, S., Gao, T., et al.. 2020. Deep multi-view image fusion for soybean yield estimation in breeding applications. arXiv:2011.07118.

Romero, G.E., Frey, K.J., 1966. Mass selection for plant height in oat populations. Crop Sci. 6, 283−287.

Rowe, D.E., Hill Jr, R.R., 1981. Inter-population improvement procedures for alfalfa. Crop Sci. 21, 392−397.

Ruan, Y., Singh, A.K., DePauw, R.M., Knox, R.E., McCaig, T.N., Cuthbert, R.D., et al., 2016. AAC Congress durum wheat. Can J. Plant Sci. 98 (2), 483−491.

Ruan, Y., Singh, A.K., DePauw, R.M., Knox, R.E., Cuthbert, R.D., McCallum, B., et al., 2019. AAC Stronghold Durum Wheat. Can. J. of Plant Sci. 99 (4), 560−567.

Russell, W.A., Eberhart, S.A., 1975. Hybrid performance of selected maize line from reciprocal recurrent and testcross selection programs. Crop Sci. 15, 1−4.

Sagaret, A., 1826. Considerations sur la productions des hybrides des variantes et des variétés en general, et sur celles de Cucurbitacées en particulier. Annales des. Sci. Naturelles, Ser. 1 8, 294–314.

Salvi, S., Sponza, G., Morgante, M., Tomes, D., Niu, X., Fengler, K.A., et al., 2007. Conserved noncoding genomic sequences associated with a flowering-time quantitative trait locus in maize. Proc. Natl. Acad. Sci. 104, 11376–11381.

Sanford, L.L., Ladd, T.L., 1983. Selection for resistance to potato leafhopper in potatoes: III. Comparison of two selection procedures. Amer. Potato J. 60, 653–659.

Sarsu, F., Penna, S., Kunter, B., Ibrahim, R., 2018. Mutation breeding for vegetatively propagated crops. In: Spencer-Lopes, M.M., Forster, B.P., Jankuloski, L. (Eds.), Manual on Mutation Breeding, Third edition Food and Agriculture Organization of the United Nations, Rome, Italy, 301 pp.

Schlenker, W., Roberts, M.J., 2009. Nonlinear temperature effects indicate severe damages to U.S. crop yields under climate change. Proc. of the Natl. Acad. Sci. 106 (37), 15594–15598.

Schnable, P.S., Springer, N.M., 2013. Progress Toward Understanding Heterosis in Crop Plants. Annu. Rev. Plant. Biol. 64, 71–88.

Sears, E.R., 1956. Transfer of leaf rust resistance from *Aegilops umbellulata* to wheat. Genetics in Plant Breeding Brookhaven Symp. Biol. No. 9, 1–21.

Sears, E.R., 1966. Chromosome mapping with the aid of telocentrics. Hereditas 2 (Suppl), 370–381.

Sedcole, J.R., 1977. Number of plants necessary to recover a trait. Crop Sci. 17, 667–668.

Sedgley, R.H., Siddique, K.H.M., Walton, G.H., 1990. Chickpea ideotypes for Mediterranean environments. Chickpea in Nineties. Proc. of the 2nd Intl. workshop on chickpea improvement. ICRISAT, Patancheru, India, pp. 87–91.

Semenov, M.A., Stratonovitch, P., 2013. Designing high-yielding wheat ideotypes for a changing climate. Food Energy Secur 3, 185–196.

Shebeski, L.H., 1967. Wheat and wheat breeding. In: Nielsen, K.F. (Ed.), Proc. Canadian Centennial Wheat Symp. Modern Press, Saskatoon, Saskatchewan, pp. 249–272.

Shebeski, L.H., 1970. Wheat and Breeding. Information Bulletin on Near East Wheat and Barley Improvements and Production Project. FAO, Rome, 7(2), pp. 19–34.

Shook, J., Gangopadhyay, T., Wu, L., Ganapathysubramanian, B., Sarkar, S., Singh, A.K., 2020. Crop yield prediction integrating genotype and weather variables using deep learning. arXiv:2006.13847

Shook, J.M., Lourenco, D., Singh, A.K., 2020. PATRIOT: a pipeline for tracing identical-by-descent chromosome segments to improve genomic prediction in self-pollinating crop species. bioRxiv 2020.10.17.343780. Available from: https://doi.org/10.1101/2020.10.17.343780.

Shull, G.H., 1908. The composition of a field of maize. Rep. Amer. Breed. Ass. 4, 296–301.

Shull, G.H., 1909. A pure line method in corn breeding. Rept. Amer. Breeders' Assoc. 5, 51–59.

Shull, G.H., 1910. Hybridization Methods in Corn Breeding. J. Hered 2, 98–107.

Shull, G.H., 1911. The genotypes of maize. Amer. Nat. 45, 234–252.

Shull, G.H., 1914. Duplicate genes for capsule form in Bursa bursa-pastoris. Z.I.A.V. 12, 97–149.

Shull, G.H., 1948. What is "heterosis"? Genetics 33, 439–446.

Siddique, K.H.M., Sedgley, R.H., Marshall, C., 1984. Effect of plant density on growth and harvest index of branches in chickpea (*Cicer arietinum* L.). Field Crops Res. 9, 193–203.

Singh, A., Pandey, M.P., Singh, A.K., Knox, R.E., Ammar, K., Clarke, J.M., et al., 2013a. Identification and mapping of leaf, stem and stripe rust resistance quantitative trait loci and their interactions in durum wheat. Mol. Breed. 31 (2), 405–418.

Singh, A., Knox, R.E., DePauw, R.M., Singh, A.K., Cuthbert, R.D., Campbell, H.L., et al., 2013b. Identification and mapping in spring wheat of genetic factors controlling stem rust resistance and the study of their epistatic interactions across multiple environments. Theor Appl Genet. 126 (8), 1951–1964 [published correction appears in Theor. Appl. Genet. 127(4):1005–8].

Singh, D.P., 2014. Genetics and breeding of pulse crops, 3rd edition Kalyani Publishers, 476 p.

Singh, A., Knox, R.E., DePauw, R.M., Singh, A.K., Cuthbert, R.D., Campbell, H.L., et al., 2014. Stripe rust and leaf rust resistance QTL mapping, epistatic interactions, and co-localization with stem rust resistance loci in spring wheat evaluated over three continents. Theor. Appl. Genet. 127 (11), 2465–2477.

Singh, A., Jones, S., Ganapathysubramanian, B., Sarkar, S., Mueller, D., Sandhu, K., et al., 2020. Challenges and opportunities in machine-augmented plant stress phenotyping. Trends in Plant Sci. ISSN 1360-1385. Available from: https://doi.org/10.1016/j.tplants.2020.07.010.

Singh, A., Ganapathysubramanian, B., Singh, A.K., Sarkar, S., 2016. Machine learning for high-throughput stress phenotyping in plants. Trends in Plant Sci. 21 (2), 110–124.

Singh, A.K., Clarke, J.M., Knox, R.E., DePauw, R.M., Wise, I., Thomas, J., et al., 2015a. AAC Marchwell durum wheat. Can J. Plant Sci 95, 189–195.

Singh, A.K., Clarke, J.M., DePauw, R.M., Knox, R.E., Clarke, F.R., Fernandez, M.R., et al., 2010. Enterprise durum wheat. Can. J. Plant Sci. 90 (3), 353–357.

Singh, A.K., Clarke, J.M., Knox, R.E., DePauw, R.M., McCaig, T.N., Fernandez, M.R., et al., 2012. Transcend durum wheat. Can. J. Plant Sci. 92 (4), 809–813.

Singh, A.K., Clarke, J.M., Knox, R.E., DePauw, R.M., McCaig, T.N., Cuthbert, R.D., et al., 2014. AAC Raymore durum wheat. Can. J. Plant Sci 94 (7), 1289–1296.

Singh, A.K., Clarke, J.M., DePauw, R.M., Knox, R.E., McCaig, T.N., Cuthbert, R.D., 2015. AAC Current durum wheat. Can. J. Plant Sci 95 (3), 589–594.

Singh, A.K., DePauw, R.M., Knox, R.E., Clarke, J.M., McCaig, T.N., Cuthbert, R.D., et al., 2016. AAC Durafield durum wheat. Can. J. Plant Sci 96 (4), 719–725.

Singh, A.K., DePauw, R.M., Knox, R.E., Clarke, J.M., McCaig, T.N., Cuthbert, R.D., et al., 2017a. AAC Cabri durum wheat. Can. J. Plant Sci 97 (1), 135–143.

Singh, A.K., DePauw, R.M., Knox, R.E., Clarke, J.M., McCaig, T.N., Cuthbert, R.D., et al., 2017b. AAC Spitfire durum wheat. Can J. Plant Sci 97 (1), 157–164.

Singh, A.K., Ganapathysubramanian, B., Sarkar, S., Singh, A., 2018. Deep learning for plant stress phenotyping: trends and future perspectives. Trends in Plant Sci 23 (10), 883–898. Available from: https://doi.org/10.1016/j.tplants.20.

Singh, D.P., 1988. Current status of mungbean yellow mosaic virus resistance breeding. In: The Proc. of Second Intl. Symp. on Mungbean held at Bangkok, Thailand, 16–20 Nov. 1987, Asian Vegetable Research and Development Centre, 1988, pp. 282–289.

Singh, D.P., 1990. Distant hybridization in genus Vigna – a review. Indian J. Genet 50, 268–276.

Singh, D.P., 1997. Tailoring the plant type in pulse crops. Plant Breed. Abstr 67 (9), 1213–1220.
Singh, D.P., Singh, A., 2005. Disease and Insect Resistance in Plants. Sci. Publs. USA and Oxford & IBH, India, 417 pp.
Singh, L.N., Johnson, L.P.V., 1969. Natural selection in a composite cross of barley. Can. J. Genet. Cytol. 11, 34–47.
Singh, R.J., Nelson, R.L., 2014. Methodology for creating alloplasmic soybean lines by using *Glycine tomentella* as a maternal parent. Plant Breed Volume 133 (5), 624–631.
Singh, R.P., Huerta-Espino, J., 2004. The use of "single-backcross selected-bulk" breeding approach for transferring minor genes based rust resistance into adapted cultivars. In: Black, C.K., Panozzo, J.F., Rebetzke, G.J. (Eds.), Proceedings of 54th Australian cereal chemistry conference and 11th wheat breeders assembly. Royal Australian Chemical Institute, North Melbourne, pp. 48–51.
Singleton, W.R., 1944. Noyes Darling, First Maize Breeder. J. Heredity 35, 265–267.
Smitchger, J., Weeden, N., 2019. Quantitative trait loci controlling lodging resistance and other important agronomic traits in dry field peas. Crop Sci. 59, 1442–1456.
Smith, S., 2008. Intellectual property protection for plant varieties in the 21st century. Crop Sci. 48, 1277–1290.
Snoad, B., Frusciante, L., Monti, L.M., 1985. The effect of three genes which modify leaves and stipules in the pea plant. Theor. Appl. Genet. 70, 322–329.
Soliman, K.M., Allard, R.W., 1991. Grain yield of composite cross populations of barley. Effects of natural selection. Crop Sci 31, 705–708.
Sperling, L., Ashby, J.A., Smith, M.E., Weltzien, E., McGuire, S., 2001. A framework for analyzing participatory plant breeding approaches and results. Euphytica 122, 439–450.
Sprague, G.F., 1955. Corn and Corn Improvement. Acad. Press, New York.
Sprague, G.F., Tatum, L., 1942. A general vs. specific combining ability in single crosses of corn. J. Am. Soc. Agron. 34, 923–932.
Sprague, G.F., Eberhart, S.A., 1977. Corn breeding. In: Sprague, G.F. (Ed.), Corn and Corn Improvement. American Soc. Agron., Madison, Wisconsin, pp. 305–362.
Stadler, L.J., 1928. Mutations in barley induced by X-ray and radium. Science 68, 186–187.
Stadler, L.J., 1944. Gamete selection in corn breeding. J. Amer. Soc. Agron. 36, 988–989.
Stakman, E.C., Piemeisel, F.J., 1917. Biologic forms of *Puccinia graminis* on cereals and grasses. J. Agric. Rec. 10, 429–495.
Stebbins, G.L., 1941. Apomixis in the angiosperms. Bot Rev. 10, 507–542.
Stebbins, G.L., 1945. The cytological analysis of species hybrids. Bot. Rev. 11, 463–486.
Stebbins, G.L., 1950. Variation and Evolution in Plants. Columbia University Press, New York.
Stelling, D., 1994. Performance of morphologically divergent plant types in direct peas (*Pisum sativum*). J. Agril. Sci. 123, 351–361.
Stevens, R.B., 1949. Replanting 'discarded' varieties as a means of disease control. Science 110, 49.
Sthapit, B.R., Joshi, K.D., Witcombe, J.R., 1996. Farmer participatory crop improvement. III. Participatory plant breeding, a case study for rice in Nepal. Experimental Agriculture 32, 479–496.
Stringfield, G.H., 1974. Developing heterozygous parent stocks for maize hybrids. Dekalb. Ag. Res. Bull., Dekalb 11.
Stuthman, D.D., 2002. Contribution of durable resistance to sustainable agriculture. Euphytica 124, 253–258.

Sunesen, C.A., 1969. Registration of barley composite crosses. Crop Sci. 9, 395–396.

Taliaferro, C.M., Bashaw, E.C., 1966. Inheritance and control of obligate apomixis in breeding buffelgrass, *Pennisetum ciliare*. Crop Sci. 6, 473–476.

Thoday, J.M., 1959. Effects of disruptive selection. I. Genetic flexibility. Heredity 13, 187–203.

Toriyama, K., 1975. Recent progress of studies on horizontal resistance in rice breeding for blast resistance in Japan. In: Gálvez, G.E. (Ed.), Proceedings of the seminar on horizontal resistance of the blast disease of rice, 9. CIAT, Cali, Colombia, pp. 65–100. Series CE-No.

Trentin, H.U., Frei, U.K., Lübberstedt, T., 2020. Breeding Maize Maternal Haploid Inducers. Plants (Basel) 9 (5), 614. Available from: https://doi.org/10.3390/plants9050614.

Tschermak, E., 1900. Uber kuinstliche Kreuzung bei Pisum sativum. Berichte der Dtsch. Botanischen Gessellschaft 18, 232–239.

Tylka, G.L., 2006. What's your type? An HG type test for SCN populations. ICM496:252–253.

Tysdal, H.M., Kiesselbach, T.A., Westover, H.L., 1942. Alfalfa Breeding. Nebraska, Agric. Exp. Sta. Res. Bull 124.

Van der Plank, J.E., 1963. Plant Diseases: Epidemics and Control. Academic Press, New York London, p. 349.

Van der Plank, J.E., 1968. Disease Resistance in Plants. Academic Press, New York London, p. 206.

Van der Plank, J.E., 1975a. Horizontal Resistance: Six Suggested Projects in Relation to Blast Disease of Rice. Horizontal Resistance to the Blast Disease of Rice. CIAT, Cali, Columbia, pp. 21–26.

Van der Plank, J.E., 1975b. Principles of Plant Infection. Academic Press, New York, p. 216.

VanRaden, P.M., 2008. Efficient methods to compute genomic predictions. J Dairy Sci. 91 (11), 4414–4423.

Vavilov, N.I., 1920. The law of homologous series in hereditary variation. In: Proc. of the III All Russian Plant Breeding Congress, Saratov, Russia, pp. 3–20.

Vavilov, N.I., 1926. Studies on the Origin of Cultivated Plants. Instt. Appl. Bot. Plant Breed, Leningrad.

Vavilov, N.I., 1951. The Origin, Variation, Immunity and Breeding of Cultivated Plants. Chronica Botanica 13, 1–366.

Verulkar, S.B., Singh, D.P., Bhattacharya, A.K., 1997. Inheritance of resistance to pod fly and pod borer in the interspecific cross of pigeonpea. Theor. Appl. Genet. 95, 506–508.

Virmani, S.S., Sun, Z.X., Mou, T.M., Jauhar Ali, A., Mao, C.X., 2003. Two-line Hybrid Rice Breeding Manual. Los Baños (Philippines): International Rice Research Institute 88.

de Vries, H., 1900. Sur la Loi de disjonction des hybrides. Comptes Rendus de. l'Academie des. Sci. (Paris).

Wallace, A.T., Luke, H.H., 1961. Induced mutation rates with gamma rays at a specific locus in oats. Crop Sci. 1, 93–96.

Wang, B., Feng, P., Chen, C., Liu, D.L., Waters, C., Yu, Q., 2019. Designing wheat ideotypes to cope with future changing climate in South-Eastern Australia. Agricultural Systems 170, 9–18.

Watson, A., Ghosh, S., Williams, M.J., Cuddy, W.S., Simmonds, J., Rey, M.D., et al., 2018. Speed breeding is a powerful tool to accelerate crop research and breeding. Nature Plants. 4, 23–29.

Watson, I.A., Singh, D., 1952. The future for rust resistant wheat in Australia. J. of the Aus. Inst. of Agric. Sci 18, 190–197.

Watson, I.A., Luig, N.H., 1963. The classification of *Puccinia graminis* var. *tritici* in relation to breeding resistant varieties. Proc. Linn. Soc. N.S.W. 88, 235–258.

Webel, O.D., Lonnquist, J.H., 1967. An evaluation of modified ear-to-row selection in a population of corn. Crop Sci. 7, 651–655.

Weinberg, W., 1909. Uber den Nachweis der Vererbung beim Menschen. Jahreshefte Verein. Naturkunde Wiirttemberg 64, 369–382.

White, M.R., Mikel, M.A., de Leon, N., Kaeppler, S.M., 2020. Diversity and heterotic patterns in North American proprietary dent maize germplasm. Crop Sci. 60 (1), 100–114.

Williams, E., Piepho, H.P., Whitaker, D., 2011. Augmented p-rep designs. Biomet. J. 53 (1), 19–27.

Williams, E.R., John, J.A., Whitaker, D., 2014. Construction of more flexible and efficient p-rep designs. Aust. N. Z. J. Stat. 56 (1), 89–96.

Witombe, J.R., Virk, D.S., 1997. New directions for public sector variety testing. In: Tripp, R. (Ed.), New Seed and Old Laws. Regulatoryreform and Diversification of National Seed System. Intermediate Technology Publication, London, pp. 59–87.

Witcombe, J.R., Virk, D.S., 2001. Number of crosses and population size for participatory and classical plant breeding. Euphytica 122, 451–462.

Woolliams, J.A., Berg, P., Dagnachew, B.S., Meuwissen, T.H., 2015. Genetic contributions and their optimization. J Anim. Breed. Genet. 132, 89–99.

Wricke, G., Weber, W.E., 1986. Quantitative Genetics and Selection in Plant Breeding. Walter de Gruyter, Berlin.

Wright, A.J., 1981. The quantitative genetics of diploid synthetic varieties. In: Gallais, A. (Ed.), Proc Fourth Meeting of the Section Biometrics in Plant Breeding. Poitiers, France.

Wright, S., 1922. The effects of inbreeding and crossbreeding on guinea pigs. U.S. Dept. Agric. Bull 1121.

Xu, K., Xu, X., Fukao, T., Canlas, P., Maghirang-Rodriguez, R., Heuer, S., et al., 2006. Sub1A is an ethylene-response-factor-like gene that confers submergence tolerance to rice. Nature 442, 705–708.

Xu, K., Xu, X., Ronald, P.C., Mackill, D.J., 2000. A high-resolution linkage map of the vicinity of the rice submergence tolerance locus Sub1. Mol. Gen. Genet. 263, 681–689.

Yan, W., Tinker, N.A., 2006. Biplot analysis of multi-environment trial data: principles and applications. Can. J. Plant Sci. 86, 623–645.

Yan, G., Liu, H., Wang, H., Lu, Z., Wang, Y., Mullan, D., et al., 2017. Accelerated Generation of Selfed Pure Line Plants for Gene Identification and Crop Breeding. Front. Plant. Sci. 8, 1786.

Yates, F., 1936a. A new method of arranging variety trials involving a large number of varieties. J. Agri. Sci. 26, 424–455. 1936.

Yates, F., 1936b. Incomplete randomized blocks. Ann. Eugenics 7, 121–140.

Zaven, A.C., Zhukovsky, P.M., 1975. The Dictionary of Cultivated Plants and Their centres of Diversity. PuDoc, Wageningen 219 pp.

Zhang, J., Singh, A., Mueller, D.S., Singh, A.K., 2015. Genome-wide association and epistasis studies unravel the genetic architecture of sudden death syndrome resistance in soybean. Plant J. 84 (6), 1124–1136.

Zhang, J., Naik, H.S., Assefa, T., Sarkar, S., Chowda-Reddy, R.V., Singh, A., et al., 2017. Computer vision and machine learning for robust phenotyping in genome-wide studies. Scientific Reports 7, 44048. Available from: https://doi.org/10.1038/srep44048.

Zhang, Y., Massel, K., Godwin, I.D., Gao, C., 2018. Applications and potential of genome editing in crop improvement. Genome Biol. 19 (1), 210. Published correction appears in Genome Biol. 20(1):13.

Zhukovsky, P.M., 1965. Maingene centers of cultivated plants and wild relatives within the territory of the USSR. Euphytica 14, 177–188.

Index

Note: Page numbers followed by "*f*" and "*t*" refer to figures and tables, respectively.

D

G

I

J

K

T